Al Mullery   Michel Besson
Mario Campolargo   Roberta Gobbi
Rick Reed (Eds.)

Intelligence in Services and Networks: Technology for Cooperative Competition

Fourth International Conference on Intelligence in Services and Networks, IS&N'97
Cernobbio, Italy, May 27-29, 1997
Proceedings

Springer

Volume Editors

Al Mullery
I.C. Europe, SARL
58 rue Glesener, L-1630 Luxembourg
E-mail: al_mullery@compuserve.com

Michel Besson
ASCOM Monetel, CICA
2229 route des Crêtes, F-06560 Sophia Antipolis, France
E-mail: besson@ascom.eurecom.fr

Mario Campolargo
European Commission, DG XIII-B/BU9
Rue de la Loi 200, B-1049 Brussels, Belgium
E-mail: mcam@postman.dg13.cec.be

Roberta Gobbi
Italtel Spa
I-20019 Settimo Milanese (MI), Italy
E-mail: roberta.gobbi@italtel.it

Rick Reed
TSE Ltd., 13 Weston House
18-22 Church Street, Lutterworth, Leics. LE17 4AW, UK
E-mail: rickreed@tseng.co.uk

Cataloging-in-Publication data applied for

Die Deutsche Bibliothek - CIP-Einheitsaufnahme

Intelligence in services and networks : technology for cooperative competition ; proceedings / Fourth International Conference on Intelligence in Services and Networks, IS&N '97, Cernobbio, Italy, May 27 - 29, 1997. Al Mullery ... (ed.). - Berlin ; Heidelberg ; New York ; Barcelona ; Budapest ; Hong Kong ; London ; Milan ; Paris ; Santa Clara ; Singapore ; Tokyo : Springer, 1997
(Lecture notes in computer science ; Vol. 1238)
ISBN 3-540-63135-6

CR Subject Classification (1991): C.2, B.4.1, D.2, H.4.3, H.5, K.4, K.6

ISSN 0302-9743
ISBN 3-540-63135-6 Springer-Verlag Berlin Heidelberg New York

Printed in Germany

Typesetting: Camera-ready by author
SPIN 10548806 06/3142 – 5 4 3 2 1 0 Printed on acid-free paper

Lecture Notes in Co

Edited by G. Goos, J. Hartma

Springer
Berlin
Heidelberg
New York
Barcelona
Budapest
Hong Kong
London
Milan
Paris
Santa Clara
Singapore
Tokyo

Technology for Cooperative Competition

The Fourth International Conference on Intelligence in Services and Networks (IS&N) has, as a theme, "Technology for Cooperative Competition". This theme is interesting because it seems to be a contradiction. Competition and cooperation are usually seen as being opposed, producing opposite requirements and results. Competition, for example, following the dictates of efficiency and market share, can lead to proprietary solutions, in which it is attempted to lock out alternative suppliers. Cooperation can lead to "open" systems that may have many levels of published interfaces, allowing suppliers to apply their solutions at any level, but where efficiency is often lost.

However, dealing with such contradictions is part and parcel of the field of IS&N. In the apparent contradiction between competition versus cooperation, the art is to find a balance between a certain level of openness, without sacrificing efficiency, and allowing competition in providing solutions that still adhere to the standards and interfaces that come with the openness.

Another, related, contradiction that arises in the work of IS&N is that of convergence versus divergence. This is particularly troublesome when applied to service architectures. The idea of an open service architecture, as such, has been at the heart of work in this area in the RACE and ACTS[1] programmes of the European Union, first raised during RACE in the ROSA and Cassiopeia projects. But other related architectures were also being developed. One can include in this list IN, NMF, TINA, and DAVIC, amongst many others. Convergence of architectures is needed for interoperability and interworking, and the divergence arises from the different applications and the two different industry backgrounds: telecommunications and computing. The actors in the industry are competing, but also need to cooperate. For cooperation between services, an agreed-upon open service architecture is essential. Each attempt at a convergence has yielded not a convergence on a single architecture but, rather, yet another architecture, increasing the order of divergence. And because there is no overwhelming market leader (yet), there cannot be de facto convergence. It may also be that none of the architectures, or at least, none of the advanced ones, has achieved sufficient simplicity to make it the obvious path to follow.

This leads to a third seeming contradiction, but one that is at the heart of IS&N — that of complexity versus simplicity. Whatever the coming "Global Information Infrastructure" (GII) turns out to be, it will present to the user an exceedingly complex ensemble of services, themselves ranging from quite simple to quite complex. Unless most potential users are isolated from the real complexity of the network or of the services (single services or an ensemble), then the GII is likely to be shelved as an interesting and valiant but wasted and irrelevant effort. Services must compete for resources, but need to cooperate for effectiveness. And experience shows that achieving elegant simplicity is not at all simple. Thus the primary challenge to IS&N is to know how to harness complexity to achieve apparent simplicity.

---

[1] Further information about the ACTS programme can be obtained from the European Commission DG XIII-B/BU9 4/82 Rue de la Loi, 200 B-1049 Brussels, Belgium. e-mail: aco@postman.dg13.cec.be tel: +32 2 296 3415, fax: +32 2 295 0654, or from <http://www.infowin.org>.

This cannot be done by magic — there is no magic, at least, not in service engineering. Improvements can often be achieved by illusion: hiding complexity to make it appear simple. The issues addressed in IS&N are complex, but no "transform" has yet been found to map software complexity into a simple world. This is not surprising because the IS&N world is becoming an integrated part of the real world, which itself is complex and full of contradictions, competition and cooperation. Of course, over the years, there have been changes, both large and small, in the way that systems are engineered, but none have been revolutionary. One idea that has recently attracted attention as perhaps being revolutionary is that of mobile agents. Unfortunately, the major use has been to cause disruption, by implanting unfriendly mobile agents in unsuspecting software systems — so-called viruses — which, if a revolution, certainly do not increase simplicity.

While you will not find magic in this volume, you will find sound and effective application of the techniques of service engineering to provide the user with an ensemble of useful, efficient, and effective communication services. The sections are arranged in the order Architecture, Service Engineering, Intelligent Networks, Network Management, Mobility, Security, and Human Factors — thus moving from what appears to be the theoretic and technocentric view to the anthropocentric view. Within any service engineering project, these two views may be seen to be competing for dominance. At the most abstract level, though, even architecture is based on an enterprise model that ultimately can only be understood in human terms, thus completing a circle. Within the domain of IS&N, competition and cooperation become not contradictory but complementary.

March 1997 The Editors

Acknowledgements

This volume could not exist without the authors of the papers. Over 100 papers were contributed. Unfortunately, many proposed papers could not be included but the authors of these papers are thanked for their efforts.

The editors would like to thank the numerous reviewers and 'mentors' of the published papers. Special thanks go to the section editors, who have also written the introductions to each section, and to Rishma Lilani and Severine Garcia who (with Michel Besson and Al Mullery) ran the review process. Also thanks to Jeanne Reed who assisted in the final editorial process, and Anna Kramer who made a final check of the book before sending it for printing.

Steering Committee

Michel Besson (Chairman), Mario Campolargo, Roberta Gobbi, Al Mullery

Programme Committee

Ordinary Members

Al Mullery (Chairman)
Jaime Delgado
Renaud Di Francesco
William Donnelly
Alex Galis
David Horne
June Hunt
Bert Koch
Gerard Lacoste
Thomas Magedanz
Rick Reed
Sathya Rao
Stelios Starzetakis
Jean-Bernard Stefani
Constantine Stephanidis
Sebastiano Trigila
Hans Vanderstraeten

Corresponding Members

Jean-Pierre Adam
Mario Bonatti
Emmanuel Darmois
Philip Dellafera
Kevin Fogarty
Yuji Inoue
Martine Lapierre
Kunji Mori
Bruce Murrill
Norbert Niebert
Roberto Saracco
Jon Siegel
Karl Ulrich Stein
George Williamson

Table of Contents

Communications Management

Mobility

Security

Human Factors

Service Architecture

Sebastiano Trigila
Fondazione Ugo Bordoni
e-mail: trigila@fub.it

In the past twenty years, many "gurus" have predicted the future of telecommunications, informatics and entertainment electronics. It suffices to browse through the first chapters of "The road ahead", a book by Bill Gates, to realise that many predictions never came true, while unforeseen phenomena have become major technical drivers in current market trends. Fax service is a small, old-fashioned example. The explosion of the WWW is an impressive, trendy example. The performance achievements of compression algorithms for still and moving images are another instance of success beyond reasonable expectation.

There are also some predictions, made a long time ago, that we cannot yet assess. For example, the prediction made in the Seventies about the convergence of technologies and services for information, telecommunications and entertainment by Nicholas Negroponte Director of Media Labs at MIT. While the "convergence conjecture" is still provocative and far from obvious, it tends to be supported by more and more signs.

We are currently in a situation where the consumers ("subscribers", in the jargon of Public Network Operators (PNOs), "audience" in that of broadcasting companies and "customers" in that of information providers) are getting familiar with a huge number of services currently provided by different technologies in terms of both network and local equipment:

- telecommunications services in their traditional form and in their multiparty, multimedia and mobile extensions (audioconference, videotelephony, video-conference, virtual private networks, GSM, UMTS, etc.)
- data services in their traditional expressions (teleprocessing, file transfer, e-mail, bulletin board systems, etc.) and in their "smart" extensions (primarily, world-wide information browsing) on a global multi-domain computer network (the Internet), encompassing hundreds of thousands of WANs and LANs across the various continents;
- traditional cable TV services and their extensions towards interactive Video on demand.

In this technological and market scenery, the roles of *consumers, connectivity providers* and *content providers* emerge with different policies and expectations.

Consumers are experiencing a number of facts: (a) many services can be offered at different conditions (such as quality of service, availability, and price), on different technological platforms or networks, by different providers; (b) no technology or provider is currently and definitely the "best" choice for all kind of services; (c) home and office can hardly accommodate the many different types of "terminals" dedicated to distinct services; (d) workstations, personal computers and laptops can reduce the number of needed terminals by offering "smart terminals" with window-based GUIs, accommodating different types of services through dedicated windows; (e) adapter cards can connect the same "smart terminal" to a number of different network types.

The distinction between the roles of connectivity providers and content providers is becoming more and more striking, also due to deregulation trends in the world of PNOs. Connectivity providers are trying to convey the widest range of possible services over their own infrastructure. Content providers would like to be reached by their customers in the most efficient and cost-effective way.

All providers are currently attempting to increase the quality, characteristics and range of services offered by their own available network infrastructures, namely:

1. Public Switched Telephone Network (PSTN), dedicated public data networks (Public: Circuit Switched Data Network -CSDN, Packet Switched Data Network - PSDN, Integrated Services Digital Network - ISDN) and forthcoming B-ISDN over ATM technology;
2. the Internet, as a world-wide federation of TCP/IP based wide area networks (including broadband segments and zones) and LANs;
3. cable-TV distribution networks.

For this reason, all players on the various platforms have been pursuing - during the past decade - standardisation activities that can be collectively tagged as the search for "architectures for control and management of services", largely independent from underlying network technology. After the pioneering studies of the European RACE Programme in this field, it has become customary to refer to those architectures as *service architectures*. The term "control" generally indicates network and terminal functionality allowing users to access and use services, by the set-up, retaining and release of necessary resources. The keyword "management" generally indicates network and terminal functionality able to create, monitor, measure, and modify the status of, resources and services. The distinction between "control" and "management" is not made in some service architectures, where "control" is identified with real-time management of resources on a per-call or service-instance basis; hence, all operations fall within the term "management".

Major examples of "service architectures" are the following.

Intelligent Network (IN) framework. The original purpose was to extend the Plain Ordinary Telephony Service (POTS) with "supplementary services" based upon service-independent building blocks chosen out of standardised "capability sets". The higher the complexity of the capability sets, the wider the range of services that can be covered by IN technology. Currently, IN aims at ensuring functional coverage of real-time personal communications and multimedia services, by suitable extension of its models at different abstraction levels ("planes").

Telecommunications Management Network (TMN) principles and architecture. Its original purpose was to provide a framework for management of network resources, via standard interfaces masking details of proprietary technology. Its powerful models were then extended to cover the management of services, both off-line and real-time. When this extension is included, TMN can be considered a remarkable service architecture.

Internet Application Protocol specifications. De facto standards emitted by Internet Engineering Task Force (IETF) related to - in historical order - FTP, SMTP, HTTP, and RSVP based services, have implicitly defined an ad-hoc service architecture. In particular, the purpose of QoS related RSVP specifications is to offer the Internet community communications applications with a pre-defined quality of service, a prerequisite for the competitive provision of telephony, video-telephony and video on demand over the Internet, according to the expectations of users today accustomed to high performance parameters on public telecommunications networks. Success in real-time communications services over the Internet is a dream of both IETF and users, who would like to be able to avoid the "costly policies" of traditional and "powerful" PNOs. A major booster of open distributed processing over the Internet is the availability of object-oriented technology and CORBA standard specifications. The main driver here is the provision of standard interfaces for client-server relations between remote computational objects residing in different operating systems and "speaking" different native languages for service and procedure invocations.

TINA (Telecommunications Information Networking Architecture). The purpose of this framework is to define the characteristics of a world-wide distributed processing environment for integrated control and management of services, beyond the paradigms offered by IN and TMN. The target is to broaden the range of services covered by the current telecommunications network and its broad-band extensions, by bringing within the PNO domain even those services currently popular over the Internet. Ancestors of TINA were the proprietary architecture AIN by Bellcore (1990-91), and the ROSA architecture (1989-92) developed within the RACE programme. Object-oriented technology has been an important input stream for TINA, which eventually adopted a superset of the OMG object model as its own computational framework.

DAVIC (Digital Audio Visual Council) specifications. Originated between 1995 and 1996, their purpose is to provide end-to-end world-wide interoperability of a vast range of interactive multimedia services and applications, across different network technologies, including cable TV networks, ATM and Internet protocol stacks. The latest offsprings of the Internet culture (like Java virtual machine) have also been taken into account.

Let us return to the "convergence conjecture". Each of the architectures mentioned above competes with the other and, in the intentions of the promoting consortia and their "addicts" in the technical and scientific community, would like to become "*the* long-term reference architecture".

Each of these architectures claims to offer concepts, building blocks, components, interaction scenarios, federation of domains, and whatever else suffices to cover all the needs of all potential services. Convergence towards a unified interface to furnish all services to users is desired by both users and providers. The actual ability of any of the above architectures to provide functionally efficient and cost-effective coverage of all types of service is still under debate. Certainly, a common service architecture has yet to be conceived. Probably, a “common service architecture” will never exist. For the moment, many experts are thinking more realistically of “interworking scenarios” between services or configurations compliant with different service architectures.

The eight papers gathered in this section of the book are representative of the different “service architecture” cultures described above.

One paper [De Zen and Faglia] deals with provision of a major multimedia service, Video on Demand, in an IN-compliant signalling context, suitably extended to work with B-ISDN.

Three papers are based on the TINA cultural stream, and deal respectively with conceptual foundations of TINA in terms of object model and interface specification language [Mercouroff and Parhar]; encapsulation of telecommunications technology (such as ATM switches, location registers in GSM networks, etc.) within a TINA-compliant context [Evans and al.]; and architecture validation experiments in the presence of broadband technology and in the perspective of provision of multimedia services [Huelamo and al.].

A paper [Sohn and al.] proposes, in support to multimedia services, an architecture aiming to provide a seamless client-server platform spanning over a federation of ATM networks.

Another paper [Marti and Delgado] exploits the latest developments in the Internet WWW technology and in TCP/IP over ATM, to present a platform independent client/server brokerage application for customers and suppliers in the publishing and electronic commerce areas.

Lastly, two papers are based on the idea of “convergence” between different platforms. The former [Limongiello et al.] presents a service architecture spawning from the idea of convergence of interactive multimedia services over ATM and Internet services and evaluates it on the basis of two prototype implementations. The latter [Choi and al.] focuses on the convergence of DAVIC standard and Internet services. In both papers, central elements of the convergence are CORBA and Java.

The subjects of the above-described papers are of immense importance in current world-wide research. They offer many exciting ideas that can stimulate further research. Any attempt to synthesise a “unified” view of service architectures would be unrealistic. The coming years will be decisive in determining when and where a “convergence architecture” will occur.

Video On Demand in an Integrated IN/B-ISDN Scenario

Giovanna De Zen, Lorenzo Faglia
ITALTEL, Central Research Laboratories 20019 Settimo Milanese (MI) - Italy
Telephone: +39.2.4388.9085, Fax: +39.2.4388.7989; E-Mail: lorenzo.faglia@italtel.it

1 Introduction

Different scenarios for the provision of the Video On Demand (VOD) service have been described in the associated literature and different standardization bodies (DAVIC [1], ATM FORUM [2] and ITU-T [3]) are working on the definition of the architecture and protocols needed to introduce this service in a broadband environment.
The scenario described in this paper foresees the provision of the service in an integrated IN/B-ISDN scenario. Unfortunately it is a long term scenario as standards for the IN/B-ISDN integration are yet in an early stage of definition.
The provision of multimedia services taking advantage of Intelligent Network (IN) facilities upon the B-ISDN infrastructure is promising from the point of view of both users and operators.
The paper will emphasize the major advantages of a service provision architecture, based on an enriched set of IN capabilities as defined in the context of the Advanced Communications Technologies and Services (ACTS) INSIGNIA (IN and B-ISDN Signalling Integration on ATM platforms) project.
In section 2 the reference network architecture is shown and the mapping of functional entities in physical elements is detailed.
The VOD static description in terms of service components is given in section 3, while section 4 deals with the dynamic evolution of a service instance, as perceived by the network elements.
The protocol stack used at the User Network Interface (UNI), Network Node Interface (NNI) and IN interfaces is indicated in section 5.
In the last section the advantages of the integrated IN/B-ISDN architecture are described when service and network capabilities evolution is considered.

2 Reference architecture and general assumptions

The IN-oriented reference architecture proposed for VOD service by ITU-T SG 13 [3] is depicted in Figure 1. Four major functional areas have been identified for the VOD service provisioning: the Customer Premises Equipment, the Access Network, the ATM Core Network and the Server System. The picture represents the identified functional areas together with the required transport, control and network management system functions. These functions are grouped in functional entities and mapped in network physical elements.
The following set of basic assumptions are considered to provide the service:

- ❐ A Broadband Intelligent Peripheral (B-IP) is present in the network architecture. B-IP and B-SCP behave as a Gateway Service Provider. B-IP acts like a specialised server for information retrieval and it is completely different if compared to traditional IP like digit receivers or tone/announcements generators.
- ❐ Access to the VOD service is made by means of a symbolic number. Therefore, number translation is a basic function required from IN. (Number translation is performed when the user accesses the service and when it must be selected by one of the Video Servers (VS) of the chosen Service Provider (SP)).

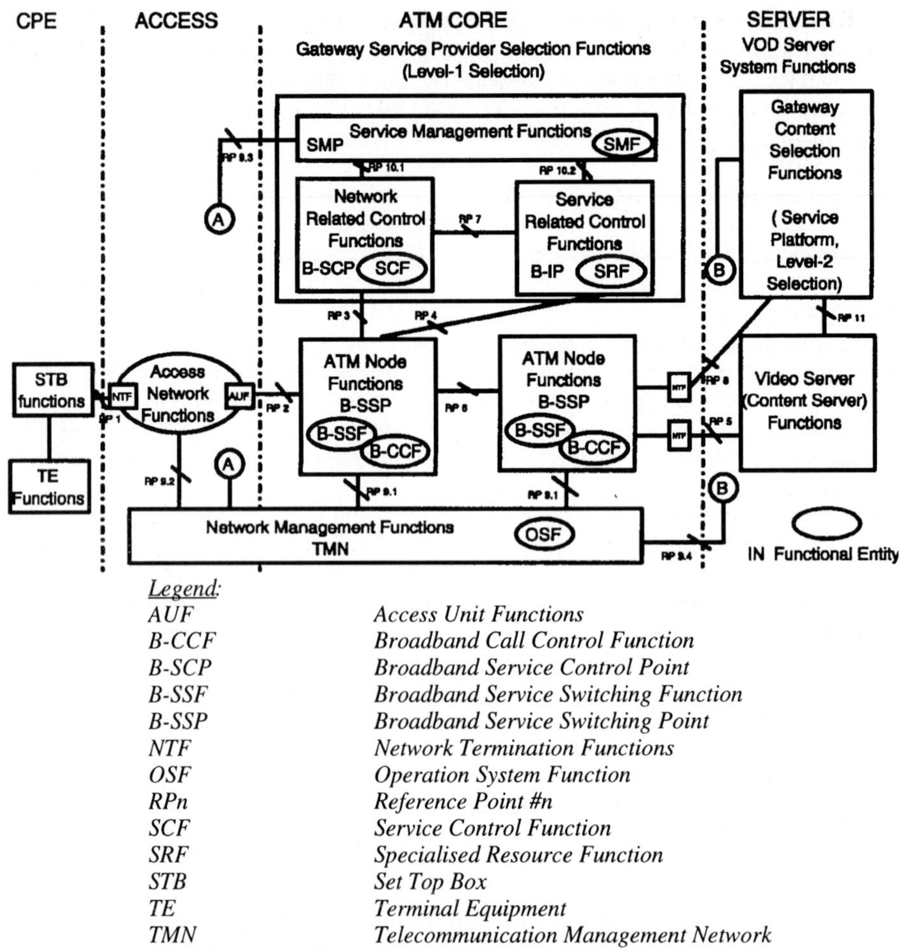

Legend:

AUF	*Access Unit Functions*
B-CCF	*Broadband Call Control Function*
B-SCP	*Broadband Service Control Point*
B-SSF	*Broadband Service Switching Function*
B-SSP	*Broadband Service Switching Point*
NTF	*Network Termination Functions*
OSF	*Operation System Function*
RPn	*Reference Point #n*
SCF	*Service Control Function*
SRF	*Specialised Resource Function*
STB	*Set Top Box*
TE	*Terminal Equipment*
TMN	*Telecommunication Management Network*

Figure 1: Video on Demand Reference Configuration

- Terminal authentication takes place in the B-SCP at the service request time (before the symbolic number translation is performed). User authorization is performed in-band after the VS selection. User's profile information is stored in the B-SCP database.
- In the first realization of the service, the B-ISDN signalling system is supposed to support only basic call/connection capabilities as foreseen by ITU-T Capability Set 1. This choice has been driven by the available signalling capabilities in the equipment of the experimental first trial of the INSIGNIA project, used to demonstrate this service architecture. This hypothesis is the most restrictive choice for the provision of a complex service like VOD, because it offers the possibility to support only point to point mono-connection calls. This implies that separate calls will be issued in order to obtain the different separate VOD information flows. In particular, one call is used in order to carry menu navigation and VCR interaction commands, while a second one carries the (MPEG-encoded) audio/video information.

- "B-SCP-initiated call establishment and release" control capability is supported. This feature can be regarded as a particular case of "third party call setup/release", where the third party is a network element and not a user.
- The term *session* is used to denote an association of calls and connections for the realization of a single instance of a service. In this context, the "session aware" network elements are the Broadband Service Switching Point (B-SSP), the B-SCP and the B-IP, but not the terminals.

3 Static service description

The VOD service is composed of three Service Components (SCs): the Audio-Video SC, the Selection Data SC and the Content Control Data SC (see Table 1).
The Audio-Video SC composes the Multimedia channel in which video and audio are multiplexed, while the Selection Data and the Content Control Data are the SCs of the Selection Path channel. Navigation commands and selection information are transported via the Selection Data SC while the VCR emulation commands use the Content Control Data SC.

Teleservice	**Interactive Multimedia Retrieval service**		*Calling Party = User* *Called Party = Service Provider*
User perspective component	**Selection Path** (mandatory)		**Multimedia Channel** (mandatory)
Service component	**Selection Data** (mandatory)	**Content Control Data** (mandatory)	**Audio-Video** (mandatory)

Table 1: VOD Service Components

The static description of the service attributes is needed to characterize the SCs, as provided in Table 2.

Mandatory service components	SC1 High quality audio-video SC2 Selection Data SC3 Content Control Data
Optional service components	None
Information transfer mode	ATM
Connection mode	Connection oriented
Traffic Type	SC1 CBR SC2 VBR, SC3 VBR
ATM transfer capability	SC1, SC2, SC3: Deterministic Bit Rate
Timing end to end	Required for Service Component SC1, SC2
Information Transfer Rate Peak Bit Rate	SC1: PCR =1.5 Mbit/s (for MPEG 1); 2-10 Mbit/s (for MPEG 2) SC2: PCR = 64 Kbit/s upstream 64 Kbit/s - 2 Mbit/s downstream SC3: PCR = 64 Kbit/s
Structure	Unstructured for every service component
Establishment of communication	On demand
Symmetry	Unidirectional (SC1) Bi-directional asymmetric (SC2) Bi-directional symmetric (SC3)
Communication configuration	Point to point

Table 2: VOD Service Attributes description

Regarding the mapping of Service Components on ATM, a connection is required for each service component.
During the communication between STB and Broker only SC2 is mandatory.

4 Dynamic service description

The VOD service is invoked like a traditional IN service by the user making a call from the terminal (STB).
The main difference with respect to the IN services in a traditional narrow-band environment is that VOD service requires more than one connection towards the same party (VS), that is one for the SC1 data flow and one for SC3. Another important issue is that the evolution of the service is determined by the user through choices on the Selection Path channel that must address actions on the Control Plane. The solutions offered by the integrated IN/B-ISDN solution is that the Service Logic residing in the SCP is able to order the SSP to establish/release the network resources needed for the progress of a service instance. To perform this task the SCP must know the resources allocated to a particular service instance: this is realized via an innovative dialogue between SSP and SCP, based on an object model able to represent the current status of the session.

4.1 Phases of the service scenario

Two different phases have been recognized during the VOD service evolution:

Phase 1: an interaction with the B-IP where the user navigates the level 1 menu and selects a VS;

Phase 2: an interaction with the VS, where the user selects a movie and the content is transferred to the STB in a real time fashion.

The first phase requires a single ATM bi-directional connection, while for the second phase two distinct calls are established, each providing a single ATM connection with specific characteristics depending on the characteristics of the transported information.
It is assumed that Phase 2 can only take place after Phase 1 calls have been successfully completed. IN takes care of performing this control.
In Figure 2 and Figure 3 a VOD service instance in terms of messages and information flows is depicted. In the figures only the successful case is illustrated.
While at the SCF-SSF and SCF-SRF interfaces the dialogue is based on exchange of Information Flows (IFs), the communication between non IN-related FEs is in terms of B-ISDN CS1 signalling messages. The detailed description of the IFs exchanged between SSF and SCF together with the actions performed by the SSF in terms of SSM state changes is contained in section 4.2.
Hereafter only a high level description of the actions sequence, regardless of the kind of IFs exchanged, is given:

Step 1) *The STB is connected to the B-IP*: A STB sends a SETUP message containing an IN-number to the CCF. When the Trigger-Detection Point "Analysed Information" in the Basic Call State Models is encountered, a Service Request is sent to the SCF. The SCF instructs the SSF to route the connection towards the B-IP. At this point the dialogue (in band) between STB and B-IP can take place.

Step 2) After the VS user selection (that is communicated by the B-IP to the SCF), the SCF instructs the SSF to *release the connection between STB and B-IP*, by dropping the B-IP party from the session.

Step 3) SCF instructs the SSF to *establish two new connections between the STB and the selected VS.*

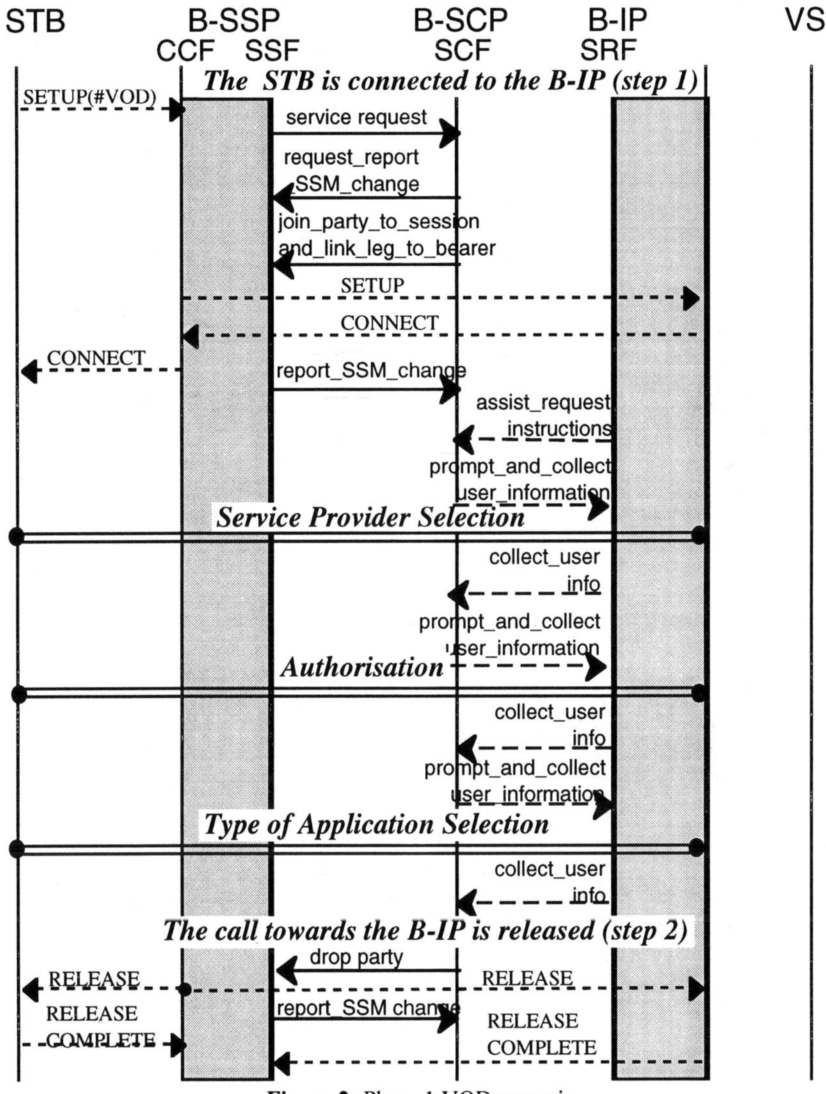

Figure 2: Phase 1 VOD scenario

The service session ends when a RELEASE message is sent by either the user or the VS. Figure 3 represents the case of a user releasing only one basic call, this is reported to the SCF and since for the service two connections are needed, the SCF requests the SSF to release the whole session.

4.2 Information flows for the service scenario

IN Information Flows defined at the SSF-SCF interface are based on the object model representing the service session configuration and the related network resources. This object model is shared by SSF and SCF and it is called IN Switching State Model (IN-SSM).

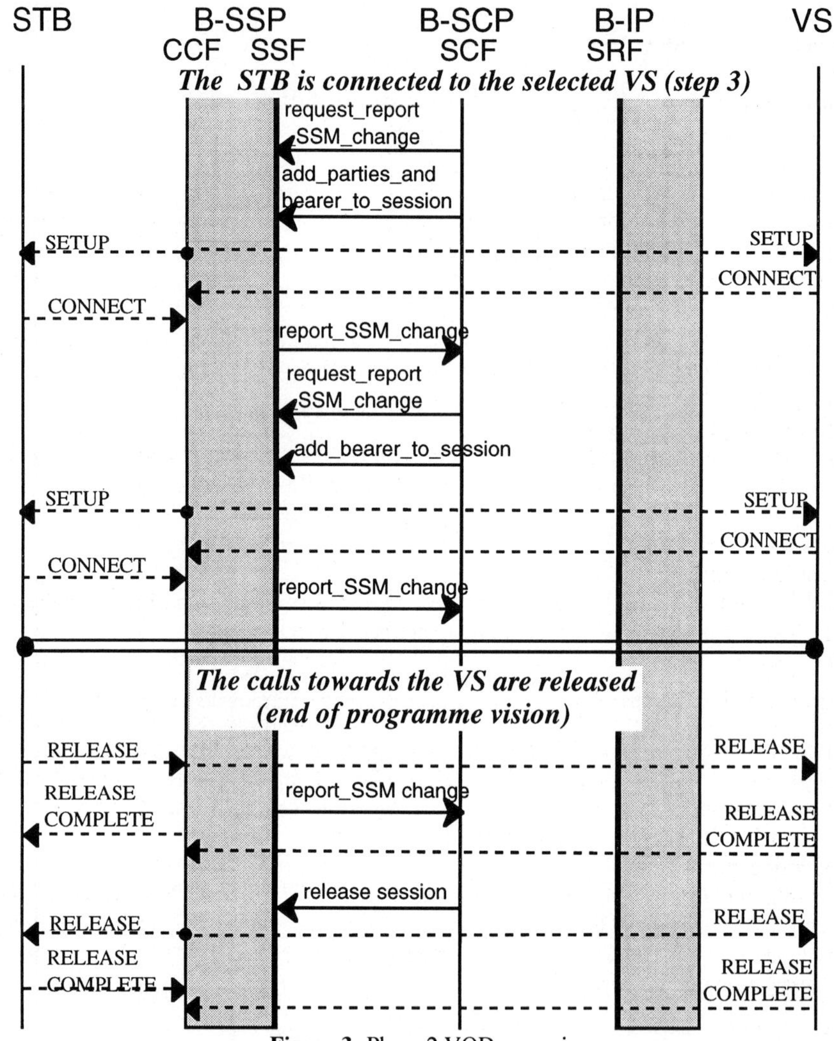

Figure 3: Phase 2 VOD scenario

The subset of Information Flows used for the realization of the VOD service are the following:

1) Service request

FE relationship: SSF to SCF

This IF is generated by the SSF when a trigger is detected at any Detection Point in the Basic Call State Model and a new IN-SSM instance is created. It means that a Service Logic Program within the SCF must be addressed.

2) Request report SSM change
FE relationship: SCF to SSF
This IF is used to request the SSF to monitor for session related events and consequently for SSM state changes, and to instruct the SSF to send a notification back to the SCF for each detected event.
3) Report SSM change
FE relationship: SSF to SCF
This IF is generated by the SSF to notify the SCF that a change in the SSM has occurred. It is sent in response to the Request report SSM change.
4) Join party to session & link leg to bearer
FE relationship: SCF to SSF
This IF is used to request the SSF to join a new party to an already established session, composed of the calling party with the corresponding leg and a bearer connection, and to link the new leg to the existing bearer connection.
5) Add bearer to session
FE relationship: SCF to SSF
This IF is used to request the SSF to add a new bearer connection and corresponding legs between existing parties within an already established session.
6) Add party & bearer to session
FE relationship: SCF to SSF
This IF is used to request the SSF to add a new bearer connection and corresponding legs, within an already established session, between the new parties specified in the IF.
7) Release Session
FE relationship: SCF to SSF
This IF is used to request the SSF to release the whole session. This implies releasing all parties, related legs and bearers and the session itself.
8) Drop party
FE relationship: SCF to SSF
This IF is used to request the SSF to drop the specified parties, from an existing session. As a consequence all the objects related to the given party are removed.
It is possible to observe that the main difference with respect to the standardized IN protocols developed for narrowband networks, is the general capability given to the SCF to order the SSF to establish or release network resources, in terms of bearer connections, parties and legs.
Regarding the information flows used between SCF and SRF, a more traditional approach is used even if the B-IP equipment is radically different from a traditional IP, because it can be regarded as a service specific server and not a digit receiver or a tone/announcement generator.

5 Protocol stack

The protocol stacks in the User Plane for the communication between the Set Top Box and the Video Server are compliant to DAVIC standards.
This means having MPEG2 for the multimedia channel and DSM-CC User to -User for the Selection Path.
The communication between the STB and the Broker is performed using a browser to navigate through HTML pages stored in the B-IP.
In the Control Plane the protocol stacks used at the different interfaces are shown in Figure 4.

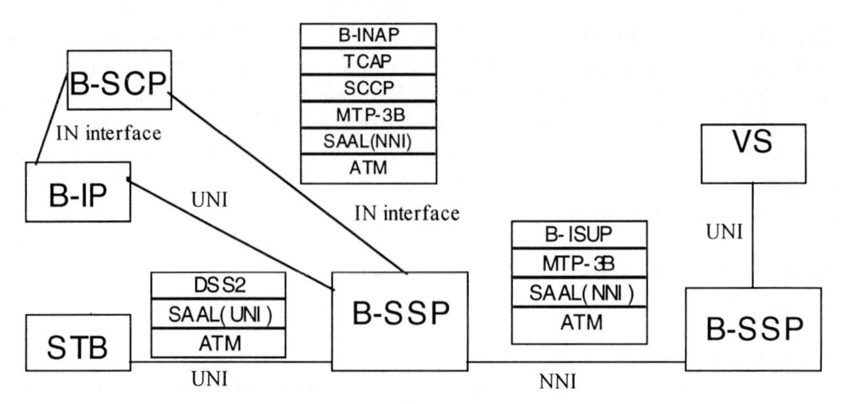

Figure 4: Protocol stacks in the Control Plane

Taking into account the implementation of the platform for the INSIGNIA trial, some considerations apply.

Regarding User Network Interface (UNI):

- the ATM layer is according to ITU-T I.361 "B-ISDN ATM layer specification"
- the SAAL layer is according to ITU-T I.363 "B-ISDN ATM adaptation layer specification", Q.2110 "B-ISDN ATM adaptation layer- Service specific connection oriented protocol (SSCOP) and Q.2130 "B-ISDN signalling ATM adaptation layer - Service specific coordination function for support of signalling at the user-network interface (SSCF at UNI)"
- the network layer is a common intersection of ATM Forum UNI, version 3.1 and ITU-T Q.2931 "Digital Subscriber Signalling System No. 2 (DSS2)- User Network Interface (UNI) layer 3 specification for basic call/connection control"

Regarding Network Node Interface (NNI):

- the ATM layer is similar to the UNI
- the SAAL layer is similar to the UNI with the only difference being the SSCF that follows ITU-T Q.2140 "B-ISDN signalling ATM adaptation layer - Service specific coordination function for support of signalling at the network node interface (SSCF at NNI)"
- the network layer is a subset of Message Transfer Part layer 3 (Q.701 and Q.704) of Common Channel Signalling System No. 7
- the basic call is handled according to B-ISUP specifications: Q.2761 "Functional description of the B-ISDN user part (B-ISUP) of signalling system No. 7", Q.2762 "General Functions of messages and signals of the B-ISUP", Q.2763 "B-ISUP - Formats and codes", Q.2764 "B-ISUP - Basic call procedures"

The IN interface applies both to the interface between B-SCP and B-IP and between B-SCP and B-SSP.

The new IN protocol named *B-INAP* lies over the same stack of the lower layers NNI interface, while for the upper layers the following considerations apply:

- the SCCP is a subset of the ITU-T Rec.:Q.711 "Functional description of the signalling connection control part", Q.712 "Definition and function of SCCP messages", Q.713 "SCCP formats and codes", Q.714 "Signalling connection control part procedures" and Q.715 "Signalling connection control part. User Guide"

- the TCAP is a subset of the ITU-T Rec.: Q.771 "Functional description of transaction capabilities", Q.772 Transaction capabilities information element definitions", Q.773 "TCAP formats and codes", Q.774 "Transaction capabilities procedures" and Q.775 "Guidelines for using transaction capabilities"

The B-INAP protocol at the SSF/SCF interface is derived from information flows described in paragraph 4.2, while for the SCF/SRF interface a backward compatible approach has been followed.

6 Service evolution

A flexible network architecture must allow the evolution of service characteristics, signalling capabilities and IN features decoupling each *domain* even in the presence of an integrated scenario.

This must be considered a key aspect of an innovative solution: as it is difficult to foresee the evolution of the information technology market, it is desirable to have an architecture that allows the capture of new requirements, without high delays due to the necessity of redesigning network elements and communication protocols.

Focusing on the control aspects, it is possible to recognize three domains:

1. *The service domain*: if the Service Provider wants the service to be enriched it is possible to modify the Service Logic in the SCF without changes in IN and B-ISDN protocols.
2. *The network domain*: if the Network Provider wants to modify the allocation of network entities without any impact on the service as perceived by the user.
3. *The customer domain*: if the terminals are able to perform new control capabilities it would be desirable to introduce them in the switches without involving a change in the service logic and in the IN protocols.

It is possible to apply this general concept to the VOD service as an example.

Starting from the description given in the previous paragraphs, a way to enrich the service characteristics is the following:

1. To allow the user to return to the level 1 menu after the viewing of the content has terminated. This implies a change in the service logic, with no modification to the protocols.
2. To specialize B-SCPs, that is allocating VOD Service Logic only in a subset of the available B-SCPs and thus modifying only IN protocols without touching service logic and signalling.
3. To introduce new signalling capabilities like multiconnection and negotiation/modification of bandwidth, that do not necessarily reflect in the service logic and in the IN protocols. For example, for service logic and IN entities it does not matter if the two user plane connections needed for the communication between STB and VS are realized via two monoconnection calls or via a single multiconnection call.

7 Conclusions

Multimedia services require enhanced call configurations, that cannot be seen by B-ISDN signalling as composing a single service instance. To solve this problem, the concept of session has been introduced at the IN level to correlate different calls and connections that compose one service instance.

In this paper a description of the provision of Video On Demand service in an integrated IN/B-ISDN scenario has been given and several advantages of this architecture have been envisaged.

First of all, the integration of IN and B-ISDN is done using a flexible approach because new IN information flows devoted to session handling have been defined such as to be independent from the B-ISDN signalling Capability Set used in the network nodes and terminals.
Another important argument is that an IN/B-ISDN platform can be used to provide a wide range of different services that share common facilities and it is not a VOD oriented solution.
The proposed model for VOD can be used even if only basic call/connection capabilities are available in the Control Plane. In fact, the service logic exploits a new signalling capability that allows the management of this complex service: it is the "Broadband Service Control Point (B-SCP) initiated call establishment and release".
This capability reflects the general concept that is the Service Logic that governs the evolution of the service, in terms of network resources establishment or release.

References

[1] Digital Audio-Visual Council "DAVIC 1.0 Specifications", December 1995.
[2] The ATM Forum Technical Committee "Audiovisual Multimedia Services: Video on Demand Specification 1.0", December 1995
[3] ITU-T draft Recommendation I.375 "Network capabilities to support multimedia services", July 1995.

TINA Computational Modelling Concepts and Object Definition Language

Nicolas Mercouroff
Alcatel Telecom,
mercouroff @aar.alcatel-alsthom.fr

Ajeet Parhar
Telstra,
a.parhar@trl.telstra.com.au

TINA Computational Modelling Concepts and their supporting language, the TINA Object Definition Language (TINA-ODL) are described in this paper. They provide a framework for the computational specification of distributed applications for telecommunications information networks. The computational specification of a distributed application describes it in terms of interacting computational entities, or program components. It specifies the structures by which the interactions occur and also specifies the semantics or behaviour of these interactions. The article concludes with the need for a strong relationship between the TINA-C community and the OMG to ensure commercial availability of future distributed processing environments supporting the TINA specifications.

The concepts and language described here have been developed in the Core-Team of the Telecommunications Information Networking Architecture Consortium. A detailed description of these concepts can be found in [1]. A complete presentation of the syntax of TINA-ODL can be found in [4].

1 TINA Computing Architecture

The TINA (Telecommunications Information Networking Architecture) Computing Architecture defines the modelling concepts that should be used to specify object-oriented software in TINA systems. It also defines the Distributed Processing Environment (DPE) infrastructure allowing objects to locate and interact with each other. These concepts are based on the Reference Model for Open Distributed Processing (RM-ODP) [5][6]. The TINA Computing Architecture defines modelling concepts for four viewpoints, used for specifying a telecommunications system:

- *Enterprise modelling concepts* focus on the purpose, scope, and policies.
- *Information modelling concepts* [2] focus on the semantics of information and information processing activities in the system. These provide a framework for the information specification of distributed systems. An information specification describes a distributed system in terms of information held by the system components and information exchanged between the components. The information may relate to the operation and use of the system, as well as to the management of the system itself. The information specification of a system focuses only on the information elements of the system and their relationships. It describes the system without regard to the distribution mechanism and does not deal with explicit interaction among the information elements.
- *Computational modelling concepts* [1] focus on the decomposition of the system into a set of interacting objects which are candidates for distribution. It provides a framework for the computational specification of distributed systems. It is the objective of the current paper to describe this framework.
- The concepts, mechanisms, and distribution transparencies assumed in the computational specifications of TINA applications have to be supported by the infrastructure, and are described by *Engineering Modelling Concepts* [3]. The execution environment for applications built according to the computational modelling concepts described here is provided by an infrastructure called the *Distributed Processing Environment* (DPE). These concepts provide a means

to specify the structure of a distributed TINA application in terms of the components that are actually distributed on a set of computing nodes, and the models and mechanisms to support their execution and interaction.

1.1 TINA Computational Modelling Concepts

To support the computational view of telecom services, TINA computational modelling concepts have been developed to satisfy the following requirements:

- *Object Structure*: The modelling concepts should address the evolutionary nature of distributed applications. Concepts to describe units of application components that can be independently released or upgraded should be defined. In addition, the modelling concepts specify functional separations that all applications must adhere to. The motivation behind these functional separations is to impose architectural rules and modularity constraints on applications additional to those implicit in object based models, such as encapsulation. The functional separations facilitate reusability of functions and interfaces. Also, they enable deployment, or configuration and flexibility in a multi-supplier environment. These functional separations are related to application decomposition and they do not have any impact on the functions and capabilities required of the DPE.
- *Object Service Description*: The modelling concepts should address both functional and non-functional aspects of distributed applications. The non-functional aspects are those related to quality of service, configuration, and enterprise policy issues. For example, accounting and security policies are non-functional aspects.
- *Object Interaction Description*: Concepts should be defined for describing interactions between application components. Such concepts should encompass both discrete communication and continuous media communication, such as audio and video communication.
- *Object Template Description*: The modelling concepts should define the framework for the specification of application interfaces including both structure and semantics. A suitable notation for computational specifications should be used. This is realized in TINA-ODL, described in Section 3.

1.2 Relationship with Existing Standards

An important requirement of TINA is to build on existing standards and results wherever possible. The Reference Model for Open Distributed Processing and Object Management Group's Common Object Request Broker Architecture (CORBA) are the main sources of influence on the TINA computational modelling concepts.

Open Distributed Processing (ODP)

The computational modelling concepts prescribed in the RM-ODP Prescriptive Model [6] have been used as the basis for the modelling concepts underlying TINA-ODL. Distributed TINA applications are composed of objects interacting with each other. Objects offer their capabilities to other objects via one or more interfaces. Interaction between objects is either by means of operation invocations and responses, or by means of stream flows where each flow is a unidirectional bit stream.

An important concept borrowed from RM-ODP is that of *distribution transparency* [5]. A central theme in the design of computational modelling concepts is that the complex details of mechanisms required for interaction between remote (and possibly independently developed) application components should be invisible in the

computational specification of a distributed application. The process of hiding the effects of distribution on system entities is known as distribution transparency.
Several dimensions of transparency have been identified in the literature (access, location, migration, concurrency, failure, replication transparency) [5], and different applications may have different transparency requirements. Applications specified in TINA-ODL implicitly use access and location transparency.
While RM-ODP has been adopted as the basis for TINA's modelling concepts, it was also felt that additions were needed. Modelling concepts that describe *collections of computational objects having properties in common* are helpful in describing the structure of large applications. These modelling concepts enable application designers to specify more explicitly the granularity at which security checks are imposed, systems management functions operate, and transparencies are selected.
In addition to the new modelling concepts, the additional concepts introduced in this paper are related to specification structures for computational specifications and their underlying syntax, TINA-ODL. In this respect, the scope of this document goes beyond the scope of RM-ODP computational model.

Object Management Group (OMG)
The OMG has developed an object model for distributed applications and has specified an architecture for the infrastructure, called the *Object Request Broker* (ORB). It supports the OMG object model ([7]), viewed as a subset of the RM-ODP computational model. This model focuses on support for object interaction. It defines a notation called Interface Definition Language (OMG-IDL) that allows the specification of the interfaces of entities supporting operation invocations. CORBA entities are seen as interfaces providing services through invocation of their operations.
The TINA Object Model extends the OMG Object Model to support the specification of telecommunications applications, where object interactions encompass discrete interaction (operation invocation), and also continuous interaction (data streams). The support for continuous interaction between objects is provided by the definition of stream flows (Section 2.2), which are additional to the OMG interaction model.
Moreover, the objectives of the TINA object model go beyond object interaction, and include support for object management. For this purpose, TINA Objects are seen as entities supporting several interfaces, providing several "views" on the service provided by an object (Section 2.1). Objects are entities that can be managed (created, activated, migrated, deleted, ...) and grouped together for collective management purposes.
Given the extended objective of TINA object model compared to OMG object model, its supporting language (TINA-ODL) has been defined as an extension to OMG IDL (see Section 3.4).

2 TINA Object Model

2.1 Objects Structure

According to the TINA computational modelling concepts, a distributed application consists of a collection of *computational objects* (hereafter simply referred to as *objects*), which may be collected into *object groups*.
An object encapsulates data (or state) and processing. It provides a set of capabilities (or functions) that can be used by other objects. This collection of capabilities is grouped into one or more subsets. Each such subset of capabilities is called a service.

Thus, an object provides one or more services that other objects can use. To enable other objects to access a service, the object that provides the service offers (or provides) a *computational interface* (hereafter simply referred to as an *interface*) which is the only means by which other objects can use the service. Thus, an object may offer several interfaces, which are the interaction points for other objects. Some of the interfaces provided by an object may be offered when the object is created while some may be offered dynamically (i.e. activated) during the execution of the object.
Objects execute concurrently with respect to one another. The activity structure within an object is not specified in this interaction model. In particular, an object may process concurrently several operation invocations and responses, and thus may have several concurrent activities within it. Depending on the semantics of an interface, concurrent use of the interface by multiple objects may or may not be possible. For example, concurrent use of an interface for file access may be allowed, while concurrent use of an interface for audio communication may not be allowed.

2.2 Object Interfaces

The specification of an interface includes both functional aspects and non-functional aspects. The non-functional aspects include performance and other quality of service parameters, some forms of distribution transparency constraints, configuration information such as the identities of physical resources encapsulated, and cost of using the service. These non-behavioural aspects are specified using *quality of service attributes* associated with interfaces.
Depending on the nature of interactions that occur at an interface, the interface is classified as being either an *operational interface* or a *stream interface*.

Operational Interfaces

The interaction that occurs at an operational interface is structured in terms of invocation of one or more *computational operation* (*operation* for short) and response to this invocation. With reference to an operational interface, the object that provides the interface is called the *server* and an object that invokes operations is called a *client*.
An object may be a client of several interfaces. Note that “client” and “server” are roles played by an object in its interactions with other objects. An object may be a client in its interaction with one object and a server in its interaction with another one. Two objects may exist such that each is a client of an interface offered by the other.
An operation in an operational interface is the interaction mechanism by which a capability (or function) provided by the server can be accessed by a client. Operations are classified into two kinds: *interrogations* and *announcements*. In an interaction via an announcement no result is passed back from the server to the client, and the client is not informed of the outcome (success or failure) of the invocation. In a blocking interrogation, the invoker waits until a response is received for the invoked operation. In a non-blocking interrogation, the invoker does not wait for the response to be received, and retrieves the response at some later time.

Stream Interfaces

A stream interface is an abstraction that represents a communication endpoint that is the source for some information flow and a sink for some other information flows. When objects interact via stream interfaces, the information exchange occurs in the form of *stream flows* between the objects, where each stream flow is unidirectional and is a bit sequence with a certain frame structure (data format and coding) and quality of service parameters. The quality of service parameters include timing requirements for different frames, and synchronization requirements between flows. With reference to a stream flow, the object that is the source of the flow is called the *producer* and the object that is a sink is called the *consumer*. An object that offers a stream interface specifies the stream flows that can occur at that interface, and specifies, for each flow, whether the object is the producer or a consumer. In order for two objects to interact by means of stream flows between them, each object has to offer a stream interface and a *binding of the two interfaces* must be realized.

2.3 Object Group

When telecommunications applications are being built, it is essential to be able to recursively construct composite components from several other components. The composite entity (object group) is viewed as a single entity for the purposes of invoking operations, and the interface of the composite entity hides its internal structure from its clients. The component objects are then said to be encapsulated. Encapsulation is an essential concept for coping with the complexity of distributed systems as it permits subsystems containing multiple composite or primitive objects to be treated as a single entity with an interface.
TINA Computational Modelling Concepts include the notion of an *object group*[8], which is intended to support hierarchical structuring of distributed applications as described in the previous paragraph. An object group can be thought of as an aggregation of objects and includes one *group manager* object, responsible for managing the other objects within the group. The group can be considered a single entity from the point of view of usage, and encapsulates the internal components which normally are not visible from the group interface. However the designer of the group can choose to make some object interfaces visible outside the group. These interfaces then becomes the group's external interfaces or *contracts*. Only the contracts are accessible to objects outside the group. At least one interface of the group manager object must be an external interface.

2.4 Object Dynamics

In a distributed application, objects may interact with each other in complex ways and engage in several activities. While many of these activities are application specific, there are certain generic management operations that can be performed on objects:

- *Object Creation*: Objects can be created. From the computational viewpoint, creation of an object is accomplished by instantiating an object template. It should not be construed that a computational object template contains all information needed for creating an instance of the template. On the contrary, it contains only the relevant information required to specify object creation from the computational viewpoint. Additional information required for object creation must be contained in additional "engineering" templates. It is a matter of system design as to which objects are responsible for the creation of which other objects.
- *Object Deletion*: Objects can be deleted. When an object is deleted, all interfaces of the object are also deleted. Depending on the application design, the deletion operation may be provided by the object itself or by some other object.
- *Interface creation* or *deletion*: The object is requested to offer an interface that is an instance of an interface template identified in the object template, or to withdraw and delete an interface that it is currently offering.

- *Interface deactivation* or *reactivation*: The object is requested to deactivate or re-activate an interface; deactivation means that no interaction is possible via that interface until the interface is activated.
- *Object migration*: During its lifetime, an object may be moved from one location (or node) to another.

Note that this is only a general list. Depending on the granularity of systems management chosen in a design, one or more of the above operations may not be applicable to individual objects. Also, the division of labour between the DPE and application level objects in the realization of these object management operations is not specified here. This is a matter pertaining to DPE design rather than a model issue.

3 TINA ODL

The concrete syntax supporting the computational specifications of object templates, interfaces templates, and object group templates has been developed, and is called, *TINA-Object Definition Language* (TINA-ODL, [4]). The design goals for TINA-ODL centre upon three objectives. These are to provide a language for:

- *Application specification and specification re-use* (at development time): When developing computational specifications, an application developer needs to be able to describe a TINA application in terms of computational modelling concepts (object groups, objects, operational and stream interfaces, and data types). In addition, development effort can be reduced if existing computational specifications of TINA applications can be reused in the specification of new applications.
- *CASE tool development*: Application specification, and the application development process, can benefit from the type of automation typically offered by CASE tools. In order to begin constructing such CASE tools, the language of specification needs to be available and supportive.
- *Application execution and interaction* (at run-time): In order to support dynamic binding and dynamic configuration management of systems, a common syntax is needed to describe the entities involved. The availability of a common language to describe operational and stream interfaces enables the definition of compatibility for interfaces, which is important for binding purposes. Interface type management needs to be extended to object types as supervisory and control systems for distributed applications become more sophisticated. Support for application execution requires the definition of run time interactions constraints, which are naturally specified in a language like TINA-ODL.

3.1 Object Type Declaration

Freedom is offered to the developer of computational specifications using TINA-ODL for independent declaration of interfaces, objects, and object groups. Each interface template in TINA-ODL may be self-contained, and may be reused in any number of object templates. Similarly, object templates may be specified as individual units, and reused in any number of object group templates.

An object template specification comprises as *header* and a *body*. The header supports the declaration of the object's identifier. TINA-ODL allows for the specifications of object template inheritance within a template header, to support specification reuse and to provide a mechanism for defining compatibility via sub-typing relationships. The object template body encompasses the following the sub-parts:

- *Object behaviour specification*, in the form of a natural language. It describes the role of an object in providing services via each of it's interfaces.
- *Object quality of service specification*, in the form of typed parameters representing quality of service attributes. The parameters are intended to allow the specification of the 'level of service' supported by a particular object instance. Such information may be defined dynamically by the management system responsible for initiating object creation. The values of these parameters may be negotiated when an object is instantiated, or changed when the parameters are altered during the lifetime of the object. Examples of such parameters are security constraints on the object location, or reliability constraints in the form of specification of the maximum admitted probability of failure.
- *Object initialization specification*, in the form of the name of an interface template, among those supported by the object, which may be used for initialization. A reference to this interface will be returned to the instantiator of the object template.
- *Required interface specification*, in the form of a list of interface templates. It specifies interface types which will be used by the object to provide its services.
- *Supported interface specification*, in the form of a list of interface templates or names of interface templates. Instances of interface types declared as supported may be offered by instances of objects being defined.

3.2 Interface Type Declaration

The specification of an interface template, which can be operational interface type or a stream interface type, comprises a header and a body. The interface header supports the declaration of an interface identifier. Similar to object template specifications, TINA-ODL allows for the specifications of interface template inheritance relationships. The interface template body encompasses the following the sub-parts:

- *Interface behaviour and usage specification*, in the form of natural language text. It describes the behaviour of the interface and the constraints on the use of its operations or streams.
- *Interface quality of service specification*, in the form of a typed parameter representing quality of service attributes. The type of the parameter is intended to allow the specification of the 'level of service' required by a particular interface instance. The value of this parameter may be set or negotiated when an object is instantiated, or may be altered during the lifetime of the object.
- *Interface trading attribute specification*, in the form of a typed parameter representing trading attributes. These attributes describe properties of an interface as used in constraint specifications when trading for interface references. For each interface type (an operational or a stream interface), it is possible to specify parametrized qualities. These quantities are typically specified by a server (or another object acting on its behalf) and used when exporting an interface reference to a trader. A client of that interface type may express its requirements to the trader in terms of this set of parameters.
- As appropriate, an *operational interface signature*, or a *stream interface signature*.

Operational Interface Specifications
An operational interface signature comprises a set of interrogation and announcement signatures, one for each operation type in the interface. An operational interface signature specifies the following information (similar to OMG-IDL): operation attribute (interrogation or announcement), operation signature (name and parameter types), and an optional exception list.

Stream Interface Template
A stream interface signature comprises a set of flow types, called *action templates*. Each action template for a flow contains the name of the flow, the information type of the flow, and an indication of whether it is a producer or consumer (but not both) with respect to the object which provides the service defined by the template. A stream interface signature specifies the following information (defining the stream server from the viewpoint of the client):

- *Signature of each stream flow* that can occur at instances of the interface type. it specifies the frame structure and coding details associated with the flow, and a producer/consumer attribute that specifies whether an object that offers the interface is a producer ("`source`") or consumer ("`sink`") of the stream flow.
- Name and type of each *service attribute* associated with each stream flow.

3.3 Object Group Type Declaration

An object group template specification comprises a header and a body. The header supports the declaration of the group's identifier and possible inheritance relationships. The object group template body encompasses the following sub-parts:

- *Group behaviour, quality of service,* and *initialization specification*: The behaviour and quality of service specifications of an object group strongly parallels that of an object. The initialization specification of an object group is slightly different to that of an object in that it specifies an object template to be instantiated instead of an interface template. In both cases an interface template is expected to be returned to the creator. It should be noted that the manager object must be one of the component objects and the returned interface, one of the contract interfaces.
- *Group Contract Specification*, in the form of a list of interface templates. Contract interfaces are the interfaces that are visible to entities outside the object group.
- *Supported objects and object groups specification*, which describes the object and object group templates that can be instantiated within the object group.

3.4 Relationships with Other Languages

A number of existing syntaxes (such as ANSA-IDL, DCE-IDL, or OMG-IDL) are capable of supporting some aspects of TINA computational specifications. Among them, OMG-IDL appears to be the most appropriate base for TINA-ODL. Its object model is compatible with that of the TINA-C architecture, and it enjoys rapid growth in industry acceptance. TINA-ODL has been developed as a super-set of OMG-IDL.

TINA ODL vs. OMG IDL
It should be stressed that OMG-IDL and other IDL languages do not fully support the concepts defined in the TINA computational model. In particular, while OMG-IDL shares most of the objectives of TINA-ODL (support to application specification, application re-use, and application interaction), OMG-IDL does not provide any support for application execution. A number of TINA computational modelling

concepts supported by TINA-ODL do not find any equivalence in OMG object model and hence OMG-IDL. TINA-ODL extends OMG-IDL in the following ways:

- At the *object level*, TINA-ODL offers support for the definition of object behaviour, object QoS requirements, objects with multiple interfaces, and object groups.
- At the *interface level*, TINA-ODL offers support for the definition of stream interfaces, interface behaviour, and operation and stream QoS requirements
- At the *operation* and *stream level*, TINA-ODL offers support for the definition of stream signatures
- At the *application structure level*, TINA-ODL supports aggregation of larger abstractions, called object groups.

TINA-ODL is a superset of OMG-IDL: the syntax defined in TINA-ODL for operational interface declaration encompasses and supports all the rules defined for OMG-IDL. It implies that OMG-IDL specifications are a part of TINA-ODL specifications (as operational interface declarations).

ITU-T SG15 work
Work is on-going in ITU-T, study group 15, for the application of ODP principles to the specification of telecommunications management systems [9]. As a support to these specifications, an ODL-like language is in the process of being defined. This language does not currently provide support to all the concepts of TINA-ODL, such as the concept of streams and object groups.

4 Conclusion

The paper has presented the main computational modelling concepts developed in TINA-C together with their underlying requirements. Given the broader objectives, and also the specific application domain of telecommunications services, extensions have been developed to existing standards such as RM-ODP or as specified by OMG. In particular, the computational specification language TINA-ODL extends OMG-IDL with concepts, such as object groups, multiple interfaces per object, quality of service requirements, and streams, supporting telecommunications applications.
Support for telecommunications applications, specified according to the concepts presented in this paper, requires the provision of a Distributed Processing Environment infrastructure. In order to see rapid commercial availability of TINA-compliant DPEs, these should be an extension of commercial CORBA platforms. This would be facilitated by the integration of TINA computing architecture concepts into the OMG standard. Contributions on behalf of TINA-C have already been injected into this group, and recognition of some of its requirements has already been achieved (see [10] or [11] for example). It is of paramount importance that the TINA-C community and the OMG community join efforts to achieve complete support for requirements and concepts underlying the telecommunications information networking architecture.

Acknowledgements

The authors would like acknowledge Barry Kitson for his substantial efforts in progressing the TINA Object Definition Language. The permission of the Director, Telstra Research Laboratories, Telstra, to publish this paper is hereby acknowledged.

References

[1] *Computational Modelling Concepts*, Document TB_NAT.002_3.1_94, TINA-C. May 1996 (available from http://www.tinac.com/deliverable/deliverable.htm)

[2] *Information Modelling Concepts*, Document TB_EAC.001_3.0_94, TINA-C. April 1995 (available from http://www.tinac.com/deliverable/deliverable.htm)

[3] *Engineering Modelling Concepts (DPE Architecture)*, TB_NS.005_2.0_94, TINA-C. December 1994 (available from http://www.tinac.com/deliverable/deliverable.htm)

[4] *TINA Object Definition Language (TINA-ODL) MANUAL*, PBL01, TINA-C. July 1996 (available from http://www.tinac.com/deliverable/deliverable.htm)

[5] ISO/IEC JTC1/SC21 N8538: *Basic Reference Model of Open Distributed Processing - Part 2: Descriptive Model*, International Organization for Standardization and International Electrotechnical Committee, April 1994.

[6] ISO/IEC JTC1/SC21 N7525: *Basic Reference Model of Open Distributed Processing - Part 3: Prescriptive Model*, International Organization for Standardization and International Electrotechnical Committee, April 1994.

[7] *CORBA Fundamentals and Programming*, Jon Siegel & al., John Wiley and Sons ed., New York, ISBN 0471-12148-7, 1996

[8] *Object Groups: Patterns in Chaos*, T.Handegard, A.Parhar, Proc. TINA'96 Heidelberg Germany, September 1996.

[9] ITU-T, Study Group 15, Draft Recommendation G-851-01, ITU, November 1995

[10] *Multiple Interfaces and Composition Request for Proposals*, OMG Document OMG/96-01-04

[11] *Streams and QoS: A White Paper*, OMG Document telecom/96-02-01

Network Resource Adaptation in the DOLMEN Service Machine

M.P.Evans[†], K.T.Kettunen[‡], G.K.Blackwell[†], S.M.Furnell[†], A.D.Phippen[†], S.Hope[§] and P.L.Reynolds[†]

[†] Network Research Group, Faculty of Technology, University Of Plymouth, Plymouth, United Kingdom.

[‡] VTT Information Technology, P.O. Box 1202, Espoo Finland.

[§] Orange Personal Communications Services Ltd., St. James Court, Great Park Road, Bradley Stoke, Bristol, United Kingdom.

In addressing the move in the telecommunications field toward Integrated Service Engineering, the paper introduces the concept of resource adaptation in a service environment. The relationship between Resource Adapter, Service Node and Service Machine is examined, along with the role of RM-ODP viewpoints in the identification and development of Resource Adapters. Further analysis of design aspects is presented and implementation requirements are discussed. Finally, the paper demonstrates the relationship between Resource Adapters and the TINA Connection Management Architecture, with an example Resource Adapter being presented.

1 Introduction

In recent years the telecommunications industry has moved away from single solution networks towards Integrated Service Engineering, the ultimate goal being the definition and development of an open, extendible network offering a wide range of fixed and mobile services.

The DOLMEN (Service Machine Development for an Open Long-term Mobile and Fixed Network Environment) project is one of a number of initiatives currently being funded by the European Commission under the ACTS (Advanced Communications Technologies and Services) programme. The core objective of the project is the development and implementation of a new telecommunications architecture, entitled OSAM (Open Service Architecture for a Mixed fixed and mobile environment), based upon the convergence and combination of concepts from the TINA[1] and OSA[2] architectural frameworks. In practical terms, OSAM makes use of a so-called *Service Machine*, which is responsible for the provision and control of telecommunications services, separating this aspect from the underlying technologies.

The Service Machine is logically distributed over a service network, which itself comprises a set of interconnected service nodes residing in identifiable physical entities. These provide access to the underlying resource infrastructure which the Service Machine uses to establish telecommunications relations between users. However, enabling the Service Machine to make use of the infrastructure is problematic, due to differences that may be encountered in terms of both service node computing platforms and telecommunications technologies involved. As such, a mechanism is required to provide an interface and resolve incompatibilities. This role is performed by a *Resource Adapter* (Resource Adapter), an engineering entity which enables the functionality of the resource infrastructure to be reachable in the service node.

More specifically, Resource Adapters are provided so that the following requirements of the DOLMEN Service Architecture can be realised :

design support should be independent of signalling and transport techniques;

design support should not be dependent upon the physical architecture;

the architecture should ensure compatibility with existing networks and services.

Examples of the resources for which adaptation may be required include transmission links, switching devices, legacy databases and existing software.
Before proceeding with the specifics of resource adaptation design, it is necessary to outline the role resource adaptation has to play in the DOLMEN Service Machine. The paper will examine the role of Resource Adapters in the Service Machine, based firstly on different viewpoints as defined in RM-ODP, and then examining less abstract views, examining where the Resource Adapters will fit into the Service Machine architecture; a closer look will also be given at the Resource Adapter's place in a given resource node. The paper will then focus on design and implementation aspects of the Resource Adapters, before finally presenting an example adapter.

2 Resource Adaptation in the Service Machine

2.1 The Service Network and Resource Infrastructure

A service network is modelled as a set of interconnected service nodes. The Service Machine is distributed over a set of these interconnected service nodes, forming the service network. A service node provides access and adaptation to telecommunications resources and resides in an identifiable physical entity. The service nodes are not involved in the propagation of streams; rather, they control the propagation through the networks formed by the interconnected resource nodes.
The Service Machine relies on the underlying resource infrastructure to actually achieve the goal of establishing telecommunications relations between IBC users. In the Service Machine, the accessible capabilities of the resource infrastructure are seen as services which are provided by resource components and are realised through the existence of a pool of Resource Adapters.

2.2 The Service Node

A service node represents a model for a collection of computing resources at one location. Each node hosts templates of a number of OSAM-components and is the location for object instances created from these component templates. These object instances are referred to as *engineering Computational Objects (eCOs)* [4], and implement computational objects of the Service Machine (see section 3).
The Resource Adapter provides the means to adapt non-OSAM interfaces of a resource (for example, an ATM switch with a proprietary management interface) to OSAM-specific objects hosted by a service node of the Service Machine. That is, the Resource Adapter adapts the resource to the Service Machine by providing the OSAM-compliant eCOs hosted by a service node with an interface to the non-compliant resource.
This, then, is the fundamental role of the Resource Adapter in the DOLMEN Service Machine: to adapt non-OSAM-compliant resources to the Service Machine such that they can be controlled and managed as if they were an integral part of it.

3 RM-ODP Viewpoints on Resource Adaptation

Section 2 gives a brief overview of the Resource Adapter's role, and its physical location in the DOLMEN Service Network. However, a Resource Adapter is a software entity and, as such, must communicate and interact with other software entities. In order to define the adapter's role and location more clearly, therefore, adaptation must be examined from a more abstract perspective; or, to be more precise, from the Reference Model for Open Distributed Processing (RM-ODP) Computational and Engineering *viewpoints*.

3.1 Computational and Engineering Viewpoints

OSAM makes use of the RM-ODP concept of different viewpoints, the most relevant in this context being the Computational and Engineering Viewpoints. RM-ODP defines the Computational Viewpoint as focusing "...on the expression of the functional decomposition of an ODP system, and of the interworking and portability of ODP functions"[5]; that is, the Computational Viewpoint focuses on *what* needs to be done. In contrast, the Engineering Viewpoint is defined as focusing "...on the expression of the infrastructure required to support distributed processing" [5]; that is, the Engineering Viewpoint focuses on *how* things are to be done.

In TINA these two viewpoints are supported by Computational Objects (COs) and engineering Computational Objects (eCOs), respectively. An object encapsulates data (or state) and processing [3] and provides a set of capabilities (or functions) that can be used by other computational objects [4]. OSAM also uses these concepts, being consistent with TINA's terminology. A service in OSAM is represented by an OSAM-component. This can be translated into a set of COs in the computational viewpoint which, together, describe what needs to be done. These COs can, in turn, can be mapped onto corresponding eCOs, which implement the functions defined by the COs.

In the computational view, a CO does not care where another CO is located; thus, it does not concern itself with the distributed nature of the system. The CO is therefore said to require *distribution transparency*.

An eCO, in contrast, does concern itself with location. An eCO encapsulates the characteristics of its related CO and must also provide the distribution transparency that the latter requires.

3.2 Adaptation as seen from the Computational and Engineering Viewpoints

Relating these viewpoints to DOLMEN, the concept of *Service Machine* has been introduced to collectively refer to the *Computational Viewpoint* of a system that is intended to provide a wide range of telecommunications and IT services over heterogeneous fixed and mobile networks. The computational viewpoint is concerned with the functional description of the system in terms of objects that interact at interfaces, thus enabling system distribution. Such a system is engineered as a Service Network. The concept of *Service Network* provides the necessary mechanisms to support system distribution, and so can be seen to represent the engineering viewpoint of the Service Machine.

With regard to resource adaptation, from the computational viewpoint, a resource provides its services to other computational objects and so its capabilities are encapsulated in resource objects known as 'resource components'. In OSAM the concept of resource component is used to denote a component that models services of resources, thus making them available in the Service Machine. As such, a resource component is an OSAM component which is statically bound to entities in the resource infrastructure.

To make a resource component available in the Service Machine, however, requires the design of a Resource Adapter which is specific to both the resource of the resource node and the computer hosting and implementing the service node functionality [3]. A Resource Adapter, therefore, is an engineering entity which makes the functionality of the resource being adapted available to the Service Machine.

Within the Service Machine, resource components are different from OSAM components with respect to their instantiation and management. Also, the roles of 'application services' (i.e. services offered by OSAM components) and 'resource services' (i.e. services of resources offered by resource components) which are available in the Service Machine are rather different: an OSAM component makes use of services provided by a resource (through resource components), while a resource component makes the services of a resource available to the Service Machine to be used by OSAM components [3].

To summarise, from the computational viewpoint, applications and services are composed of COs, describing what needs to be done without regard for the location of either themselves, or other COs. If a CO requires the services of a resource, it communicates with that resource's resource component. The functionality of each CO (and thus the application or service described by it) is realised through the use of equivalent engineering COs. These are deployed COs in the engineering viewpoint, which each encapsulate the functionality of their corresponding CO, but which provide mechanisms to implement the required distribution transparency.

An eCO is a distributed object that forms part of a distributed application, sitting within a Distributed Processing Environment (DPE), on top of the Distributed Processing Platform (DPP). If an eCO requires the service of a resource, it must communicate with that resource's adapter (acting, effectively, as a *deployed* resource component). This performs the same functions as the resource component, but from the engineering perspective. It is an engineering entity which is comprised of two parts: one part that resides in the DPE, and provides a unified, generic interface to the type of resource being adapted; another which is specific to the technology of the resource being adapted, and which is comprised entirely of objects sitting beneath the DPE.

The services offered by a resource can come in many flavours, reflecting the types of resource available to the Service Machine. For each type of resource there is a corresponding adapter. Note that from a computational viewpoint, all these resource services are modelled as computational objects and reside in the Service Machine as resource components. However, it is the engineering viewpoint in which the Resource Adapters actually provide these services to the Service Machine.

4 Toward Resource Adapter Development

While the above presents the overall requirements of resource adaptation from different viewpoints, in order to demonstrate the OSAM architecture, the Resource Adapter requirements need to be developed into a more concrete definition.

We should consider the central role of Resource Adapters as adapting non-OSAM compliant resources to the Service Network, such that the resource is visible and accessible to the Service Machine. We have two considerations; where the Resource Adapters fit in the Service Machine, and at a lower level, where the Resource Adapters fit in a given Service Node.

4.1 Identifying Resource Adapter Locations from a Service Machine Perspective

Resource Adapters may reside in a number of areas within the Service Machine, as indicated in figure 1. The location of the first "type" of Resource Adapter lies between applications and Transport Network. This adapter's use lies in an application requiring services from a resource that could not be managed by the Service Machine (in the case of legacy databases, etc.). The second "layer" of resource adaptation is concerned with the management of network resources to establish physical connections in the transport network. It is this level of resource adaptation that would be required in an ATM switch adapter (see section 6.2).

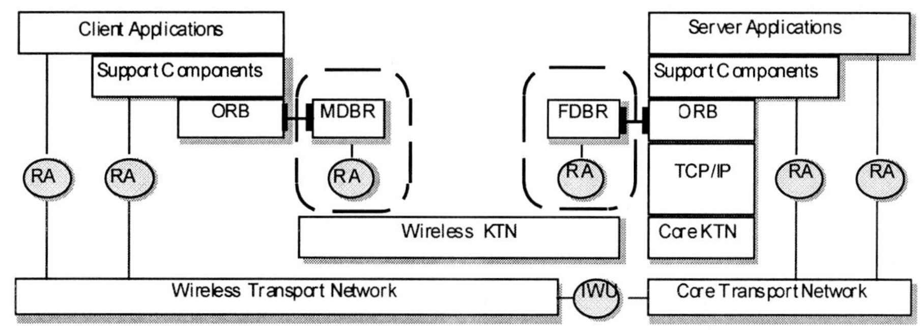

Figure 1: Resource Adapter location in the Service Machine

Finally, we are concerned with enabling signalling information, the kernel Transport Network (kTN), between the Service Machine's fixed and mobile domains. In a CORBA-based DPE such as the DOLMEN Service Machine, we can consider signalling information to be inter-object communications. If current technology is used without adaptation, the restrictive bandwidth of the mobile link would considerably slow down the kTN. Therefore we require an adaptation of the object communications to enable a fast transmission of the signalling information over a mobile connection. An interoperability mechanism between the ORB in the fixed and the ORB in the mobile domain is established based upon the bridging mechanisms defined in [4], with a lightweight interORB protocol (LW-IOP) developed to enable a stripped version of the inter object call to be transmitted. Bridging between the Fixed Domain Bridge (FDBR) and the Mobile Domain Bridge(MDBR) will strip down and rebuild the object call to allow transmission using the LW-IOP. The role of the Resource Adapter in this situation is again management of any resource (probably in the domain of mobile network management) and also it is envisaged that the Resource Adapter will provide a multiplexing mechanism to allow both signalling (kTN) and user (Transport Network) information to be transmitted over the same radio link.

4.2 The Resource Adapter in a Service Node

Following the identification of where resource adaptation is required, we need to focus on the overall requirements of a given Resource Adapter. We can consider the central role of a Resource Adapter to be the adaptation of non-OSAM compliant resources to the Service Network, such that the resource is visible and accessible to the Service Machine. As such, a Resource Adapter must contain two interfaces (see figure 2): a technology-specific *Lower Interface*, which connects onto the resource being adapted, or to a management interface of such a resource; and a general *Upper Interface*, which should make the resource's functionality available to the Service Machine. The generality of the upper interface should be such that the Service Machine component can control any resource of a given type (for example, an ATM switch) using the same interface and function calls. The Resource Adapter has a requirement to take the function call from it's upper interface and translate the into a function call specific to the technology of the underlying resource.

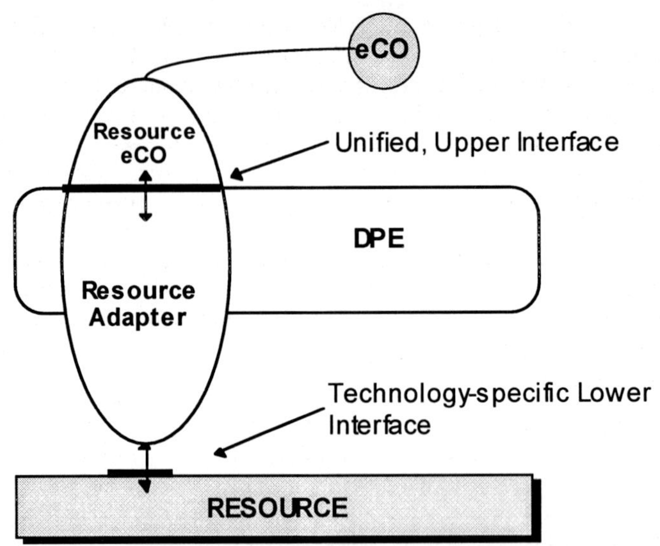

Figure 2: Resource Adapter Interfaces

5 Adaptation to Mobility

A major aspect of the DOLMEN project is the issue of both personal and terminal mobility in the Service Machine. Clearly the fact of terminal mobility means that the topology of the network(s) involved is subject to change as mobile terminals detach from one access point and attach to another - with possible periods of inaccessibility between. Quite apart from the issue of restricted bandwidth, as referred to above, this raises questions of connection reliability and (most significantly) routing to-from mobile terminals. Such considerations apply equally to both the kernel Transport Network and the Transport Network.

Whilst not realisable in a single software entity, adaptation to terminal mobility is a real issue for which a defined strategy and support structures are required. The nature of those structures, and their interrelation, is part of the current development in DOLMEN; further information on that development will be released in future publications.

6 Resource Adapter Design Considerations

If we now consider the implementation aspects of the Resource Adapters, the architecture and adapter analysis places a number of requirements in the final design and implementation of sample Resource Adapters:

- the adapters should be designed in such a way that the Service Machine does not need to be aware of the technology specifics of the underlying resource (accessed through a generic upper interface);
- the architecture is intended to be a long term, therefore existing adapters should be able to provide a framework for the derivation of new adapters. New Resource Adapters should be able to re-use a large amount of existing code;
- as the environment is one of a very real time nature, resolution of lower level function calls (the technology specifics) should be dynamic. It is of no good to the environment to merely resolve the technology specifics at compile time. This would create an entirely non-adaptive environment.

The realisation of these requirements is aided by the Object Oriented (OO) nature of the implementation and the exploitation of the OO "triad" of encapsulation, inheritance and polymorphism. It is envisaged that, at each level of Resource Adapters (see below) abstract objects will define the required "functionality" for a given object, objects derived from the abstract classes will implement the functionality. Therefore, the uppermost object of a Resource Adapter can be provided with the required functionality at design time, and this functionality can be realised in the run-time environment in a dynamically polymorphic manner. Interface derivation can also be used to provide "templates" for new resources adapters based on, but not identical to, those for existing resources (for example, if a new generation of ATM switch were developed with added functionality, a new interface could be provided at the upper level by deriving from the existing ATM switch interface and adding the functionality in the new interface).

6.1 Initial Object Design for Resource Adapters

Following requirements definition it is now necessary to consider software internals which would make up a Resource Adapter. There is a need to identify the required object classes in order that abstract classes can be defined to provide an "adapter framework" from which actual adapters can be derived.

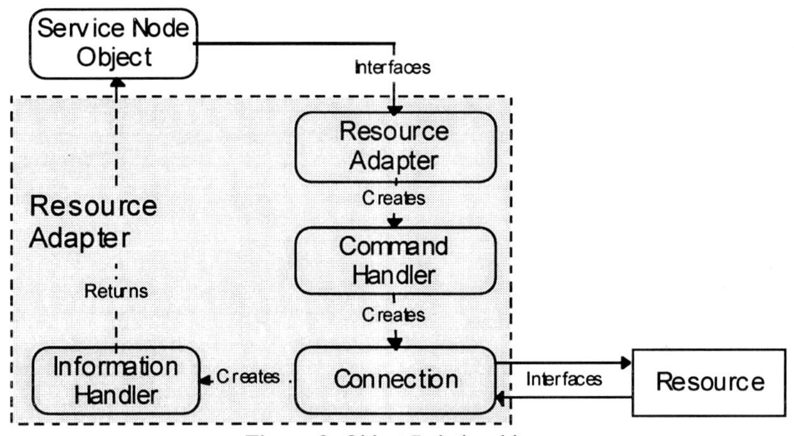

Figure 3: Object Relationships

A useful starting point for the object design lies within protocol adaptation needs on the World Wide Web (WWW). Whilst the complexity of DOLMEN's needs may be greater than that of a WWW application, the underlying mechanisms, of utilising a specific resource in a common way, can be seen to be very similar. Object structures defined and discussed in [7] provide a basis for object definitions. By applying the DOLMEN requirements to the discussion, four objects types have been defined. These should be considered the abstract base object from which "real" objects can be derived. Figure 3 illustrates the relationships between the objects:

ResourceAdapter: Can be considered as the *Upper Interface* of the Resource Adapter for DOLMEN for implementation purposes. The *ResourceAdapter* object is the only object in the Resource Adapter that the Service Machine will interact with and is responsible to lower level object creation based on calls from the given connection performer object in the Service Machine; it should resolve which *CommandHandler* object should be created, based on resource type, etc.

CommandHandler: A handler object that needs to be implemented to allow for protocol specifics, and handle command resolution. The CommandHandler is responsible for creating the appropriate *Connection* object to deal with lower interface command resolution.
Connection: The *lower interface* in DOLMEN terminology. An object that can interface with the resource in a way the resource understands. In the event of information being passed back to the upper *ResourceAdapter* object, the Connection object should create the appropriate *InformationHandler* object.
InformationHandler: This object will take resource-specific information from the *ResourceConnect* object and structure it in a way that will enable it to be passed back through the upper interface to the connection performer in a common format.

7 Resource Adapters and the TINA Connection Management Model

The following section is included to demonstrate the relationship between the TINA Connection Management Architecture (CMA) and Resource Adapters. A sample Resource Adapter is then illustrated to bring together this and the previous section, demonstrating how the principles discussed in this paper so far can be implemented.
It is the functionality provided by the TINA CMA that manages the switching elements of the resource infrastructures external to the DOLMEN Service Machine and allows the establishment of the stream between the end-user applications.
The CMA, adopted by DOLMEN, is a framework to model transport network connection establishment from a TINA environment. It is a service application, used by OSAM, which will run in the OSAM service environment and comprises the functionality for setting up, maintaining and releasing telecommunication connections (streams) on a connection-oriented basis [3]. From the CMA, the Communication Session Manager (CSM) has the task of setting up a stream, but if the resource is non-OSAM -compliant, this can only be achieved via a Resource Adapter.

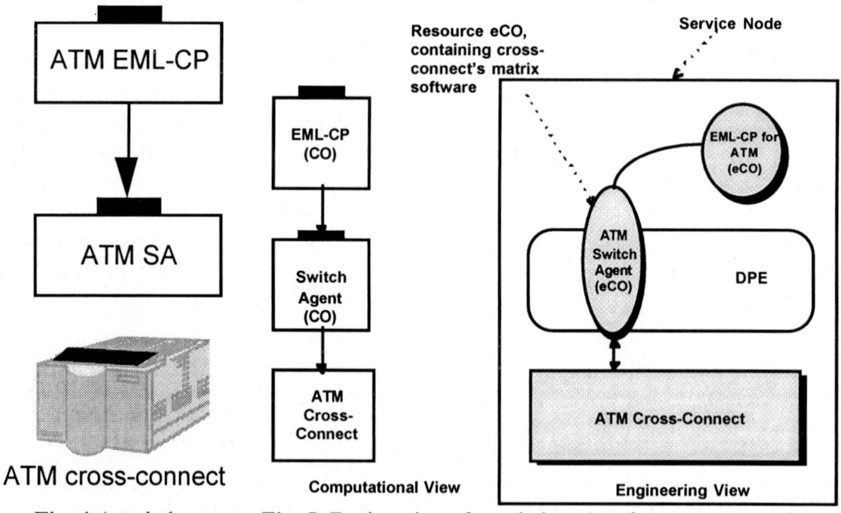

Fig. 4:A switch agent offering services to a connection performer

Fig. 5: Engineering of a switch agent of a cross connect is a service node

The ability to establish a cross-connection within a single ATM cross-connect is required in the lower part of the Connection Management (CM) service. CM is modelled as any other service, as a set of interacting computational objects (running 'on top' of the DPE). The CMA defines a hierarchy of objects involved in communications management, with the CSM being the top level object, and an object called the Element Management Layer-Connection Performer (EML-CP) being the bottom-level object. This object is responsible for a network element; in this example, the network element is the resource being adapted, i.e. the ATM cross-connect. From [1], the EML-CP will interact with an object acting as a proxy for the cross-connect. As the resource is a switch in this example, such an object is called a Switch Agent (SA), and its services are requested by the EML-CP. Figure 4 depicts this relation. Note that the EML-CP is technology dependent, hence it is an EML-CP for ATM, and the SA is vendor dependent.
The EML-CP is the last component in the TINA Connection Model, and is part of the Service Machine. The ATM-Cross-Connect is the physical network resource. TINA views all resources as being TINA-compliant, and if the ATM cross-connect in this example were TINA-compliant, then simply implementing the SA would give the EML-CP the required functionality to control the resource. For the ATM cross-connect to be TINA-compliant, however, it would have to be a service node, with the matrix software implemented by an eCO corresponding to the Switch Agent CO. Figure 5 depicts the engineering of a cross-connect with service node software.
However, no cross-connect is currently available that is also a service node, i.e. where the matrix software is encapsulated in an eCO. This is quite logical, keeping in mind that (de-facto) standardised service node implementations, e.g. CORBA, have not been well established. What is needed, is a Resource Adapter to adapt non-OSAM-compliant (or non-TINA-compliant, the terms are interchangeable in this specific example) resources to the Service Machine.

7.1 ATM Cross Connection Resource Adapter

If we take the example discussed above, we can develop it to show how the Resource Adapter would be built into the resource component. If we refer back to figure 4, we see an EML-CP connected directly to the ATM switch agent. However, in reality a service node that is also a cross connect does not exist. Therefore we require a means to interface with the cross connect. Nowadays, cross connects offer services through network management protocols such as CMIS and SNMP. If we are to take SNMP for this example, what is required is a Resource Adapter that can interface with the SNMP manager of the cross connect while providing the resource's services to the service node. Diagramatically, we have a situation as illustrated in figure 6.:

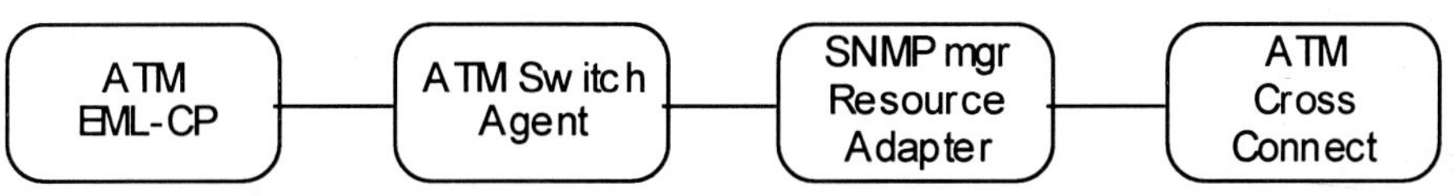

Figure 6: Resource Adapter Location in ATM Cross Connect Resource

If we combine this information with the object structure discussed in section 5, we have the necessary design for the required Resource Adapter:

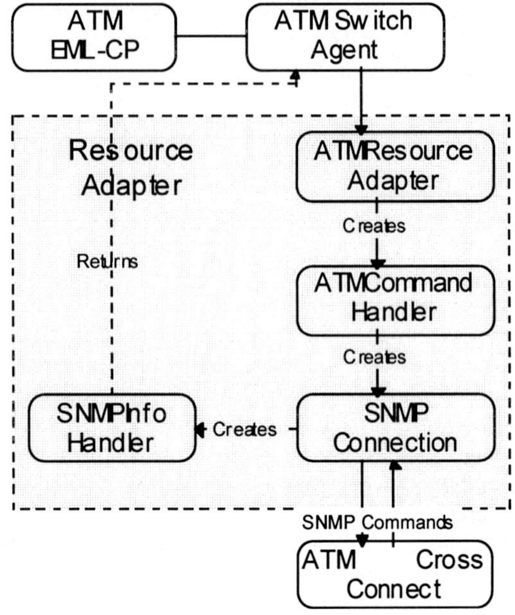

Figure 7: Design for an SNMP manager Resource Adapter

8 Conclusions

Resource adaptation provides us with a means of exploiting resources external to the DOLMEN Service Machine in an open and long term manner. The design of the Resource Adapters should be carried out in such a way to help realise the long term viability of the DOLMEN architecture.

Future work in the DOLMEN project will further develop the resource adaptation principles and refine development and implementation of adapters for use in a comprehensive trial in the final year of the project. These will be developed in such a way that the implemented objects will provide a framework for new adapters. Technologies that will be adapted to as part of the trial include ATM, GSM and Wireless LAN.

References

[1] TINA-C 'Overall Concepts and Principles of TINA'. Public Deliverable February 1995.

[2] CASSIOPEIA. 'Open Services Architectural Framework for Integrated Services Engineering'. Deliverable R2049/FUB/SAR/DS/P/023/b1. http://www.fokus.gmd.de/step/cassiopeia/

[3] TINA-C. 'Computational Modeling Concepts'. Public Deliverable February 1995.

[4] TINA-C. 'Engineering Modeling Concepts' Public Deliverable December 1994.

[5] Open Distributed Processing Reference Model. 'Part 3: Architecture'. February 1994. http:// www.iso.ch:8000/RM-ODP/part3/

[6] OMG. 'The Common Object Request Broker: Architecture and Specification', Revision 2. July 1995. http://www.omg.org/public-doclist.html

[7] Niemeyer and Peck (1996). 'Exploring Java'. O'Reilly. ISBN 1-56592-184-4

A TINA based Prototype for a Multimedia Multiparty Mobility Service

Javier Huélamo & Hans Vanderstraeten, Alcatel Telecom
Juan C.Garcia, Telefónica I+D
Paolo Coppo, Centro Studi e Laboratori Telecomunicazioni
Juan C.Yelmo, Universidad Politécnica de Madrid
George Pavlou, University College London

1 Introduction

The Telecommunication Information Networking Architecture (TINA) is being defined by an international consortium (TINA-C) constituted by over 50 companies including the major network operators, computer suppliers and network equipment vendors. TINA is aimed to enable the efficient creation, deployment, operation and management of a wide range of services; taking full advantage of the most recent distributed processing and object orientation technologies; and paving the way for the emerging global multi-supplier multi-operator telecommunication sector. TINA-C was formed at the end of 1992. The period from 1993 to 1995 was devoted to the definition of this open distributed architecture. In 1996, a great effort had been invested to validate and consolidate TINA with the target of achieving a seamless and consistent reference architecture.

VITAL (Validation of Integrated Telecommunication Architectures for the Long term) is both an ACTS project and a TINA auxiliary project whose main objective is to develop, demonstrate and validate a collection of reusable components based on TINA for developing, deploying, managing and using services composed of multimedia, multiparty and mobility service features. The project has adopted a phased approach where three European wide field trials will be carried out involving heterogeneous national host infrastructure. In each phase, a set of service features, stakeholders, subsystems, components, and transport infrastructure elements will be selected to validate several architectural aspects taking as main input the last version of the TINA deliverables available in the design stage. The validation results of each trial will provide feedback to TINA to refine and complete the architecture incorporating these improvements in new versions of the TINA deliverables which will be the base for the next trial. This phased approach not only allows a rapid first validation of the architecture, but will also evaluate its extendibility and reusability, both in terms of providing additional service features and using heterogeneous underlying transport technology. At the end of each phase feedback will be collected from all the relevant stakeholders.

This paper is based on the results [1] of the VITAL 1st trial (VITALv1) which was finished at the end of October 1996. The following TINA business roles are supported: consumers, a retailer and a connectivity provider. This trial has been installed and demonstrated in the Spanish and Italian national hosts. After presenting the overall architecture defined for VITALv1, all the developed subsystems and their components are described in detail mentioning the main refinements and extensions with respect to TINA.

2 Overall Architecture for VITAL 1st Trial

The TINA based architecture defined for VITALv1 consists of components in the consumer or private domain, and in the provider or public environment which is composed of one retailer domain and one connectivity provider domain. In VITALv1, there are two types of terminals: end-user terminals and operator terminals. The end-user terminals in the consumer domain contain the end-user applications (EA) and a set of components which interact with other ones located in the retailer domain. Two end-user applications have been implemented in VITALv1: a session control application and a desk top video-conference application. These applications support the interaction with end-users; the interaction with subscribers will be incorporated in the 2nd trial (VITALv2). Therefore the subscription data is introduced by the retailer using an operator terminal. The operator terminals contain two types of applications: subscription management application (SA), which allows the interaction with the retailer, and resource configuration management application (CA) for the connectivity provider.

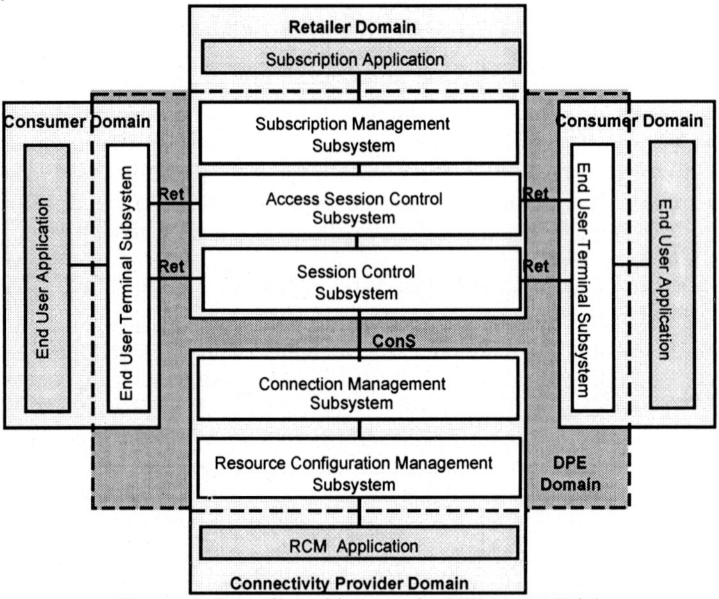

Figure 1. Overall Architecture for VITAL 1st Trial.

The architectural components developed are distributed units or computational objects (COs). These components are over a Distributed Processing Environment (DPE) installed in the terminals and in computing nodes within the retailer and connectivity provider domains. The DPE selected for VITALv1 is the IONA's Orbix 2.0. which is CORBA compliant. All the terminals in VITALv1 will be based on workstations with DPE. The DPE kernels are interconnected by means of a kernel transport network over TCP/IP. The transport network used is an ATM network provided by the Spanish National Host (NH) and the Italian NH. These NHs will not be interconnected in VITALv1.

The components of VITALv1 are grouped in the following subsystems: End-User Terminal Subsystem (EU), Access Session Control Subsystem (AS), Session Control Subsystem (SC), Subscription Management Subsystem (SM), Connection Management Subsystem (CM), and Resource Configuration Management Subsystem (RC). Figure 1 depicts the developed applications and subsystems within the corresponding domains. Currently two TINA reference points (RP) are partially supported: the Ret-RP between the consumer domain and the retailer domain, and the ConS-RP between the retailer domain and the connectivity provider domain. New components such as the Provider Agent will be incorporated in the next trials to align the architecture with the 1996 TINA deliverables .

Figure 2 presents the main COs and their interfaces in the consumer and retailer domains. The EU subsystem of the consumer domain contains the following COs: User Application (UAP), Generic Session End Point (GSEP) and some COs associated with the connection management in the terminal.

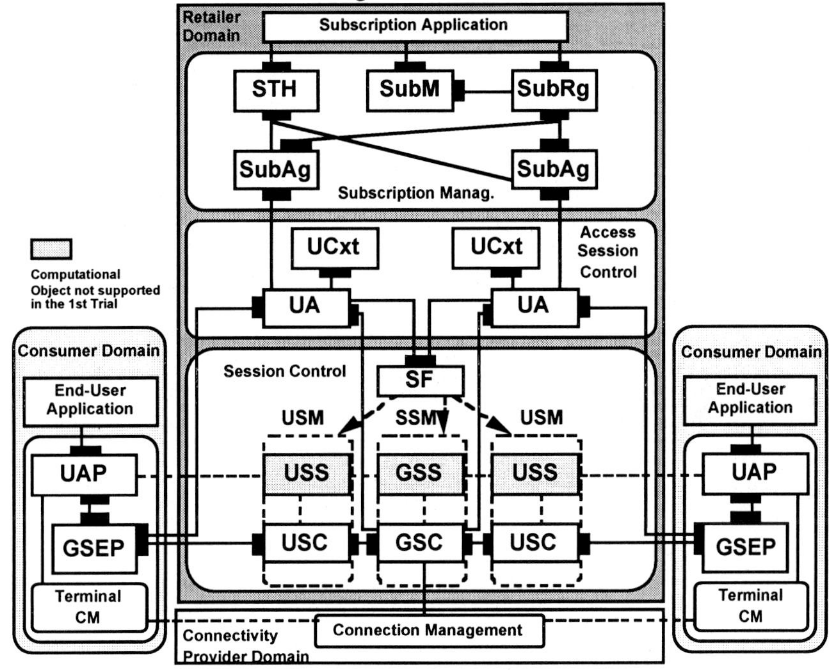

Figure 2. Main COs in Consumer and Retailer Domains

The retailer domain consists of the three subsystems: the SM subsystem which interacts with the SA application, the AS subsystem which interacts with the EU subsystem in the consumer domain, and the SC subsystem which interacts with the CM subsystem in the connectivity provider domains.

The SM subsystem is composed of the following COs: one Subscription Agent (SubAg) per User Agent, Service Template Handler (STH), Subscriber Manager (SubM) and Subscription Register (SubRg).

The AS subsystem is formed by the following COs: one User Agent (UA) and one User Context (UCxt) per end-user. The UA interacts with the Service Factory to create instances of user sessions and service sessions.

The SC subsystem consists of the CO Service Factory (SF) and two types of building blocks: User Session Manager (USM) and Service Session Manager (SSM). These building blocks contain service specific COs and service independent COs. The User Service Segment (USS) belonging to the USM and the Global Service Segment (GSS) belonging to the SSM are service specific COs, these components have not been developed in VITALv1. The User Service session Control (USC) belonging to the USM and the Global Service session Control (GSC) belonging to the SSM are service independent COs. The UAP will interact via the GSEP with the USC and GSC in VITALv1.

The main COs and their interfaces in the connectivity provider domain are shown in Figure 3. This figure also presents the main information objects handled in each interface of the CM subsystem. This domain is formed by the CM subsystem and the RC subsystem.

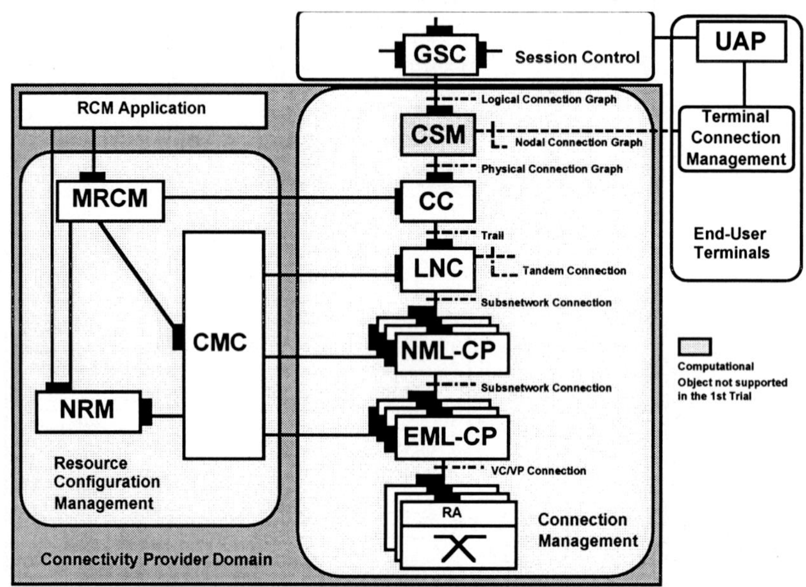

Figure 3. Main COs in Connectivity Provider Domain

The CM subsystem contains the following COs: Communication Session Management (CSM) which converts logical connection graph requests to physical connection graph requests and it will not be developed in VITALv1; Connection Coordinator (CC) which converts physical connection graph requests to trail requests in the corresponding layer network and it will interact directly with the GSC in VITALv1; Layer Network Coordinator (LNC) which converts trail requests in sub-network connection requests; Network Management Layer-Connection Performer (NML-CP) which converts sub-network connection requests in children sub-network connection requests, LNC could request tandem connections to other LNC via its federation interfaces although this feature will not be supported in VITALv1; Element Management Layer-Connection Performer (EML-CP) which converts sub-network connection requests in VC/VP connections in the corresponding network element via its associated Resource Adapter (RA).

The RM subsystem is composed of the following COs: Management Resource Configuration Manager (MRCM), Connection Management Configurator (CMC) and Network Resource Map (NRM). The RCM Application interacts with the MRCM to configure the network resources in the NRM and to launch one CMC per layer network. The CMC is in charge of launching the corresponding LNC and CPs via their configuration interfaces. The MRCM also launches the CC (if it has not yet been launched) and tells it the references of all the LNCs.

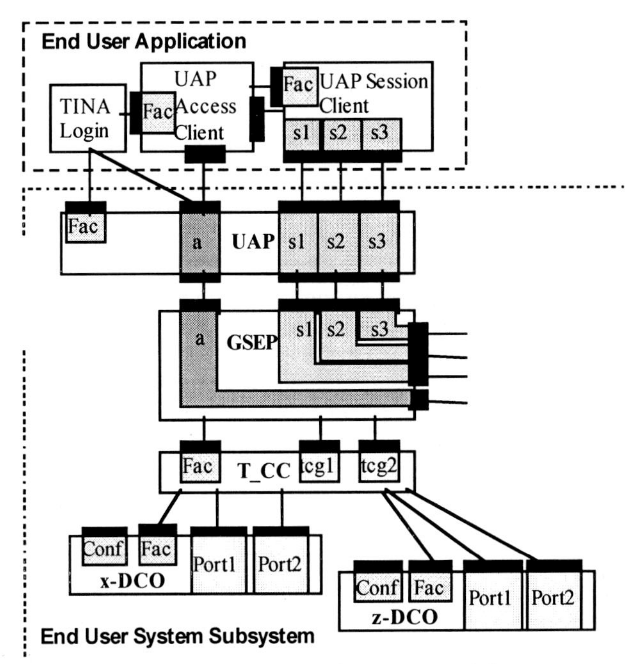

Figure 4. EU Computational Model

3 End User Terminal Subsystem

This subsystem consists of the components in the consumer domain. The computational model of this subsystem is shown in Figure 4. This figure depicts the situation where the terminal is involved in 3 active service sessions (s1, s2 and s3). UAP and GSEP have been taken over from [2]. However, while TINA models UAP as a completely integrated application, VITALv1 models UAP to provide a more high-level API on top of which actual applications can be more easily developed.

The TINA Login component allows users to login/logout as TINA users. This component is application-independent and creates the UAP Access interface passing the received authentication data which is checked within the retailer domain and if successful, the request and indication interfaces of the UAP Access Client are created.

The UAP Access Client component is also application-independent. It implements the abilities to use TINA access facilities (e.g. request a new session, request to join an existing session, receive an invitation to join a session). The UAP Access Client provides the interfaces: Request interface towards the UAP Session Client, Indication interface towards the UAP, and Factory interface towards the TINA Login component to allow creation of Request and Indication interfaces.

The UAP Session Client component implements the abilities to use the Session Control capabilities of the UAP. This component is application-specific. It can decide which session control changes can be initialized by the user, respond to indications on session control changes, and decide which of the information on session control changes will be presented to the user. Furthermore, this component can also request for a new session (via the UAP Access Client request interface) and respond to incoming invitations to join a session. The component provides the Session Control

interface (s*) towards the UAP; and Factory interface towards the UAP Access Client to allow creation of Session Control interfaces.
The UAP provides a higher layer API to enable a faster application development. It can also be considered as a gateway to non-CORBA applications. Its "server interfaces" could be developed using e.g. UNIX sockets, enabling an access to applications which are not implemented as CORBA objects. The UAP provides the Access interface (a) towards the UAP Access Client and GSEP; Session Control interface (s*) towards the UAP Session Client and GSEP; and Factory interface is used by the TINA Login to allow creation of the UAP Access interface.
The GSEP is the single point of contact in the consumer domain for both the access and service session. It offers an Access interface towards the UAP and UA; and a Service Control interface towards the UAP and USC, different Session Control interfaces are created for different sessions. It has been developed in accordance with [2]. This means that it also acts as the contact point for the access session, while in [3] the Provider Agent has been introduced.
VITALv1 introduces the Terminal Connection Coordinator (T_CC) and the Device Controller Objects (DCO) to manage connectivity within the consumer domain. The T_CC coordinates the technology layers in the terminal, similar to the CC in the network. It should negotiate with the peer T_CC and the CC about the selection of the technology layers. In VITALv1, this selection is pre-programmed. The T-CC provides Terminal Connection Graph interfaces towards the GSEP and a Factory interface to create TCG interfaces. A port within a TCG represents the termination of a user information stream within the consumer domain. Such a termination is typically implemented as a protocol stack. Every protocol stack element represents a technology layer modeled as a separate DCO (e.g. MJPEG DCO, AAL5 DCO). A DCO provides Port interfaces to handle a specific protocol layer; a configuration interface to configure the corresponding protocol layer; and a Factory interface to create Port interfaces.

4 Subscription Management Subsystem

This subsystem (see Figure 2) provides to the retailer, via the corresponding subscription application, functions to subscribe/cancel subscribers, to collect/modify/show subscription information, to authorize/bar service use for Subscription Assignment Groups (SAGs) associated with subscribers, and to determine the subscription information available to a user during service access.
The SM information model considers the following information objects: Service Description, Service Template, Subscription, SAG, Service Profile, Subscription Contract, and Subscription Portfolio.
The SubRgs allows to collect/modify/show/cancel subscription information. A single SubRgs handles the subscriptions of one service. The SubRgs maintains the subscription contract, subscription and service profile information objects.
The SubAg allows its clients (UA) to get subscription information about a particular user during an access session. In particular, the SubAg provides a list of services available and a description of a service to be executed for a user. Subscription operators interact with SubAg through the SA application to assign/de-assign Service Profiles to SAGs. The SugAg is created by the corresponding UA. There is a SubAg per UA.

The STH handles one or more service templates that represents service capabilities provided by the service retailer, and verifies subscription characteristics on request of the corresponding SubRgs.
The SubM manages the information of subscribers, subscription portfolios and SAGs. The SubM CO is independent of any particular service.
The SM subsystem is based on [2, 4] although some extensions and refinements have been done in order to make them suitable to VITALv1 requirements. The most relevant are: common definitions have been updated and extended; GSEP can interact directly with SubRg to get subscription data on behalf of the end-user; and COs that create dynamically other COs export their references to the trader.

5 Access Session Control Subsystem

This subsystem allows the access to the services offered by the retailer providing a limited personal mobility functionality where the user can register on any terminal to receive invitation of a particular service and can request to establish service sessions from any terminal. Some authentication mechanisms are also available. This subsystem is composed of two type of COs: UA and UCxt (see Figure 2).
The UA represents a user in the retailer domain. The UA is the single point of contact to access the available services within a retailer domain. The UA provides some capabilities to handle sessions allowing a user: to create a new session, to join in an existing session, to be invited to participate in an existing session, to leave from an existing session, to delete an existing session. The UA also provides the list of services available in a retailer domain via the SubAg and the list of existing and public sessions via the SF. The UA allows the user to be authenticated. Personal mobility is also provided by being able to register on a terminal for incoming communication. The registration information is handled in the corresponding UCxt.
There is one UA per user in the retailer domain and each UA has associated one UCxt. The UAs are created by the SubM when a new user is defined in the SA Application. When a UA is created, the corresponding UCxt and the SubAg are created by this UA. The UAs interact with the SF to create or delete service sessions and user sessions handled in the GSC and USC respectively. The GSEP and the GSC are the clients of the UA interfaces. The UA interact with a Authentication Server, USCs, GSEPs and SubAgs. The UAs are the clients of the UCxt interfaces.
The AS subsystem is based on [2, 4] although some modifications have been introduced. The main changes in the IDL definitions of the information objects are: the session Id has been included in the session description, the reference included in the service execution information is a USC reference, and several new information objects have been introduced (terminal configuration, user configuration and terminal registration). In the computational model, the main changes have been: the Personal Profile has not been developed; the Terminal Equipment Agent has been replaced by information objects; and the Provider Agent and Initial Agent proposed in [3] has not yet been considered.

6 Session Control Subsystem

This subsystem is in charge of handling the user and service sessions associated with the services available in the retailer domain. In VITALv1, there is only one service provided by the retailer. This service offers the operations to manage the Session Graph (SG) concept. This functionality is performed by service independent components according to TINA specification. An overview of the TINA Service

Session model is represented in Figure 5. The elements in gray will not be implemented in VITALv1.

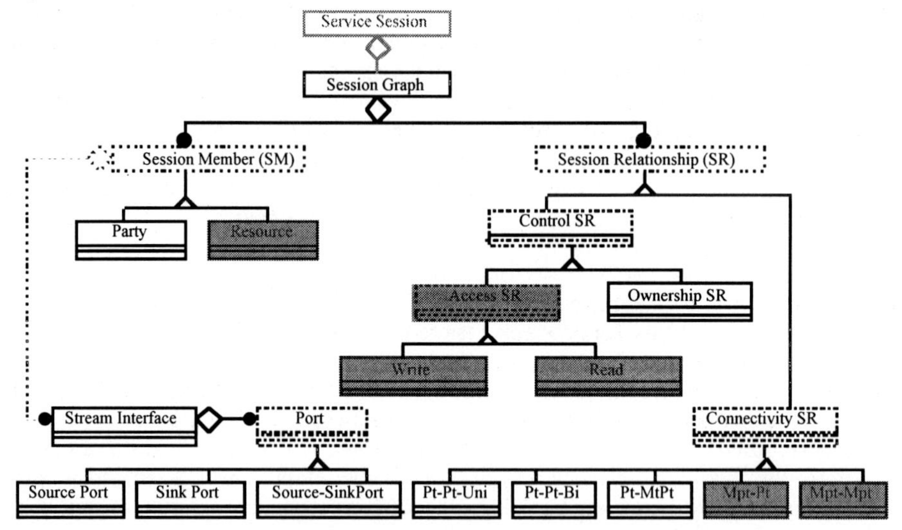

Figure 5. Service Session Graph Overview

The Service Session (SS) contains one -and only one- SG object. An SS is composed of Session Member (SM) and Session Relationship (SR).

The SM is an abstract class used only to show the relationships and aggregations on a generalized level in the model. The SM is further specialized as Party Session Member (PSM). All the SM properties are inherited by PSM, including all the relationships the SM has in the model with other classes. The PSM represents a negotiating entity taking part in the session. The user takes part in the session by defining him/herself as a PSM in the SG. A user invites another one in the session by requesting such a PSM to be created for that invited user. The SM is also defined as an aggregation of Stream Interfaces (SI). The SI represents a grouping of related Port objects. A Port represents a source, sink or source+sink of a stream of user information. A Port is composed of the attributes: information type, direction, port description and port address.

The SR is an abstract class used only to show the relationships and aggregations on a generalized level in the model. The SR is further specialized as Control SR and Connectivity SR. All the SR properties are inherited by Control SR and Connectivity SR, including all the relationships the SR has in the model with other classes. The Connectivity SR models, from the Service Session point of view, a stream binding (expressing the request for user information exchange). The Connectivity SR contains the following attributes: information type, topology and connection description. The Control SR is also specialized into the Ownership SR (OSR). The OSR associates a PSM to any other SG object instance in the session graph.

In VITALv1, this subsystem is composed of the following COs: SF, USC and GSC (see Figure 2). The USC and GSC implement the service common parts of session control. The service specific parts are not implemented in VITALv1. The SF's usage interface is invoked by UAs. The USC provides interfaces towards the SF, GSEP and GSC. The GSC provides interfaces towards the SF and USC. The USC and GSC session control interfaces represent operations on the Session Control information model defined by the SG. The SC subsystem is partially based on [3]. TINA allows more than one SG object in one Service Session. This is not supported in VITALv1. The following SG elements are not supported in VITALv1: resource session member, write and read access control relationships, mpt-pt and mpt-mpt connectivity relationships, and SM and SR groups. The SG objects do not implement a scheduling mechanism. The following restrictions apply to the OSR: the owner in the OSR can only be a PSM, not a PSM Group; the ownership type is always equal to the semantics of "owned-shared" as specified by TINA; and the only voting rule supported by the ownership relation is 100% unanimity. The Port and Connectivity SR classes are not further specialized, but an attribute is introduced to distinguish between the conceptual sub-classes.

7 Connection Management Subsystem

This subsystem provides connectivity capabilities to the retailer domains. The CM information model is based on the TINA Network Resource Information Model (NRIM) [5] which started by considering the object classes identified in ITU-M3100 and some extensions have been introduced. The NRIM is decomposed in several fragments.

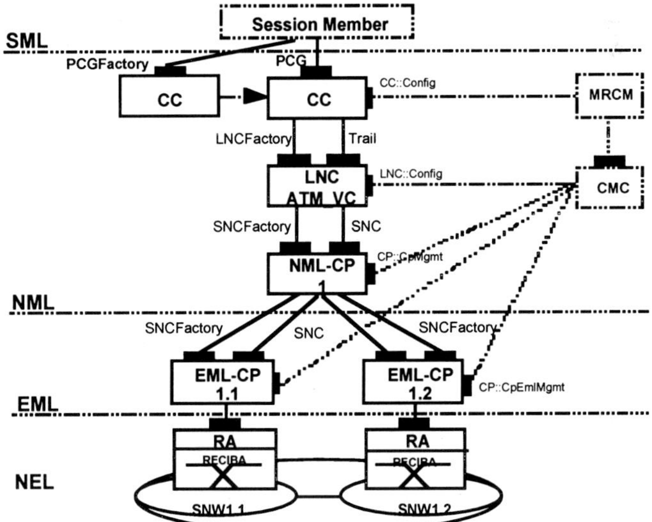

Figure 6. CM Computational Model

For CM computational model, TINA defines two types of COs: those which are concerned with offering an interface to service components independent of network structure and technology (CSM and CC) and those which are concerned with the connection management within a layer network (LNC and CP). The VITALv1 CM computational model is depicted in Figure 6.

For a layer network partitioned in several connection management domains, there is an LNC, for each of these domains, which serves the requests for trails to the clients of the layer network with access to the domain. The LNC relies on lower level entities, called CPs, which are able to provide sub-network connections for the sub-network in their scope. These entities are called NML-CPs. Each NML-CP relies on other NML-CPs, and finally on lower level entities, called EML-CPs which control directly network elements. The LNC and CPs are concerned with connections

in a single layer network. They are specialized for the technology that is characteristic for that layer. When a trail spans different domains, an LNC interacts with other LNCs in the neighbor domains, using a federation mechanism, which is defined in terms of tandem connections (not supported in VITALv1).
A Sub-network Connection (SNC) corresponds to a CP interface. In other words, operations concerning creation, manipulation and interrogation of information represented by a SNC are defined on the CP. An operation on a CP is propagated along the CP hierarchy.
Interworking of different layer networks is provided by the CC. The CC offers its clients one interface, defined in terms of a Physical Connection Graph (PCG), which allows the specification of end-to-end connectivity between points which can be associated with different layer networks. This specification of the connection topology is independent of the technology and topology of the underlying networks. The CC will take the client's request, match the specified endpoints on compatibility, select the appropriate layer networks, and request the corresponding LNCs for the further detailed connection set-up.
A higher level of abstraction of the connectivity between service components is provided by the CSM. It provides for end-to-end connectivity among stream interfaces defined at the computational viewpoint.
The CM subsystem is based on [5, 6] with some modifications. In the following, there is a summary of the main changes.
Only a part of the information model is used, attending to the VITALv1 requirements. A specialization of ports and lines has been included in the information model taking into account the allowed connection types (point-to-multipoint, point-to-point unidirectional and bi-directional).
The connectivity services in VITALv1 are offered by CC, the CSM has not been developed. A new interface has been defined to configure the CC. It provides an operation to set the LNCs associated to the CC.
A configuration interface for the LNC has been incorporated. It is used to set the top NML-CP supporting the Layer Network and to set the LN access points. EML-CP and NML-CP management interfaces have been refined and extended in order to allow for a more flexible configuration.
The CM subsystem has the capability for selecting a termination point in case only the network address (or termination point pool) is specified by the user. In this case, it returns the identifier of the selected termination point. In VITALv1, the network address is an E-164 address and is associated to a pair <ATM port, VPI>. The termination point identifier indicates the particular VCI used in this pool. Groups of lines, trails or SNCs are not considered in VITALv1.
A new interface, the Resource Adapter, has been defined in VITAL. This interface encapsulates the particular switch vendor implementation and associated switch control protocols. All the network elements will offer this common control interface independently of the particular implementation and protocols they offer. So in VITAL, the EML-CPs are technology dependent but vendor independent.

8 Resource Configuration Management Subsystem

The Resource Configuration Management (RCM) [7, 8, 9] is responsible for managing the configuration of the components of the Resource layer. The resources can be network resources (e.g. LNW. SNW, trail), service resources (e.g. UA, TA, session), computing resources (e.g. node, capsule), and management resources (e.g.

LNC, CP). The RCM is applicable not only to network resources, but also to other resource type. In [8], it is recognized the fact that within the Configuration Management domain, there is Configuration Management of the Network, Service and Computing, although the concept of Management Configuration Management is not explicitly mentioned. For example, the Connection Management Configurator (CMC) is in fact a Resource Configuration Manager for Connection Management resources (LNCs, CPs). In a similar way, although not yet defined by TINA, there could be a Fault Management Configurator or Performance Management Configurator. All of these Management Configurators belong to the RCM domain. The configuration management functional area can be applied to all types of resources in TINA. This analysis leads to the conclusion that a new domain of Configuration Management is needed namely Management Configuration Management which contains one or several managers responsible for the "meta-management" of the TINA management architectural components.

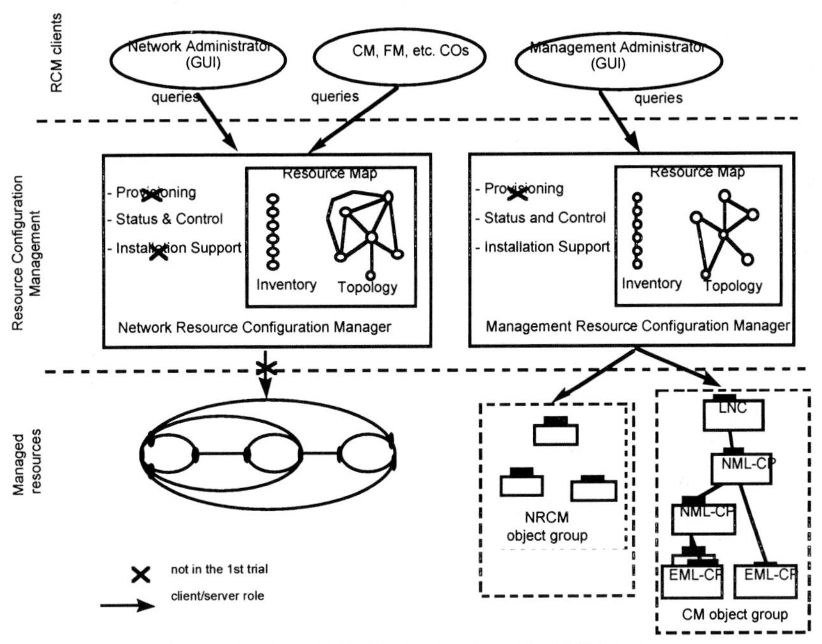

Figure 7. Overall RCM Architecture for VITALv1

A result of this is that Connection Management Configuration is distinct from Network Resource Configuration Management. They are both Resource Configuration Managers, but the former manages the configuration of Connection Management resources (CPs, LNCs) while the latter is responsible for the configuration of Network Resources (LNWs, SNWs, TPs, etc.). A benefit of this view is that it now becomes clear where to position CMC. Previously it has always been unclear whether CMC should be part of Connection Management or part of Resource Configuration Management. The answer is that it is part of RCM but not part of Network Resource Configuration Management.

Figure 7 shows the initial proposal for the overall architecture of both NRCM and MRCM for VITALv1. The figure is informal and deliberately does not map directly to an information, computation or engineering model. The figure is divided into three

layers. The center layer represents RCM, and contains both NRCM and MRCM which are not decomposed into their constituent COs. Although it is assumed that NRCM will be composed of several COs, only the Network Resource Map (NRM) has been developed for VITALv1. An initial analysis reveals that the MRCM block contains the CMC, FMC etc, plus some other objects for bootstrapping. The other objects are responsible for: configuring and managing the NRCM; registering for notifications from NRCM as network resource objects are created; creating the CMC, FMC, etc., parts of MRCM; and other roles are for further study. In VITALv1, the MRCM block contains a Management Resource Configuration Manager and a CMC.
The lowest layer contains the managed resources. The figure clearly shows that NRCM and MRCM operate on different managed resources. NRCM manages the configuration of (static) network resources, while MRCM manages the configuration of the management architecture itself.
The upper layer in the figure contains the clients of RCM. Both NRCM and MRCM are assumed to act as servers to administrators - the network administrator and the management administrator respectively. The administrators are responsible for issuing queries and updates to the RCMs. NRCM acts as a server to CMC in the sense that it provides network topology information which CMC uses to configure the CM object group(s). An RCM application, for both network administrator and management administrator, has been developed for VITALv1.
In short, the RM subsystem is composed of MRCM CO, CMC CO and NRM CO.
MRCM CO is responsible for the initialization of the management COs of the Connection Management. It launches the NRM, learns how many layer networks exist in the NRM and how many sub-networks for each layer network, launches one CMC per layer network (the CMC launches the corresponding LNC at this stage), passes to each CMC the sub-network names it is responsible for. The CMC reads a configuration file or accesses the NRM to find out the details of that sub-network; it subsequently launches and initializes the corresponding CP and informs the LNC about it. MRCM learns the corresponding LNC reference from each CMC. Finally MRCM launches the CC (if it has not been launched already) and tells the CC the references of all the LNCs. The Management Configuration Management (MCM) domain is entirely new to TINA. It has been an architectural extension that was considered necessary to manage resources of a TINA system. As such, the MRCM CO is completely new in TINA.
NRM CO provides the network resource map of NRCM. CMC queries the network topology information to configure the CM COs. Topology information (SNW, TPPool, TP, etc.) are represented as managed objects internal to the network resource map. The NRM CO for VITALv1 provides only the query facilities, no updates, and provides a CMIS-like IDL interface. This IDL interface offers the front-end to clients such as CMC, and acts as a gateway to the resource map. The latter is constructed using an OSI technology platform, and thus the resource map is an OSI Management Information Base (MIB). NRM CO is part of the NRCM domain and represents the network resources. This CO has not yet been defined in TINA.
CMC CO provides the configuration and management of the CPs and the single LNC in a Configuration Management Domain. For this purpose the CMC needs information on access points, subordinate sub-networks and links that constitute each sub-network in a layer network. CMC is initialized by the MRCM CO and fetches information from the NRM CO.

9 Conclusions

VITALv1 has allowed us to refine and extend TINA specifications to meet the requirements of a service covering multimedia, multiparty and personal mobility aspects. We can conclude that it has been a very important step to validate TINA and contribute to the definition of two TINA RPs Ret and ConS. In this paper, a detailed description of all the developed subsystems for VITALv1 has been presented indicating the main changes introduced with respect to TINA.

In the next VITAL phase, the VITALv2 will be developed based on the last version of TINA specs., and on the results and experience gained in VITALv1. The new features to be incorporated in VITALv2 have been selected, considering their architectural impact and the possibility to demonstrate the suitability of TINA to develop and deploy complex services, as the ones required for a teletraining application, considering the enterprise model defined in TINA with multiple business roles. VITALv2 will contain multiple retailer domains. Each retailer domain could provide one or several services to the consumer domains where there will be the first version of a teletraining application. These services (e.g. white board, slide presentation) will consist of the corresponding service specific components which will interact with the updated version of the service independent components developed in VITALv1. The retailer domains will request connectivity service to four connectivity provider domains, one per NHs. There will be four NHs in VITALv2. These connectivity provider domains will be interconnected via federation interfaces at the LNC level. The accounting management will be included as a new subsystem. On-line subscription, customization, management of multiple layers (ATM VP, VC) are some of the additional features to be supported. In VITALv2, the TINA RPs Ret and ConS will be aligned with the most recent TINA specifications and two new RPs will be added TCon and LNFed.

10 References

[1] ODTA Validation 1st Trial, VITAL D05 Deliverable, 31.10.96.
[2] Service Architecture, TINA-C Baseline, 31.3.95.
[3] Service Architecture, TINA-C Baseline, Version 4.0, 28.10.96.
[4] Service Component Specification, TINA-C Document, Draft, 31.3.95.
[5] Network Resource Information Model, TINA-C Document, 12.94.
[6] Connection Management Specification, TINA-C Document, Draft, 6.3.95.
[7] Resource Configuration Architecture, TINA-C Document, 12.93.
[8] Network Resource Configuration Management, TINA-C Document, Draft, 4.96.
[9] Report on Fault Management & Resource Configuration Management, TINA-C Doc., v1.0, 1.95.

Architecture of Multimedia Service Interworking for Heterogeneous Multi-Carrier ATM Network

Sung Won Sohn*, Jong Soo Jang*, and Chang Suk Oh**
*Intelligent Network Department
Electronics and Telecommunications Research Institute(ETRI)
**Department of Computer Engineering
Chungbuk National University, Korea

In this paper, we propose the concept of middleMEN (middleman for Multimedia service Environment) to provide an infrastructure for building multimedia service interworking platforms that support interactive multimedia applications in distributed heterogeneous network environments. We describe the main techniques developed to achieve the service interworking, and also the functional architecture of middleMEN in heterogeneous ATM network environments.

1 Introduction

Thousands of multimedia services are being developed for use over existing public networks and high-speed communication networks. Because various network technologies and services are proposed, information transport capabilities and service control schemes for each network vary widely. However, the design and engineering activities of all these emerging services, networks, and terminals are currently fragmented across many companies and organizations without any explicit regulatory mechanism that could ensure end-to-end service accessibility.

In the multi-carrier network, seamless integration of terminals and services is essential in order to offer an end-to-end and user-independent service. This will enable users to place multimedia calls anywhere, anytime, and to anyone without worrying about connectivity, compatibility, or coverage. The interworking and interoperability become the key requirements for integrating and interconnecting various local and remote multimedia service networks.

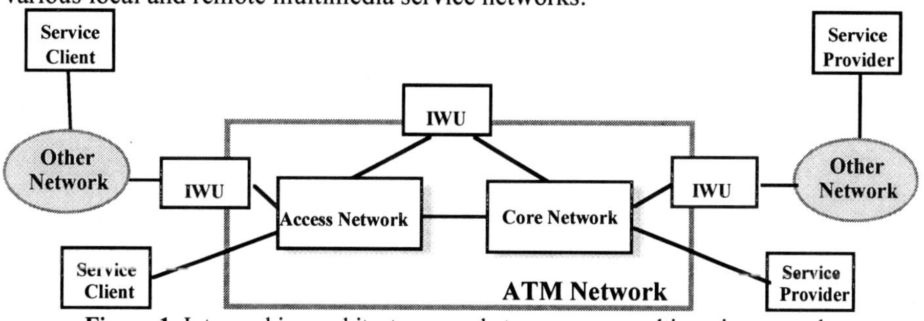

Figure 1: Interworking architecture over heterogeneous multi-carrier networks

Advances in computer and communication technology have resulted in faster and more powerful interworking systems. ATM is the preferred interworking protocol between multimedia systems if the bandwidth and quality of service requirements of real-time and interactive multimedia applications are considered. The Interworking Function(IWF) may be located in the access point of a carrier's network, or it may be located in the carrier's network. Figure 1 is an illustration of the IWF to inter-network between an ATM network and the other network.

Based on an ATM network, the interworking architecture over heterogeneous multi-carrier networks contains five segments: service client, service provider, access network, core network, and interworking units which are independent of the network topologies and protocols. The interworking architecture is characterized by: an access network that is capable of transporting information over ATM channels; different types of clients connected to the access node through the various networks; a core network based on ATM transport; an interworking unit connected to the access or core network; and, a service provider that is directly connected to an ATM switch or the other networks [1].

In this paper, we focus only on the topic of the service interworking in a heterogeneous environment with the ATM technology as the underlying technology that will unify the various networked services. First of all, we present the overall requirements of service interworking based on the ATM network and identify the key technical issues. Second, we propose four interworking scenarios on multi-carrier ATM network environment. Third, according to the scenarios, we define the concept of middleMEN for realizing the service interworking brokerage with a scheme for transport and service control scheme. The primary goal of middleMEN is to provide an infrastructure for building multimedia service interworking platform that support interactive multimedia applications dealing with synchronized, time-based media in heterogeneous distributed environments. Finally, we study the functional requirements and the functional architecture of middleMEN for multimedia service interworking in heterogeneous network.

2 Interworking Requirements

As the networked service has increased at an explosive rate in the Internet and multimedia services, the number of end users to access the different networks has also grown. ATM technology has been gaining tremendous industry support, and is maturing to become a dominant networking technology that will support various services, each requiring a different quality of service. In order to respond to these changes, it is important to establish appropriate connections between users, network, and service provider, to integrate networks and service equipment for the networked interactive applications, communicating both an ATM network and existing networks. Therefore, it will be necessary for different types of networks to inter-network each other at the network and service level.

Interworking can occur at various layers of the Open Systems Interconnection (OSI) layering model, and various locations in the network environment. By employing the interworking at various layers, the demand to achieve these essential objectives can be met quickly and easily, whenever, wherever, and in whatever form is required. Each user will be able to choose a network that suits their own purpose and uses their own existing network.

There are two types of interworking - network interworking and service interworking. The aim of the service interworking is to define a network architecture that satisfies the following requirements: seamless service provision across existing network by integrating with B-ISDN; flexible service provision that allows different service providers to customize their service by interacting with a service gateway component; and, global service provision through virtual integration of service provider.

Many issues are associated with service interworking in heterogeneous network environments. As a great number of terminals are involved, it is necessary to communicate all service terminal and servers using all protocols, even those that are

not necessary for the user, and very long lead time is required to introduce new services. In resolving the issues, the technologies required to accomplish seamless connection and to construct the transparent network have come into the focus in many discussions.
Multimedia service requires support for seamless interworking among various networks in which users need to access different types of media. For instance, communication service should be possible between groups of people in two different locations. Such interworking requires a robust architecture and protocol specifications so that applications can be built to satisfy the needs of users. Users supported by interworking function should be able to communicate from wherever they may be and with whatever capabilities they may have. The key features of the service interworking concept are as follows:

- Platform independence. Users that need to communicate should be able to use different desktop platform, such as PCs, workstations, televisions, and cellular phones.
- Access independence. Users should be able to logically connect through a variety of network access technologies, such as B-ISDN, N-ISDN, PSTN (Public Switched Telephone Network), wireless network, and LAN.
- Media independence. Users should be able to use various capabilities they have access to, such as audio, video, and data information. Further complications arise if transcoding or live speech translation needs to take place to achieve friendly and useful collaboration models.
- Service and application independence. Users should be able to participate any type of service and application with whatever they may require to join the service session.
- Service control and management independence. Users should be able to access any service counterpart to which they need to communicate in spite of different types of service control and management scheme.
- Place Independence. Users that need to communicate should be able to collaborate among themselves at different times and from different places.

Several interworking architectures have been proposed and implemented [2, 3, 4]. These support only certain requirements stated in earlier descriptions. So, we propose a set of architectures which cover a wide range of available solutions, such as distributed architectures, user signalling and proxy signalling, session management and connection management.

3 Service Interworking Scenarios

Two types of interworking platform architecture can be distinguished by their functionalities shown in Figure 2. One network interworking platform has access functionality, while the other has service gateway functionality. The interworking platform with access functionality, called IWF-element, is located in the access network domain, being the point of flexibility for providing access to network and services, and offering proxy server functions. The Interworking platform with service gateway functionality, (called a Service gateway element) is located in the network provider domain, and is the point of flexibility for providing access to services, offering value-added functions, and supporting communication capabilities to users in different service domains. The Service gateway element plays a number of roles. One role is the Level 1 gateway functions that can support equal access to any service provider within various network environments. Another role is connection and session management. The service gateway element can act as a proxy server in order to provide Internet-like services, and can offer enhanced functionality to the user such as managing users and server profile, authentication, etc. It provides functionality that relates to service collaboration between systems in different service domains. Another important feature is supporting the communication capability of a system for several users with different quality of service requirements.

The class of services whose access is facilitated by a service gateway element is called MMOD (Multimedia on Demand). The services in this class are VoD, tele-shopping, tele-conference, multimedia retrieval service, and so on. In the standards bodies such as ITU-T, ATM Forum, DAVIC, etc., it is not yet clear how multimedia services will be provided within the ATM network and the existing networks. However, several interworking scenarios can be sketched based on the level of functionality offered by interworking platform. Considering the conceptual architecture depicted by Figure 2, we identified four interworking scenarios as summarized in Table 1.

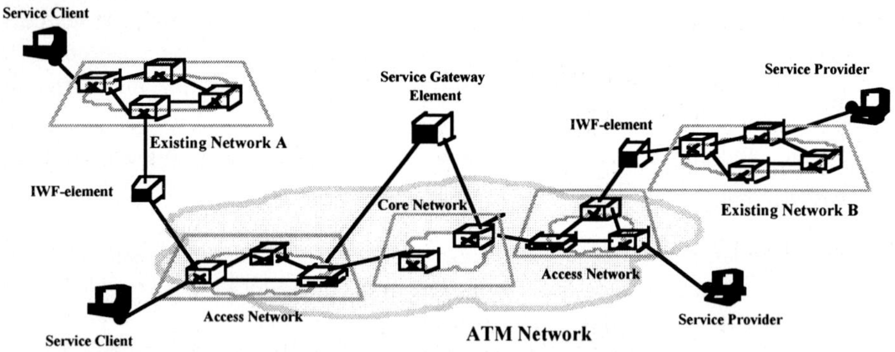

Figure 2: Interworking Platforms on heterogeneous multi-carrier networks

	interworking scheme	interworking location	interworking functionality
Scenario 1	Transparent ATM Connection	Network Adapter, IWF-element	- physical adaptation
Scenario 2	Network Interworking	IWF- element	- physical adaptation - PVC, SVC connection control
Scenario 3	Service Interworking for Equal Access	IWF- element or Service Gateway element	- PVC, SVC connection control - access support - service guide - media conversion - virtual server
Scenario 4	Service Integration	IWF- element and Service Gateway element	- access support - service guide - media conversion - virtual server - capability support - service integration

Table 1: Interworking scenarios on ATM network

In these scenarios, interworking functions can be performed by an IWF-element or service gateway element. Such an interworking function may be in a separate platform or can be integrated in a network node. Separate IWF systems connecting terminals to a network are usually called network adapters, and IWF systems connecting two networks are generally IWU.

Scenario 1: This scenario is based on the improvement of currently existing systems, such as N-ISDN terminal and Ethernet card in the PC. In this scenario, existing infrastructures could be reused and the service provision method is similar to the existing networked service. Although this approach is simple, it would be limited by the constraints on service connectivity that is provided by PVC connection management.

Scenario 2: This scenario provides the basic network interworking by implementing physical adaptation, via the PVC/SVC connection control function. In this scenario, the interworking function is performed by an IWF-element in IWU.

Scenario 3: In case of adding interworking functionality such as session control, access support, service guide, media conversion, and virtual server function to the previous scenario, we can achieve the service interworking. Interworking function can be performed by an IWF or service gateway element.

Scenario 4: We complete our interworking scenario by adding capability support and service integration function to the previous scenario in order to achieve the service integration. This interworking function is also performed by an IWF or service gateway element.

Scenario 1 and 2 could support communication level interworking across an ATM network. These scenarios are not considered because they use transport level interworking, not to service level interworking. Scenario 4 is based on a wide collaboration and integration of service end-systems. It is not feasible in a short or medium time frames. Therefore, we have chosen scenario 3 for the definition of the service interworking, because it is flexible enough to support multimedia service, it is based on ATM protocols, and it can be supported by existing network terminals. Furthermore, scenario 3 can be supported by different network capabilities.

4 Functional Architecture

According to the scenario above mentioned, we define the concept of middleMEN for realizing the service interworking brokerage in a scheme of transport and service control and the middleMEN platform is a network node with service interworking functionality. The definition of middleMEN derives from the concepts of a service gateway and a Level 1 gateway. It conforms to the ATM protocol specifications for network aspects, and the specifications of DAVIC (or another standards body) for service aspects.

4.1 Functional Requirements of middleMEN

MiddleMEN is a platform supporting control and management of communication capabilities between service providers and clients, and brokerage capabilities that are offered by the network to user applications. MiddleMEN can also provide various ability and functionality to applications. It provides an user's initial access to multimedia services. It represents the first contact point for the users to the service environment.

To support the service interworking in various networks, middleMEN has to validate the user identification, calculates an acceptable QoS, sends the negotiated QoS with user identification to the client and with server identification to the server. It also detects the differences between different service capabilities of clients and the server, selects the associated interworking assistant function supporting to the server or clients. It has the limited capabilities to play and control the service, and performs the interworking service functions for the server and clients.

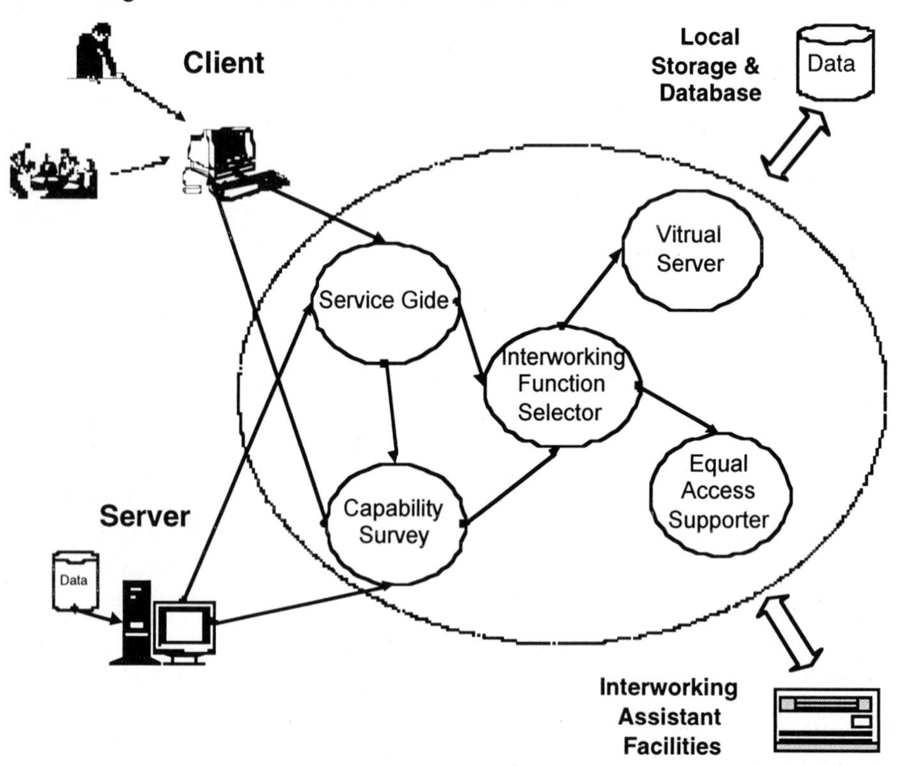

Figure 3: key functions for service interworking

Figure 3 shows conceptual view of domains of service interworking functions in a general network architecture. The architecture of service provision across various networks provides an infrastructure for delivering real-time multimedia services, one of which may be video on demand (VOD). Based on an ATM network, the service interworking functions are independent of the network topologies and protocols. The service interworking functions are used by network provider to communicate with service providers and end users for end-to-end service provision. The key elements of this architecture are Service Guide, Capability Survey, Interworking Function Selector, Equal Access Supporter, and Virtual Server.

As depicted in Figure 3, several functions of middleMEN can be introduced. Functionality associated with middleMEN is as follows:

- service guide and navigation in heterogeneous service environments,
- capability survey that monitors and compares users' and servers' service capabilities,
- function support for equal access,
- server networking in distributed server environments for making a virtual server.

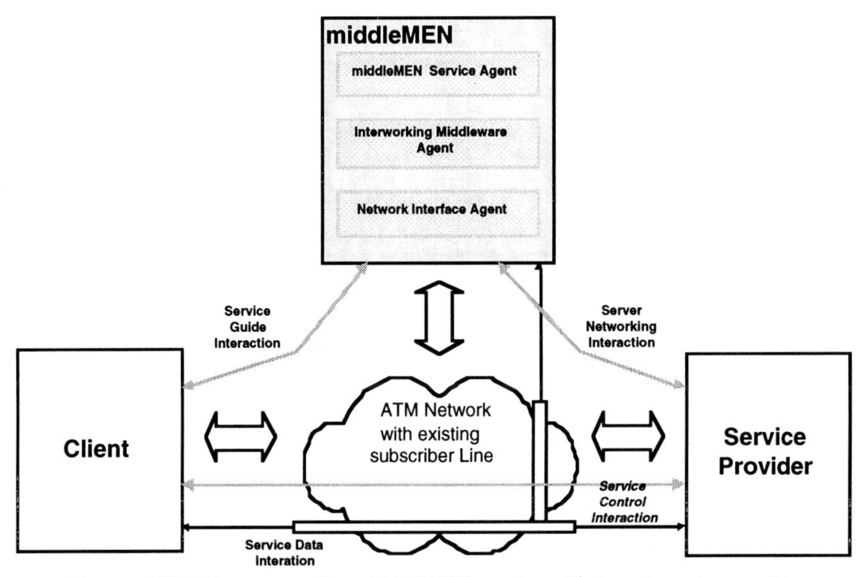

Figure 4: Relation among the middleMEN, service client, and service provider

Figure 4 represents the high-level relation of the network architecture with middleMEN, service clients, and service providers. It shows two interactions supporting the service interworking based on ATM network: between middleMEN service clients, and between middleMEN and service providers.

The service guide and navigation support user interactions with middleMEN for the purpose of selecting a service provider. Using this capability, the middleMEN provides the user's initial entry point to the service and it guides the user in service selection, thus users can search different program titles or previews of the program titles simultaneously.

Capability survey has an application feature that detects whether the capability of the user matches its server point of presence. In particular, a set of functionality for managing the service protocol and associated data has been considered. This means that middleMEN can collect; service subscription data; user access data; information related to all of the services the user can select and use; accounting and fault information. MiddleMEN may support associated service capability that the user wants to use. To support clients with different presentation capability, middleMEN may provide data conversion of new kinds of devices and media types.

MiddleMEN performs the access support functions for equal access between different service terminals using profiles of clients and servers. It must ensure that all service providers have equal access to users through middleMEN. Equal access operates in one of two ways, depending on who is taking the initiative: equal access for users to various service providers; equal access for service providers to any users. In addition, it is necessary to provide interworking function supporting disability users and platform.

MiddleMEN provides essential services to the server, that is a networking function conceptually making one virtual server with the service gateway capability in the distributed server environment. MiddleMEN also offers specific interfaces to servers. These interfaces enable middleMEN to retrieve the configuration information of the distributed server. A user generally wants to use the same menu for various service during service navigation. The virtual server function provides unique a service menu for users to commonly access to middleMEN through these interfaces .

4.2 Functional Architecture of middleMEN

With the above functional requirements, middleMEN provides the Level 1 gateway and service guidance functionality necessary to satisfy multimedia service requirements for equal access to any service providers, using common communication API and standard protocols, if available. It is important for users to be able to access multimedia services without needing to browse to find the service provider. MiddleMEN can be realised using:

- A common application environment using common communication API (CC-API). The CC-API is a set of APIs that provide a high-level interface for exchanging data between middleMEN applications and the underlying communication systems. It presents applications with a choice of standard communication objects, such as session, connection, and port.

- Two common protocols for seamless service interworking. These are a set of communication protocols between various clients and servers, and a set of gateway protocols between different service domains, which is necessary to mediate policies for collaboration, media conversion, and other feature between clients and servers.

The middleMEN services provide the service interworking functionality to users and servers. The middleMEN services include service guide, navigation, virtual server function, and access support for disability systems and users.
Based on the above mentioned methods, we can suggest the functional structure of middleMEN as shown in Figure 5. It consists of three types of functional agents: middleMEN Service Agent (MISA), Interworking Middleware Agent (INMA), and Network Interface Agent (NIFA).

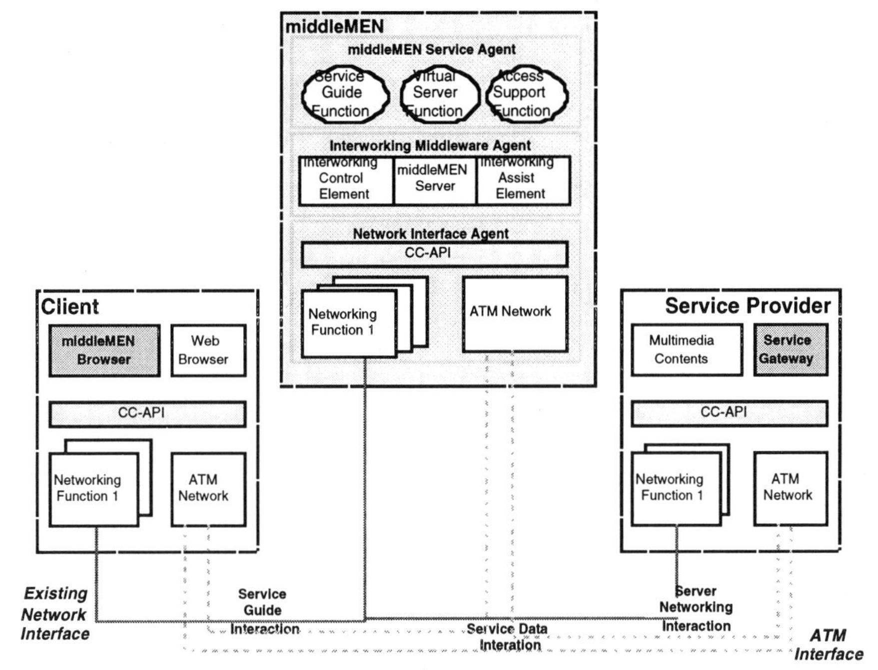

Figure 5: Functional structure of the middleMEN

The NIFA provides the communication infrastructure to connect the non-ATM network and ATM network. It has the CC-API that provides a common interface for exchanging data between the INMA and NIFA. This function is composed of interface function for ATM network and the non-ATM network, and the CC-API handling function. The networking function provides the protocols to communicate with B-ISDN and other network elements.

The INMA consists of three functional elements. A description of the main operations follows. The Interworking Control Element manages both client's and server's profile. It provides session and call/connection management functions using Level 1 gateway functionality. The middleMEN Server includes a demon program for communicating between middleMEN and clients, and between middleMEN and servers. The Interworking Assist Element performs the media conversion function from input information to valid information for client. MiddleMEN communicates with other network elements using the appropriate session layer protocol such as DSM-CC, as determined by DAVIC. INMA also maintains a profile data base, including pre-selected service information and any viewing restriction set by service gateways. This function is also used to support functions of the middleMEN Services Agent.

MISA performs applications for service interworking, such as the service navigation and service guide, virtual server function for contents binding, and access support for disability systems and users. Using its subscriber service capability, the middleMEN is an end user's initial entry into the network and guides the end user in service selection. This provides the Level 2 gateway role mandated by DAVIC for service networks. For multimedia retrieval service and interactive service, middleMEN offers a service directory (also called a service guide). The service guide function provides navigation service allowing choice of service providers and listings of services

provided. Its detailed functions are service selection, proxy server functions, provision of the directory and naming service for menu, billing, measurement, security function for the middleMEN, and so on.
According to the needs of the middleMEN service function, the MISA can use an associated interworking function supported by INMA for service interworking provision.

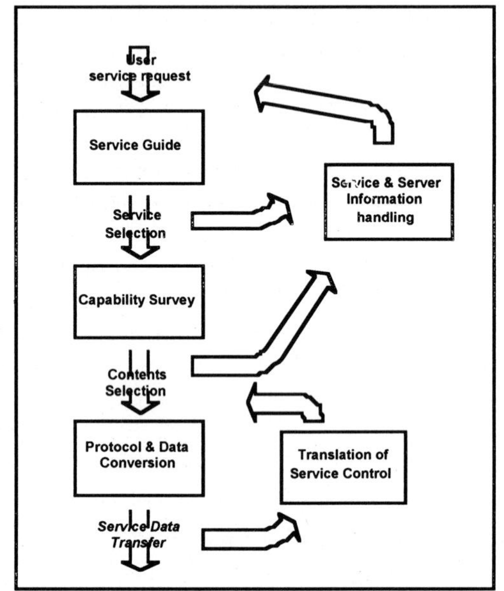

Figure 6: Operation flows of the middleMEN

5 Service Operations

Generally, service operations in a single service domain are as follows: user identification, service navigation, and service data transfer. But, in middleMEN, some more operations are required for supporting service interworking in various networks. Figure 6 depicts an example of service operation flows in various multi-carrier network environment with middleMEN. Five main operations are: Service Guide, Capability Survey, Protocol & Data Conversion, Translation of Service Control, and Service & Server Information Handling.

Service Guide operation: If a user needs to receive a service from a certain server located in an anonymous remote network, the client sends a service request with its user profile. MiddleMEN verifies the service request, and manages the user's profile. It guides a user in service selection and allows the user to choose a service from those available.

Capability Survey operation: The Capability Survey procedure compares internally the service capabilities of a client with those of a server, and selects the required interworking assist function to support the selected service. MiddleMEN maintains the user identifier and location of the required interworking assist function to support the selected service, and informs the user of the required interworking assist function.

Protocol & Data Conversion operation: During transfer of service data, middleMEN supports data conversion of new kinds of protocols and media types. For examples, if there is MPEG-2 data format in some platform and H.261 data format in the other platform, the middleMEN can support the communication between them through the conversion function.

Translation of service control operation: During transfer of service data, middleMEN supports communication between other network elements using different session layer protocols (such as DSM-CC, HTTP) by translating service control schemes. Where standards do not exist, middleMEN supports the appropriate protocols for service client and service providers.

Service & Server Information Handling operation: During the Service Guide operation, middleMEN receives content lists from the selected servers via the virtual server function. The middleMEN also informs to the selected service provider which is requested by client.

MiddleMEN can be implemented in ATM networks and Internet service environments for seamless service provisioning. We identify the typical applications of middleMEN in various viewpoints. For example, Figure 7 shows the general network configuration for service interworking in various heterogeneous network environments. In this figure, middleMEN platform can be used typically for interworking in ATM network and Internet service environments, and also for interworking in ATM network and the various wireless service networks (called middleMEN-mobile).

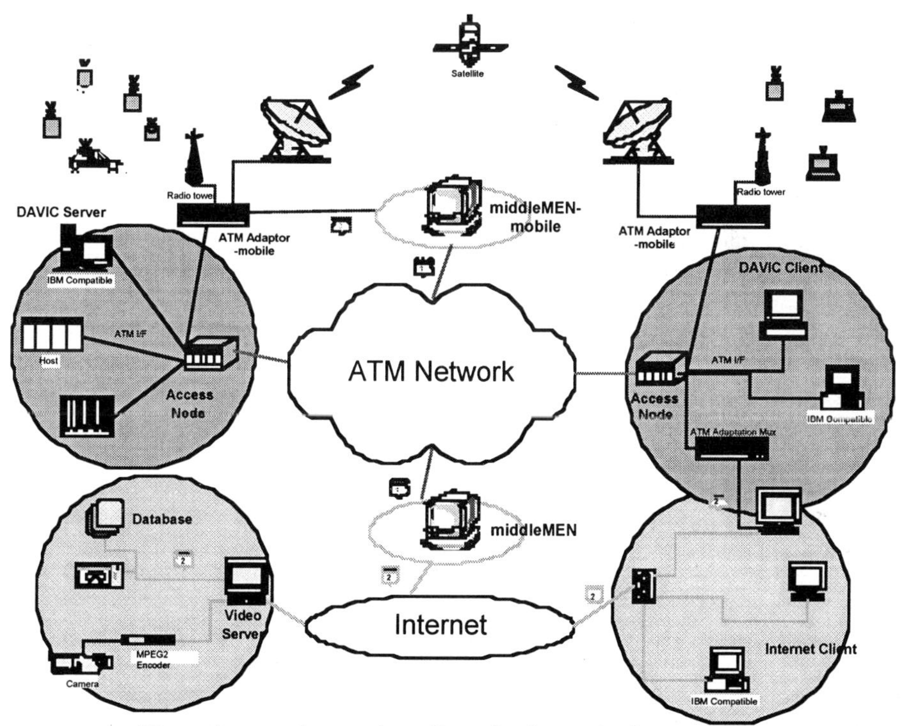

Figure 7: general network configuration for service interworking

6 Conclusions

At present, there are no specifications available on how the service interworking procedures between the various existing network and the ATM network should be done. However, we identified the service interworking scenarios and architecture to provide a seamless service in heterogeneous network environments. MiddleMEN has also proposed to be an effective architecture for service interworking of some sophisticated multimedia applications that are well integrated with existing networks and the B-ISDN network. We are currently exploring more sophisticated application areas, such as Internet access over B-ISDN. MiddleMEN will continue to provide the functionality needed as the market, standard, and new services continue to evolve.

References

[1] Hegering, S. Abeck, and R. Wies, "A Cooperate operation Framework for Network Service Management," IEEE Comm. Mag., Jan. 1996

[2] R.E. Libman, M.T. Midani, and H.T. Nguyen, "The Interactive Video Network: An Overview of the Video Manger and the V Protocol," AT&T Tech. Journal, Sep/Oct., 1995

[3] H.J. Fowler, J.W. Murphy, Network Management Considerations for Interworking ATM Networks with Non-ATM Services," IEEE Comm. Mag., June, 1996

[4] G.P. Balboni, and P.G. Bosco, "A TINA Structured Service Gateway," TINA'96 conference proceeding, Heidelberg, Germany, Sep., 1996

Use of WWW Technology for Client/Server Applications in MULTIMEDIATOR

Ramon Martí, marti@logiccontrol.es
Jaime Delgado, delgado@ac.upc.es, delgado@logiccontrol.es

This paper presents the design and implementation of a platform independent Client/Server Brokerage Application based on WWW Technology. This work has been done in the context of the MULTIMEDIATOR Project (ACTS 096).

This document is structured in three parts. The first one gives an overview of the MULTIMEDIATOR system, the second one explains the implementation of a Client/Server Model for a specific access method (forms access) in MULTIMEDIATOR, and the third one explains how the MULTIMEDIATOR system implements some brokerage features.

A key innovation in the forms access mechanism of the MULTIMEDIATOR system is the re-use of existing Internet technology and software for developing a Client/Server Distributed Application following the DOAM (Distributed Office Applications Model, ISO/IEC 10031) Model.

Keywords: Architecture, Brokerage, Client, Intelligent, Interfaces, Java, MULTIMEDIATOR, Networks, Server, Services, WWW

1 The MULTIMEDIATOR Project

1.1 Introduction

MULTIMEDIATOR (Multimedia Publishing Brokerage System) is an ACTS project that plans to demonstrate the use of an intelligent multimedia brokerage service for customers and suppliers in the publishing and electronic commerce areas. Services offered will include specialised video-on-demand, hypervideo, videorating, conventional publishing services, and electronic commerce on virtual gallery and education. For that purpose, the project will integrate existing technology and its own project developments. A working application over ATM will be demonstrated by piloting video services, conventional publishing services and some electronic commerce services.

The procedure for the customers to obtain services or to buy information consists of a sequence of steps such as to look for suppliers, to select the ones that fit into their needs, to negotiate the job or purchase, to order it, to control the work progress, and to accept it. In parallel, accounting and billing is to be performed.

Three access mechanisms are used simultaneously and co-operatively to provide the services in the MULTIMEDIATOR system: Forms, Remote store access and Interactive access. Nevertheless, this document is mainly concerned with the Forms access mechanism.

In the Forms mechanism, the Broker Machine provides a form to the user (customer/supplier). Each specific service has its own form that allows specifying constraints or options of the service. The form is filled in by the user and sent back to the Broker Machine, which may respond using any of the possible mechanisms (not necessarily with another form). This mechanism is useful for queries.

Figure 1 shows an overview of the MULTIMEDIATOR system.

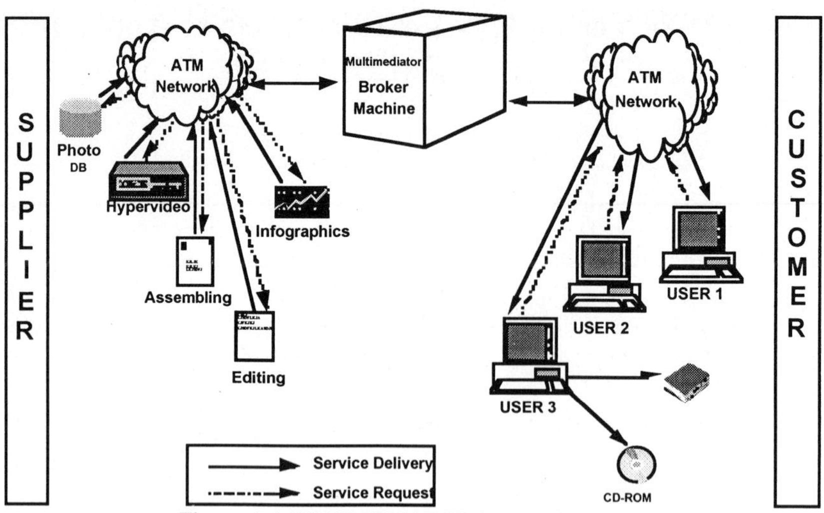

Figure 1: MULTIMEDIATOR System Overview

1.2 Architecture

The architecture of the MULTIMEDIATOR service includes:

- A Broker Machine offering "interactive" (order negotiation, ordering, quality control, etc.), and non-interactive services (document transfer, MM-information remote access, etc.); intelligent automatic processing of information, like formats conversion, is also provided.
  Information about Customers, Suppliers, Data and Services is stored in a DFR (Document Filing and Retrieval, ISO/IEC 10166) server accessed through the Remote Store Handler module. Additional information for static catalogues is stored in HTML pages in a WWW server.
  A Smart Search Engine able to combine standard Search Engine and DFR searches is also included in the Broker Machine.
  Management information, including Accounting and Billing, is also stored in the Broker Machine database and accessed through a Management Handler.
  An IPR module will also be included to manage the Intellectual Property Right issues.
  An ATM communication platform will be included for communication with Customers and Suppliers.
- Customers and Suppliers multimedia machines to dialogue with the Broker Machine, which will include whiteboards for multimedia production.
  An ATM communication platform will be included for communication with the Broker Machine. Any communication between Customers and Suppliers will always be through the Broker Machine.

Figure 2 shows an overview of the architecture of the MULTIMEDIATOR system with the modules mentioned above.

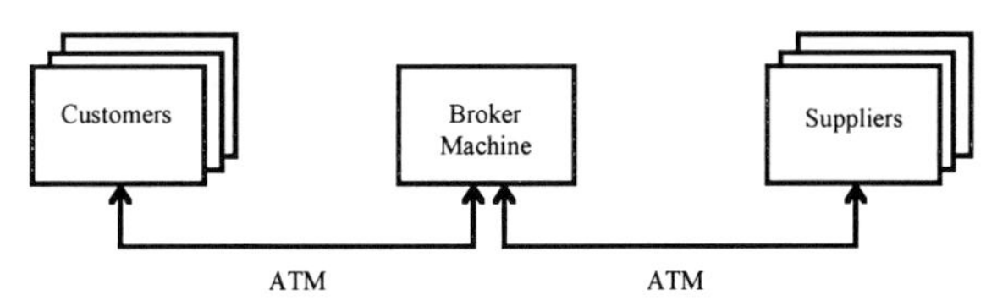

Figure 2: MULTIMEDIATOR Architecture Overview

A more detailed architecture of the Broker Machine is presented in figure 3.

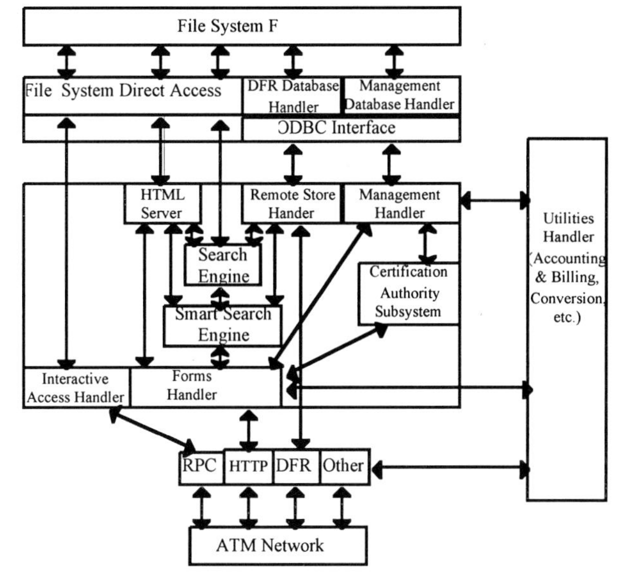

Figure 3: MULTIMEDIATOR Detailed Architecture

1.3 User Requirements

In MULTIMEDIATOR, the User Requirements have already been identified. The most relevant ones for this paper are:

- Distributed Client/Server Application
- Reuse of existing commercial software and technology
- Platform Independence
- Multiple Users Network Access
- Security
- Modularity

2 Client/Server Applications in MULTIMEDIATOR

2.1 Distributed Applications: DOAM Client/Server Model

DOAM (Distributed Office Applications Model, ISO/IEC 10031) standard [DOAM] specifies a Distributed Application Model with the following modules:

- *User*: The user of the Application. This user can be a User Interface able to present information to a Human User and interact with him.

- *Client*: The client part of the Application. The user interacts with the Client to use the service provided by the Application.
- *Server*: The server part of the Application.

Between the three modules, the following communication protocols are identified:

- User Interface <-> Client: It is usually implemented as a local protocol, although it could also be remote.
- Client <-> Server: It is usually implemented as a remote protocol.

Figure 4 shows the relationship between the different DOAM modules and the communication protocols:

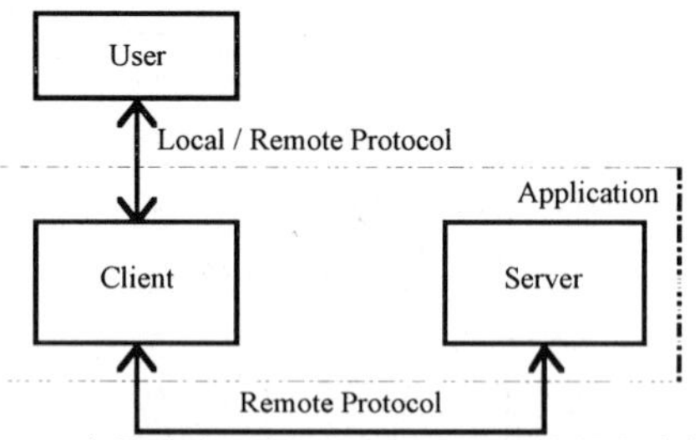

Figure 4: Relationship between DOAM Modules

2.2 DOAM Model in MULTIMEDIATOR

MULTIMEDIATOR Customer, Supplier and Broker Machine modules are structured following the DOAM Client/Server Model.

Since the functionality of the Broker Machine is to give the MULTIMEDIATOR Service to Customers and Suppliers, it will correspond to the Server module. Then, Customer/Supplier machines will correspond to both Client and User modules.

The architecture of the MULTIMEDIATOR system for the Forms mechanism (introduced in 1.1) is discussed in the next sub-clauses, and the final structure is shown in figure 5.

2.3 Reuse of Existing Software and Technology: Internet

Re-use of Internet Technology, such as HTTP and HTML, has been selected in MULTIMEDIATOR. This means that TCP/IP communications and related protocols will be used, and reuse of software already available for the Internet will be possible. In any case, MULTIMEDIATOR will not be a global Internet application, but a "private" Internet and OSI application running on the ATM network.

2.4 User: User Interface, WWW Browser and HTML

The User Module has to be present in both the Customer and the Supplier modules, since it is the User Interface in charge of presenting the information to the human User. In MULTIMEDIATOR, the information is related to the Negotiation of the services and to the presentation of the multimedia data (text, audio, video, etc.) to the User.

HTML [HTML] format gives all the needed functionality, since it is able to handle forms, for the negotiation phase, and to present any kind of multimedia data to the user through a WWW Browser. Then, the use of HTML as interchange format and a WWW Browser as a User Interface has been chosen in MULTIMEDIATOR, at least for the Forms access mechanism.

2.5 Communication Protocol: HTTP

A Communication protocol is needed between User <-> Client and Client <-> Server modules following the architecture decisions already taken.
Once the use of a WWW Browser has been decided and HTML for the User module, the chosen communication protocol for the MULTIMEDIATOR User <-> Client and Client <-> Server interactions has been HTTP [HTTP].

2.6 Client: Local Proxy

From the previous decisions, the Client module has to be an intermediary program that acts as both a server and a client from and to the WWW Browser and the Broker Machine. Requests are serviced internally or by passing them, with possible translation, on to other servers. It has to interpret and, if necessary, rewrite a request message before forwarding it.
The previous definition corresponds to a "proxy", so the following architecture decisions were taken for the MULTIMEDIATOR System:

- Implementation of the Client module as a "proxy".
- Use of the proxy in the same machine as the WWW Browser is running (instead of the proxy placed in a machine remote to the one where the WWW Browser is).

2.7 Server: WWW Server and WWW Local Proxy

The Server module corresponds to a part of the MULTIMEDIATOR Broker Machine. The fact that the User module uses HTML and HTTP, implies that the Server will receive HTTP requests with HTML Input Forms from the clients (Customer / Supplier), will analyse them, if needed will look for response forms, and will send HTTP responses with HTML Output Forms back.
The MULTIMEDIATOR solution for the Server module (Broker Machine) is to implement it with two modules: a local proxy, to provide the MULTIMEDIATOR Broker Machine brokerage features, and a WWW Server to provide HTML forms to the proxy.

2.8 Distributed Application Functionality: HTML Forms Management

In the MULTIMEDIATOR system, as in usual Client/Server Application, the functionality of the application is distributed between the Client and the Server modules.
Since most of the interactions between the user interface and the application are done through HTML forms, the functionality will be mainly based on the Management of these forms.

2.9 Platform Independence: Java

For the User Module and the WWW Server, there is no hardware platform problem, since WWW Browsers and Servers are already available in almost all the existing operating systems, so it can be considered platform independent.
The problem with programming platform independence for the Client Proxy and the Server Proxy modules has been solved by developing these modules using Java as the programming language.

2.10 Multiple Users Network Access: Remote User <-> Client Protocol

With the solution of developing the Customer and Supplier machines as a WWW browser and a proxy server, a network solution with a single client (WWW proxy) module accessed by multiple users (WWW Browser) is also possible without any modification. In this case, the User <-> Client interaction must be implemented as a remote protocol.

2.11 System Implementation

With all the considerations presented in this clause, the final architecture of the MULTIMEDIATOR system following the DOAM Client/Server Model is shown in figure 5.

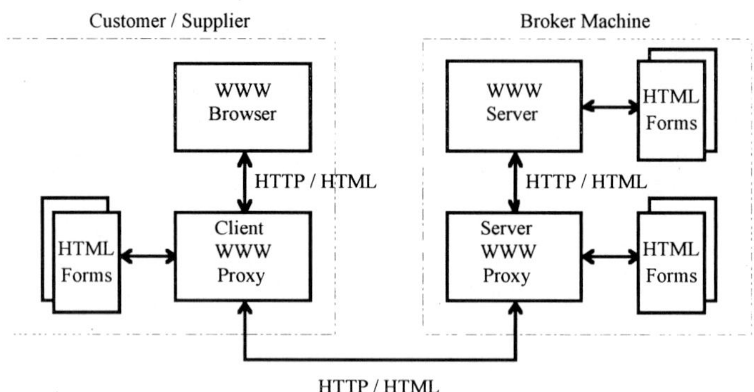

Figure 5: MULTIMEDIATOR Forms Access Architecture

3 Brokerage Features in MULTIMEDIATOR

3.1 Suppliers Information

Each Supplier is able to store in the Broker Machine information about Data and Services that it provides, in order to be consulted by the Customers. Two types of information can be stored in the Broker Machine by the Supplier:

- A set of WWW linked pages with general information to create catalogues. A main meta-catalogue with links to the catalogues of all the different Suppliers is also included in the Broker Machine.
- DFR [DFR] store information, with detailed description about Services and Data provided by each supplier. When consulting this store, the information is also presented to the Customer as HTML pages, but, in this case, dynamically generated by the Broker Machine.

With this implementation, users can access the information that is stored in the Broker Machine in different ways, leading to different visions of the structure of the information.

3.2 Remote Operations: Real Operations and Forms Operations

Two kinds of Remote Operations have been implemented in the MULTIMEDIATOR system to access the two different types of Suppliers information:

- Real remote operations: To access the DFR store.
- Forms operations: The operations are done through forms, and it is used for Client <-> Broker Machine dialogue. The request is an empty form, while the returned filled form can be considered as the operation response. The information returned in the filled form is analysed by the Forms handler in the Broker Machine.

3.3 Smart Search: Service and Data Search

The Service and Data requests in the forms operations are analysed by the Forms handler in the Broker Machine and sent to the Smart Search Engine module, that implements the Smart Search facility in the MULTIMEDIATOR system (see figure 3). Three kinds of Service and Data requests are possible in the system.

Unknown Type of Service, Data or Supplier

When the user does not know the type of Service, Data or Supplier, two options are provided by the MULTIMEDIATOR system:

- Using the Forms access, but filling the form only with the known fields. The Smart Search Engine is in charge of searching the information by making a call to the Search Engine to get a first approach of the results, and then refine them with a call to the search operation in the DFR module.
- Navigation through the database information: This option is also supported by the MULTIMEDIATOR system. In this case, the Broker Machine is in charge of dynamically generating the HTML pages to navigate through the information from the user requests.

Known Service or Data

When the exact service or data is known, MULTIMEDIATOR provides the "Remote store access" method. In this case, only DFR store access protocol is used without passing through the Smart Search Engine module.

Abstract Requests

The Search Engine is the module introduced in the Broker Machine to allow the system to make abstract requests. Since the user only interacts with the Smart Search Engine, this module is the one in charge of making a call to the Search Engine, and returning the results to the user. In this case, the Smart Search Engine makes no call to the DFR module.

3.4 Security

A Certification Authority with X.509 certificates, and SSL (Secure Sockets Layer) are used to provide security services both to the Customers and to the Suppliers, with the following features:

- Confidential transmission and the integrity of the information: It is obtained by encrypting the information.
- Authentication of the identity of the Customers and Suppliers: It is obtained by adding electronic signature to all the transactions.
- Non Repudiation.

Since the Broker Machine can be considered as a trusted party for both the Customers and the Suppliers, it can be in charge of managing certificates. Therefore, a trusted Certification Authority to generate and manage certificates and certificate revocation lists will be included in the Broker Machine.

3.5 IPR (Intellectual Property Rights)

MULTIMEDIATOR will integrate an ECMS (Electronic Copyright Management System). To do this, support is received from the COPEARMS ESPRIT project. COPEARMS is in charge of assisting other European projects on IPR issues, and will help our project on integration.

Moreover, the concept of volatile and non-volatile space will also be implemented inside the MULTIMEDIATOR project for the Customer <-> Broker Machine <-> Supplier communication. In this way, copyrighted information will be protected.

3.6 Management Services: Modularity

Since the Broker Machine is implemented in a modular way, the management functionality is distributed in a set of modules: Accounting and Billing, Management Handler, Remote Store Handler and Interactive Access Handler. In any case, new management modules can always be added.

4 Conclusions

The implementation of the MULTIMEDIATOR system includes several new technologies and trends into a single system:

- It follows the DOAM Client/Server Distributed Application model
- Use of Internet technology: HTTP, HTML and TCP/IP.
- Use of WWW Browsers for the user interfaces.
- Implementation of the remote application with two proxies (Client and Server Modules) and a WWW Server.
- Client/Server platform independence using Java as programming language.

From the services and networks point of view, the key innovations in MULTIMEDIATOR are:

- DFR information and HTML pages allow storing and generating static and dynamic WWW data and services information pages.
- Two kinds of remote operations are defined: Real remote operations and HTML forms operations.
- The Smart Search Engine module in the Broker Machine allows the user to search for abstract and unknown services and data.
- SSL and a Certification Authority dealing with X.509 certificates are used to provide security to the service.
- IPR Management and protection will be included in MULTIMEDIATOR.

5 References

[DOAM] ISO/IEC 10031-1:1991, Information technology - Text and office systems - Distributed-office-applications model - Part 1: General model.

[DFR] ISO/IEC 10166-1:1991, Information Technology - Text and office systems - Document Filing and Retrieval (DFR) - Part 1: Abstract service definition and procedures
ISO/IEC 10166-2:1991, Information Technology - Text and office systems - Document Filing and Retrieval (DFR) - Part 2: Protocol Specification

[HTML] RFC1866 - T. Berners-Lee, D. Connolly, "Hypertext Markup Language - 2.0", 1995/11/03

[HTTP] RFC1945 - T. Berners-Lee, R. Fielding, H. Nielsen, "Hypertext Transfer Protocol – HTTP/1.0", 1996/05/17.

[X.509] ITU Recommendation X.509, The Directory: Authentication Framework (ISO/IEC 9594-8), 1993.

An Experimental Open Architecture to Support Multimedia Services based on CORBA, Java and WWW Technologies[1]

A. Limongiello (*), R. Melen (**), M.Rocuzzo (*), V. Trecordi (+) and J. Wojtowicz (+)
(*) TELECOM Italia Headquarters, Roma - (**) Politecnico di Milano
(+) CEFRIEL, Via Emanueli 15, 20126 Milano MI, ITALY,
Ph.: +39-2-66161261, Fax: +39-2-66100448, E-mail: vit@mailer.cefriel.it

The convergence of interactive multimedia services and Internet services characterises the new services scenario and also highlights the important role of the terminals and their intelligent capabilities. Thus the scenario moves from the classical telecom setting where the “intelligence” is concentrated and embedded into the network (switches) towards a computer networking scenario in which more “intelligence” is located at the periphery (hosts, computers and terminals). The primary objective of the ORCHESTRA project was to demonstrate by fast prototyping an experimental control architecture capable of complying with a large number of the emerging requirements for the support of advanced services in broadband networks, with particular emphasis on emerging technology integration and interoperability.

This paper describes the principles according to which the architecture was developed. The fast technological innovation rate has led to two successive implementations of the prototype. The prototype is currently installed at Telecom Italia premises in Rome and at CEFRIEL in Milan and testing across the Italian high speed ATM backbone, named SIRIUS, is currently being performed.

1 Introduction

We are presently witnessing a convergence process of Telecom-style interactive multimedia services and Internet services (e-mail, file transfer, access to multimedia information, etc.) leading to a new services scenario in which Telcos, Cable TV (CATV) operators, Internet Service Providers (ISPs) and content providers share an interest.

The approach taken by the various players interested in this future multimedia communication business obviously depends on their background, but, in spite of their differences, recently a significant convergence can be seen on some of the fundamental technical elements of future networks, such as high-speed backbones based on the Asynchronous Transfer Mode (ATM) [1] technology and MPEG video coding standards. Moreover, emerging services require, in order to be implemented efficiently, technical elements coming from various fields (nomadic access to data networks by means of Personal Digital Assistants is one example).

In this scenario, the field where more questions are still unresolved is probably the control architecture. By control we mean the set of functions which allow users to request access to services and to reserve the resources necessary for their support. It includes functions which span a wide range of abstraction levels and its implementation is subject to severe performance and scalability requirements.

Existing proposals range from the classical telecom setting where the "intelligence" is concentrated into the network (switches) towards a computer networking scenario in which more "intelligence" is located at the periphery (hosts, computers and terminals). Presently various initiatives are carried out on advanced control models,

[1]Research work carried out at CEFRIEL, under the sponsorship of TELECOM Italia.

both in industrial and research environment; among others Telecommunications Information Networking Architecture Consortium (TINA-C) [2], Internet QoS architecture (RSVP [3]), Xbind [4], Active Networks [5], QoS-A [6] can be quoted. The present paper describes an experimental architecture addressing these control issues, with particular emphasis on emerging technology integration and interoperability.

2 Project Motivations and Reference Model

ORCHESTRA is a prototype software architecture for the support of services in broadband networks. The primary objective of the ORCHESTRA project was to demonstrate by fast prototyping an experimental control architecture capable of complying with a large number of the emerging requirements for the support of advanced services in broadband networks. The project was committed to investigate the effectiveness of emerging technologies, such as WWW, Common Object Request Broker Architecture (CORBA) [7] and Java [8] and the power of promising software engineering paradigms such as software agents [9], to tackle the complexity of the requirements of future-proof service-oriented networks. The study aimed at investigating also the possible alternatives of distribution of intelligent functions between the terminal equipment and the network nodes, thus trying to reconcile the telecom based view and the computer based view of the network. Two key concepts of the design are the capability to offer a unique user interface to access all the services in an advanced broadband network and to propose a layered model of the network functions targeted at supporting the independence of service and information transport. This latter property results in a control architecture open to a variety of possible future business scenarios and market segments.
The architectural paradigm we investigate introduces a distribution of both control and management applications. This means that different and/or autonomous software components (agents) possibly located at different points in the network (client terminals, switches, servers) co-operate, within a distributed software environment (middleware), in order to offer services to end users and/or service providers.
Flexibility, distribution and autonomous software components are the keywords which suggest the applicability of the Agents technology. Agents are interpreted language programs which can be dispatched from a client computer and transported to a remote server computer for execution.

A general reference model for the kind of control architecture investigated is depicted in Fig. 1. It consists of three main layers: the Transport, the Middleware and the Service. The interaction between the layers is achieved by means of well-defined interfaces. The information transport layer is at the bottom of the reference model. This layer accounts for the functionalities devoted to carry information across a network.
The core of the model is the Middleware layer which hides the details and the complexity of the underlying network and includes the framework for supporting the design, the rapid introduction and execution of advanced services in an heterogeneous network environment. The Middleware Layer can be divided into two sub-layers: the Distributed Processing Environment (DPE) and the Service Independent Layer. The latter defines the set of basic mechanisms used by all services and consists of four fundamental components:

- Broker, that is the definition of the mechanisms for mapping service requests to service instances;
- Session Control, that is the definition of the mechanisms which rule a service session;
- User Control, that is the definition of mechanisms for keeping user profiles and locating users;
- Resources Control, that is the definition of the mechanisms for opening, controlling and closing the communication channels as well as allocating and releasing special resources.

Finally, the Service Layer defines how a specific service (distributed application) employs the available mechanisms (e.g. for connection set-up, service management etc.) offered by the underlying infrastructure.

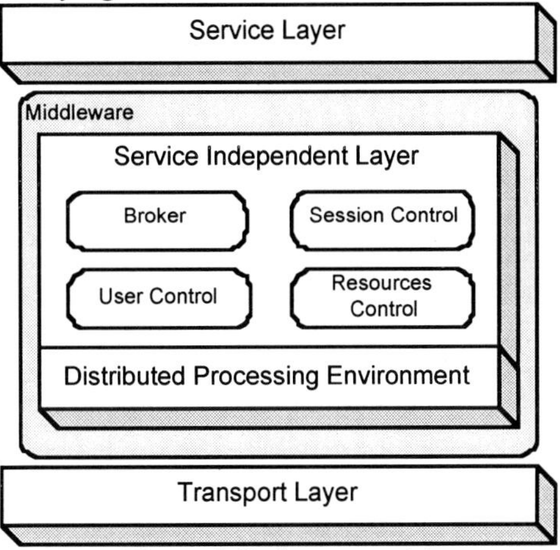

Figure 1: Reference model

3 The ORCHESTRA Architecture

Fig. 2 describes the main elements of the ORCHESTRA architecture, which can be seen to be consistent with the model outlined in the previous section.

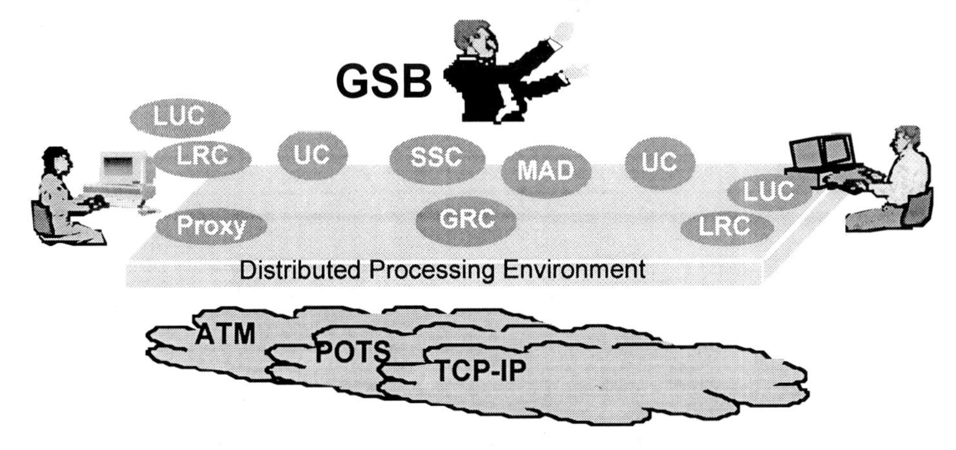

Fig. 2. ORCHESTRA architecture

The Service Independent Layer components define the distributed infrastructure of the whole experimental architecture. Hereafter the distribution of those components among different logical domains is described.

- User Domain components, Local User Controller (LUC) and Local Resource Controller (LRC), ensure the interaction between the user and the infrastructure. Those components can be installed in the user terminal equipment or can be dynamically downloaded from the access provider when user logs-in. The LUC controls the authentication process and presents to the user its graphical interface which allows service selection and access. The LRC executes the service logic and controls the local resources available on the terminal.
- Network Access Domain components, User Controller (UC), Proxy Local User Controller (PLUC) and Proxy Local Resource Controller (PLRC), allow the access to the services. The UC maintains all the data related to a single user as user profile, subscribed services, list of suspended sessions etc. Moreover, it authenticates the user access requests coming from the LUC. In order to support the broadest range of terminals with different computational capabilities, the proxy components have been introduced (Proxy LUC and Proxy LRC). In case of the simplest terminals (as ordinary telephone), the proxies bridge the capabilities of the terminals and those of the infrastructure.
- Service Broker Domain components, Global Service Broker (GSB), Meta-Application Distributor (MAD), Service Session Controller (SSC) and Global Resource Controller (GRC), control all the aspects related to the service execution. GSB, on user request, locates the services offered by different service providers, contacts the MAD, which controls the storage and distribution of the meta-applications, and instances the SSC, which maintains and controls the state of a single service session. GRC manages the resources necessary for service execution.

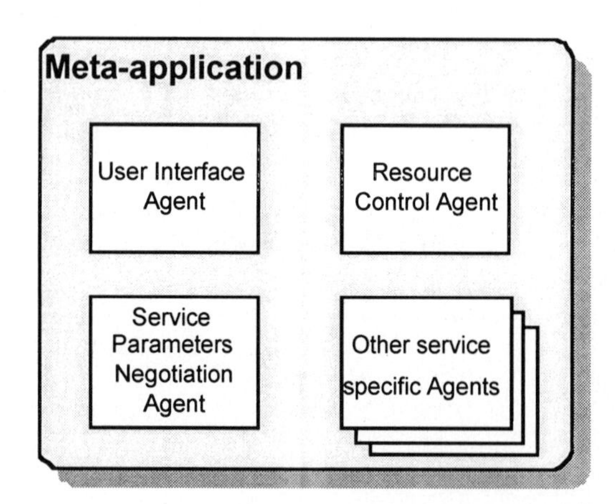

Figure 3: Meta-application's components

From a conceptual point of view, the Service Layer is composed of a set of *meta-applications*. The meta-application concept (see Fig. 3) has been introduced as a means to define the service logic. A meta-application is a logical entity that consists of various Agents that can be transferred across the network and executed by

appropriate run-time interpreters to carry out service specific functions (including local program activation). According to the remote execution paradigm of Agents, each Agent is implemented as a script and an interpreter of the scripting language is supposed to be available in each terminal equipment. Proxy interpreters are supposed to be available as a service for those terminals that are not equipped with an interpreter.
The meta-application components are transferred between software entities of the Service Independent Layer by using standard mechanisms of the DPE and are executed by the appropriate daemons installed on the user terminal and the network nodes. The use of Agents for defining the service logic allows to keep the service independent on the changes of the underlying network infrastructure. The dynamic interaction between system components based on the remote execution paradigm, ensures higher flexibility.

4 Technology Integration and Experiments

A first release of a prototype following the ORCHESTRA principles used OSF DCE [10] as the DPE layer. In particular the basic service used was the Remote Procedure Call (RPC), released in public domain in autumn of 1994, ported to Solaris 2.4. Tcl/Tk [11] was the language for the implementation of Agents as well as of the user interface to access to the services provided by ORCHESTRA. In particular the communication from terminals to the GSB was carried out via RPC, while the communication from the GSB to the terminals was based on Agent remote execution.
A second release of the ORCHESTRA prototype is currently available and the core technologies have been upgraded as follows. An implementation of CORBA 2.0, namely Black Widow, by Post Modern Computing[2], has been used as the DPE. All the entities envisioned in the proposed architecture have been implemented as Java objects whose services are traded by the CORBA broker by means of well known object interfaces. An Internet Web browser has been used as a mean to access the services made available by ORCHESTRA, i.e. the user interface has been embedded within the data space of the browser, thus achieving transparent portability on a wide range of hardware platforms. The interaction between the user and the middleware has been mediated by a gateway between WWW and CORBA, the Gatekeeper developed by Post Modern Computing. New features have been considered in the new release of the prototype, such as session mobility, i.e. the capability to suspend an active service session and to resume the session from a different terminal equipment, and the interworking with terminal equipment with low-end capabilities, such as standard telephone sets.
The development of applications is outside the objective of our work. In order to demonstrate sample services within our prototype implementation we chose a set of multimedia applications freely distributed in Internet for use in Mbone [12]. We chose two applications: VAT 3.4 (Visual Audio Tool) for audio and VIC 2.6 [13] (VIdeo Conference) for video. Since we wanted to experiment the delivery of data directly over ATM, we ported VIC over ATM by using Fore System's API[3]. We implemented two meta-applications in our ORCHESTRA prototype. The first one realises a Videoconference service and the second one a Video on Demand service (VoD).

---

[2] The Post Modern Computing joined Visigenic Inc. and Black Widow is now available as Visibroker for Java.
[3] Our ATM-based VIC implementation is part of the current VIC publicly available releases.

The whole experimental effort was aimed at the fast development of a prototype, capable to evolve rapidly in response to the technology changes. This objective was pursued by the integration of innovative technologies characterised by a widespread industrial support. We are now in a position to evaluate the effects of the choices we made in this development.
On the positive side we obtained a high level of portability of the architecture, in particular we were able to demonstrate it in a multiplatform client environment; this was mainly due to the adoption of Java as the basic programming tool. A further advantage lies in the positive correlation that the system capabilities exhibit with respect to the parallel evolution of the component technologies. An example is reported in the following which shows how an improvement in Java implementation led almost directly to a significant advantage for the performance of our prototype.
There are several disadvantages, however, which can be tracked down to our implementation strategy or to the specific choices we made.
A first problem is related to the dependence of the performance on the behaviour of commercial Operating Systems which tend to be not particularly well suited to real-time support. In various cases a performance degradation followed OS upgrades. Moreover, we are not in a position to take precisely in account the OS scheduling characteristics when determining the grade of service that an application instance is going to receive.
A second consideration regards the adoption of a CORBA-compliant platform as the foundation of our distributed system; we realise that, at present, the complexity of CORBA does not payoff in terms of added flexibility or portability, and that this solution to interoperability is probably too heavyweight for a large part of our application environment. The substitution of the client implementation with a simpler, Java-based software could be rewarding at the present state of things.
The choice of innovative technology brought together some low-level technical problems which must be mentioned, mainly consisting in system hiccups, non-standard behaviour and unpredictable performance across various platforms (this is the case of the Java virtual machine). This class of problems, however, is expected to become less relevant in the future.
Finally, we faced various performance problems due to the complex interaction of the various components of the architecture. In particular the first experiments led us to pay a greater attention to the performance issues involved in the agents downloading. We identified as a specific problem the necessity for separately loading each class of the Java applets in Java 1.02, coupled with the stateless nature of the current HTTP implementations, and verified a significant performance improvement when resorting to new Java implementation (1.1) which allow the downloading of an entire applet with a single file transfer [14].

5 Conclusions

In our opinion, the ORCHESTRA prototype has successfully proven the viability of a distributed and open approach (to the control and management of a telecommunication network) that uses emerging technologies and focuses on interoperability.
The next step will be to investigate the feasibility of this approach in a real network environment. In order to do so, ORCHESTRA architecture needs to be validated against crucial aspects of a real telecommunication network, such as scalability and performance.

In any case we believe that the use of general purpose software technologies inside network elements is only a matter of time and maturity; openness, lightweight and interoperability will be the keywords in future network environments.

References

[1] A. Alles, ATM Internetworking, http://cell.relay.indiana.edu/cell-relay/docs/ cisco.html, May 1995.

[2] M. Chapman and S. Montesi, "Overall Concepts and Principles of TINA", TINA Baseline TB_MDC.018_1.0_94, TINA-C, 17 February 1995.

[3] L. Zhang et al., Resource ReSerVation Protocol (RSVP), IEEE Network, 1993.

[4] Lazar, A.A., Bhonsle, S. and Lim, K.S., "A Binding Architecture for Multimedia Networks", Journal of Parallel and Distributed Systems, Vol. 30, Number 2, November 1995, pp. 204-216.

[5] D. L. Tennenhouse and D. J. Wetherall, "Towards an Active Network Architecture", Multimedia Computing and Networking (MMCN 96), 1996, San Jose, CA.

[6] Campbell, A., Coulson, G., and D. Hutchison, "A Quality of Service Architecture", ACM SIGCOMM Computer Communication Review, April 1994.

[7] OMG, "The Common Object Request Broker: Architecture and Specification", Technical Report, Object Management Group, 1991, Rev. 1.1.

[8] J. Gosling and Henry McGilton, "The Java Language Environment: A White Paper", Sun Microsystems, 1995.

[9] C.G. Harrison and D.M. Chess and Aaron Kershenbaum, "Mobile Agents: Are they a good Idea?", IBM Research Report, RC 19887, March 1995.

[10] D. Fauth, "Remote Procedure Call: Technology, Standardization and OSF's Distributed Computing Environment", Technical Report, Open Software Foundation, 1992.

[11] J.K. Ousterhout, "Tcl and Tk Toolkit", Addison Wesley, Reading, MA, 1994.

[12] H. Eriksson, "MBONE: The Multicast Backbone, Communications of the ACM", 37(8):54-60, August 1994.

[13] S. McCanne and Van Jacobson, "Vic: A Flexible Framework for Packet Video", ACM Multimedia, November 1995.

[14] http://www.javasoft.com/products/JDK/1.1/index.html

DAVIC and The Internet Convergence: A Convergence Architecture Using WWW, JAVA, and CORBA

Taesang Choi, Eunyoung Son, Kyeong-yeol Yu
Multimedia Communications Section, ETRI
161 Kajong-Dong, Yusung-Gu, Taejon City, 305-350 Republic of Korea
Tel: +82-42-860-5628/ Fax: +82-42-861-1342
{choits,eyson,yky}@etri.re.kr

As the world is being rapidly connected by a global information network, distributed interactive multimedia services are attracting significant attention from various sectors of the world community. DAVIC (Digital Audio Visual Council), whose importance we acknowledge, was established in June 1994 to provide end-to-end interoperability of a wide range of audio visual services of applications across countries and applications/services.

This paper proposes our system architecture to implement DAVIC specification 1.1 [1] compliant audio visual services such as Video on Demand (VoD) and teleshopping, especially focusing on our approach to integrate WWW, JAVA and CORBA technologies for providing user-friendly and familiar navigation of services, yet, retaining the power of high quality real-time interactive multimedia services. Convergence with the Internet is raised as one of the most important recent issues in the DAVIC community.

1 Introduction

Digital media will play much different and important roles in the coming 21st century. It will become the core means of providing new working, shopping, and leisure environments. The HAN/B-ISDN (Highly Advanced National/B-ISDN) project has started to build a communication, information, and service infrastructure to meet these challenging requirements in Korea. Some of the important researches done or planned in this national scale project include ATM core switches, ATM access network equipment, various transmission equipment, B-ISDN terminals, PCS (Personal Communications System), FPLMTS (Future Public Land Mobile Telecommunication System), and high quality land/mobile multimedia services. The development described in this paper is part of the HAN/B-ISDN project and will serve as a platform to provide high quality distributed interactive multimedia services over B-ISDN. Our experimental services developed so far are VoD and teleshopping. Realization of these services requires integration of a huge number of audio and visual technologies including real-time transmission capabilities. In other words, it is essential to reach agreements on open and interoperable interfaces to glue them seamlessly and with scaleability which can incorporate future innovative technologies.
DAVIC (the Digital Audio Visual Council) was established in June 1994 to provide end-to-end interoperability of a wide range of audio visual services of applications across countries and applications/services. Its specifications address the complete spectrum of vertical and horizontal aspects of its goal. DAVIC specification 1.0 published in December 1995 includes interfaces, protocols, and tools to interconnect content provider systems, service provider systems, delivery network systems, and service user systems. It was developed based on important principals, that is, adopting as many existing official or industry-based standards as possible and one solution for one functionality to avoid unnecessary conflicts. We closely followed this activity and our recent developments conform to DAVIC specification 1.0 [2].
While DAVIC builds a foundation to provide high quality interactive multimedia services, a new digital information service paradigm is spreading with enormous

speed and supports throughout the world. That is, the Internet is approaching closer and closer to the public. Especially, thanks to the WWW, the Internet seems to be becoming the global information infrastructure. All kinds of innovative services are under testing or under operation already. Due to the number of technical problems as of today, however, the quality of services does not fully meet the expectation of the majority. DAVIC considers this interesting situation as an opportunity to speed-up the convergence between two worlds to harmonize and share advantages. Thus the recently published DAVIC specification 1.1 provides a few interesting tools such as the Java virtual machine and direct Internet access tools. The main focus of this paper is to propose an architecture to integrate the Internet services and DAVIC-based services over our development platform. This platform will be our cornerstone not only to apply it to our currently developing services but to build future creative integrated services.
The next section briefly describes the overall architecture of our development system. Our integration architecture and an example scenario are provided respectively in the following sections. We will conclude our work and discuss future works to be addressed in the final section.

2 Overall Architecture

Our development system, called IMPRESS (Interactive Multimedia exPRESS), consists of ATM switches which are used as a core network and CANS (Centralized Access Node System) that has a stream duplication and channel selection control function which is used as an access node. For the server side, there are four kinds of servers: movies on demand server, switched video broadcasting server, Internet server and multimedia conference server for multipoint videoconferencing service. An MPEG-2 (Moving Picture Experts Group-2) set-top box, a PC based MPEG-2 set-top box, and a PC based desktop videoconferencing terminal are being developed as service user terminals. All the service user terminals are connected with 155Mbps STM-1 interface. Table 1 is a list of the service equipment that will be used for the service platform. Protocol stacks of the current service platform are compliant to DAVIC 1.0 and 1.1 specification. [1,2]

Core network	Public ATM switch
Access network	CANS (Centralized Access Node System) with stream duplication and channel selection control function
Server	Movies on demand server, Switched video broadcasting server, Internet server for ISAP (Internet Service Access Point), Conference server
Terminal	MPEG-2 STB (Set-Top Box) PC based MPEG-2 STB PC based DVT (Desktop Videoconferencing Terminal)

Table 1. List of Equipment in Service Platform

The plan is to build the following basic services over the platform next year: movies on demand, Internet access services, multipoint videoconferencing service, and switched video broadcasting. Among these services, we have implemented MPEG-2 quality movies on demand, home-shopping, and Internet access service based on the DAVIC specification 1.0 and 1.1. The main objective of this development is establishing a new service developing an environment for integrated interactive multimedia services combining telecommunications, broadcasting, and computer services. Figure 1 illustrates an overall architecture of the IMPRESS.

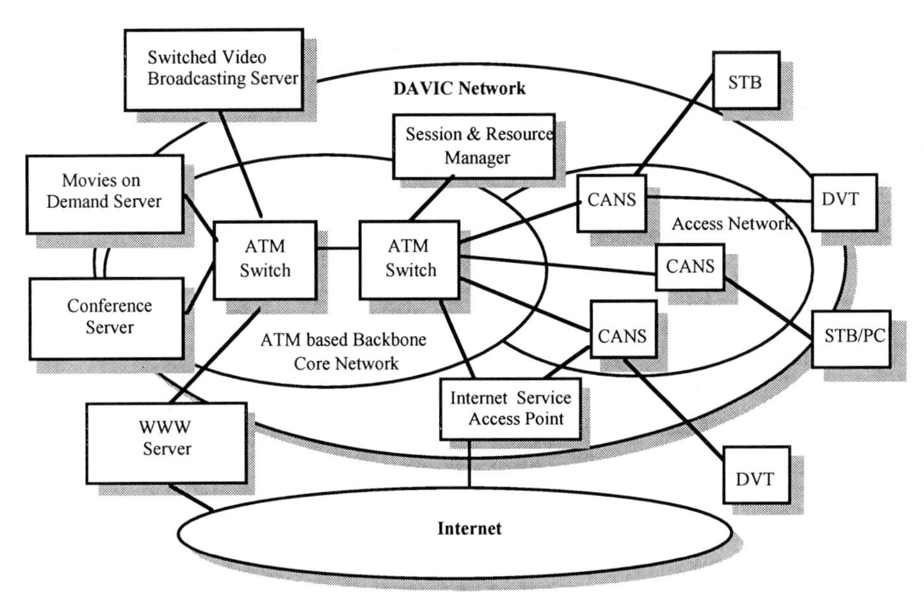

Figure 1. Overall Architecture of the IMPRESS

As mentioned above, our current system supports on-demand interactive services such as VoD and teleshopping. In the server side, an application server and a video pump together provide high quality video server's functionality. The video pump can handle a number of MPEG-2 encoded movies and a number of audio/video clips and a large-scale content server is under development. An application server which can provide service gateway, stream control, and application control functionality is designed to be scaleable to accommodate future expansion since it is a CORBA-based object-oriented server. For the service user terminal, STB (Set-Top Box) was developed as a one add-on board for PC system hardware. MINIBA (Media and Network Interface Board Assembly) is the main unit that performs the MPEG-2 decoding and ATM network interface function. MINIBA is divided into two modules: ANIM (ATM Network Interface Module) and MPM (Media Processing Module). ANIM provides termination of physical signal, ATM and AAL processing, and Q.2931 UNI signalling function. MPM provides demultiplexing of MPEG-2 TS (Transport Stream), decoding of MPEG-2 AV (Video and Audio), and AV output function. STB software operates in the Windows NT environment. They are classified as the CB (Connection Block) that controls the low layer connection between STB and server, SM (Session Manager) for session management, and ACB (Application Control Block). Figure 2 shows the basic system architecture of the STB.[9]

These high quality multimedia contents are transferred between content providers and service user systems through broadband networks. Public ATM switches and access devices such as CANS are being developed for this purpose.

Towards our ultimate integration architecture, we are expanding our current development platform to support multipoint videoconferencing and switched video broadcasting during this year. Absorbing the Internet into this architecture has significant effects in terms of user-friendliness, flexibility, and marketability as will be explained in the next section.

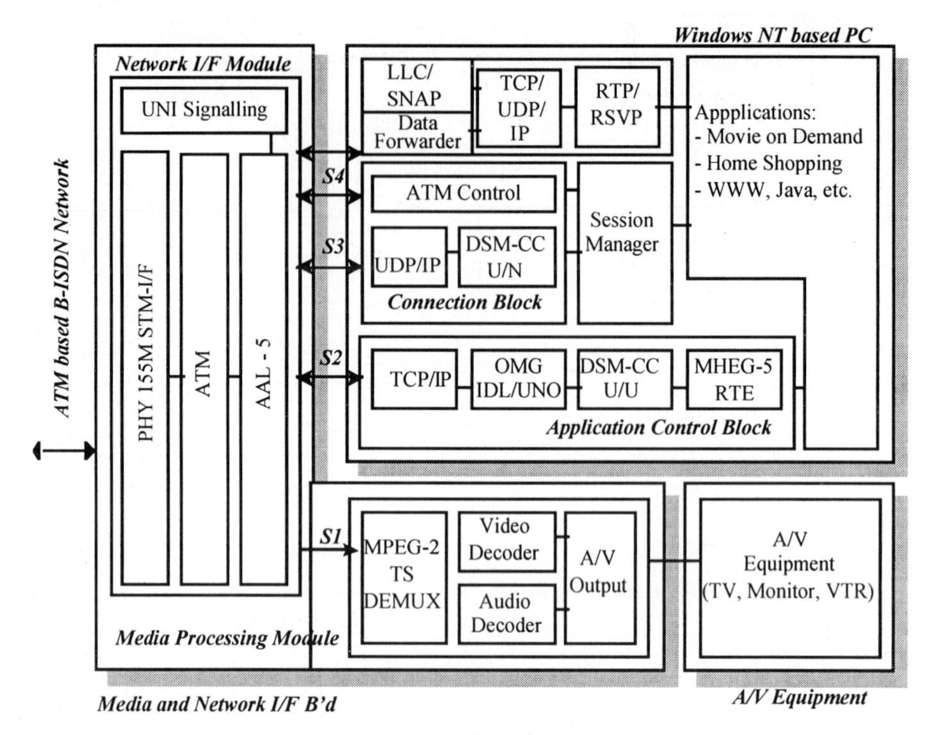

Figure 2. Detailed View of the STU

3 DAVIC and Internet Convergence Architecture

In this section, we focus on our approach to harmonize the Internet and DAVIC services by integrating WWW, Java, and CORBA technologies. Figure 3 below illustrates our proposed architecture.

In DAVIC specification 1.0, end-to-end user controls for interactive on-demand types of services use the DSMCC-UU (Digital Storage Media Command & Control - User-to-User) protocol and navigation presentations depend on MHEG-5 (Multimedia and Hypermedia Experts Group-5) objects and engine. Typically, MHEG-5 objects reside in a service provider system and an MHEG engine is in an STU. These MHEG-5 control events such as navigating a particular service or choosing a movie map to corresponding DSMCC-UU primitives. Mapped requests from an STU are transferred to the service provider system which has a DSMCC-UU server through CORBA 2.0[6] object brokers over IIOP (Internet Inter-ORB Protocol). An end-user who wants to receive a VoD service, for example, needs an STU with MHEG-5 based navigation tool and an engine. However, our proposed architecture utilizes WWW, JAVA, and CORBA to distribute a DSMCC-UU client as a JAVA applet. Thus, end-users can navigate multimedia services interactively using their own familiar and user-friendly web browsers. Some of the advantages of this approach are as follows:

- an STU no longer needs DSMCC-UU client software;
- platform independence due to the Java virtual machine;
- a single user-friendly interface to access existing Internet and high quality multimedia services;

- scaleability and expandability not only for future DAVIC and Internet services but also for existing back-end legacy systems such as Database systems, network management systems, etc;
- a foundation to build new innovative Internet and DAVIC integrated services.

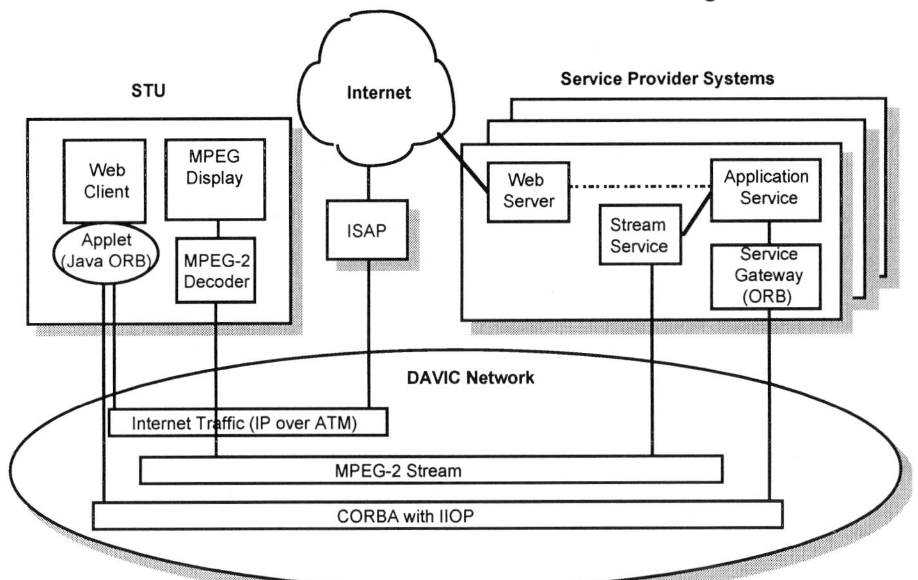

Figure 3. DAVIC and the Internet Convergence Architecture

As shown in figure 2, our STU has ATM, MPEG-2 decoding, and direct Internet access capabilities. The JAVA ORB which can handle JAVA based CORBA clients is added with JAVA virtual machine [7]. When a DSMCC-UU applet client is downloaded, JAVA ORB transparently relays requests and replies to/from DSMCC-UU server in a service provider system. One interesting scenario we can think of is that an end-user can search for a football player's profile of his/her interest via Internet search engine while watching a high quality football game through the same web browser. Also, there is a module called, Data Forwarder, in an STU. With 10-Base T physical port, it relays IP traffic from/to an attached PC. That is, an STU can be used as a very high speed modem to access the Internet. No modifications are required in PCs, thus any existing TCP/IP and its application software can be used.

For direct Internet access, an ISAP(Internet Service Access Provider) relays IP over ATM traffic from/to an STU. It is a typical router which has ATM and IP LAN (e.g., ethernet, token-ring, FDDI, etc.) interfaces. Any commercial off-the-shelf . of this category will be fit for its purpose. DAVIC specification 1.1 allows one ISAP per STU, but it is extending the scope for supporting multiple ISAPs of the user's choice and other alternatives to access Internet [8]. Our architecture will incorporate future extensions. For our development, a UNIX workstation based gateway is used instead and will be replaced in the future.

In the service provider system, an HTTP server is installed with existing application service modules. It maintains navigation presentation objects and DSMCC-UU JAVA applets. Another important role is to advertise its services to not only DAVIC customers but also non-DAVIC Internet users. It can be an excellent means to establish and expand its customer base. We are planning to add RTP(Real Time Protocol) to support the quality of multimedia services for non-DAVIC Internet users. CORBA IIOP is a transparent delivery vehicle between DAVIC clients and DAVIC service providers. The service gateway plays a server role for the Java based DAVIC client to provide various services such as application service, delivery service, and stream service. JAVA ORB and IIOP allows existing DAVIC services to be accessed with no or minor modifications.

With this architecture, future high quality multimedia service users, no matter which services they are using, can download necessary clients which is a part of their familiar user interface on demand. More importantly, all this client software maintenance and version control is done by service providers. That is, this architecture incorporates future network-oriented software rental concepts instead of letting every customer worrying about the maintenance of their client software. The marriage between DAVIC for high quality multimedia service framework and the Internet for the existing and fast-growing huge customer base will lead to a leading-edge position in this highly competitive market.

4 A Sample Scenario: Video on Demand

As mentioned before, the convergence with the Internet is not limited to the interactive on-demand types of services such as VoD, PPV (Pay Per View), Teleshopping. It will be applied to the rest of multimedia services such as below and communicative services (such as multipoint videoconferencing). However, we will illustrate a scenario based on the currently developed service: VoD.

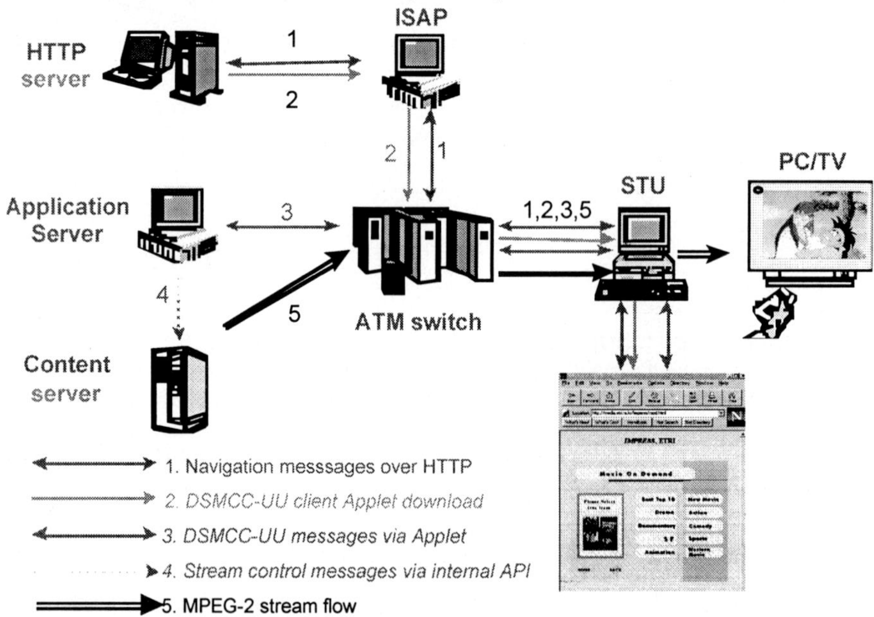

Figure 4. A Sample Scenario: Video on Demand

When an STU starts up, a configuration setup module initiates a dialogue with its DAVIC network provider's configuration setup module. As a result, the STU receives information about the ISAP to connect to the Internet access. After its connection with the ISAP, an end-user starts a web browser. While navigating through the Internet, the user finds an interesting multimedia service site. During further surfing, the user can decide to accept the service provider's contractual terms and select high quality DAVIC VoD service. Through the negotiation, the service provider gets enough information about the user and the user's STU. After authentication verification, an applet is downloaded to the customer. This applet initiates resource allocation via the STU's DSMCC-UN(Digital Storage Media Command & Control - User-to-Network) protocol. After successful resource allocation, the VoD service navigation menu with a DSMCC-UU client applet is downloaded and the user selects a favorite movie. Finally, the movie is delivered to the user through MPEG-2 downstream channel. Figure 4 illustrates interactions involved with this scenario pictorially. In the picture, it is assumed that a DAVIC session between an STU and an ISAP is established in advance. While watching the movie, the user not only can control the stream interactively but also can navigate the Internet at the same time.

5 Conclusion and Future Work

We proposed a system architecture to implement DAVIC 1.1 compliant audio visual services such as Video on Demand (VoD) and teleshopping, especially focusing on our approach to integrate WWW, JAVA and CORBA technologies for convergence with the Internet. Well harmonized convergence has a great strategic importance for both worlds and for the success of next generation multimedia services.

Our development platform, IMPRESS, has been progressed significantly. STU's hardware, stream control flow, DSM-CC user-to-network, and user-to-user control flow are ready. However, the DSMCC-UU JAVA applet for VoD and other multimedia services is still under development. Validation and interoperability testing will be followed when the prototype implementation complete and our experience will be incorporated into future versions of the paper.

References

[1] DAVIC specification 1.1, Geneva, Switzerland, September 1996.

[2] DAVIC specification 1.0, December, Berlin, Germany, December 1995.

[3] ISO/IEC DIS 13818-6, Information technology - Generic Coding of Moving Pictures and Associated Audio - Part 6: Digital Storage Media Command and Control (DSM-CC), 1995.

[4] ISO/IEC DIS 13522-5, Information technology - Coding of Multimedia and Hypermedia information - Part5: MHEG Subset for Base Level Implementation.

[5] RFC 1577: M. Laubach, Classical IP and ARP over ATM, Hewlett-Packard Laboratories, January 1994.

[6] CORBA 2.0, OMG Technical Document PTC/96-03-04

[7] Sun Microsystems Computer Corp., The Java Virtual Machine Specification, Release 1.0 beta DRAFT, August 21, 1995.

[8] Digital Audiovisual Council, DAVIC 1.1 Specification Internet Access Tools Baseline document #28 Revision 3.0, Geneva, September, 1996.

[9] J.H.Park, J.S.Choi, S.J.Kim, Y.D.Park, "Multimedia Terminal in ATM Network", KICS Proceeding Vol.14, No.2, 1995.

Service Engineering

Thomas Magedanz
GMD FOKUS / TU Berlin, magedanz@fokus.gmd.de

This section of the book is devoted to service engineering. Service engineering has been defined as the art and science of building software for communication services. In this, it deals with methods, techniques, languages, tools, architectures, interfaces, and common components by which service software can be conceived, specified, designed, built, tested, deployed, run, maintained, improved, and retired. More abstractly, anything anywhere in the life cycle of service software that can be applied to more than one service can be considered within the domain of service engineering. Furthermore technological progress, such as the shift from function-orientation to object-object-orientation or the emergence of agent technology add additional diversity in this domain. Hence the spectrum of issues present in the service engineering domain is a big one.

This section of the book features ten papers, which are grouped in accordance with the hot topics of the telecom services environment into three major parts: *Intelligent Networks, Object-oriented Middleware / Service Platforms* and *Mobile Agents.*

Within the last decade, the *Intelligent Network (IN)* concept has coined the face of the telecommunications environment. Rather than just a network architecture, the IN represents a complete framework for the creation, provisioning and management of advanced communication services. The first IN architectures were introduced in the 80s in order to provide a network independent and service independent service platform. The basic goal was to provide more flexibility in service delivery than is possible in traditional service environments. Since the beginning the IN has been the subject of research and ongoing evolution in order to meet emerging market demands and take into account technological progress.

Three papers of this section are related to IN issues. The application of IN principles on top of broadband networks represents a major research issue. The paper of Hussmann *"Intelligent ATM Networks: Services and Realisation Alternatives",* presenting the results of the ACTS project INSIGNIA, discusses several realisation alternatives for network intelligence in order to provide a broadband VPN and a broadband videoconference service in ATM networks. Another area of ongoing research in the IN world is the feature interaction problem as a consequence of combining multiple services which have been designed independently. The paper of Kimbler et.al. *"Goal-based Filtering of Service Interactions",* presenting results of the EURESCOM project P509, introduces a concept of interaction filtering, which identifies interaction-prone combinations by analysing the relations between services and user`s goals. Finally, the introduction of an IN within a specific network environment requires careful analysis. The paper of Makhrovskiy et.al. *"Intelligent Networks Planning Supported by Software Tools",* presents a complex tool for IN dimensioning and planning in order to maximise efficiency.

In the beginning of the 90s, research on object-orientated distributed service platforms, also referred to as "middleware", has received increasing attention, since object-orientated software development has been considered to provide advantages in the context of software structuring and reuse of components. Therefore the integration of service control (i.e. IN) and management (i.e. TMN) platforms into a unified architecture taking into account emerging Open Distributed Processing concepts has represented and still represents a fundamental research issue for future telecommunications. In this domain the *Telecommunications Information Networking*

Architecture (TINA) and the underlying distributed processing environment (DPE) technology, such as OMG`*s Common Object Request Broker Architecture (CORBA),* have evolved as defacto standards during the last three years.
Therefore we will recognise these architectures within most of the five papers within the second part of this section. Object-oriented platforms require corresponding service development tools. The paper *"Adopting Object-oriented Analysis for Telecommunications Systems Development"* by Martin reports on initial experiences with the adoption of object-oriented analysis for the development of telecommunications systems. Dede et.al. introduce in their *paper "OSAM Component Model - A key concept for the efficient design of future telecommunication systems"* a basic modelling concept for the consistent and effective analysis, design, use and management of services, developed within the ACTS project DOLMEN. On the other hand effective techniques for the rapid integration of service and management systems over multiple organisational domains are required in face of increased competition, complex service provision chains and integrated service offerings. Furthermore, multimedia services are considered as the drivers for the new service platforms. Therefore appropriate mechanisms have to be provided by the platform. Qian and Campbell propose in their paper *"Extending OMG Event Service for Integrating Distributed Multimedia Components"* a timed event service, representing an extension of the standard OMG event service with temporal factors so that the system can deliver large volume events, like video frames, in real time. Finally, the paper by Newcomer et.al. *"STDL as a High-Level Interoperability Concept for Distributed Transaction Processing Systems"* describes how the concept of STDL is used in the ACTS ACTranS project for portability and interoperability of heterogeneous Distributed Transaction Processing Systems, including support of object-oriented TP systems.
In 1994 another new paradigm has gained momentum considered as an alternative to traditional Remote Procedure Calls in client/server systems: *Mobile Agents*. Mobile Agents enable the delegation and automation of specific tasks to autonomous software entities which could move between network nodes in order to perform their tasks. This enables on demand service provision and network computing. Dedicated "Agent execution environments" form the prerequisite for agent-based services taking care for the secure execution, communication and transfer of agents. Today scripting languages, such as Java, are considered most appropriate for implementing mobile agents. Rather than considering Mobile Agents and Remote Procedure Calls as exclusive approaches, maximum flexibility will be achieved by hybrid middleware platforms.
Two papers are present in this final part of the book section. The paper of Park et.al. *"JAE - A Multi-Agent System with Internet Services Access"* introduces a Java-based Agent Environment, a multi-agent system comprising mobile agents communicating with a fixed infrastructure of agent servers making services provided through the Internet accessible. JAE will be used within the ACTS project OnTheMove. In their paper *"AgentSpace: A Framework for Developing Agent Programming Systems"* Silva et.al. concentrate on the development of distributed applications based on agents and their support programming systems. The idea is to consider agent-based applications spanning through three levels: agent framework, programming system and application level itself.
In summary it can be stated, that the outlined papers present a good snapshot of the current diversity of research issues within the service engineering domain.

Intelligent ATM Networks: Services and Realisation Alternatives

Heinrich Hussmann
Siemens AG, Public Communication Networks
Heinrich.Hussmann@mch.scn.de

In order to fully exploit the potential of broadband networks for various user groups, network intelligence is required. Network intelligence functionality has to be considered in the context of telecommunication services as well as in the context of network architectures.

This paper is based on results of the ACTS project AC068 INSIGNIA which realises an ATM-based Intelligent Network. The paper describes the services which have been chosen by INSIGNIA for prototyping. A general perspective is taken, and several realisation alternatives for network intelligence in ATM networks are discussed.

Keywords: Intelligent Network, ATM, Broadband Services, INSIGNIA

1 Introduction

In order to fully exploit the potential of broadband networks for various user groups, network intelligence is required. Network intelligence in this context means additional network-related functionality which is not necessary to establish the pure transport of information from source to sink, but which provides added value for application developers and users. Examples for intelligent network functions are flexible routing and billing, support for user groups, or management of complex communication sessions.

The ACTS project AC068 INSIGNIA (IN and B-ISDN Signalling Integration on ATM platforms) aims at adding network intelligence to B-ISDN networks. INSIGNIA will provide trials of experimental services in a Europe-wide testbed in spring 97. INSIGNIA comprises a well-balanced selection of telecom operators, equipment manufacturers and research institutions.

In the early phases of the INSIGNIA project, a general discussion of the issue of network intelligence for ATM networks took place. This paper summarises these discussions and briefly describes the decisions taken by INSIGNIA.

Two issues can be distinguished here which should be discussed separately: the services offered to users of the network and the architecture chosen for realisation of the services. INSIGNIA has decided to implement a selection of three prototype services. Two of these services are used as running examples throughout this text. INSIGNIA implements the selected services on an architecture closely resembling the classical Intelligent Network architecture as defined by ITU-T [2]. In this paper, however, the relationship between services and architectures is analysed in a more general perspective, covering also alternative realisation options to the architecture chosen by INSIGNIA.

The structure of this paper is as follows: After a brief introduction into the topic of network intelligence, the services selected by INSIGNIA are described briefly in section 2. Section 3 introduces several alternative realisation options, stressing the distinction between client-server solutions and network-based solutions. The realisation of the example services on these alternative options is discussed in section 4, giving indications which implementation alternative fits best to specific service

---

This work was partially funded by the European Union through the ACTS programme within the project AC068 INSIGNIA.

classes and user groups. Section 5 gives a summary of the observations and tries to identify common functionality which is independent of the chosen network architecture.
The general assumption of this paper is that a *pure ATM network* connects all end systems, and that all end systems have access to an ATM switched bearer connection service through UNI signalling. In the short and middle term, this assumption will hold only for business customers, except for a few trial areas where experimental ATM infrastructure is available to private customers. However, section 3 below includes one architectural option which can be used also in networks composed of several network technologies (internets).

2 Network Intelligence: Services

2.1 Application and Network Services

Several different classes of services can be distinguished. In the following, a distinction between *application services* and *network services* will be used.
If a user participates in a telecommunication service, he/she always uses some kind of user interface to handle the service. In a classical telephone service, the user interface simply consists of the telephone handset and the functions it offers. In this case, there is more or less a one-to-one mapping between functionality at the user interface (e.g. buttons for call transfer, call redirection etc.) and the features of the network. In a more advanced environment, the user may use a modern graphical user interface through a computer connected to the network. In this more complex case, the user perceives a telecommunication service as a unit which is defined by the user interface, without even knowing about the detailed features on an underlying network. We call such a service an *application service* since it is often realised as an application program running on the user's computer. Examples for application services are Electronic Mail, World Wide Web, or Videoconferencing.
Any application service uses functionality of the underlying network. In the simplest case, this is just establishment of bearer connections and/or transmission of information. However, the network may offer additional functionality, i.e. network intelligence, towards application services. In any case, the functionality offered by the network is called a *network service.* So an application service uses network services.
Using this terminology, we can distinguish between two kinds of network intelligence.

- *(Standalone) Intelligent network services*: These services are so close to the core functionality of the network and so generic that they are meaningful for many or all kinds of application services. A famous example from classical Intelligent Network is the Freephone service (800 numbers). For a complex application service, it is easy to use this kind of network service by just using special numbers when setting up connections. So for example, a database access application may use an 800 number to connect to its server.
- *Intelligent application services:* These services are complete application services which are realised by application programs running on the end systems but which may contain network-oriented features. An example for a network-oriented feature (network intelligence) is a generic user authentication service which is valid for all users for the network and which may be used within a videoconference application service as well as in a database access service.

For this paper, two services will serve as examples, one being a typical network service and the other a typical application service. Both services are realised within the INSIGNIA network trials.

2.2 Example: Broadband VPN

Broadband Virtual Private Network (B-VPN) is an example of a standalone network service which can be combined which any application service, provided the application service is based on the B-ISDN User-Network Interface (UNI) and uses the Q.2931 protocol for establishing Switched Virtual Connections among end systems.
B-VPN is a special package of network features to support distributed groups of users collaborating over long distances. For this purpose, a virtual network can be constructed which appears to the user group similar to a local network. For example, the group has the possibility to administer addresses (private numbering plan) and to define certain access and egress rights for the communication with other users outside the group.

2.3 Example: Broadband Videoconference

Broadband Videoconference (B-VC) is an example of an application service which is visible for the user as an application program running on the local workstation. The videoconference service provides a directory of potential participants in conferences and a directory of actual conferences (participant lists). The user can create a new conference by defining a list of participants together with appropriate privileges. Afterwards, the conference is still inactive but stored in the conference database. At any point in time (and also repeatedly), such a stored conference can be activated by any user who is authorised for this operation, leading to actual connections being set up among the participants (each to each). Of course, many other useful functions are necessary (e.g. dynamic change of conferences or handling of invitations to conferences) in order to provide a user-friendly application service.

3 Network Intelligence: Architectural Options

We distinguish and describe here two different approaches for the provision of network-oriented features in broadband communication services which are:

- the end-system-based approach (client/server solutions); and
- the approach based on the Intelligent Network (IN) architecture.

This very general distinction essentially describes two different options for the function split between end systems and network: In the first option the network is not involved into the services beyond its pure transport functionality; in the second option the network provides part of the service functionality.
It is obvious that most popular application services of today (e.g. Electronic Mail, World Wide Web) follow the end-system-based approach, since they are realised on networks using the IP (Internet Protocol) suite. However, for future high-performance multimedia services on B-ISDN networks, it is very likely that applications will become "aware" of the underlying ATM/B-ISDN infrastructure (see [1]). In this case, there will be an explicit interface between the application and B-ISDN signalling software according to the ATM UNI (User Network Interface) specifications. The remainder of this paper restricts its focus only to applications which are aware of underlying B-ISDN and make use of signalling. For this class of application programs, the above-mentioned options from realisation alternatives will need to be discussed.

3.1 End-System-Based Approach

In the end-system-based approach to service realisation, the network itself is not involved in the provisioning of the service at all; it is just used as a means for information transport. The application resides on end systems of the network. In sophisticated services, a distributed client/server architecture can be found, where some end systems act as servers and some as clients. Such service solutions can (and will) be extended to make use of ATM UNI signalling, in order to benefit from the features of an all-ATM network (e.g. well-defined quality of service). However, the ATM network itself does not contain any specific implementation of the service.

Figure 1 gives a picture for this approach indicating the locations of service control and interaction resources. Service control resides within end systems, and is distributed among them; in this way part of the overall service logic is provided by the client. Moreover, interaction resources (like audio/video encoding and transmission components) are found at the terminal equipment (TE) and at servers.

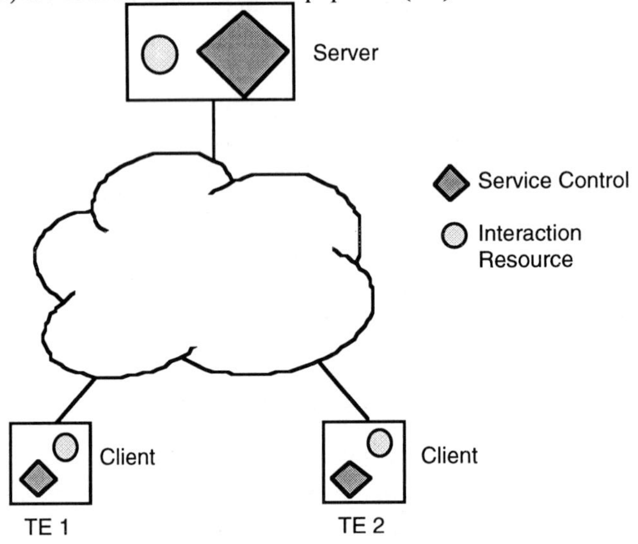

Figure 1: End-System-Based Approach

The main advantage of this solution is that new services can be easily provided without affecting the network itself. New services are deployed just by connecting the clients and servers to the (unmodified) network. The drawbacks of this solution are:

- The compatibility of end systems (clients and servers) is problematic. There are special classes of end systems which can interwork, and others which cannot.
- The network is not able to take any specific measures to make the service more secure or more reliable. So for example, an arbitrary end system on the network is able to put a high load on a server for such a service, simply by placing frequent calls to the server.
- The role of the network operator is just to provide reliable transport of information, without specific "added value" targeted for service users. This makes the business of network provision less profitable.

3.2 IN-Based Approach

In the IN-based approach to broadband services, as it is studied by INSIGNIA, the service implementation is tightly integrated with the network infrastructure. The services are realised on special network elements called Service Control Point (SCP) and Intelligent Peripheral (IP). Due to the use of most flexible (and therefore long-living) interfaces between SCP, IP and the underlying switching network, it is possible to add and modify service logic without affecting the software in the switching elements of the network.

In figure 2, it is pointed out that the service-specific control information is centralised at the SCP. The network itself contains only generic (i.e. service-independent) control functionality which is realised by a Service Switching Point (SSP). Neither does the network itself provide any interaction resources. Interaction resources required for the service logic (e.g. for the SCP) are located in the IP element[1]. The end systems contain special software components to communicate with each other and with the IP.

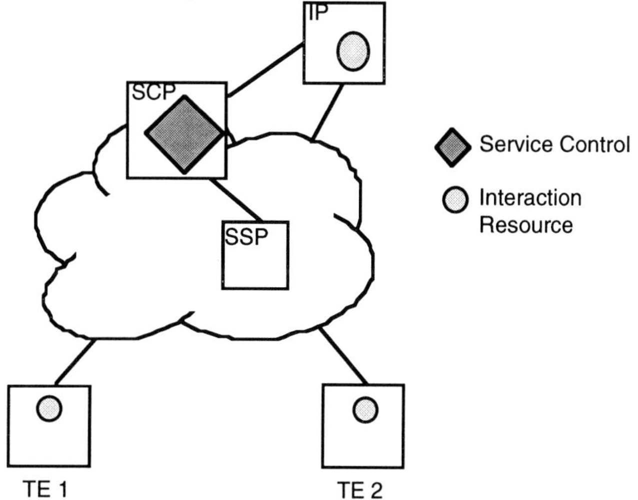

Figure 2: IN-Based Approach

When comparing figure 2 with figure 1, it becomes apparent that a quite drastic change in the implementation of services is required in order to move from an end-system based solution to an IN-based one. Experience with the realisation of the INSIGNIA trial services has shown that only minor parts of existing end-system-based applications (mainly the interaction resources and parts of the user interface) can be re-used when migrating towards an IN-based solution. An existing end-system based solution can be taken as a starting point for developing an IN-based service realisation, but the control logic of the service has to be re-implemented on the SCP platform.

The main advantages of the IN-based approach are:

- From the consumer point of view, the network becomes more powerful. Additional network features become available for many applications in a homogeneous style.

[1] In an implementation using the so-called User-Service Information features of IN, a limited form of interaction resources may also reside on the SCP.

- Since the network has the responsibility for IN-based services, well-developed mechanisms for high reliability, overload control and load sharing are applicable to the Broadband IN services.
- From the network operator point of view, the network operator can differentiate itself from other network operators by means of value-adding services.

On the other hand, the major disadvantage of IN-based solutions is that services have to be developed specifically for an IN framework. Since standardisation of Broadband-IN cannot be expected for the near future, there is a danger of introducing proprietary interfaces, and therefore limiting the wide-spread use of services.
The next section illustrates the two approaches by briefly discussing alternative realisations of example services. Two examples are given for the two services which were introduced above (B-VPN and B-VC), covering both an end-system based and an IN-based approach.

4 Realisation of a Broadband VPN Service

4.1 IN-Based Realisation of B-VPN

Examples for application scenarios of the B-VPN service are as follows:

- A user may want to indicate for a call that it belongs to the VPN, for instance to have it billed onto a company account or for achieving special access privileges. For this purpose, the end system has to transmit an identifier for the Virtual Private Network which is addressed and a local user identification (private number) within this network. Of course, the VPN service will apply additional security checks, e.g. request for a PIN.

- A user may want to place a call into the VPN (from outside or inside the VPN). For this purpose, again the VPN identification and the private number within the VPN have to be transmitted. The VPN service applies screening rules for such a call and performs a translation into a physical address, which can be made dependent on parameters like day, The Broadband Virtual Private Network service is a typical pure network service. The core functionality of the B-VPN can be invoked just by choosing special numbers as addresses for a call setup. time, or origin of the call.

These examples show clearly that the main functionality of B-VPN can be achieved by using the normal interface from an application program towards UNI signalling, just by using special numbers. For instance, the fact that a number is an IN number can be expressed by using a special prefix, followed by sequences of numbers indicating the individual VPN instance and the private number within the VPN. If this number format is used for the *calling party address*, it indicates that the call is meant as coming from the VPN; at the position of the *called party address* it indicated that the call goes towards a VPN destination.
So there is a natural way for mapping the service invocation onto the interface which is already offered by the B-ISDN network in the form of UNI signalling. The natural choice for implementation therefore is an IN-based solution where the service is realised within the network, as it is indicated schematically in figure 3. The "application" shown here can be an arbitrary program using switched ATM connections, for example a videotelephony or file transfer application.

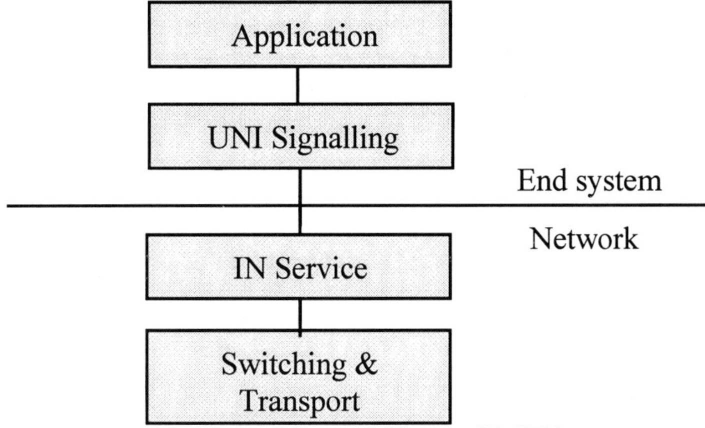

Figure 3: IN-Based Realisation of B-VPN

4.2 End-System-Based Realisation of B-VPN

Basically, it is possible to realise the functionality of a VPN in a pure client-server style, such that from the network only basic switched connections are expected. To achieve this, a special server has to be realised which provides the centralised resources for the service like number translation database. On the client systems, specific software is required which interacts with the VPN server before attempting a call setup. The overall architecture of such an approach is indicated in figure 4.

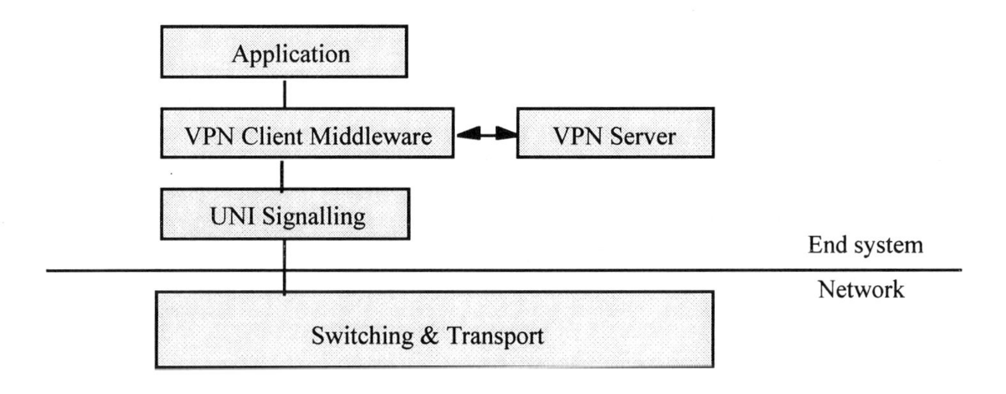

Figure 4: End-System-Based Realisation of B-VPN

The relatively complex architecture alone is a first indication that this realisation alternative is less attractive than the IN-based solution. Other problematic issues with such a client-server realisation are the following:

- Communication between client and server software has to be established before the service can start to work. This is additional overhead to the IN-based solution where the existing signalling channel is used.
- VPN client software has to be available on all kinds of involved end systems in such a way that it can interwork with the underlying signalling software ("middleware"). Therefore, a huge variety of hardware/software platforms have

to be supported by this software (whereas the signalling interface used in the IN-based implementation is readily available from the manufacturer of the interface card).

- Service-specific communication protocols are used between VPN client and server. This communication may be realised by a standard Distributed Processing Environment (DPE).

To summarise, the best choice for B-VPN is the IN-based realisation.

5 Realisation of a Broadband Videoconference Service

5.1 End-System-Based Realisation of B-VC

Examples for application scenarios of the B-VPN service are as follows:

- A user wants to create a new conference. For this purpose, the participants are picked from a user directory and rights assigned to them (e.g. the right to add users, the right to close the conference). The conference description is stored in a conference database.
- A user wants to activate a conference which is stored in the conference database. For this purpose, the conference is selected out of the list of available conferences and activated. The service checks the permission for this action and then tries to contact all participants, by establishing the required audio/video communication channels.

These examples show that the B-VC service (being an application service) has a significantly different structure compared to B-VPN. There is a central database as in B-VPN, but in B-VC the users need on-line access to this database during service runtime. Moreover, a management of communication sessions is required which sets up the necessary connections (in a star or mesh configuration) and monitors any events for them (e.g. loss of a single connection). The interface offered by the B-ISDN network (UNI signalling) is not a natural way to express all invocations of B-VC service functions. In particular, an extensive (multimedia) dialogue between the service logic instance and the user is required.

These are the reasons why it is natural to realise this service in a client-server style, as it is shown in figure 5.

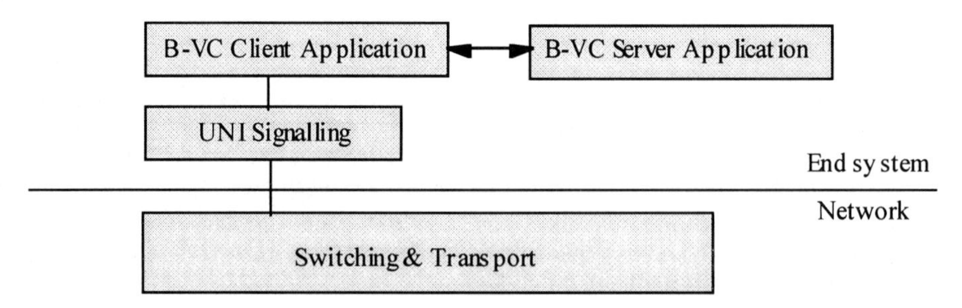

Figure 5: End-System-Based Realisation of B-VC

In the end-system-based solution, the B-VC client and server applications define their own application-specific protocol for internal communication. The server application is responsible for keeping the database and for the management of communication sessions, whereas the client application establishes the required connections through UNI signalling.

5.2 IN-Supported Realisation of B-VC

The end-system-based solution which was sketched above is adequate for the B-VC service. However, there are a few aspects where a client-server implementation can be improved. For example, the session management is a function which is very close to the network, so it makes sense to integrate it in a more close way. The mechanisms of an Intelligent Network can be used to provide an effective monitoring of connections in all their states. Moreover, some generic functionality should be reused in several services, like profile information for users and end systems.
This leads to the idea of extending an end-system-based solution with some features realised by the Intelligent Network, as indicated in figure 6.

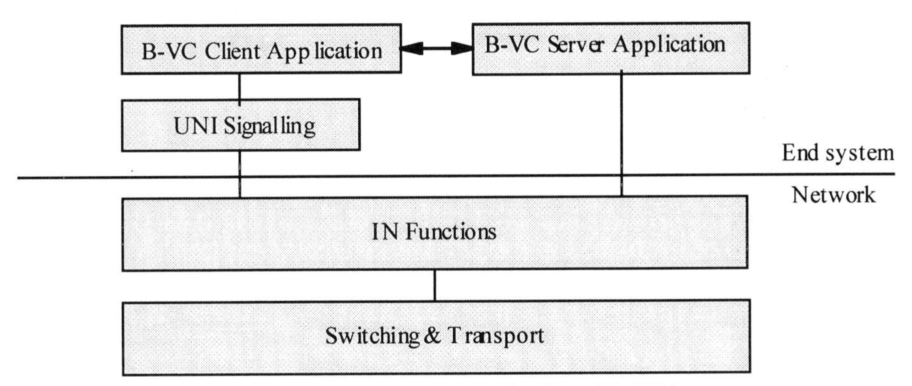

Figure 6: IN-Supported Realisation of B-VPN

This picture shows that a part of the server functionality has been moved into the IN part of the network (in particular communication session management). Of course, there is a need for interaction between the IN functions (which are realised as Service Logic Programs running on an SCP) and the B-VC server application. This can be achieved by redefining the B-VC server as an enhanced variant of the traditional Intelligent Peripheral network element defined for the IN architecture.

6 Conclusion

Two alternative solutions for provision of broadband services were discussed, the end-system-based approach and the IN-based approach. Basically, both approaches can be used to realise any service. However, the end-system-based (client/server) approach seems to offer a natural solution for application services (at least in the short and middle term), whereas an IN-based approach seems to be natural for network services. This has been exemplified on a rather abstract level for two example services, but it can be assumed that these observations hold in general.
For application services, a possible compromise between the two paradigms has been outlined where part of the server functionality in a client-server solution is supported by IN.
In the ACTS project INSIGNIA, the consequences of the above discussion are drawn, and a B-VPN service is realised in a pure IN-based way, whereas a B-VC service is realised in a client-server style supported massively by IN functions.

References

[1] M. T. Jeffrey, Signalling and the Broadband Internet, In: Proceedings Signalling for Broadband, IEE Colloquium Series Vol. 240, 1995, pp. 5.1-5.5.
[2] ITU-T Recommendation Q.12xx series.
[3] Draft ITU-T Recommendation I.375 (SG13), Network capabilities to support multimedia services.
[4] H. Hussmann, Th. Theimer, J. Totzke, G. v.d. Straten, An IN-Based Implementation of Interactive Video Services, In Proceedings ICC 1995, Seattle.

Goal-based Filtering of Service Interactions

Kristofer Kimbler, Niklas Johansson, Johan Slottner
Department of Communication Systems, Lund University
Box 118, 221 00 Lund, Sweden
tel. +46 46 2229008 fax. +46 46 145823
e-mail: chris@tts.lth.se

In order to assure short time-to-market and high quality of new services, effective and efficient pre-deployment service interaction handling is necessary. Service creators and service providers need mechanisms for assessing the impact of newly introduced services on the existing ones. In particular, efficient methods for detecting undesired interactions are required. To satisfy these needs, the EURESCOM[1] Project P509, "Handling Service Interactions in the Service Life Cycle" created the concept of interaction filtering which is a step up towards interaction detection. Filtering identifies service and feature combinations which are interaction-prone and to eliminate those which are unlikely to cause any interactions, which simplifies the actual interaction detection. Filtering uses a number of complementary methods which analyse different aspects of services. The paper presents and evaluates one of such methods called goal-based filtering. The method identifies interaction-prone combinations by analysing the relations between service properties and user's goals.

Keywords service interaction, service creation, detection, filtering, goals

1 Introduction

Rapid service creation and provisioning was one of the main rationales behind the concepts of IN and TINA. Unfortunately, there are several obstacles to achieving this goal. The service interaction problem is commonly regarded as one of them. Undesired interactions can significantly deteriorate the quality of provided services. Therefore, in order to assure short time-to-market and high quality of services, service providers need effective and efficient methods for pre-deployment service interaction handling.

A service interaction occurs when the behaviour of one service is affected by the behaviour of another service or another instance of the same service. Such situations happen because new services and features exceed the original design limitations of the underlying networks [1], and because new services are developed independently upon different, often inconsistent assumptions.

We usually talk about an *interaction problem* when a service interaction is unexpected or undesired from the point of view of an actor. It should be stressed, that the interaction problem is not limited to telecommunications. It is an intrinsic problem of any large and distributed system which evolves in time.

The interaction problem in telecommunication systems has many dimensions and can be addressed on different levels of abstraction and in different phases of the service life cycle. In the current industrial practice, pre-deployment detection of service interactions is usually done by means of costly and time consuming functional testing. There are efforts to develop network mechanisms for on-line interaction resolution in service operation, such as the Feature Interaction Manager or the Negotiating Agent Model [2]. Also, new concepts of future service architectures, such as TINA [3], are investigated. They provide for separation of concerns, e.g. calls from connections,

---

[1] EURESCOM stands for *The European Institute for Research and Strategic Studies in Telecom*. The institute was established in 1992 to perform cooperative R&D in pre-competitive areas. 26 Public Network Operators in Europe are EURESCOM shareholders. The views represented in this paper do not necessarily reflect the views of the shareholders.

services from calls, users from terminals, etc., which would eliminate certain categories of interactions existing today, but not the problem as such.

The main R&D efforts are focused, however, on methods for interaction detection during service creation. The proposed methods are often based on formal modelling of services and underlying networks in SDL, LOTOS, Z [4, 5], or different logic [6, 7], combined with automatic or manual verification of service properties. Unfortunately, none of the detection methods proposed so far proved to be efficient when applied to a large set of services. The complexity of the interaction problem which grows exponentially with the number of services in a network as well as high costs of formal modelling and verification could be blamed for this. Indeed, the thorough formal analysis of all service combinations would be probably too time consuming and could cause unacceptable delays of service creation process.

To create services quickly without compromising their quality, the process of interaction detection must be fairly simplified. There is a need for a simple but effective mechanism which could be applied prior to the actual interaction detection in order to identify service combinations which are interaction-prone and to eliminate those which are unlikely to cause any interaction problems.

The EURESCOM Project P509, "Handling Service Interaction in the Service Life Cycle", developed a concept of *filtering* as a step up towards interaction detection [8]. Filtering uses a number of complementary methods which analyse different service aspects, such as concepts, user's goals, invocation conditions or temporal relations between service features and Basic Call Processing.

The goal-based analysis presented in this paper is one of the filtering methods created and evaluated by the EURESCOM Project P509. The method identifies interaction prone combinations by analysing the relations between service properties and user's goals. The method focuses only on interactions on the requirements level and does not consider any design and implementation aspects of services and features.

In the subsequent sections, we will present the idea of filtering, the principles of the goal-based filtering method including the applied goal model, and the results of method evaluation in the extensive case-study performed by Project P509.

2 The Concept of Filtering

2.1 What is Filtering?

The filtering activity can be seen as a step up towards interaction detection, or as a kind of interaction detection "pre-processing". The main task of filtering is to make rough and quick evaluation of how interaction-prone particular combinations of services and features are. Filtering is not supposed to spot actual interaction problems, but rather to identify service combinations which might cause undesired interactions. In this way, the combinations that are unlikely to result in any interactions can be promptly eliminated, leaving less work for the actual interaction detection.
The filtering uses a few informal and quasi-formal methods which analyse different service aspects, such as concepts, user's goals, invocation conditions or temporal relations between service and feature invocations and Basic Call Processing (BCP) [9]. The key point is to select mechanisms which are simple to use and efficient, and which together cover all the important technical and non-technical aspects of services. The different methods are run in parallel to increase the time efficiency. Their results are then combined to assess for each service combination, how prone it is to interaction.

2.2 Service Interaction Handling Process

The EURESCOM Project P509 positioned the filtering activity as an integral part of the *Service Interaction Handling Process* (SIHP) [10]. The SIHP is an *independent expert* responsible for detecting interactions between services and features, as well as for finding optimal solutions for these interactions. The SIHP is designed to serve several instances of the Service Creation Process simultaneously.

The SIHP is run every time a need for interaction analysis occurs. It is applicable not only for new services and features, but also in a variety of other cases, such as modification of existing services and features, service platform upgrading, purchasing of service software or service nodes from external vendor, or resolution of interactions detected after service deployment.

The SIHP is subdivided into five sub-processes: Preparation, Filtering, Detection, Solving, and Support, which in turn consist of activities. The first four sub-processes contribute to interaction detection and solving, whereas Support is an auxiliary sub-process which manages the data used by the other four sub-processes as well as provides communication with external processes, e.g. the Service Creation Process. A schematic description of the SIHP is shown in Figure 1.

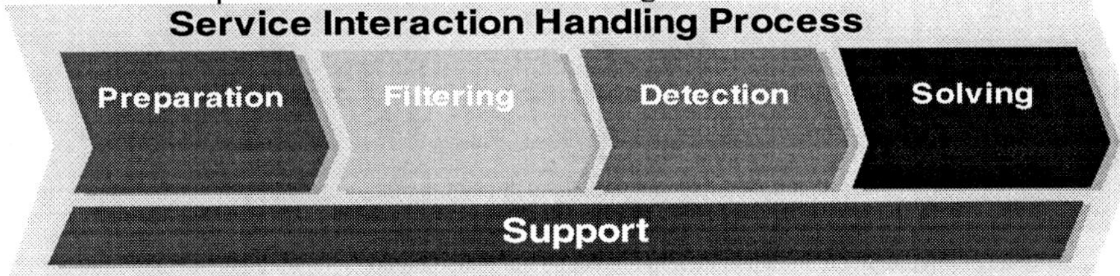

Figure 1. The structure of the SIHP

Every run (invocation) of the SIHP follows a sequence of actions.[2] First, Support informs other sub-processes about the service and platform modifications that have taken place since the last run. Then, Preparation sub-process produces/updates the service and feature descriptions and decides which service and feature combinations have to be analysed in the current run of the SIHP. Subsequently, Filtering performs a rough and quick analysis of the selected combinations to assess which of them are interaction-prone; Detection makes a deep and detailed analysis of the interaction-prone combinations to identify those that may cause real interaction problems; and eventually Solving selects possible solutions for detected interaction problems, evaluates them and recommends optimal solutions.

3 Goal-based Filtering Method

The goal-based analysis originates from a simple observation that interacting services and features have often conflicting or similar goals, whereas the combinations of services whose goals are "orthogonal" usually do not cause any interaction problems. Conflicting goals as a reason of undesired interactions are rather obvious. On the other hand, similar goals might not be so self-explanatory.

Deeper studies of the interaction problem show, however, that many interactions are actually caused by services which are used to realise the same or very alike goals but use different technical means to achieve them. For instance, Call Waiting (CW) and

[2] Resolution of an on-line detected interaction is an exception from this pattern. In such a case only the Support and Solving sub-processes are involved.

Call Completion on Busy Subscriber (CCBS) aim at increasing the user's reachability, but different combinations of these services cause interaction problems.

3.1 The Goal Model

The Goal Model used by the mechanism does not consider any implementation aspects of services and features. It focuses only on interactions on the requirement level. Therefore, there can exist interaction cases which are not identified by this method. Those, however, are covered by the other filtering methods.

There are two abstraction levels of goals considered by the Goal Model: objectives and properties. The *objectives* represent the user's goals, whereas the *properties* capture the functional characteristics of services and features as such, i.e. they describe how the objectives are obtained by technical means.

The objectives are more abstract than properties, and do not directly correspond to any particular service or feature. They explain what the services and features are used for by their users. The same user's goal may be reached by means of different services. The same property may support one objective while obstructing another. We can thus define the *Support and Obstruction Relation* (SOR) between properties and objectives.

The objectives themselves can also be interrelated. One objective can be regarded as specialisation of another, e.g. *to receive a message* contributes to objective *to increase reachability*. On the other hand, objectives may conflict with each other, e.g. *to constrain access* is in contradiction with *to increase reachability*. We can thus define *Contribution and Contradiction Relation* (CCR) between objectives.

The structure of Goal Model is illustrated in Figure 2 below.

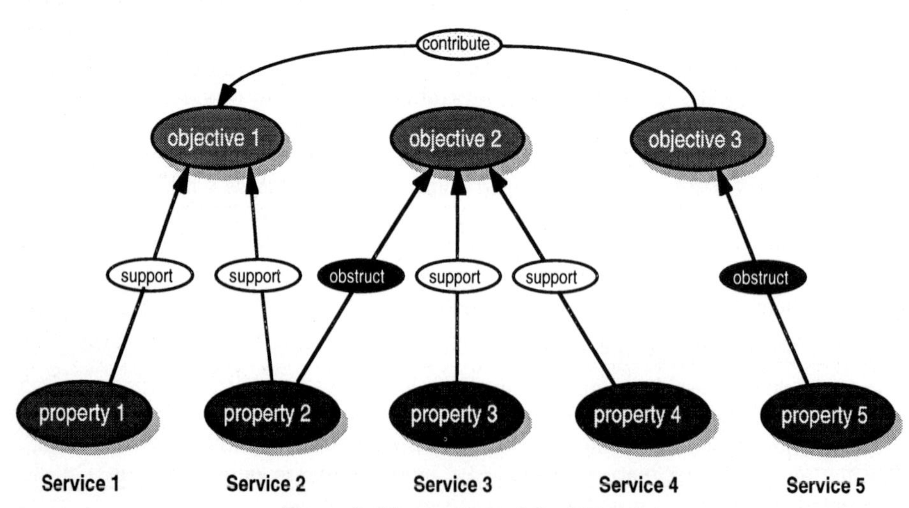

Figure 2. The structure of Goal Model

The Goal Model is represented as four tables, two defining goals and two defining relations between them. The first and second tables contain a list of objectives and a list of properties. The third and the forth tables contain definitions of the Support and Obstruction Relation (SOR) and the Contribution and Contradiction Relation (CCR), respectively. The particular items are discussed in the next sub-sections.

The first table contains a list of objectives and its main purpose is to enumerate these objectives. The objective table contains numbers and definitions of objectives as shown in Table 1 below:

no.	Objective
1	to increase my reachability
2	to be reached on the line where I am
3	to receive messages while being unable to answer incoming calls
4	to answer more than one call simultaneously
5	to allow access to service only for authorised users
6	to constrain access to my line

Table 1. Example of a list of objectives

The second table contains a list of services and features with their associated properties. Each service and feature may have one or several properties. The property table may contain only core (main) features, or optional features, related to the services, as well as core features. This depends on the desired scale of the analysis. The property table is illustrated in Table 2 below.

no.	service	feature	Property
1	CFU	core	to forward all incoming calls to another number
2	CW	core	to notify subscriber about another incoming call
3	UPT	core	to allow the subscriber for personal mobility,.
4		AUTV	to control service access through key code authentication
5	VM	CRA	to notify the calling user by announcement that VM is reached

Table 2. Example of a list of properties

The third table of the model (illustrated in Table 3) contains the Support and Obstruction Relation (SOR). The table has rows enumerated by properties and columns enumerated by objectives. A '+' or a '-' in a table entry means that the row property supports or obstructs the column objective respectively. Since a service is built up by features, each service is related to all the objectives these features' properties are related to.

Properties \ Objectives	1	2	3	4	5	6
1	+	+				+
2	+			+		-
3					+	
4					+	
5	+					

Table 3. Example of SOR table

The fourth table of the goal model contains a definition of the Contribution and Contradiction Relations (CCR). Note that the two relations are not symmetrical. The CCR table is square with columns and rows enumerated by the objectives. A '+' or a '-' in a table entry means that the row objective contributes to or contradicts with the column objective respectively. The diagonal is marked with '+' since an objective contributes to itself. This is illustrated in Table 4 below:

Objectives Objectives	1	2	3	4	5	6
1	+					-
2	+	+				-
3	+		+			
4	+			+		-
5					+	
6	-	-	-	-	+	+

Table 4. Example of CCR table

3.2 Goal Analysis

The method analyses the Goal Model presented above to find interaction-prone service feature combinations. Every two service features are analysed for all possible combinations of objectives that they are related to.

A general conclusion is that *an interaction might occur whenever two features (or rather properties) support or obstruct (SOR) the same objective or two different, but interrelated objectives (CCR)*. Since every objective contributes to itself, the first case can be treated as a special instance of the second one. In other words, the case when two properties related to the same objective is interpreted as if the properties were related to two contributed objectives. In the following, we will thus not explicitly consider the first case.

In Table 5 below all the possible cases to be considered are listed. There are two objectives, namely X and Y. They may have a contribution or contradiction relation to one another, as explained above, or no relation at all. The properties may support or obstruct these objectives, as explained above, or not be related to them at all. Property 1 relates to objective X, and property 2 relates to objective Y yielding nine different cases.

Relation	Contribution		Contradiction		No Relation	
Objective	X	Y	X	Y	X	Y
Case 1	+	+	+	+	+	+
Case 2	-	-	-	-	-	-
Case 3	+	-	+	-	+	-
Case 4	-	+	-	+	-	+
Case 5	+		+		+	
Case 6		+		+		+
Case 7	-		-		-	
Case 8		-		-		-
Case 9						

Table 5. Interaction Cases

If the two objectives are not interrelated or at most one property is related to its objective then no interaction may occur and is therefore not considered in the rest of the analysis. Thus, only the four first cases (shaded in Table 5) might cause any interactions. In each of these cases, we distinguish two instances corresponding to the first and the second column in Table 6. In the first instance one objective contributes to the other, whereas in the second they contradict each other. These four interaction-prone cases are explained and exemplified below.

Case 1 & 2

In the first instance of these cases, two properties (features) support (case 1) or obstruct (case 2) two contributory objectives (a '+' in the CCR table) and they probably do this by different technical means. A network problem with selecting one of the features (means) might occur, e.g. when one is selected the other one can be inhibited.

An example of this instance (case 1) is shown in Figure 3 below. Two services *Call Waiting* and *Call Forwarding Unconditional* support two different objectives. Please notice that due to contribution relation between the objectives, *Call Forwarding Unconditional* supports indirectly the objective *to increase reachability*.

In the second instance, two properties (features) support (case 1) or obstruct (case 2) two contradictory objectives (a '-' in the CCR table). If the features are invoked at the same time or even sequentially, there could be a possible interaction since at least one of the properties obstructs the user's goal.

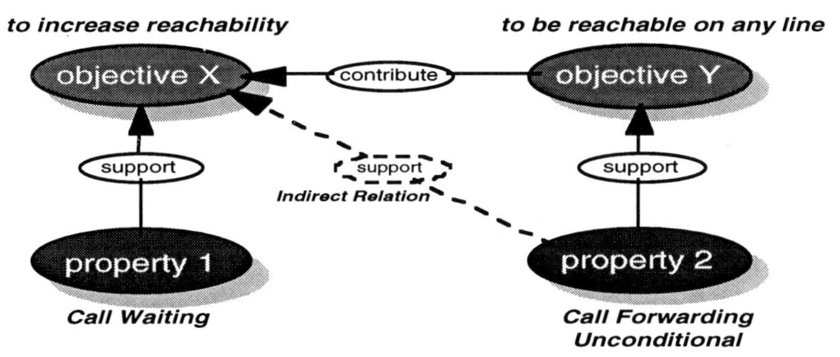

Figure 3. An example of goal-based filtering, case 1

Case 3 & 4

In the first instance of these cases, one property (feature) obstructs and the other property (feature) supports two contributory objectives (a '+' in the CCR table). If the two features are invoked at the same time or even sequentially, there is a possible interaction since at least one of the properties obstructs the user's goal.

In the second instance of these cases, one property (feature) obstructs and the other property (feature) supports two contradictory objectives (a '-' in the CCR table), i.e. both obstruct the contradicted objective, they probably do this by different technical means. A network problem with selecting one of the means can occur, e.g. when one is selected the other one can be inhibited.

An example of the second instance is shown in Figure 4 below. Two services *Do Not Disturb* and *Screen Incoming Calls*, the first one obstructing and the other one supporting two different but interrelated objectives. Due to contradiction relation between the objectives, *Screen Incoming Calls* obstructs indirectly objective *to increase reachability*.

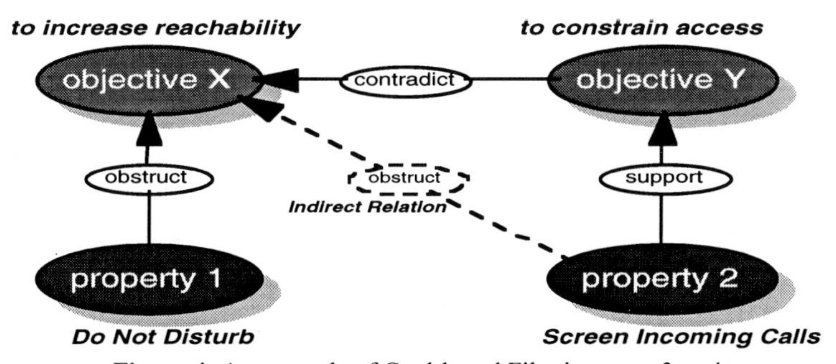

Figure 4. An example of Goal-based Filtering, case 3 or 4

Filtering Algorithm

The goal-based filtering method can be expressed by the following filtering algorithm:

```
for each A_property do
    for each B_property do
        for each A_objective do
            if entry (A_property,A_objective) in the SOR table is marked then
                for each B_objective do
                    if entry (B_property,B_objective) in the SOR table is marked then
                        if entry (A_objective,B_objective) in the CCR table is marked then
                            entry (A_property,B_property) is marked in the interaction table
```

Each possible interaction found, is marked in the interaction table indicating an interaction-prone combination. These combinations are subject to a more thorough investigation. Please notice that the interaction table is symmetric.

Properties Properties	1	2	3	4	5
1	X	X	X	X	X
2	X	X	X	X	X
3	X	X	X		X
4	X	X		X	
5	X	X	X		X

Table 6. Interaction table

4 Evaluation of the Method

The goal-based method was evaluated during a case study carried out by the EURESCOM Project P509. The case study aimed at evaluating the effectiveness and efficiency of the filtering and detection method proposed by P509 [4].

In the first step of the case study, a set of services and service features was selected to be applied by all the tried methods. Both supplementary services and IN services were considered. In total 5 IN services, such as VCC, FPH, VPN, VM and CONF (using the abbreviations recommended by ETSI), 12 switch-based services, such as CFU, CFB, CFNR, CW, CCBS, CB, SICB, CLIP, CLIR, DND, WUC and TKCS and about 40 features of these services were selected.

Descriptions of the services and features were compiled from definitions given by the standardisation bodies (mainly by ETSI). It was promptly noticed that short textual descriptions were not sufficient as they only address the main functionality of the service in an informal way and leave place for very different interpretations on how the services should work. Thus it was decided to produce *templates* for service and feature descriptions. Then, the services and features were described using these templates.

Specification of user's goals and service/feature properties were one of the elements of service and feature descriptions. They were used as the basis for producing the Goal Model that forms an input to the goal-based filtering method. The Goal Model created during the case-study covered 70 properties and 29 identified objectives. The goal analysis was first made manually. Later a simple tool support was created which made subsequent applications of the method more efficient. The Goal Model has been improved during this process.

The results obtained during the case study were evaluated against the P509 Interaction Benchmark [11]. The benchmark was based on the interaction classification also

developed by project P509, and contained 30 representative cases of interactions between the services selected for the case study.
In Table 7 below the interactions presented in the benchmark are marked with 'x', and the combinations pointed as interaction-prone by the method are shaded. Since the table is symmetric only the upper half of it is considered.

	CFU	CFB	CFNR	CW	CCBS	CB	CLIP	CLIR	DND	WUC	FPH	LGS	CD	LIM	UPT	VCC	CRA	VPN	CUG	SLD	TKCS	CRL	PNPI
CFU	X	X	X		X	X					X								X	X			
CFB				X							X								X	X			
CFNR										X	X								X	X			
CW													X		X								
CCBS									X							X							
CB																							
CLIP								X								X							
CLIR																							
DND																							
WUC																							
FPH															X	X						X	
LGS																	X						
CD																							
LIM														X									
UPT																							
VCC																					X		
CRA																							
VPN																					X		
CUG																							
SLD																							
TKCS																							X
CRL																							
PNPI																							

Table 7. Interaction and interaction-prone combinations

As can be seen above all of the 30 interaction cases included in the benchmark are pointed as interaction-prone by the goal-based method. This confirms the effectiveness of the method.

Table 8 indicates that "over-coverage" of goal-based filtering is very high. This could be explained by the fact that the combination of services used in the case study and the benchmark (especially the supplementary ones) are very interaction-prone by their nature. Moreover, the benchmark contains only a small subset of all the possible interactions between these services. Therefore, if we consider all the interaction cases between these services, the "over-coverage" would be probably much lower.

The results from the full case study are presented in Table 8 below.

Type of services	Combinations filtered out
Supplementary vs. supplementary	10%
Supplementary vs. IN	20%
IN vs. IN	35%

Table 8. Percentage of combinations filtered out

5 Summary and Conclusions

The goal-analysis presented in this paper is one of the filtering methods developed by the EURESCOM Project P509. The method identifies interaction-prone combinations by analysing the relations between IN and supplementary services and features and user's goals. The method was applied and evaluated in an extensive case study which proved both its efficiency and effectiveness. The method works on requirements level, and thus is not implementation-dependent, which makes it applicable also to services provided on other platforms, such as the Internet or TINA.

The original goal of filtering in general and goal-based analysis in particular was to simplify and speed up interaction detection by eliminating "harmless" service combinations and select those which should be a subject to a more thorough analysis. The result of the case study indicated, however, that not many of those combinations can be really filtered out.

Nevertheless, application of the goal-based filtering provides valuable guidelines for the detection mechanisms, and in that sense it can simplify and speed up interaction detection and make it more structured. Moreover, it gives a quick and early assessment of interaction-pronity of different service combinations, and thus indicates the potential problem areas.

The future work related to the goal-based filtering should focus on improving the *filtering criteria* used by the method, so that the accuracy of the filtering grows and more irrelevant non-interaction-prone combinations are filtered out. Beside it, a more advanced tool support should be developed as well.

References

[1] N. Griffeth, Y. Lin, Extending Telecommunications Systems: The Feature Interactions Problem. In IEEE Computer, August 1993.

[2] N. Griffeth, H. Velthuijsen, The Negotiating Agent Model for Rapid Feature Development. In Proceedings of the Eight International Conference on Software Engineering for Telecommunication Systems and Services, March 1992.

[3] M. Chapman, P. Farley, R. Minerva, A. Oshisanwo, ROSA - A Service Architecture for TINA. In Proceedings of TINA'93, 1993.

[4] P. Combes, M. Michel, B. Bernard. Formal Verification of Telecommunication Service Interactions using SDL Methods and Tools. In SDL'93: Using Objects, Elsevier Publishers, 1993.

[5] A. Lee, Formal Specification - a Key to Service Interaction Analysis. In Proceedings of the Eight International Conference on Software Engineering for Telecommunication Systems and Services, March 1992.

[6] A. Gammelgaard, J. E. Kristensen. Interaction Detection, a Logical Approach. In Feature Interactions in Telecommunications Systems, IOS Press, 1994.

[7] C. A. Middelburg, A Simple Language for Expressing Properties of Telecommunication Services and Features. In Proceedings of the FORTE'94, October 1994.

[8] C. Capellmann, J. Jonasson, K. Kimbler, P. Samaras, J. Pettersson: The Concept of Filtering - A Step up Towards Interaction Detection, Submitted to FIW'97.

[9] C. Capellmann, J. Jonasson, P. Samaras, K. Kimbler: Service Description and Interaction Analysis using a Multiple Actors' View on the BCSM, In Proceedings of IEEE IN'97, May, 1997.

[10] H. Velthuijsen, K. Kimbler, T. Nauta: Integration of Service Interaction Handling with Telecom Business Processes, In Proceedings of IEEE IN'97, May, 1997.

[11] K. Kimbler, H. Velthuijsen, C. Capellmann, T. Nauta, J. Ruiz: Benchmarking Feature Interactions, Submitted to FIW'97.

Intelligent Networks Planning Supported by Software Tools

O. Makhrovskiy, V. Kolpakov, V. Shibanov, Yu. Soloviov, I. Tkachman
RUBIN Research Institute
Kantemirovskaya st. 4
197342, St. Petersburg, Russia
Phone: + 7 812 245 44 23
Fax: +7 812 245 95 52
E-mail: rubin@sovam.com

Now that Intelligent Networks (IN) are becoming reality, some new problems have to be considered while implementing these new architectures in real network structures.
One of the main problems is the need for new (advanced) Software (SW) Tools to optimize the dimensioning and planning process of the IN structures.
The objective of this paper is to present a complex of SW tools named POSET that is related to the IN structures dimensioning and planning. POSET allows modelling of the economics of the transition towards an IN. These tools also assist a study of network modification to improve the efficiency of the existing network.
The proposed paper is further development of work started three years ago by the Rubin Research Institute on modernization of existing networks and deployment of IN within the Russian Federal Programme of Advanced Telecommunications Networks Creation in the St. Petersburg region.
Keywords: Intelligent Networks, Service Demands, Software Tools, Planning Process, Performance Analysis, Access Network, Cable, Radio.

1 Introduction

The implementation of IN is regarded as a necessity by most of the operators of public telecommunications networks. Some of them have already established a first solution supporting a few services on a relatively small scale. In the future the number of services provided by IN architecture will continue to grow significantly, with open service architecture as a long term evolution of IN. The evolution of the current network infrastructure towards advanced IN includes a smooth evolution of existing services and expansion of networks, rapid development and smooth deployment of new services, reduced cost of managing networks and services, etc. These factors emphasize the need for comprehensive Personal Computer (PC) based interactive vocational SW Tools for dimensioning and planning of future IN structures.
This paper presents some current results of the project started three years ago by the Rubin Research Institute on modernization of existing SW tools and developing of advanced SW tools for assisting the IN planning process in the St. Petersburg region within the Russian Federal Programme of Advanced Telecommunications Networks creation. The final objective of this Programme is to establish guidelines for the introduction of advanced communications networks and multimedia services in a competitive multi-service environment. The St. Petersburg region project results will address different user communities within both residential and business market segments, including both fixed and mobile communications. The main outputs of the project are techno-economic guidelines and recommendations to IN Planners, Service Providers and Equipment Manufacturers.
The paper is organized in the following way: section 2 reviews the main features of the IN dimensioning and planning process based on previous investigations [1], section 3 describes the complex of advanced SW tools with emphasis on two newly developed modules, and section and contains conclusions.

2 The IN Dimensioning and Planning Process

The description of IN includes the division between the physical elements and the functional components. The physical elements dealt with in the dimensioning and planning process are the exchanges, trunk lines, signalling links, Intelligent Peripheral (IP) equipment (answering machines, voice message machines, etc.), Signalling Transfer Points (STP), and Service Control Points (SCP). More are defined in [2].
In this paper the term "network dimensioning and planning" means to decide the telecommunications network structure for the whole planning horizon given. That is, to find the locations and the capacities of the elements involved in handling IN services as well as the routing related to these elements at the each stage of the planning horizon.
When performing the network dimensioning, we have to establish a measure of the quality of a given network structure (usually the cost of the involved equipment) and a set of requirements must be fulfilled.
The task can be formulated as an optimization problem:

MINIMIZE NETWORK RESOURCE COST
with respect to
PERFORMANCE MEASURES $\leq$ REQUIRED VALUES
for a given set of service demands and when the possible alternative network structures are defined for a given time stage.

In order to perform such optimizations, software tools are needed [3].
The required input data to the dimensioning and planning process are mainly of three categories: the service demands, the required values for the IN performance values and the possible network structures for each stage of the planning horizon.
The set of service demands must be given for a certain geographical area based on the service forecast with the level of details appropriate for the purpose of the dimensioning and planning. The performance requirements may be identified as described in [4].
A number of alternative network structures for each stage of the planning horizon is given by the implementation similar to [5].
For a long-term planning, major changes of the IN structure can be foreseen and some cost estimate for the alternative solutions must be established. Two aspects are related to the long-term planning. First, one can define the IN target structure for the end of the planning horizon. That is, a close to optimal structure according to some measure that fulfils the set of planning requirements. The second aspect is related to the evolution scenarios that are being suggested for future INs.
The target structure for an IN can be searched for without considering the equipment that is available at present, but may be foreseen according to a planning horizon, e.g., 10 or 15 years. The quality of a network structure is given by some economic measure. Therefore, one should establish some cost function used as the objective function in the optimization problem. As we stated above, the cost function is considering the cost of the network resources. Several other aspects of the cost could be taken into account. But, as usual, there is a trade-off between using more exact relationships and the complexity that the problem turns out with. The cost of the network elements (e.g. exchanges, transmission lines etc.) could be related to the purchase, installation, operation, etc. or some combinations of these factors at a time.
When a network structure is examined, the characteristics of each of the elements should be captured by more coarse relationships in order to limit the dimensions of

the problem. This is in line with the hierarchical modelling approach which is applied to large systems by the use of queueing networks.
Figure 1 shows the structure of an IN. One may recognize some similarities with the subscriber access network [6].

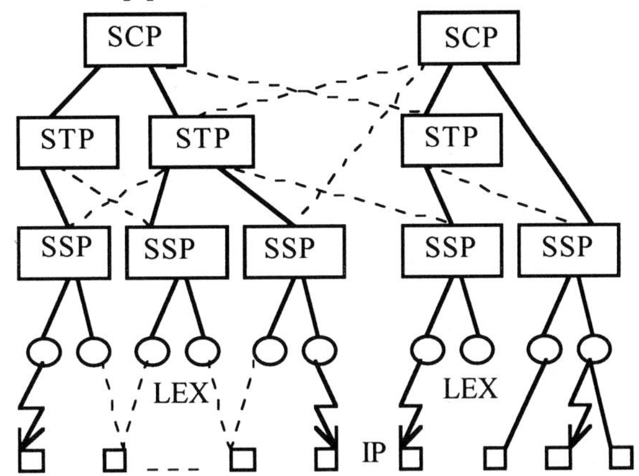

Figure 1. Typical IN structure

More complex configurations of IN can be present, depending on the architecture of the signalling network, the number of SCP's available for each Service Switching Point (SSP), and so on.
After an IN structure has been found for each stage of the planning horizon, the various transition scenarios towards a target IN structure are identified.
The idea of our approach is that structures found in each stage of the planning horizon are examined as alternative solutions in one dynamic optimization process [7].
The implementation of this IN dimensioning and planning process may have separate algorithms or one could strive for a common optimization framework for dealing with dimensioning and evolution problems. In the latter case variants of this problem could be handled by fixing some of the variables compared to the most general case.

3 Intelligent Networks Dimensioning & Planning POSET Complex

We have developed a complex of IN planning support tools integrating the planning and design process. The key element of this complex is a system named POSET - Plans Optimization by Selection of Embedding Times.
The economical part of the complex has as its basis the SW tools for strategic telecommunications network planning elaborated early [3, 5] and broadened by a set of new capabilities.
POSET is a mathematical model of the equipment replacement decision-making process that takes into account traffic models, equipment parameters related to the type of environment, and evolution aspects that are driven by changing requirements over time and economic aspects. The model takes information about costs and revenues and their trends in time associated with existing and emerging technologies and services. From this information there is a calculation of a point in time for technology replacement to minimize the Net Present Value (NPV) that characterizes embedded plant and its replacement. We also use an additional optimization criterion - the Internal Rate of Return (IRR) that may be essential if the discounting rate is unstable and badly predictable. Thus through bi-criterial optimization on the (NPV,

IRR) - plane, the model reveals the zones of risk, the zones of steady incomes etc. and further carries out the detailed optimization in the zones.
In the POSET model it has become possible to set the initial data with their probability distribution and correspondingly to get the distributions of optimal embedding times.

3.1 General organization of POSET

The POSET SW tools have three main subsystems, as shown in Figure 2 by grey blocks.

- *Display Subsystem* - which displays information on maps showing roads, bridges, rivers, railways, area development plans. The map also shows social and economic statistics of the region, traffic flows, current and planned facilities.
- *Network Dimensioning and Planning Subsystem* - provides demand analysis, traffic forecasting and selects Alternative Configurations concerning access and transit segments. Forecasting the demand for various services: telephone, data on demand and other newer telecommunications services. e.g. multimedia, is very important for the economical and effective construction of IN.
- *Network Evaluation and Optimization Subsystem* - provides evaluation and selection of Transition Scenario from Initial (current) Network to future IN Architecture - Network Targets, during the horizon given.

Furthermore, there is the *Data Translation Subsystem* which translates data from external database to other subsystems.
Two important advanced SW modules of POSET complex are described below in subsections 3.2 and 3.3.

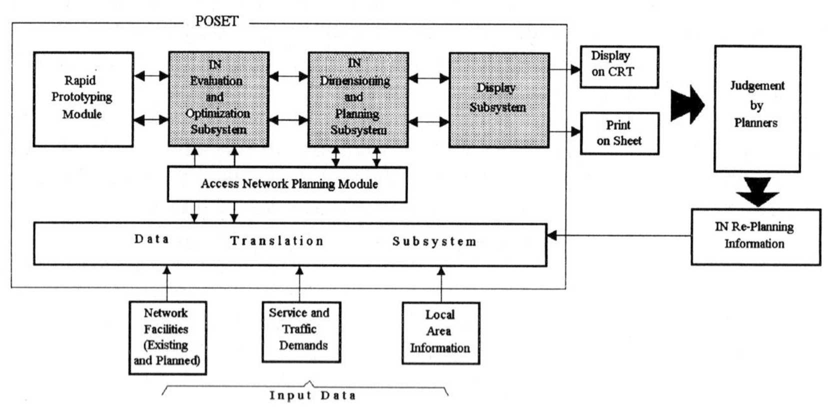

Figure 2. Structure of POSET complex

The complex of SW tools uses geographical information to efficiently plan the telecommunications infrastructure. These tools can display the information needed visually on maps to support the planned solution under the uncertainty of future demand and technology, and the diversification of customer services.
The software package is designed as modular, fully integrated in an overall planning process and able to incorporate future developed modules.

3.2 QoS Parameters Estimating & Rapid Prototyping of Services Interaction

The POSET module described below belongs to the IN Evaluation and Optimization subsystem and analyses some technical parameters of INs including the estimating of Quality of Service (QoS) parameters.

When a user of IN wants to get a service (e.g. exchange of multimedia with an other user with acceptable values of QoS parameters), the user first establishes a service association with the provider, then utilizes this service, and then finally disconnects the association. When utilizing the service the IN mechanisms first establish a network connection, then fulfil the information exchange and then finally disconnect the network connection. In this way, each service usually has a multi-layered structure, that is, it uses other services (maybe of other service providers) in its execution. So from the point of view of QoS characteristics such as delay parameters (mean delay and delay variation, for example) the (stochastic) time of service execution in IN contains many such components that are not easily determined. This situation is not met in simpler kinds of networks where the time of service execution consists almost completely of the inter-user information transmission time.

Thus the functioning of the service in IN may be represented as an oriented graph with stochastic times of (partially simultaneous) execution of its arcs. These times are mutually dependent and connected with an ensemble of queueing systems (it resembles a Petri net but is more complicated). The time dependencies are mostly unknown, therefore one has to build suitable bounds on the necessary values often using extreme (in a sense) cases. A convenient approach here appears to be an apparatus of coordinated stochastic values developed by us based on the notion of associated stochastic values [8]. In combination with other more conditional theoretic approaches, it gives a means to obtain sufficiently good estimates of the delay QoS parameters in IN. The module that includes these means for estimating of IN QoS parameters is not yet completely developed.

Another essential part of the POSET complex is a module for rapid prototyping of services interaction (see Figure 2). The application includes prototyping the interaction of services (or services features) in the IN. In future we shall consider services implying the service features also.

Certain unforeseen service interactions can lead to grave consequences such as deadlocks and other similar disturbances breaking the network functioning. To discover them the usual proper tools are sometimes recommended such as Petri nets [9]. Regretfully, they are not well adapted to prototyping the estimate efficiency characteristics, and for this purpose other approaches are more suitable.

Deterioration of the QoS can arise from interaction of services connecting one end-user to another (e.g. Call Rerouting Distribution - CRD [10]) and having valuable delays (both explicit and implicit ones). The implicit delays are contained in many services dealing with handling of IN databases. The explicit delays are contained (for example) in Selective Call Forwarding on Busy/Don't Answer - SCF [10]. Some services bring stochastic elements into the interaction model, and certain service features are capable of directing calls into queues (Call Queueing - QUE - or Consultation Calling - COC).

Prototyping a combination of services in IN is made even more complicated by taking into account the influence of the networks resources limitations, that determine the behaviour of the whole system that can be appreciably different from its "ideal" version. For the rapid prototyping of the services and the service features in IN (taking into account their interaction) various simulation tools are now being developed by a team of developers.

Simulation languages were and are being developed for use in estimating the efficiency parameters of entities, probably verifying entities, and to assist the conversion of the entity informal model into the formal. These languages are usually

based on the queueing network paradigm (unlike formal specification languages based on the paradigm of extended finite-state machine - EFSM - such as SDL). We have developed such a language named SIMQUENT (SIMulator of QUEueing NeTworks). It was implemented as an external class of language SIMULA and thus it is object oriented. It has the graphical-tabular presentation (D-SIMQUENT) and the programming one (P-SIMQUENT). The D-description of a scheme is converted into its P-description almost element-by-element.
SIMQUENT has only 3 basic types of elements and several types of derivative elements. Moreover it has powerful means of aggregation of these elements.
SIMQUENT has shown good results at prototyping of algorithms and protocols in Russian regional telematic systems and communication networks over several years. On the basis of SIMQUENT, a new language PROTOSINT (PROTOtyping of Services In NeTworks) is being developed as a graphic-oriented high-level system implemented at PCs.

3.3 Access Network Planning Module

To provide a connection with the telephone network for new users, new links are laid or any available ones may be used. The users may be at different distances from the existing or contemplated Local Exchanges (LEXs) in the region. The aggregate of such links and IPs, and other facilities forms the so called "subscriber access network" of IN.
Throughout the world various physical layers of communication links are used, such as copper pairs, coaxial, fibre, radio, satellite. When the size of city/rural areas or their parts are relatively small, satellite links are expected to be disadvantageous for access, but other physical layers may be applied successfully. A real region has individual properties of terrain, buildings, users needs and locations, existing access network loops, capacities and throughputs of switches as well as varying capabilities of Network Designers, Operators and Service Providers. In this case differing physical lines and combinations are used, for parts of a particular communication route.
Thus the creation of a regional access network is a multi-parameter and complex problem, i.e. it involves a set of some sub-problems. To solve it, a universal approach for the short term (3-4 years) planning of access network is being developed, taking into account all the above factors and peculiarities of the real region given.
Short term planning proposes to involve the following overall issues:

1. Forecast and identification of users in the region studied.
2. Planning of users needs and services demand scenarios.
3. Definition of types and number of subscriber terminals and their distribution over region area.
4. Estimation forecast of future user traffic in the region.
5. Definition of rational locations for User Network Interfaces.
6. Generation of possible topological network structures based on the various layers which are may be suitable.
7. Forming of alternative access network structures for the region.
8. Assessment of investments required to build the access network as well as the revenue expected.

1-4 issues are the preparation part of this work. When the main part is carried out (i.e. issues 6-8), the set of the criteria and constraints are specified by giving the following parameters:

- minimum cost for network building;
- non-negative (maximum, if possible) revenue;

- required level of service provisioning.

The module of POSET complex which realizes the planning issues above is named ACCESS.

The ACCESS module contains the following functional sub-modules:

- creation and representation of region map model;
- planning of cable routes;
- planning of radio coverage;
- evaluation of planning parameters;
- control and interaction of above modules.

As the planning process conception does not envisage the precise final solutions which are required for the direct design, the computer tools use some modelling simplifications. They are of sufficient precision for planning results, and so are valid.

The software gives the Network Planner an opportunity to plot and display on a computer monitor the locations and dimensions of buildings, streets, roads, hills, rivers and others obstacles and terrain undulations. A scanner is used to support these activities as well as a computer mouse. A relevant "weight" co-efficient is given to each geographical object so that the streets, fields or another authorized zone where the cables are laid usually, have the minimum values of co-efficient. The Planner can correct them during the planning process and revise the location, dimension of real terrain undulation or any other parameter, if desired. A SW menu allows selection of a region map.

During the process of wire/cable network planning, SW analyses the weight co-efficients, and then sets out the lines of routes for various cable links (coax, copper, fibre), and finally shows them in the PC monitor screen by different colours. This algorithm is based on the classic theoretical routing method, the "Travelling Salesman Problem", modified by the authors for the Access Network Planning purposes. The SW presents to the Planner, the planning route lines connecting the premises of region users (houses, offices etc.) with the nearest LEX as well as the generic, calculated values of cable parameters. These are the lengths, capacities and costs.

The length and capacity are broken into three relevant parts of cable:

- inside of building;
- between building and underground manhole;
- between manhole and LEX.

In addition all the above planning results may be used as initial data for the Network Designer to choose the real types of cables.

When being planned, the access radio network assumes the hypothesis of optical radio wave propagation, i.e. radio links are supposed to be "line of sight". The SW tool identifies how to place rationally the minimum number of the base radio stations in the region area to involve the most number of users.

The results of such a planning are shown in Figure 3.

If any obstruction (e.g. building) is present, the Planner should increase the height of one or both antennas using the computer mouse. When the straight line between antennas does not touch the obstruction the access radio network may be considered as planned finally. During this process software tools will recalculate the new planning parameters. The locations of the base radio station and relevant LEX are supposed to be the same or they are to be located in close proximity to each other.

Each of the planning parameters may be viewed in the relevant Tables (see e.g. presented in Figure 4).

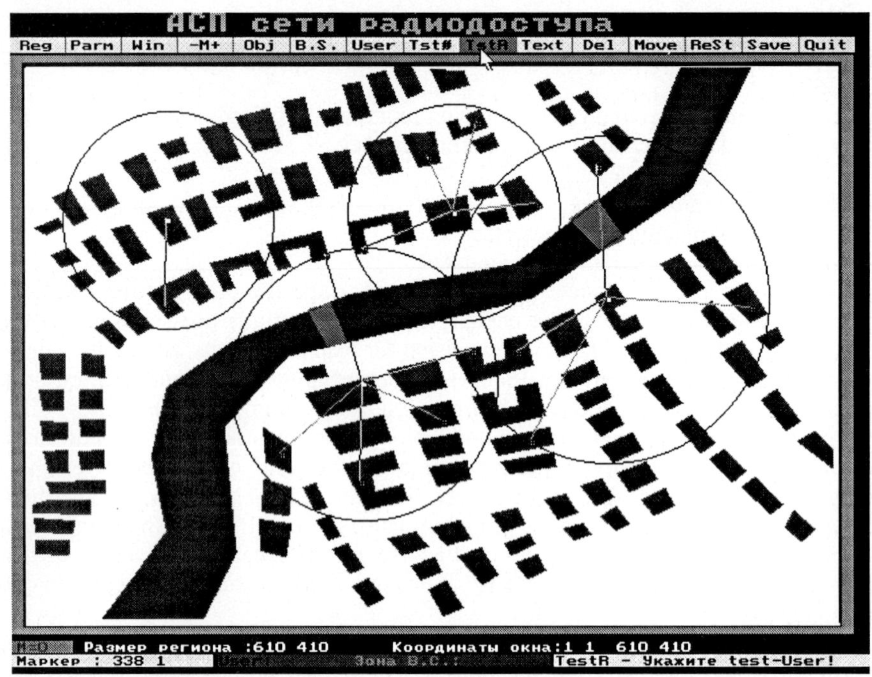

Figure 3. Rational allocation of radio coverage over region area

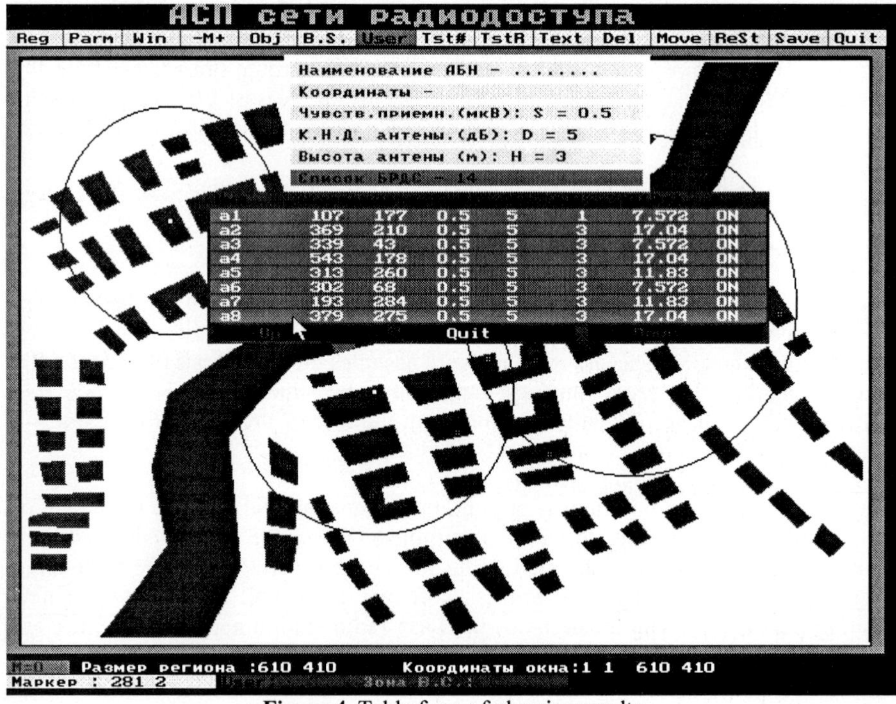

Figure 4. Table form of planning results

4 Conclusions and Further Research

The well known B-ISDN concept is now one of the major trends of the telecommunications industry. However, in accordance with forecasts transition towards full B-ISDN will take a sufficiently long time period in Russia, so that the realization of additional services on the basis of PSTN by means of Intelligent Networks conception has became more attractive.

A crucial part of the IN creation problem is to have methods and SW tools that support the finding of good IN evolution paths.

The complex of SW tools named POSET related to the IN structures dimensioning and planning is now being developed at the Rubin Research Institute for the purpose of the accelerating a smooth migration from existing networks to the future network architecture.

The main objective of this paper is to give the description of the POSET complex to plan IN structures, with emphasis on access network segments (e.g. cable and wireless) out of the whole area the tools cover.

At present the POSET is being used successfully for IN planning and dimensioning in the region of St. Petersburg. In the future, the proposed complex will be incorporated into the framework of a large-scale Engineering Planning System with the integration of fixed network and FPLMTS based on IN technology.

References

[1] Makhrovskiy O., Kolpakov V., et al, "Optimal Network Evolution Planning on the Base of the Demand Model in the Region of St.Petersburg", Proc. 3rd Intl Conference on Intelligence in Networks, Bordeaux, France, 11-13 October, 1994, pp. 227-231.

[2] ITU-T Recommendations Q.12xx Series.

[3] Kolpakov V., Kronina L., Makhrovskiy O., et al. "New Software Tools for Network Planning and Design", Proc. 3rd St. Petersburg Intl Conference for Regional Informatics, St. Petersburg, Russia,10-13 May, 1994,pp.132-142.

[4] Jensen T., "On the Dimensioning of Intelligent Networks", Proc. Intl Conference on Informational Networks and Systems, St. Petersburg, Russia, 24 -28 October, 1994, pp.417-423.

[5] Bonatti M., Gobbi R., Makhrovskiy O., et al.: Regional information networks strategic planning. "Telecommunications and Radio Engineering", 1995, No.8.

[6] Calhoun G. Wireless access and the local telephone network. Artech House, Boston, 1992, 597p.

[7] Makhrovskiy O., et al.: Design of the regional information systems based on the intelligent networks. "Telecommunications and Radio Engineering", 1995, No.5.

[8] Barlow R.E., Proschan F., Statistical theory of reliability and life testing. Probability models. Holt, Rinehart & Winston, Inc. 1975.

[9] Kim H. et al. "Modelling and detecting feature interaction using Petri-nets". Proc. 3rd Intl Conference on Intelligence in Networks, Bordeaux, France, 11-13 October, 1994, pp. 385-386.

[10] ITU - T Recommendation Q.1211 (1992), Annex B.

Adopting Object Oriented Analysis for Telecommunications Systems Development

Declan Martin
Broadcom Eireann Research, Ltd.
Kestrel House, Clanwilliam Place, Dublin 2, Ireland,
ph: +353-1-604-6000, fax: +353-1-676-1532, dmn@broadcom.ie

It has been accepted for some time that errors introduced in the early phases of software development are significantly more expensive to correct that those introduced at later stages. Therefore, development teams should emphasise the analysis phase to ensure that system requirements are captured correctly and documented unambiguously. This paper reports on initial experiences with the adoption of *object oriented analysis* (OOA) for the production of telecommunications systems. The purpose of the paper is to report the experiences gained, to offer some recommendations for those attempting to introduce object oriented analysis into their organisation, and to remind developers of important issues that are sometimes overlooked. Three main conclusions are drawn: adopting a methodological approach to object oriented analysis is difficult but beneficial, using existing telecommunications specific models within the context of an OOA method causes problems, and additional tool functionality supporting telecommunications systems development would be appreciated.

Keywords: Object Orientation, Analysis, Migration, Lessons learned.

1 Introduction

Competitive pressures within the telecommunications industry have led to the advocation of the use of object oriented technologies for software development. Indeed, the benefits of object orientation (OO) are well documented and proven within other software development domains [11]. These benefits include better management of complexity, closer mapping to problem domains, and increased productivity, reuse, and evolutionary capabilities, due to the use of the same paradigm from requirements right through to implementation and testing. Despite these benefits, however, completed software systems still contain errors and often do not satisfy their requirements. Indeed, it has been widely reported that the majority of computer systems failures are the result of incorrect or poorly specified requirements [1, 15], and that the cost of correcting such failures can be as much as 200 times greater than the cost of correcting failures due to errors introduced later in the development process [3]. It is important, therefore, that software development teams pay attention to the analysis phase of the development process to ensure that the requirements are correctly understood and documented before design and implementation begin.

The output of the analysis effort should be a correct, complete, consistent, and unambiguous requirements specification. The requirements specification is a representation of the proposed system that is intended to describe all the relevant properties of that system. It should describe the goals of the system, the functional and non functional requirements, and any design and implementation constraints on the system. A good quality specification should also be understandable, testable, and implementation independent. Producing good quality specifications, however, is not a simple task that can be achieved (repeatedly) using an ad-hoc approach. To facilitate the creation of specifications a number of software engineering researchers and practitioners have developed methods providing guidelines and heuristics on how to approach the analysis and specification effort, and notations for the resulting system

models. The goal of these methods is to facilitate the production of specifications possessing the quality characteristics mentioned above.
The objective of this paper is to summarise some experiences gained adopting object oriented analysis for telecommunications systems development. A number of development projects were studied in compiling the report. The paper also draws on the author's experience on a number of funded research projects, and on some ongoing work in the Network Management Forum (NMF).

2 Background

The following experiences and recommendations are based on the work of six telecommunications projects. The development projects upon which the findings are based range in duration from 2 man months to 12 man months. The systems being developed were all telecommunications related: some service focused, some management system focused. The results were collected through review, observation, questionnaire and interview.
Two separate organisations were involved in the study. The first organisation had several years experience in the use of OO technology but had not previously emphasised the analysis phase of development. Training consisted of a three day course, given by an external consultant, involving a real world example modelled using the OMT [16] and Use Cases. No pilot project was used during the migration. The second organisation was unfamiliar with OO technology but followed a defined software process that included analysis. The initial OOA process mainly used the OMT, with Use Cases used for requirements capture. These were supplemented with elements from Booch [4] and Fusion [6] and incorporated into the Rapid Applications Development (RAD) development approach DSDM [7]. A pilot project followed training.

3 Lessons Learned

The experience gained and recommendations are discussed under the following headings: education and training, analysis effort and benefits, using OOA methods with telecommunications specific modelling approaches, tool support for analysis, use cases, and general experiences and impression of OOA use in the industry.

3.1 Education and Training

When attempting to adopt OOA there are often two difficulties that must be faced: the disruption of the development process, and the paradigm shift that is often required for practicing software engineers who are not familiar with object oriented development. The first thing to do to overcome these problems is to educate staff in the benefits and importance of analysis (see below), and to provide both managers and developers with adequate training in the analysis method chosen. Training should be given by someone experienced in both development and training. If it is not possible to get a trainer who is familiar with the telecommunications domain then the one chosen should have considerable experience in full life-cycle OO development (including OOA) within some other domain.
Classroom training should be followed by hands-on experience in a realistic development setting (such as work on a pilot project that allows developers try out their newfound knowledge). On-going training should also be provided throughout the development process, however, for those staff members not familiar with the new development paradigm. Often this is best achieved with adequate access to an OOA "guru" who is on hand to provide assistance and guidance when needed. Moreover,

additional training in moving from analysis models to design models should also be provided as developers have difficulty with this step.
It is also important to provide developers with examples of previous OOA work both within the domain and from other domains. In the opinion of one developer:
"*OOA is in an early adoption phase in the telecommunications domain - it should be supported by more specific examples related to the analysis, modelling and design of telecommunications systems*".
Indeed, this is where the work performed by various industry sources, such as the EC funded RACE and ACTS projects, can prove useful. Many of these projects provide deliverables containing examples of OOA models applied to the telecommunications domain. Moreover, ongoing OOA modelling work in the Network Management Forum (NMF) using initially the OMT and (when it stabilises) the UML (Unified Modelling Language) [17] will also provide examples of the use of OOA. Organisations should try to avail themselves of this work.

Experience	Recommendation
Training from someone experienced in both (full life-cycle) development and training works well	Make sure that the trainer is experienced in both aspects - development experience is as important as training ability
Classroom and hands-on training are needed	Allow developers test their understanding in a real setting
Access to previous domain related examples is important	Search for as many examples as possible - books won't contain sufficient examples
Initial training was performed before projects began but no on-going training took place	Train staff before projects begin and provide on-going training throughout the projects. Mentoring from a guru is useful

3.2 Analysis Effort and Benefits

It may seem strange, but many practicing software engineers (and managers) today often don't realise the value of time spent analysing system's requirements before embarking on implementation. There is still a tendency to rush through to coding in order to have something concrete to show managers and clients. Indeed, when asked, many of the developers involved in the projects studied for this report admitted that in previous projects little or no time was spent on analysis. Developers and managers must be educated on the benefits of analysis effort (for example, by citing the figures mentioned above regarding the costs of correcting failures due to poorly specified requirements). Having done this, however, it is still very difficult to get developers to produce analysis models unbiased by implementation details. It was found that the models produced were often directly translatable into the developers favourite programming language (developers freely admitted this). One way to avoid this is to perform regular reviews of the models produced to ensure that they remain as implementation independent as possible. It helps to have someone who is used to producing and reviewing analysis models on hand to review the models. Developers are, of course, sometimes careless regarding supporting documentation (such as the production of data dictionaries, and the storage of even intermediate models). To avoid this it is important that these items are explicitly checked during the reviews.
Having educated developers on the importance of the analysis phase and trained them in the chosen OOA method the next challenge is to allocate time within the development plan for the extra effort required to model requirements. This time spent analysing the domain and producing relevant models proves extremely useful to

developers in further development projects - this is where the real benefit of OO can be felt. Organisations attempting to adopt object oriented technologies for software development often do so because they are seduced by the promise of reuse. In general, developers often tend to think of code as being the only reusable artefact from development. It has been our experience, however, that provided they are stored in an effective way there are many artefacts that can be reused. OOA models in particular can be reused with appropriate modification. It is recommended, therefore, that reuse within telecommunications organisations is focused initially on the analysis phase of development.

Experience	Recommendation
Domain analysis is needed	Train staff and allocate time
Developers unfamiliar with analysis produced models that were implementation oriented	Educate staff on the purpose of the analysis activity. Perform regular peer reviews of models produced
Reusing analysis models (or parts thereof) saves effort and aids understanding	Focus reuse strategy on the analysis stage Store results in reuse library
Developers are careless regarding supporting documentation	Explicitly check supporting documentation during reviews

3.3 Using OOA Methods with Telecommunications Specific Modelling Approaches

One of the features of telecommunications software development that distinguishes it from development in other domains is the existence of existing information models produced by various standards bodies, and industry fora (such as those mentioned above in Section 3.1). Current OOA methods are weak in the area of handling existing models. Within the telecommunications domain these existing models are described using a variety of notations and languages such as Guidelines for the Definition of Managed Objects (GDMO) notation, the Object Management Group's (OMG) Interface Definition Language (IDL), and the TINA-C's Object Definition Language (ODL)[1]. In fact these models are more design and implementation oriented than analysis oriented. Developers need training in both OOA and telecommunications modelling approaches. This can cause confusion, however, for those unfamiliar with the OO paradigm (because of concept and naming conflicts, for example). It was also found that developers have difficulty using these models when attempting to follow an industrial strength OOA method. The following quote from one interviewee bears out this point:

"*Methodologies should be extended to prescribe and optimise the use of OOA within the telecommunications domain. Specifically the use of existing external models should be catered for*".

There exists a wide body of knowledge on the successes and pitfalls of using OOA but not on the use of OOA in conjunction with existing telecommunications modelling approaches. Analysis and design patterns, and frameworks specific for telecommunications systems would be useful.

Compounding this problem is of course, the fact that often software developers don't understand exactly how they are supposed to use these existing models during development. Figure 1 shows briefly how some of these models should be used. GDMO models are used to indicate the information that must pass over a TMN interface. However, developers are not forced to use the objects in the models exactly

[1] While the intentions of these models differ the problems created by their existence are similar.

as they appear. It is possible to extract other objects (classes) from the GDMO models. IDL/ODL models, on the other hand, are directly translated into code. When faced with existing models given in these notations (that have been produced external to the project) developers should attempt to reengineer OOA models in the notation of their chosen method from the existing models. This will allow development to be derived from the OOA models. To do this effectively, of course, tool support is needed. Indeed, recently the NMF InfoMod team (whose task it is to decide on the method and tool used for NMF modelling purposes) has issued a Request for Proposal (RFP) [13] to vendors regarding their support for telecommunications system development, including the reengineering from the models mentioned. It may be some time, however, before this is supported fully.

One of the development efforts studied attempted to create its own method based on elements of a number of well known industrial methods from the outset. This was found to slow development down as developers needed constant clarification on terminology and concepts. It proved a particular problem for those not familiar with OO development. Because of this, it is felt that at the outset at least developers should not mix methods. That is not to say that access to written material and examples from other methods is not needed. On the contrary access to such material can prove valuable in clarifying concepts. It is important, however, to base development on a single well defined method for developers unfamiliar with OOA.

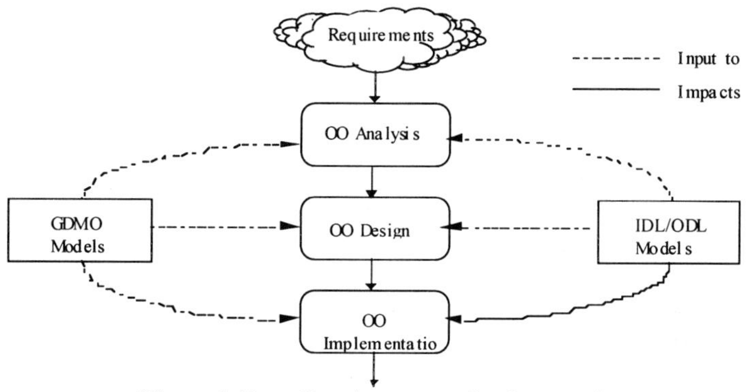

Figure 1: From Requirements to Implementation

Experience	Recommendation
Developers should be familiar with both the OOA method and existing notations (IDL, ODL, GDMO, ..)	Training in both the OOA method chosen and telecommunications modelling approaches is needed
OOA methods and support tools are weak at handling existing models	OOA methods and tools should be extended to handle existing models especially those within the telecommunications domain
Developers found it hard to harmonise existing models with OOA methods	Represent existing information models in the notations of the OOA method being used. Don't spend too much time on this though
Mixing methods can confuse a team and lead to delays	Don't mix and match methods too soon. Make sure the team members are familiar with OOA before mixing

3.4 Tool Support for Analysis

It has been reported that the use of a CASE tool is essential to a successful transition to OO development [8]. The analysis phase in particular requires tool support since all resulting development effort is dependent on the output of this phase. However, tools have two important shortcomings: firstly, they don't support methodologies - only notation, and secondly they encourage isolated design, performed by developers working on their own [10]. It was found that much useful work can be performed using a whiteboard and marker (or similar non automated approach) before attempting to use the tool. When agreement has been reached, the models produced can then be recorded using the selected tool. It is important, however, that staff exploit the tools capabilities (that is, not just use it as a diagramming tool). Therefore, training in all its capabilities should be given.

The choice of tool will be related to the method chosen [2]. Indeed, it is best to select a tool that supports more than one method so that if the method used is changed the tool does not become obsolete. A good place to start looking at tool features is the Web (for example, a selection of tools can be found at [18]). Often demo versions of the tools can be downloaded for trial use. It is also worth monitoring the work of the NMF InfoMod team in this area, as mentioned above.

In relation to development of TMN systems in particular the following quote makes a valid point:

"*There is such a gap between OOA/D tools and TMN development environments. A recommendation is that either the OOA/D tools be extended to support TMN development, or TMN development environments be extended to support OOAD based graphical modelling using an accepted notation such as that of the OMT*".

Therefore tool vendors should work closely with each other to determine the exact needs of developers who wish to follow a software engineering approach using an general OOA method. Data interchange between tools should be provided.

Experience	**Recommendation**
White boards and paper are useful analysis tools	Allow developers freedom to choose what to use to start but ensure that models are recorded correctly
Tool support is necessary to record models	Provide tool support early
Training in the full use of the tool is necessary	Train staff in the complete range of the tool's capabilities (e.g. configuration management, syntax checking, and so on)
The Web can provide useful information on CASE tools	Download demo versions for trial use before committing to a particular tool
The NMF's RFP on tool support provides information on candidate tool requirements	Monitor the work of the NMF on tool support
It is hard to use OOA/D tools and TMN development environments together	Tools vendors should cooperate to provide (at least) data interchange between tools

3.5 Use Cases

Nowadays it is hard to find anyone involved in software development who has not heard of Use Cases. Unfortunately Use Cases are one of the most misused models in software engineering. Everyone has heard of them, but everyone has their own concept of what they are and how they should be used. This understanding has often derived from word of mouth conversations. Unfortunately, the great strength of Use Cases is also one of their weaknesses. Because of their informality they are easy to produce and understand. This informality, however, often gives developers a false sense of security (this was also a problem with a previous software engineering favourite, the Data Flow Diagram). When using Use Cases as the basis for software development this can lead to poorly understood systems. It was found that developers use their own variant of Use Cases based on their understanding of them. It is important, therefore, to give developers a good grounding in Use Cases before allowing them to proceed. In addition, it is important to have a company wide standard for what a Use Case consists of, the conventions used, and guidelines for their production. Candidate Use Case formats can be found in [9].

Use Cases provide an excellent means of eliciting requirements and modelling external system behaviour. However, there are a number of other uses to which Use Cases can be put: testing and validating the finished system, managing complexity, providing a means for end users to participate in requirements analysis, and identifying candidate objects. In practice it was found that developers usually only use them to capture requirements and to identify candidate objects. To overcome this a description of how Use Cases can be used should be defined at each stage of the development process (where appropriate).

Experience	**Recommendation**
Use Cases were used heavily and were helpful for identifying objects and eliciting requirements	Start requirements capture using Use Cases
Developers have their own understanding of what a Use Case should contain	Produce company wide standards for Use Cases
Developers don't exploit Use Cases to their full potential	Exploit Use Cases throughout the development process

3.6 General Experiences & Impression of OOA use in the Industry

Developers often tend to take a bottom up approach to learning object orientation (that is, they start by learning a particular programming language - almost invariable C++). This generally does not lead to a true understanding of the benefits and usefulness of object orientation. That is not to say that knowing a particular object oriented programming language does not help understand the principles. However, using a programming language that supports a particular paradigm neither ensures good quality code nor good quality systems. In fact this is one of the myths of object orientation; it is just as easy to develop a poor quality system (in terms of maintainability, extendibility, reusability, and so on) using the object oriented paradigm as it is using the procedural paradigm. One way to avoid this is to select a set of analysis and design quality metrics. A collection of object oriented metrics can be found in [5]. In reality applying such metrics is difficult but developers should be made aware of their existence and should be encouraged to use them. Initially only a small number should be selected and applied.

Developers have a tendency to learn from mistakes at a low level (that is, coding experiences). It is also important to learn at a higher level in the process. This can be achieved by the application of a continuous improvement program such as at the Software Engineering Institutes (SEI) Capability Maturity Model (CMM) [14] for software development. In addition to storing artefacts and models, decisions, pitfalls and shortcuts should be also recorded for projects and used to improve the development process.

Developers find it difficult to apply a pure OO approach. This may have something to do with the implementation language used and the bias towards solutions that are based on knowledge of this. It was found that developers who are unfamiliar with OO

learn the concepts easier if they learn OO analysis before OO design or OO programming.
Despite the fact than many organisations, standards bodies, and relevant fora advocate its use, and are active in promoting OO, a full methodological approach to its use is often not followed. Many funded research projects claim to follow a particular methodology (currently OMT is the "flavour of the month") but in reality when one attempts to find out exactly how these methodologies have been used one finds that invariably only the notation of the method has been used. Drawing a diagram using the notation of a particular method will not by itself help engineer quality systems.
There may be several reasons why full methodologies have not been used. One of them could be the background of the people in the industry. Many people involved in development projects do not have a proper grounding in good software engineering principles. Developers in the industry come from such diverse backgrounds as physics, electronic and electrical engineering or other backgrounds. Therefore, all those involved in development should be provided with software engineering training.
There seems to be an emphasis within the telecommunications industry on *information models*. Unfortunately there is often disagreement as to what an information model is (for example, is it an Entity Relationship Diagram (ERD), an object model, or a collection of object, dynamic, and functional models?). It is worth pointing out that an information model (in a general software engineering sense) is only one part of a *requirements specification* (they are also included in other specifications as well, of course). Perhaps more focus should be placed on requirements specifications (which include information, dynamic, and functional models, as well as other system related information) briefly described at the beginning of this paper.
Finally, it appears that industry fora and have not adequately addressed the problems of migrating to the use of object orientation and the peculiarities of the industry with this regard. This issue is starting to be addressed for the wider object oriented community [12] but there are certain characteristics of the telecommunications industry that need to be considered. One of these is the existence of information models mentioned earlier. This issue needs more attention.

Experience	**Recommendation**
Design metrics were not used	Know and use metrics - select a small number and apply them
Developers often need a grounding in good software engineering principles	Adopt a continuous improvement program such as the SEI's CMM
Object orientation is difficult to apply completely	Teach OO analysis before OO design or OO programming
Information models and requirements specifications are not the same thing	Information models are part of requirements (and other) specifications
The industry has not addressed migration problems adequately	The peculiarities of the industry (diverse backgrounds, etc.) should be addressed by fora and research organisations.

4 Conclusions

While many telecommunications industry fora, standards organisations, and research initiatives advocate the use of object orientation for software development, and while these groups have produced standards, examples, and documents on the subject, adopting object oriented analysis remains difficult for organisations who have either not emphasised the importance of the analysis phase or who are only familiar with

legacy developments approaches. There are two major reason for this. First, the existence of the models produced by various industry sources makes it difficult to adopt a methodological approach. While following a methodological approach has proven beneficial it can prove particularly difficult for organisations not familiar with object oriented development. To overcome this problem it is recommended that methods and tools be extended to cater for existing telecommunications models.
The second reason why adopting object oriented analysis is difficult in telecommunications is because industry fora have not adequately addressed the problems of migrating to the use of object orientation and the peculiarities of the industry with this regard. This issue is starting to be addressed for the wider object oriented community but there are certain characteristics of the telecommunications industry that need to be addressed. To overcome this problem organisations can learn from the experiences of others who have attempted to migrate. This paper is a first attempt at disseminating such experiences.

References

[1] Basili V.R., and Perricone B.T., "Software Errors and Complexity: An Empirical Investigation," *Communications of the ACM*, Vol. 27, No. 1, pp 42 - 52, January 1984.

[2] Bell R., "Choosing Tools for Analysis and Design", IEEE Software, May 1994, pp 121 - 125.

[3] Boehm B.W., *Software Engineering Economics*, Prentice Hall, Englewood Cliffs, NJ, USA, 1981.

[4] Booch G., *Object-oriented Analysis and Design*, 2nd Edition, Benjamin/Cummings Publishing Co., California, USA, 1994.

[5] Chidamber S.R., and Kemerer C.F., "Towards a Metrics Suite for Object Oriented Design", in OOPSLA '91, Proceedings of the 6th Annual Conference on Object Oriented Programming, Systems, Languages and Applications, ACM Press, 1991, pp 197 - 211.

[6] Coleman D., *et al.*, *Object-oriented Development: The Fusion Method*, Prentice Hall International, 1994.

[7] DSDM Home Page URL: http://www.dsdm.org/

[8] Fayad M.E., T.S. Tsai, and M.L. Fulghum "Transition to Object Oriented Software Development" in Communications of the ACM, Vol. 39, No 2., pp 108 - 121, February 1996.

[9] Harwood R.J., "Use Case Formats: Requirements, Analysis and Design", in Journal of Object Oriented Programming, January 1997, pp 54 - 57.

[10] Johnson D., "The Important things are Always Simple", Object Magazine, September 1996, pp 66 - 70.

[11] Lewis J.A., Henry S.M., Kafura D.G., and Schulman R.S., "An Empirical Study of the Object Oriented Paradigm and Software Reuse", in OOPSLA '91, Proceedings of the 6th Annual Conference on Object Oriented Programming, Systems, Languages, and Applications, pp 184 - 196, ACM Press, 1991.

[12] McGibbon B., *Managing Your Move to Object Technology*, SIGS Books, 1996.

[13] Network Management Forum (NMF), 1201 Mt. Kemble Avenue, Morristown, New Jersey 07960, Request for Proposal for Object Oriented Analysis and Design Tools, June 1996.

[14] Paulk M., "Capability Maturity Model for Software Development, Version 1.1", Tech. Report CMU/SEI-93-TR-24, Feb 1993.

[15] Posten R.M., "Preventing Software Requirements Specification Errors with IEEE 830," *IEEE Software*, Vol. 2 No. 1, pp 83 - 86, 1985.

[16] Rumbaugh J., Blaha M., Premerlani W., Eddy F., Lorensen W., *Object Oriented Modelling and Design*, Prentice Hall, 1991.

[17] The Unified Modelling Language (UML), Version 1.0, Rational Software Corporation, January 1997.

[18] Collection of OOA tools can be found at
URL: http://www.csse.swin.edu.au/manfred/allpages.html#oo_ooa_ood_tools

OSAM Component Model

A key concept for the efficient design of future telecommunication systems

Aggeliki Dede[1], Spiros Arsenis[1], Alessandro Tosti[2], Ferdinando Lucidi[2], Richard Westerga[3]
[1]National Technical University of Athens (NTUA), {adede, arsenis}@telecom.ntua.gr
[2] Fondazione Ugo Bordoni (FUB), {alex, nando}@fub.it
[3] KPN Research, r.s.westerga@research.kpn.com

In this paper, guidelines for the efficient design of telecommunication services are presented. The driving force is the OSAM Component, a basic modelling concept for the consistent and effective analysis, design, use and management of services. Based on the OSAM Component Model, a service machine is built out of components integrating both mobile and fixed network services.

Keywords. OSAM, Component Model, Service Design, CORBA, Distributed Objects

1 Introduction

The current telecommunications industry is facing the growing need of making telecommunication services more versatile, easier to develop, interoperable, consistent, manageable and independent of the underlying network. The demand for new sophisticated services, such as universal personal telecommunications, mobile, multimedia and broadband services, is on the increase. These services require more flexible access, management, and charging mechanisms.

To meet the needs of future telecommunications, new architectural frameworks have been proposed (IN, TMN) and others are currently under development (TINA, OSAM), that provide the means to build services and a service support environment. One of the major goals of these new architectural frameworks is to assist and constrain designers in the complex process of service creation. During this effort, software portability, interoperability and reuse will be of prime importance. Services are increasingly being realised as software modules which may reside in general purpose computers attached to a network. Therefore, the introduction of the *component* notion, as the basic modelling concept of services, seems to be the best solution to face, in a integrated broadband communications environment, emerging issues related to service nesting, service reusability and integrated design.

This paper presents the OSAM Component, stemming from the OSAM service architecture as this has been elaborated in the DOLMEN ACTS project[1], a basic modelling concept for the description, the specification and the implementation of the system entities-services. The OSAM Component is mainly derived from the OSA Component concept, as defined in the RACE project CASSIOPEIA [1], however worked out and refined in a number of aspects that have only been sketched for the OSA Component. A basic contribution of the OSAM Component, beyond the efficient description of the functional aspects of the system entities, is the definition of the dynamic behaviour of the system entities.

The paper is organised as follows: Section 2, presents the structure of the OSAM Component, based on the functional separation of the service capabilities, as well as a generic state model for structuring the OSAM component internal behaviour. Guidelines for applying the OSAM component model throughout the service

[1] The aim of DOLMEN ACTS 036 project is to develop, assess and promote an Open Service Architecture for an integrated fixed and Mobile environment (OSAM).

development process are provided in Section 3. An application example is presented in section 4, demonstrating the use of the OSAM component model in the description, specification and implementation of service components realising the DOLMEN service machine. The paper concludes with evaluation remarks and recommendations for future work in section 5.

2 OSAM Component Model

Analysis and design should be guided by a well defined modelling concept enabling a better understanding of the information and behaviour of service entities to be built. Such a modelling concept, already adopted by both OSA [1] and TINA-C [2] architectures, as well as service creation environments such as SCORE [3], is the concept of *component.* Services can be expressed and built as discrete systems made out of components, which can eventually be mapped onto the service support environment. Similarly, services of the service support environment, as well as the resources of the underlying resource infrastructure may also be modelled in terms of components and thus be accessed by other components, such as application related components.

Component modelling ensures that services and components are designed integrating different aspects such as usage, management and logic. The component model allows services to be designed, specified, built, and managed in a modular fashion, thus promoting consistency, reuse, and simplification of management.

The OSAM Component is defined as an entity providing a set of services and might be realised in an environment for service deployment and service provision. The services offered by an OSAM Component comprise the OSAM Component Functional Model that describe all the capabilities which represent the reason for existence of the component, as well as capabilities that are of subordinate interest but necessary for the component to be realised in a particular environment.

Nevertheless, for the full specification of the system services, beyond the definition of the entities functional model, the definition of the entities dynamic behaviour is required. A basic contribution of the OSAM Component is the efficient definition and the formal description of the component dynamic behaviour consisting of the OSAM Component Dynamic Model.

2.1 OSAM Component Functional Model

Based on the OSAM component model (Figure 1), all services are modelled as consisting of a *mission* and *ancillary facets.*

A mission describes a service as to its primary value to a user. It represents the reason for existence of the service i.e. the core service capabilities. The ancillary facets represent a set of constraints on the intended usage, on the supporting environment and on the used resources. They are further classified into the following groups: *usage* supporting capabilities enabling the user to exploit the component functionality as this is provided by its mission. It mainly provides information about the component itself as well as instructions on how to access and use the component in a given provision environment; *life* supporting management capabilities and capabilities related to the component's life-cycle, mainly derived by the FCAPS categorisation of management activities; *resource* dealing with capabilities that the component requests by its environment, in order to correctly perform its functionality.

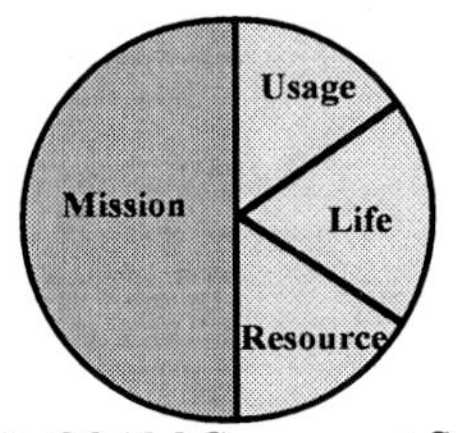

Figure 1: OSAM Component Structure

This separation reflects the difference between service capabilities that are essential for a service to fulfill its goal (mission), and those of subordinate interest, yet necessary for the service to exist in a particular service provision environment. As the amount of development resources today go into capabilities ancillary to what really makes a service stand out, one can think of how much economy of design is possible for services differing only in their mission while reusing ancillary facets (e.g. by inheritance).

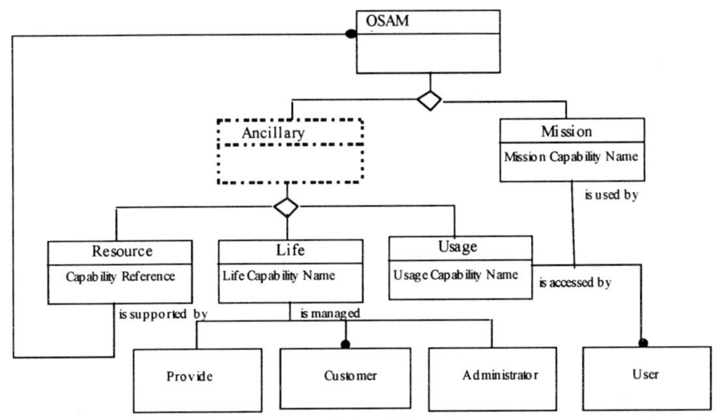

Figure 2: OMT diagram of the OSAM component structure

Moreover, the OSAM Component structure fulfills the objectives of the ODP Enterprise Viewpoint analysis and the ODP Information Viewpoint structuring. In Figure 2, the OMT[4] notation is used as a meta-model to express the semantics of the OSAM component structure as a composition of objects. The aggregation tree shows the partition of the component and the different actors of the surrounding environment that access the component according to their role in the service execution scenario.

2.2 OSAM Component Dynamic Model

Another OSAM component basic characteristic is the component internal behaviour, describing the component's *way of being*. Through a state model we define the different type of states governing the whole life cycle of a component. This state model will guide the components detailed specification. The component's state is composed of a set of values, characterising the component successive states of availability and operation during the component's life. We can distinguish three generic state types:

- The *mission state*, reflecting the use that we make of the component. The mission state values and their transitions describe the component's internal functional behaviour (the way in which component capabilities are performed) by indicating if the component is idle or active as well as on which step of its operation is found in a specific moment,

- The *life operational state*, reflecting the component's internal behaviour concerning the component resources availability (behaviour pre-coded on the component level and not initiated by the administrator). The life operational state indicates if the component is physically installed, if it possesses the necessary resources (memory, transfer capacity, etc.) to operate as well as if it is operating properly.
- The *life administration state*, reflecting the management applied on the component by the administrator. The life administration state indicates the permission or the interdiction of the component use, imposed by the administrator. The life administration diagram states the administration policy (interdiction of use of system services for certain users) or the administrator reaction to specific operation problems.

These states combination give the complete image of the component functional and management treatments. Using this triad of values (mission, life operational and life administration values), we manage to track and control the component's behaviour during its whole life cycle. Moreover we guarantee the component's functional and management autonomy as a distributed entity [5]. The life operational state expresses this need for management autonomy, by delegating part of the control activity on the component level. The life administration state structures the actions performed by an administrator on a component and the effect that these actions have on the component internal behaviour.

The OSAM component state model allows the exact modelling of the component's internal behaviour. The transition from one state to another is initiated by the occurrence of an important event for the components operation. The *mission* and the *life operation* states give an indication of the component internal activity. Their values are modified directly by the component during its life-cycle. They can only be consulted and they cannot be modified by a direct action from the component service user or the administrator. This fact reflects the nature of the components as autonomous entities.

During the component's operation the states become a powerful *control tool* allowing at any instance the monitoring and the control of the component behaviour as well as the component behaviour degradation. Through a state operation mode, the coherence (execution step by step) of the component's behaviour is enforced. Furthermore, using this state information, management mechanisms as fault tolerance can be easily implemented.

3 Component Construction

In this section, the application of the OSAM component model for the development of the DOLMEN service machine components is presented. The OSAM component model provides a continuous support from analysis to implementation. The component structure (mission, ancillary facets) is preserved throughout the design process, however different modelling concepts, techniques, tools and languages are used to express the component semantics in each phase.

An overview of the component driven development process is depicted in Figure 3. A 3-step approach is followed in the component construction: component description, component specification, component implementation. The purpose is, starting from the component's conceptual structure to be able to derive a target implementation consisting of C++ objects executing in a CORBA distributed processing environment.

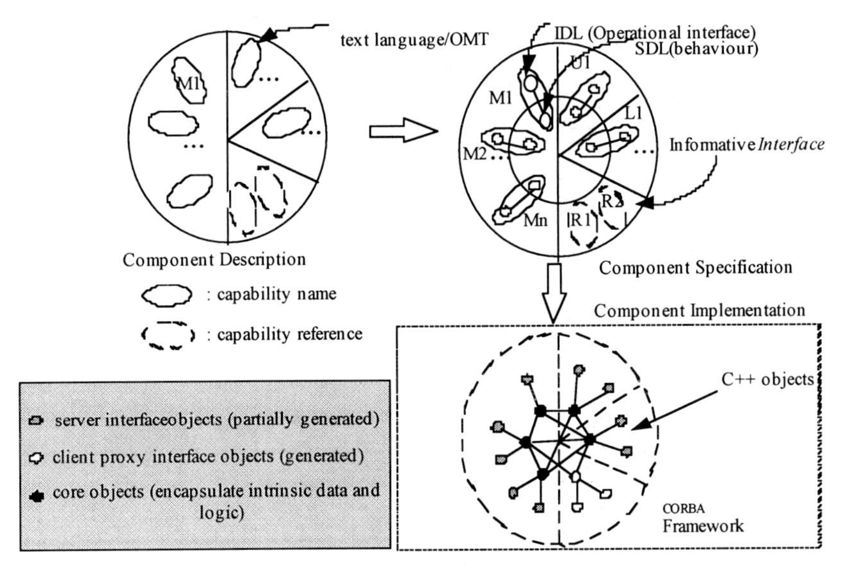

Figure 3: Component development process

The development process for the service machine components starts with a description of the component at a certain abstract level. At this stage service capabilities to be provided through the different facets of the component functional model are described by means of text, tables or OMT diagrams.

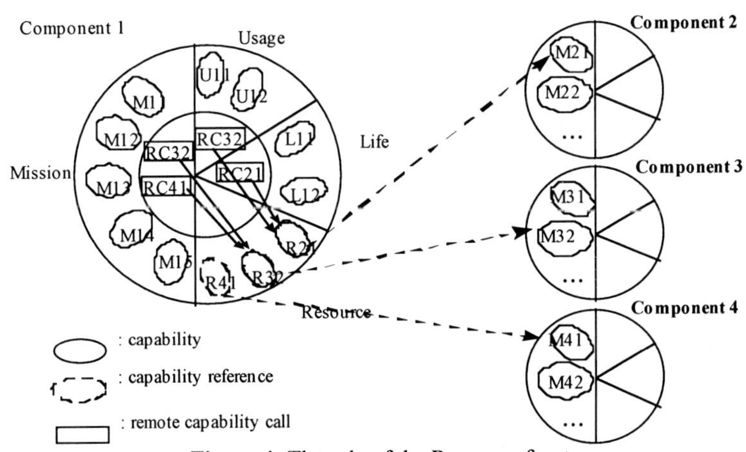

Figure 4: The role of the Resource facet

Based on the above description, the second step is the specification of static and dynamic aspects of an OSAM component. The OSAM component specification phase should be based on client/server and object oriented paradigms, for service

specification in distributed systems. At this stage, a second layer of separation is introduced in the component structure taking into account computational aspects [6]. A service is specified as a composition of computational objects, interacting by means of well defined interfaces in order to achieve the service goal. Therefore, an OSAM component is viewed, at the specification stage, as a specialised computational object that supports the interfaces as derived from the semantics of the component structure: Mission, as well as Usage and Life operational interfaces. For the Resource only an informative interface is used while behaviour does not exist. Through the Resource facet, other components' interfaces can be invoked as it is shown in Fig. 4.
The component specification starts with the interface descriptions of the computational model by using the standard OMG Interface Definition Language (IDL) [7]. For the detailed specification of the component's dynamic behaviour, SDL has been chosen [8]. Figure 5 shows how SDL can be used in the development process.

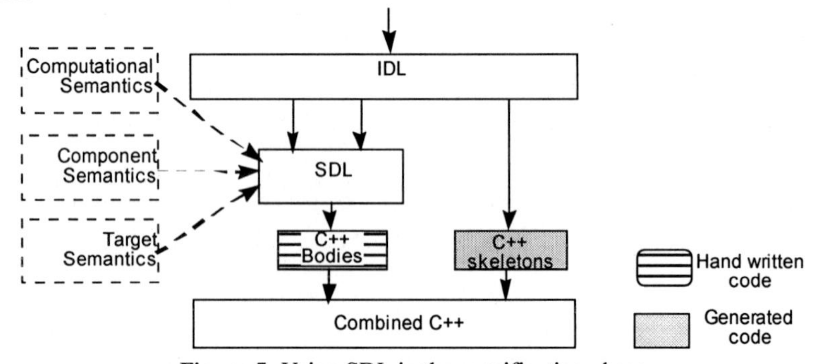

Figure 5: Using SDL in the specification phase

From the IDL interface definitions of the component, an SDL specification is created. Adding the necessary detail in SDL, the goal is to generate C++ code. The basis for the SDL model stems from the computational semantics, component semantics and the target guidelines. So far there is no tool that supports automatic C++ code generation. However, taking into account the above aspects during specification, we will easily render to a CORBA compliant C++ hand-written code which, combined with the C++ skeletons automatically generated from IDL, can be compiled and run on a CORBA-compliant platform.
The OSAM component model will guide the components detailed specification using SDL. The mapping of the component semantics to the SDL concepts, is presented in Table 1. The component entity splits up into several processes in order to meet the computational and engineering requirements. Nevertheless, to conserve the component's behaviour atomicity and coherence during its implementation and execution, the dynamic state model properties must be observed.
Finally, the implementation phase is considered, though not explicitly outlined as it is up to the individual designer's freedom to explore his/her own creativity and mechanisms for the most promising implementation. By taking into account computational and engineering requirements in the SDL specification, we can easily derive a structure of C++ objects comprising the component (Figure 3). There are objects implementing *server* type interfaces as well as *client* proxy objects, a *container* object encapsulating the data specific to the component and a *core* object providing the means to glue the structure collection of objects in order to form a single concrete entity [9].

Component Concepts	**SDL construct**	**Remarks**
Component	Structure of SDL processes: Core and Interface processes	The structure will be clarified in the application example (section 5)
Mission Facet	SDL process	Evident
Life Facet	SDL process	Evident
Usage Facet	SDL process	Evident
Resource Facet	SDL Data	The Data contains references to interface processes which is a collection of capabilities. The datatype will be SDL Process Identifiers (PID)
Component Dynamic Behaviour	SDL Data + Controls on the SDL procedure	The Component State variables are stored as data in the Core SDL process. Before the execution of any capability the component state variable is checked to avoid the object's incoherent operation
Capability	Exported procedure	A capability is modelled as a procedure where the actual behaviour of the capability can be specified.
(Remote) Capability call	Remote procedure call	Since the capabilities are exported procedures, other components may call remote capabilities by using this SDL concept.
Internal component communication	Remote procedure call	All data is stored in the Core SDL process. If an interface needs data it has to retrieve it from the core process. The mechanism is modelled using also remote procedure calls.

Table 1: Mapping of component concepts to SDL

4 Application example

In this section, the component construction complete guidelines are demonstrated by using a simple example based on the access session of the OSAM service machine. The component chosen is the User Agent (UA), which plays a key role as a contact point for a user to access various services in the provider domain. The description of the UA basically consists of a capability list, given as a combination of text and tables.

Mission Capability	**Rational**	**Usage Capability**	**Rational**
access request	initial access and authentication to the service environment	access list	provide list of authorized users
service request	create a new session for a specific service	usage info	provide information about the component
Life Capability	**Rational**	**Resource Cap. Ref**	**Provided by**
component initialize	initialize a component instance	invite a new user	Provider Agent
component shutdown	shutdown a component	check subscription	Subscription Agent

Table 2: UA component capabilities

Table2 shows a list of capabilities that the UA should offer, grouped into Mission, Usage and Life, as well as the ones that the UA requests from other components (Resource Capability References), e.g. Provider Agent (PA). For the sake of the example, it is assumed that only a limited capability list is supported by the UA.

From the above descriptions, an IDL specification can be derived such as:

```
module UA_Component {
// data declarartions   ...
// interfaces
 interface i_UAMission {
  boolean UA_access_request(in string PIN);
  boolean UA_request_service(in long UserID, in long ServiceID);
 };
 interface I_UAUsage {
        void UA_access_list( out string Users );
        void UA_info( out string Description, out long ProviderID );
 interface i_UALife {
        void UA_initialize( );
        void UA_shutdown( );
 };
// the Resource is a client type interface that appears only in the
implementation
}; // end of UA module
```

The next step concerns SDL guidelines and their application to the UA component. In order to better illustrate the structure of the SDL system, we use an example of two access session components, the UA and the -other side of access session interactions-Provider Agent (PA). Considering the engineering view of the ORBIX - CORBA compliant - platform [10], these reside in two different capsules in the provider domain and end-user system respectively. The notion of capsule here resembles the notion of the ODP cluster [11].

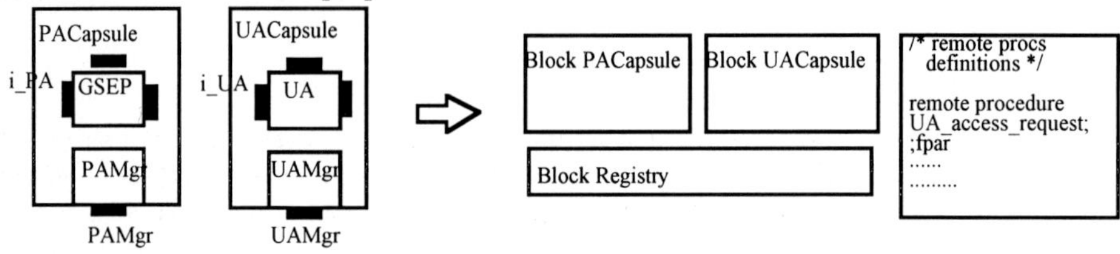

Figure 6 : Mapping of engineering to SDL concepts

Objects are clustered in a capsule and every capsule has a capsule manager (CMgr) that takes care of managing the objects inside the capsule by creating and deleting object instances. The CMgr interface is the only interface visible outside the capsule and the one that the ORB binding function can provide a reference to. Other interfaces inside the capsule are provided by the CMgr. Figure 6 shows how engineering concepts are translated into an SDL system. Capsules are modelled as blocks. The registry block models the ORB that offers the possibility to find components in a capsule. Remote procedure definitions are used for every exported procedure such as the *UA_access_request.*

The next step is the structure of the UA component within a capsule. Figure 7, illustrates the SDL diagram of the UA capsule. The CMgr is an SDL process responsible for the instantiation of the Core process. The Core process encapsulates all data manipulated by the component and is also responsible for the instantiation/deletion of the component Interface processes. The capabilities supported by the Interface processes are modelled as Exported Procedures. The figure shows the instantiation of the component's UAMission, UAUsage, UALife interface processes. The Resource facet is realised as SDL data stored within the Core process and is actually referenced to other components. Those data are manipulated by means of GetData and SetData procedures.

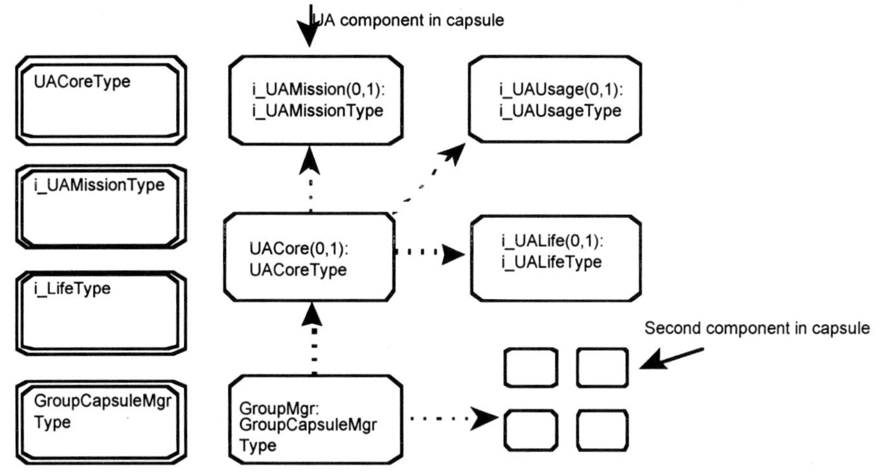

Figure 7: UA component specified as a capsule

Using this model the OSAM component semantics are fulfilled together with the target environment requirements. SDL can be used, through its powerful tool support (e.g. SDT, GEODE), to simulate and validate our system in order to ensure the coherent behaviour of OSAM components.

5 Conclusions

In the DOLMEN project, the OSAM Component has been proposed as the modelling entity that guides the description, the specification and the implementation of system entities. The OSAM component conceptual structure enforces the designer to implement all necessary features for the successful realization and operation of a service. Reusability is promoted by the separation between mission and ancillary facets. The OSAM Component facilitates the smooth integration of both functional (interface model) and dynamic (state model) aspects of a service. Moreover, the coherence (execution step by step) of the component's behaviour is guaranteed, and the functional and management autonomy of the component entities is enforced by delegating part of the control activity on the component level. Finally, the model adheres to existing modelling techniques and standards, though being decoupled from specific tools and languages

Concerning the difficulties of applying the OSAM Component Model, some effort is required for the designers to get familiar with the concept and the semantics of the OSAM component facets, especially the Ancillary Facets, in order to fully exploit its design capabilities. The Life support facet poses certain requirements to the underlying distributed system technology. Telecommunications management aspects are not yet fully incorporated within the current distributed objects technology, while platforms dedicated to management are missing object-oriented, distributed processing aspects. Therefore, a close consideration of the target execution environment is necessary in all phases of the component development process as it helps to decrease the designer's overheads.

Acknowledgements

This work was supported by the ACTS DOLMEN project; however, it represents the view of the authors. The authors would like to thank their colleagues in the project for their contribution and the fruitful discussions.

References

[1] RACE 2049 Cassiopeia, "Open Service Architectural Framework for Integrated Service Engineering", R2049/FUB/SAR/DS/P/023/b1, March 1995

[2] TINA-C Deliverable, "Overall Concepts and Principles of TINA", Version 1.0, TB_MDC.018_1.0_94, Feb 1995.

[3] RACE 2017 SCORE, "The SCORE Component Model and it Framework", R2017/SCO/WP1/DS/P/024/b2, Dec 1994

[4] J.Rumbaugh et al., "Object-Oriented Modelling and Design", Prentice-Hall International Inc., ISBN 0-13-630054-5, 1991

[5] S.Arsenis, N.Perdigues, N.Simoni, "Distributed Applications and Networks Integration: from Modelling to Implementation", TINA Conference, Brisbane, Australia, Feb 1995

[6] TINA-C, "Computational Modelling Concepts", TB_A2.HC.012_1.2_94

[7] OMG, "The Common Object Request Broker Architecture", The Object Management Group, Inc., Version 2.0, Jul 1995

[8] ITU-T Recommendation Z.100-annexF.3 "SDL Formal Definition: Static and Dynamic Semantics" March 1993.

[9] H.Korte and R.Westerga, "Guidelines for Computational Modelling in CORBA Environments", KPN Research, Accepted to the ISADS'97 Conference

[10] IONA, "The Orbix Architecture", IONA Technologies Ltd, ARCH.DOCv1.3, Jan 1995

[11] ITU-T, "Recommendation X.901 - Reference Model of Open Distributed Processing - Part 1", ISO/IEC/JTC1/SC21/WG7, Jun 1995

Extending OMG Event Service for Integrating Distributed Multimedia Components

Tin Qian and Roy Campbell
Department of Computer Science
University of Illinois at Urbana-Champaign
1304 West Springfield Avenue, Urbana, IL 61802, USA
{tinq, roy}@cs.uiuc.edu
Fax: (217) 333-3501 Tel: (217) 333-7937

Developing multimedia applications, such as video and audio applications, is a difficult task because of the stringent requirement on system resources and the great diversity of multimedia standards and devices. Many existing multimedia systems are monolithic and extremely complex. It is hard to extend and reconfigure those systems. Easy extensibility and reconfigurability is desirable since multimedia research and development is one of the fastest changing fields in computer science. On the other hand, event-based systems seem to be the right solution to these software engineering problems by allowing software integration in a loose and flexible way. Several emerging standards for distributed integrative environments, like the Common Object Request Broker Architecture(CORBA) from the Object Management Group (OMG), have defined standard event service interfaces. However, most existing event services in those systems cannot provide the Quality of Service (QoS) that multimedia applications need. In this paper we address this problem by extending standard OMG event service with temporal factors so that the system can deliver large volume events, like video frames, in real time. A new type of event service called timed event service is proposed and deployed in constructing the high-performance event services. To demonstrate the feasibility of our design, we have prototyped this fast CORBA-compliant event service in the Distributed System Object Model (DSOM).

Keywords. Event, Multimedia, Interface, Asynchronous, Real-time

1 Introduction

Multimedia research and development has recently become one of the hottest areas in computer science. People have developed many state of the art multimedia systems and applications. However, constructing multimedia applications, especially video/audio ones, remains a difficult task because of stringent requirements on system resources and the great diversity of multimedia standards and devices. Many existing multimedia systems are monolithic and very complex. It is hard to extend and reconfigure those systems. Easy extensibility and reconfigurability are often required by the fast evolution of multimedia technology. On the other hand, recent studies in event-based software integration [3, 12, 5] show that event-based object oriented frameworks can be a promising solution to these software engineering problems because of the loose and flexible way in which they integrate software. Several emerging standards on integrative environments for distributed object computing have also included event services as the basic object facility. For example OMG has defined a standard interface for event services in CORBA [1, 6]. However the performance issues in these event-based integrative environments have been largely neglected. In addition, different applications require different QoS for monitoring, processing, and notifying events. For example, dynamic multi-point (DMP) applications [14] and video conferencing applications require timely delivery of large quantities of continuous media while interactive applications, like many GUI toolkits, have a much less stringent performance requirement for event services. Moreover high-performance event services usually impose high demands on system resources.

How to guarantee the resource requirement by these high-performance applications while still retaining the ability to serve other traditional applications is an open issue. In this paper we propose an adaptive and efficient framework for constructing CORBA-compliant event services. an extension of the OMG event service with a real time event delivery mechanism. The prototyping implementation is done in the context of DSOM, a CORBA-compliant distributed object-oriented programming environment from IBM. The benefits of using CORBA is due to its support for strong interfaces in the Interface Definition Language (IDL), and for distributed object oriented technologies, such as sub-typing to extend and specialize functionalities, thus encapsulating the diversity of underlying hardware platforms and system services.
The remainder of this paper is organized as follows: Section 2 gives a brief description of event-based software architecture and OMG event services. In section 3 we present the architectural design of the extended OMG event services for integrating multimedia components. Section 4 describes our prototyping implementation based on the design in previous sections. Related work is presented in section 5. Finally we conclude our work in section 6.

2 Background

In an event-based object model, the communication between objects is decoupled. The objects producing event data are called *suppliers* and those processing event data *consumers* The systems supporting event services will deliver events from suppliers to consumers. To handle events efficiently, the event services often allow users to specify the types of events they are interested in through event registration mechanisms. Then later when this certain type of events is generated, they will notify the interested objects and dispatch the events to them. Therefore, allowing applications to be constructed in such a loose and flexible way, the event-based model enables applications to avoid premature commitment to platform or application dependent mechanisms. This deferred binding can enhance software engineering perspectives, like portability, reusability and extensibility, as well as performance, by tailoring the services for application behaviours. Because of these advantages, the incorporation of the event-based model into the conventional object oriented programming paradigm is believed to be one of the most promising and novel ways to solve many difficult problems arising in software packaging, such as the extensibility of long-run applications. As one of the most basic services used by applications, it is the major factor in the overall system performance in event-based architectures. However, few studies have been conducted on improving its performance. The traditional event service mainly focuses on supporting single-threaded event notification mechanisms [13]. It has been widely used in interactive applications such as graphical user interfaces, windowing systems, etc. The emerging DMP applications and multimedia applications make it necessary to devise high-performance event handling mechanisms [14]. These applications usually produce large volumes of event traffic continuously, such as satellite telemetry processing systems, video conferencing, real-time data analysis systems, and others.
OMG defines a set of standard event service interfaces for supporting decoupled, asynchronous communication between objects [1]. It supports both *pull* and *push* event delivery models. In the *pull* model a consumer of events requests event data from a supplier while the *push* model allows a supplier to initiate the transfer. To allow multiple suppliers to communicate with multiple consumers OMG event services have also defined *EventChannels* as standard objects. In figure 1 an *EventChannel* has two administrative objects, *ConsumerAdmin* and *SupplierAdmin.*

The former is a factory for adding consumers while the latter is for adding suppliers. To add a consumer, which is traditionally an event registration process, the consumer has to obtain a proxy supplier from *ConsumerAdmin*. It then connects to this proxy supplier. The consumer and the supplier can use either the *pull* model or the *push* model. In this way, new suppliers and consumers can be dynamically attached to an *EventChannel*. Although the OMG event services do not specify the quality of service (QoS) of those *EventChannels* we can support different QoS requirements and extend basic event services by sub-classing the basic *EventChannel* class. Various filtering functionalities can also be introduced by chaining several *EventChannels* together.

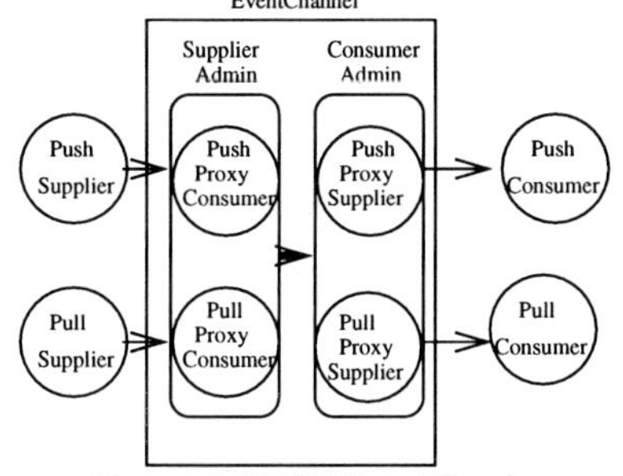

Figure 1. OMG Event Service

3 Architectural Design

In the traditional event-based model, the time between generating an event and consuming this event is quite long. We usually consider this type of event model as a pure asynchronous model. However for high-performance applications, like multimedia applications, time constraints become a major factor in getting a satisfactory outcome. Although these types of applications can still be implemented in an asynchronous way to achieve a great-degree flexibility, timing factors have to be considered in order to guarantee their QoS requirement. Therefore, we call this type of event services *timed event services*. The reason why we believe it is better to implement the communication of those application in the asynchronous way is because it can:

- enable dynamic configuration and easy extension of these systems
- fit naturally into the underlying network communication model, e.g. group communication
- provide a generic and consistent event-based framework to integrate diverse applications

The essential idea of our design is to introduce temporal factors into the traditional event model by integrating many low level system services, such as light-weight thread scheduling, fast cross domain data transfer and hardware multicast. Meanwhile, a runtime mechanism is provided to customize the system adaptively by selecting the most appropriate event service according to the static and dynamic application behaviours.

The key in our approach to building high-performance event services is to exploit parallelism among services. Differing from most other designs, a pipe-based event processing model is adopted. Event consumers and event suppliers, even EventChannels in our design, can hence all run in parallel.

Because of the similarity between event services and communication networks, we borrow many ideas from research in the field of network communication when designing fast transport facilities for delivering and dispatching events. There are two major types of *EventChannels*: inter-process and inter-host. For inter-process *EventChannels* the main issue is how to provide fast data movement across protection boundaries. Recent advances in operating system research and development makes it possible to devise real time event delivery mechanisms. New OS facilities, like shared memory, user-level threads and schedulers, support high-performance event services for applications running as different processes.
As to inter-host *EventChannels* underlying communication network facilities are fully utilized to enable fast data transfer. For many shared medium LAN technologies, such as Ethernet or Token Ring, the hardware supported multicast or broadcast capability can be used as an efficient way to supply event data to multiple consumers. In a high-speed network like ATM, these *EventChannels* can be constructed using the special point-to-multipoint connections. To process events timely, every event that subscribes to a *Timed EventChannel* is associated with a time stamp, and a deadline-based event dispatching mechanism [11] is used in these *Timed EventChannels* to ensure that event data is processed within the time constraint. Basically, the time stamp indicates the deadline before which this event needs to be delivered. The algorithm will assign high priority and resources including CPU time, memory and network bandwidth to those events with closest deadlines.
The increased concurrency may cause high contention for system resources. Therefore it becomes critical that resources are managed dynamically and adaptively such that both high performance for certain resource intensive applications and high utilization of system resources can be achieved. In our design this task can be carried out by a *performability manager* [9] which can reconfigure the system at runtime by monitoring application resource access pattern and availability of system resources. Besides managing resources, its functionality can be extended to support failure recovery, future extensions, and automatic tailoring to underlying hardware platforms. All these features can be provided with no system down time which is essential to many mission critical applications like banking and stock exchange packages.

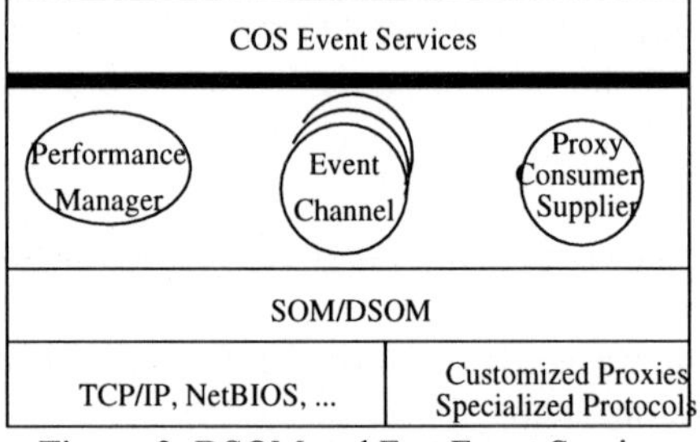

Figure 2. DSOM and Fast Event Service

4 Implementation

Based on the design described in previous sections, we implemented a prototype system in IBM's Distributed System Object Model (DSOM). Figure 2 illustrates the relationship between DSOM and our extended OMG event service. The most distinctive feature here is that we implemented a specialized event dispatching protocol by customizing proxy objects of event service, which will be described in detail in the rest of this section. The idea is essentially to set up a region of shared memory as an EventChannel among event suppliers and event consumers. Therefore the implementation of the event service is totally distributed among applications in the form of libraries, without running any centralized event servers. The general structure of this customized proxy based implementation is shown in figure 3.

4.1 Distributed Proxies

As shown in figure 3 *ConsumerAdmin* and *SupplierAdmin* are customized using DSOM's proxy object interfaces so that their corresponding proxy objects reside in client spaces and implement a specialized protocol for share-memory-based event dispatching. One interesting feature of this protocol is that it enables a fully distributed implementation of OMG event services. For example, when a supplier registers to an *EventChannel* it obtains a *proxy consumer*, which is in the same domain as the supplier, from the *SupplierAdmin*. There is no centralized server running in the system, so the event service can be easily scaled and reconfigured.

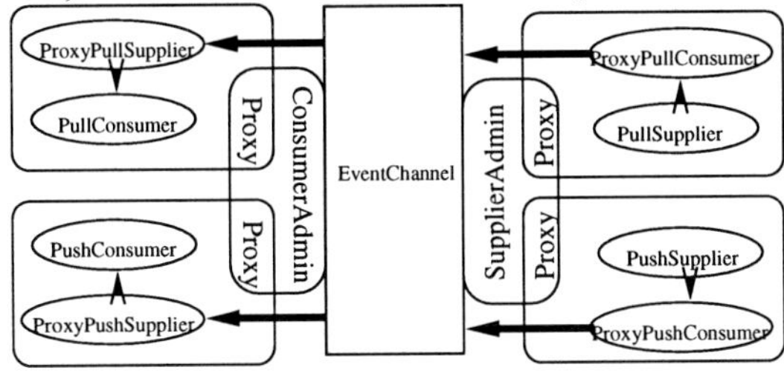

Figure 3. Design

However, when obtaining *proxy consumers* or *proxy suppliers*, we require a synchronization mechanism to coordinate the access to the shared *EventChannel*. In our implementation, semaphores are employed. An *EventChannel* is also a CORBA object, so we can use other services, like name service, to locate an existing *EventChannel*. Since the current DSOM has no such service available, finding existing EventChannels has to be done through some external mechanism in our implementation. For example, a file containing externalized object references, or a well-known server managing a directory of existing EventChannels can be used.
The interfaces between *EventChannels* and user-supplied *EventSuppliers* or *EventConsumers* are realised by *proxy suppliers* and *proxy consumers*. Depending on the event model they use, there are four different kinds of proxy objects: *ProxyPullSupplier*, *ProxyPushSupplier*, *ProxyPullConsumer* & *ProxyPushConsumer*. DSOM allows users to customize object proxies[1] by overriding system-provided ones. For example our implementation customizes the transport protocols used by *EventChannel* by re-implementing *EventChannel__Proxy*. *EventChannel__Proxy* inherits *EventChannel*'s interface.

4.2 Sample Control Flow

To better understand how those proxy objects described in the previous section can be customized to support real-time multimedia streams, we illustrate a sample control flow as following:

[1] The proxy objects, like *proxy consumers$* and *proxy suppliers$*, described before are actually DSOM objects. Here the object proxies refer to object stubs linked to user programs so that they can access the real objects independent of their locations.

1. Create an *EventChannel* or find an existing one:
   When a new *EventChannel* is created, a shared memory region is created. Meanwhile semaphores are also created for this shared memory to synchronize its access. All these are done through overriding *somDefaultInit* methods of *EventChannel* and its proxy *EventChannel__Proxy*. For the sake of simplicity and performance, the shared memory region is mapped to the same address across all domains using it. We implement each *EventChannel* as a separate UNIX process running in parallel with other DSOM server objects.
2. getting *ConsumerAdmin* or *SupplierAdmin*
   Since the proxies for these administrative objects are customized, user-supplied object proxies will be dynamically loaded in when object references for them are returned. If they are for newly created *EventChannel* they will be created and associated with the new *EventChannel*. But for existing *EventChannels* some external sources, such as a file containing object references or a name server, will be consulted to locate them.
3. Get *Proxy Consumers* or *Proxy Suppliers*
   With customized *ConsumerAdmin* and *SupplierAdmin* these proxy objects will be created locally.
4. Connect client-side consumers and suppliers with the proxy objects
   If the proxy object is *ProxyPullConsumer*, it will end up in an infinite loop to pull event from client-provided *PullSupplier*. It is the same situation with *ProxyPushSupplier*.
5. Pull or Push Events
   On the suppliers side, *ProxyPullConsumer* keeps pulling event data from the pull supplier it connected with. Then it stamps it with the desired deadline for delivery and puts them into the shared EventChannel. On the other hand *ProxyPushConsumer* is a passive object. It will be given event data whenever the connected push supplier has generated an event. Then it also stamps it and puts the data into the shared EventChannel. Meanwhile, on the consumers side, *ProxyPushSupplier* is in an infinite loop of pushing event data to its connected push consumer. *ProxyPullSupplier* is also a passive object which will provide event data out of the shared EventChannel whenever it is invoked by client-side consumers. The EventChannel will dispatching these events based their deadlines.

We can easily tailor the above flow into many existing multimedia application scenarios. For example to develop a video on demand (VOD) system, we could model the video service provider as a *PushSupplier* and the service users as *PushConsumers*. So a service user can send a request to the service provider to start a video. The EventChannel will deliver the video clips within the desired time period to assure real-time performance. The advantage of modelling it in this way is that it fits with most exiting VOD communication model. Therefore little changes need to be made to the existing software.

4.3 Future Work

When EventChannel gets (provides) event data from (to) user-supplied suppliers (consumers), it can do either memory mapping or data copying. For large quantity events, it is faster to map event data to and from the EventChannel. However, for events with small amounts of information, it is more efficient to copy the data directly, since the overhead of memory mapping is relatively larger for small data

exchange between processes. Therefore we will devise an adaptive mechanism to automatically choose the right *EventChannel* for different event types.
Because the limitation of the current implementation platforms, the overhead of context switching is quite high due to the increased number of processes. Without life cycle service support, in the current demo sample supplier and consumer run as parent and child processes so that they can share the same *EventChannel*. We plan to port the prototype to other platforms with better system support. We are interested in conducting performance studies on new generation operating system support [15] for high performance event services.
The current implementation of *EventChannel* can only support one event dispatching scheme. However, we plan to use a generic event container which will allow different strategies to coexist e.g. priority-based and deadline-based scheduling policies. In [9] Leonard Franken et. al. presented an integrated approach to model and evaluate various functional and quantitative aspects of distributed systems. In future we will construct a performability manager for our system based on their approach so that it can dynamically reconfigure the system to guarantee user-requested QoS even in a failure-prone environment.

5 Related Work

Recently, much work has been done on developing a unified event-based framework for software integration [5, 12, 3]. For example implicit invocation [10] is a programming abstraction for event-based object invocation. It includes event declaration, event binding, event naming and event dispatching. Based on the same framework, the group invocation commonly seen in DMP applications and video conferencing can be naturally modelled as software multicast [4]. In those studies, it has also been pointed out that many add on functionalities can be easily encapsulated within *EventChannels*. For example, when too many event consumers request the same type of event data, replication can be made to reduce memory and network bandwidth contention. Events can also be logged by making them persistent. Many other event processing abilities like filtering, transforming and even complicated statistical analysis and reporting can be added without affecting existing clients.
The *fbuf* mechanism discussed in [7] provided a novel OS facility to transfer and manage buffers cross domains efficiently. *fbuf* only supports static configuration of data flow. The operations considered in *fbuf* are mainly segmentation and reassembly while event services will support event filtering and message transformation in addition to these operations. Although it is mainly targeted at supporting high-speed networks, its idea of early demultiplexing can also be applied in designing other communication facilities like EventChannels.
In [14] Douglas Schmidt presented an object-oriented framework for high-performance event filtering based on CORBA. He extended CORBA IDL to enable the declaration of filter expressions so that event filters can be automatically generated and optimized. Although we believe extending CORBA IDL will cause interoperability problem, the optimization techniques used in his work can be employed to reduce the filtering overhead.
In recent literature, there are some efforts on software architecture for adaptively and dynamically binding and configuring multimedia components [16, 8, 2]. However none of the work has addressed the issue of providing QoS in the context of standard event services. Because of the general RPC approach most of them had adopted, their systems could not support the loose integration enabled by event-based systems. Thus less flexibility and extensibility could be achieved.

6 Conclusion

In this paper we extended OMG event services for integrating large distributed applications especially those resource-intensive applications like multimedia applications in a CORBA-compliant environment. A new type event service called *timed event service* is proposed and deployed in constructing high-performance event services. We implemented a prototype system in IBM's DSOM using a shared memory based event dispatching mechanism. The overall system structure is fully distributed and highly scaleable. Therefore it provides both flexibility and efficiency to the resource intensive and mission critical applications.

References

[1] Common object services specification, volume i., Technical Report OMG Document Number 94-1-1, Object Management Group, 1994.

[2] Shailendra K., Bhonsle Aurel A., Lazar and Koon Seng Lim., A binding architecture for multimedia networks., In *Proceedings of the International COST 237 Workshop Multimedia Transport and Teleservices*, pages 24--33, Vienna, Austria, November 1994.

[3] Brian Oki, Manfred Pfluegl, Alex Siegel, and Dale Skeen., The information bus - an architecture for extensible distributed systems., *Proceedings of the Conference on Object-Oriented Programming Systems, Languages and Applications*, pgs 58-68, December 1993.

[4] Chen Chen, Elizabeth L. White and James M. Purtilo., A packager for multicast software in distributed systems., 1993.

[5] Daniel J. Barrett, Lori A. Clarke, and Peri L. Tarr., An event-based software integration framework., Technical Report Version 3.0, Laboratory for Advanced Software Engineering Research, Computer Science Department, Uni. of Massachusetts, May 1995.

[6] Digital Equipment Corporation, Hewlett-Packard Company, HyperDesk Corporation, NCR Corporation, Object Design, Inc., and SunSoft, Inc., The common object request broker: Architecture and specification., Technical Report Revision 1.2, Object Management Group and X/Open, 1993.

[7] Druschel and L. L. Peterson., Fbufs: A high-bandwidth cross domain transfer facility., In *Fourteenth ACM Symposium on Operating Systems Principles*, pages 189--202, Dec 1993.

[8] Posnak, H. M. Vin, and R. G. Lavender., Presentation processing support for adaptive multimedia applications., *Proceeding of Multimedia Computing and Networking*, Jan1996.

[9] Leonard J. N. Franken and Boudewijn R. Haverkort., The performability manager., *IEEE Network*, 8(1):24--32, January/February 1994.

[10] Govindan and D.P. Anderson., Scheduling and ipc mechanisms for continuous media., In *Proc. Thirteenth ACM Symposium on Operating Systems Principles*, pages 68--80, California, USA, October 1991.

[11] David Garlan and Curtis Scott., Adding implicit invocation to traditional programming languages, *Proceedings of the 15th International Conf. on Software Engineering*, 1993.

[12] James M. Purtilo., The polylith software bus., Technical Report UMCP-TR-2469, Computer Science Department and Institute for Advanced Computer Studies, University of Maryland, College Park, 1990.

[13] Douglas C. Schmidt., Reactor: An Object Behavioral Pattern for Concurrent Event, Demultiplexing and Event Handler Dispatching., August 1994.

[14] Douglas C. Schmidt., Scalable high-performance event filtering for dynamic multi-point applications., In *the 1st International Workshop on High Performance Protocol Architectures*, pages 1--8, Sophia Antipolis, France, December 1994.

[15] See-Mong Tan and Roy H. Campbell., μChoices: An Object-Oriented Multimedia Operating System., In *Fifth Workshop on Hot Topics in Operating Systems*, Orcas Island, Washington, May 1995.

[16] Tin Qian, See-Mong Tan, and Roy Campbell., An integrated architecture for open distributed multimedia computing., In *the proceedings of the IEEE Workshop on Multimedia Software Development*, March 1996.

STDL as a High-Level Interoperability Concept for Distributed Transaction Processing Systems

Eric Newcomer
Digital Equipment Corp
Systems Engineering
110 Spit Brook Road ZKO1-3
Nashua, NH 03062, USA
newcomer@miasys.enet.dec.com

Hartmut Vogler, Thomas Kunkelmann
Darmstadt University of Technology, ITO
Alexanderstr. 10
D-64283 Darmstadt
{vogler, kunkel}@ito.th-darmstadt.de

Malik Saheb
INRIA Rocquencourt BP 105
F-7831 Le Chesnay Cedex
* ENST, 46 rue Barrault
F-75634 Paris Cedex 13
Malik.Saheb@inria.fr

The Structured Transaction Definition Language (STDL) is a language-based programming interface to transactional protocols and runtime systems. STDL isolates within the language transaction processing features, allowing an implementation to hide underlying communications mechanisms, like TxRPC and CORBA/OTS, from the programmer. Because of its design centre in distributed processing, STDL already includes many features of object-oriented systems. Completing the transformation to an object-oriented language provides a migration path from procedure-oriented TP to object-oriented TP and simplifies the substitution of object-oriented communication managers.

This paper describes how the concept of STDL is used in the ACTranS project for portability and interoperability of heterogeneous Distributed Transaction Processing Systems, including the support of object-oriented TP systems.

Keywords: Distributed Transaction Processing, STDL, OTS, X/Open DTP, Interoperability, Portability

1 Introduction

The *Structured Transaction Definition Language* (STDL) [BGW93], recently specified by X/Open as its high-level language [XOP96b] for *Distributed Transaction Processing* (DTP) [XOP96a], is a procedure-oriented language designed specifically for distributed transaction processing, and to resolve the industry problem of incompatible TP monitor programming interfaces. STDL provides communication subsystem independence for transactional applications and includes a foundation on which to build a migration path from the procedure-oriented transaction processing environments of today to the object-oriented transaction processing environments of the future.

In the context of the *ACTS* (Advanced Communication Technology and Service) research programme of the European Commission, the *ACTranS project* (A Transaction Processing Toolkit for ACTS, AC081) [Vog96] demonstrates the interoperability of different DTP systems in heterogeneous environments. ACTranS has achieved an interoperability of the two DTP standards of X/Open and the OMG by using a half bridge. To realize a global view on the different DTP concepts for the development of transactional applications, we use STDL as a high-level portability concept in the ACTranS project. For the end user of the ACTranS toolkit it is transparent if he develops transactional applications either for a single DTP standard, or beyond the boundaries of different standards, as the underlying protocol differences are hidden by the language.

This paper describes the possibilities for increasing the level of transforming STDL into an object-oriented language and the integration into an object-oriented communication platform. Section 2 gives an overview over STDL, in section 3 the usage of STDL in an object-oriented transaction processing environment is described.

The sections 4 and 5 describe the interoperability of the ACTranS project and how STDL can be used in the project to achieve an overall interoperability concept. The migration path of STDL into an object-oriented description language, suitable for OTS applications, is shown in section 6. Section 7 concludes this paper.

2 The Structured Transaction Definition Language

STDL was specified to resolve the incompatibility of vendor-specific TP monitor programming interfaces. Using STDL for the development of TP applications for multiple platforms allows programmers to concentrate on business solutions rather on the complex notation of programming interfaces. The specification of STDL has been implemented independently by different vendors and publicly demonstrated at Telecom '95 [Tele95]. Figure 1 demonstrates how STDL integrates TP monitors of different vendors, based on different operating systems.

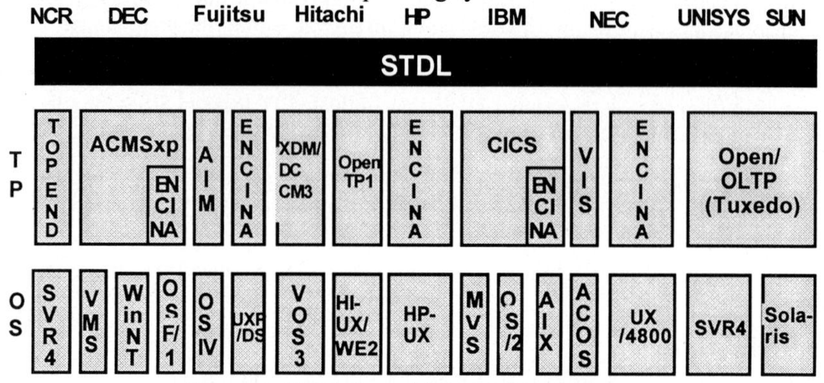

Figure 1: STDL for multiple vendors platforms

The STDL language achieves multi-platform portability and modularity by encapsulating TP functionality not available in the standard C, COBOL or SQL languages. The TP functionality interfaces (e.g. transaction control, exception handling, and communications management) are incorporated within STDL syntax. STDL applications consist of *STDL tasks*, which demarcate a transactional context, and from where calls to standard C or COBOL functions with embedded SQL commands can be made. STDL task procedures are also callable from C and COBOL procedures.

Figure 2 illustrates the STDL three-group model, in which application procedures are arranged according to the type of work they perform: user or device access (*presentation procedures*), transactional flow control and error handling (*STDL task procedures*), and data access (*processing procedures*). A separate interface definition is created for each group of procedures and one procedure calls another via this interface definition. STDL does not define a protocol for the procedure calls except for remote task calls, which use the X/Open TxRPC [XOP95] protocol. In addition, the remote procedure call (RPC) of the *Distributed Computing Environment* (DCE) [Sch93] can be used to call an STDL task from a client external to the TP system or for remote non-transactional task calls. The focus of the STDL design is the concept of the RPC. Consequently STDL includes an interface definition language (IDL) that is used to create interface definitions separately from the procedures themselves. The STDL IDL is called a *task group specification.* This IDL represents *encapsulation*, the fundamental attribute on which the integration of STDL and object-oriented technology is based.

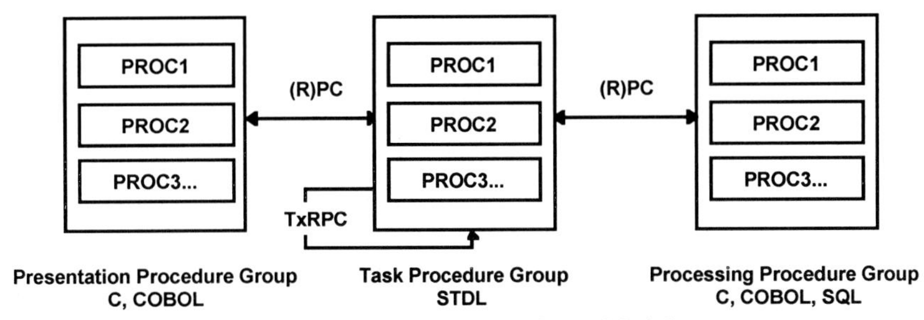

Figure 2: STDL Three-Group Model

The concept of an IDL is also central for both DCE and the *Common Object Request Broker Architecture* (CORBA) [OMG95]. Although the IDLs of CORBA, DCE and STDL differ, simple type mappings are possible. A mapping between DCE and CORBA IDLs is realized in [VoGr95]. The mapping from STDL to X/Open TxRPC and DCE RPC is part of the STDL specification (including data type mappings).
A possible mapping of STDL data types to CORBA IDL data types is shown in the following table:

STDL Data Type	CORBA IDL Data Type
TEXT CHARACTER SET SIMPLE LATIN SIZE	char[x]
TEXT CHARACTER SET SIMPLE LATIN-1 SIZE	char[x]
TEXT CHARACTER SET SIMPLE LATIN-2 SIZE	octet[x]
TEXT CHARACTER SET KATAKANA	octet[x]
TEXT CHARACTER SET KANJI	octet[x]
INTEGER	long
DECIMAL STRING	char[x+1]
OCTET	octet
ARRAY	type id[n]
ARRAY DEPENDING ON	sequence
RECORD	struct

The major problem for a real interworking is the migration from a procedure-oriented system like DCE towards an object-oriented approach, which is characteristic to CORBA environments. Here STDL can help because it provides a *language-based* approach to TP instead of a *system-service* approach. The language-based approach has the advantage of allowing a language compiler or precompiler to be introduced as an intermediate step in creating a TP application (the system service approach relies instead on calls embedded within another language). STDL compiler options are one way to change underlying communications managers without impacting application source code. Language extensions for complete OO features can be added in an upwardly-compatible way, providing a smooth migration for application programs.
Today, sending a message to invoke a method within an object can be seen as calling a procedure within a procedure group. An STDL task group is essentially the same as a network object, and the tasks within it are essentially the same as methods.

3 Using STDL for Object-Oriented TP

Figure 3 illustrates a first step in the migration path towards object-oriented TP from current STDL, which is an integration of the two worlds between a CORBA client and an STDL server. The proxy accepts a call from the CORBA client as a CORBA Server and repackages the message into a TxRPC to invoke the task procedure in the STDL server, now acting as a STDL client. In this case, the CORBA domain is accessing STDL as an external client. The implementation of this proxy can be realized in different ways, e.g. as a gateway or as a bridge. An example for a gateway is Digital's implementation between *ObjectBroker* (CORBA-compliant) and *ACMSxp* (STDL-compliant TP monitor) [BCDW95].

Figure 3: Integration of STDL with CORBA Client

The result is an object-oriented front end to a TP system. Once the call reaches the STDL server, a transaction is started and data from that point onward in the execution of the request is transactional. From the point of view of the client, the object-oriented interface encapsulates the STDL task within an object invocation, providing location transparency for the server.

In the second step of the migration path from STDL to object-oriented TP the STDL client stub accepts the CORBA protocol directly, eliminating the need for a proxy. In STDL terms, this implies specifying the language bindings for the additional *Internet Inter-ORB Protocol* (IIOP) and recompiling the application to generate the stubs for a new communications manager.

The net effect from the application point of view is identical to the first approach; the STDL task is still encapsulated as a network object and invoked as an object method. The advantage of mapping the protocol directly to STDL is that one communications hop is eliminated, most likely providing better performance.

The final migration step is to transform STDL itself into an object-oriented language, analogous to the way that C was transformed to C++ or COBOL extended with object-oriented capabilities. Starting with the foundation of the existing IDL, additional object-oriented aspects can be considered as extensions, enhancements, or modifications to STDL itself. Among these are stronger data typing, classes, inheritance, and adding C++ procedures/classes for the processing and presentation functions. STDL can also be extended to generate CORBA IDL in place of the DCE IDL that some STDL vendors already generate.

4 The Interoperability concept of ACTranS

Although STDL provides a framework at the programming level, the problem of incompatibilities among various communication protocols is still not resolved. To get from closed transaction systems to an open transaction processing environment some new and existing concepts for the interoperability must be realized. An important issue is to close the gaps in different specifications for the sake of a useful and practicable realization. The ACTranS project [Vog96] closes this gap by developing new tools for the interworking between different specifications.

Today's available transaction systems are based on the procedural interfaces of the X/Open standard and the OSI TP protocol [ISO92]. Due to their complexity and their

widely dissemination they can only incrementally be converted to use object-oriented systems, like the *Object Transaction Service* (OTS) [OMG94] of the OMG for CORBA. For this reason some systems for the interoperability between OMG OTS and X/Open DTP domains must be developed [KVT96].
Consider the fact that the interoperability of DTP and OTS implicitly contains an interoperability between DCE and CORBA [YaVo96]. Such an interoperability between heterogeneous environments can normally be achieved by a *gateway* or a *half bridge* [SUZ96]. For the interoperability of OTS and X/Open the half bridge concept seems to be meaningful, because the TxRPC *Communication Resource Manager* (CRM) in the X/Open model is very well suited for the half bridge kernel.

- In the **client-domain** the TxRPC CRM allows a DCE client to initiate a transaction, to call a TxRPC and to end a transaction with commit or rollback. By means of the TxRPC the TM in the client domain can propagate a commit or rollback to the subordinator domain.
- In the **server-domain** the TxRPC CRM enables a DCE server to receive a TxRPC. By means of the TxRPC the TM in the server domain can receive a commit or rollback from the superior domain and acknowledge it.

In Figure 4 the half bridge concept used in the ACTranS project is illustrated. The implementation is based on the X/Open compliant CRM of the ACTranS toolkit [ACT96].

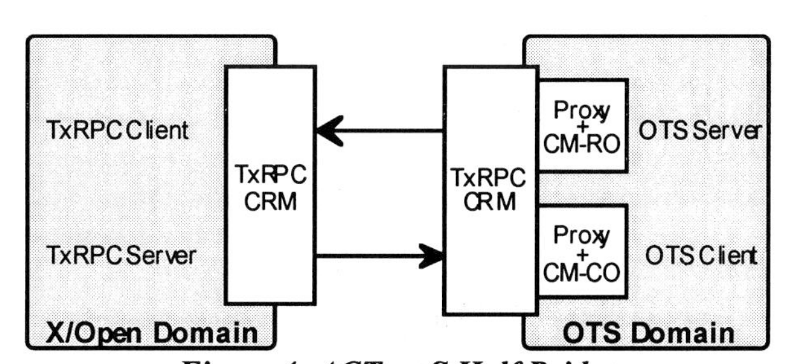

Figure 4: ACTranS Half Bridge

In [YaVo96] the interoperability between DCE and CORBA is described, using different bridging concepts (static, on-demand and dynamic bridge). For the interoperability between distributed TP systems in ACTranS only a static bridge is planned for the present due to the fact that no dynamic behaviour is expected, and because performance aspects must be considered. For the conversion of the different DCE IDL and CORBA IDL a translation like in [VoGr95] is sufficient.
As the TxRPC CRM forms the kernel of the half bridge, no changes at the X/Open side are required. Considering the integration into the OTS domain we have to distinguish whether the OTS domain or the X/Open domain is the initiator of the transaction. For this reason we have to introduce two new modules into the OTS system for the two different directions, a *C-bridge* and an *S-bridge*.

- Coming from the OTS domain and calling a server in the X/Open , the C-bridge performs as a proxy, which acts both as a recoverable object and as a TxRPC client. Besides this the module contains a *communication manager resource object* (CM-RO), which provides a resource object interface to the superior OTS coordinator.

- Coming from an X/Open and client calling a recoverable object in the OTS domain, the OTS-S bridge performs as a proxy, acting as both a TxRPC server and an OTS transactional client, and a component called the *communication manager subordinator coordinator object* (CMSub-CO), which address the subordinate OTS by the interposition mechanism.

Both C-bridge and S-bridge communicate with the X/Open domain using DCE and OSI TP.

A more detailed description of this half-bridge concept for the interoperability between OMG OTS and X/Open DTP can be found in [KVT96].

5 Using STDL in the ACTranS Project

With the half bridge described above the ACTranS toolkit achieves an interoperability between the procedural X/Open DTP system and the object-oriented OTS. The current solution is rather a patchwork than a global concept for a distributed TP environment. For ACTranS a higher-level framework is missing. Introducing STDL will bring this framework into the project.
The current specification of STDL is based on the usage in the X/Open DTP domain. As already mentioned, several implementations of STDL exist for X/Open DTP. At the moment there is no mapping of STDL to OTS available. With the migration of STDL to object-oriented TP described in the second section this mapping can be achieved. For the ACTranS toolkit it is necessary to start with the same STDL description files, which have to be mapped to an TxRPC IDL file and an OMG IDL file. Those IDL files can be used to build a client/server application in the X/Open domain and in the OTS domain as well as for an inter-domain application.
Fig. 5 shows how STDL can be used to build a transactional application where the client is located in the OTS domain, and the server resides in the X/Open domain. The scenario in the other direction, with the client in the X/Open domain and the server in the OTS domain, can be constructed in an analogous way.

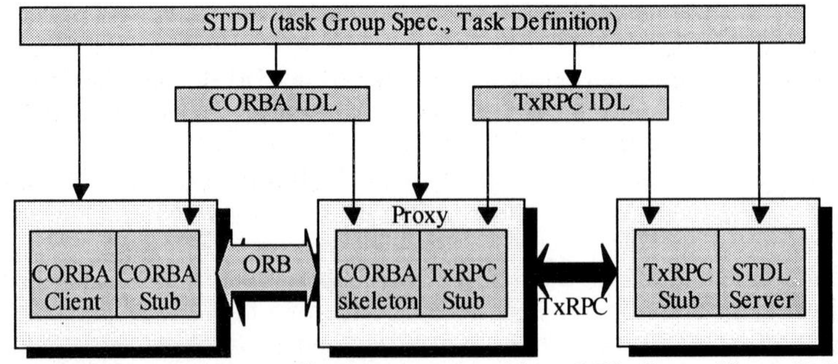

Fig. 5: Using STDL to build an application scenario with an OTS client calling a TxRPC server

Based on the *task group specification* in the STDL file a CORBA IDL and a TxRPC IDL file are generated. The mapping of a *task group* onto TxRPC IDL is part of the STDL specification. For the mapping onto CORBA IDL a *task group* is translated into two CORBA interfaces, one for the non-transactional service, which corresponds to the non-composable task, and one for the transactional service, which corresponds

to the composable task of STDL. With these IDL files the respective skeleton and stub routines for the client, the proxy and the server can be generated.
Besides the different IDL files, the STDL compiler generates C files for the task implementation.. The C-files generated from the *task definitions* are used for building the server application. A detailed description of this step can be found in [BCDW95]. Additionally, the STDL compiler generates C code for the task processing [XOP96b], which can be used for the client and the proxy implementation.

6 Transforming STDL into OO STDL

An original design goal of STDL was compatibility with distributed processing environments. The language assumes therefore that communications protocols are available on the implementation platform and that any given implementation includes support for distributed application development. Another original design goal was modularity – in a procedure-oriented environment the ability to call other procedures and accept calls from other procedures, both local and remote, again transparently to the program source code.
These original design goals of STDL result in the foundation for object-oriented extensions. This section first explains the existing foundation for OO features, and then briefly describes three of the most important transformations needed to complete the OO TP support. Because STDL contains no callable services, and is entirely a language-based approach
The STDL *task group,* because of its separate *task group specification* (STDL's version of an IDL), is similar to a C++ class definition or Java class definition. Within the task group, an STDL *task* is similar to a C++ method. STDL includes a data structure called a *shared workspace*, which allows data sharing among multiple tasks in a task group. The STDL shared workspace is similar to a static, private C++ member variable, and maintains the shared state in STDL applications.
In addition, some limited amount of polymorphism is supported within the language because different task groups can be given the same task group name. Task group names are additionally qualified by their network destination names, allowing the possibility of different task groups containing tasks with the same name.
A typical implementation of STDL includes the ability to generate communications stubs (client and server) based on some form of an IDL. Within the STDL source code translation operation (compiler or interpreter), the STDL specification language is mapped to the IDL associated with the communications manager at hand, for example DCE RPC or OMG IIOP.
Among the main additional features needed to complete the transformation of STDL into OO STDL are:

- Inheritance
- Full polymorphism
- Dynamic objects

These are described in the following subsections. These descriptions are indicate the major areas of STDL OO transformation for STDL and what the changes may look like. A draft specification of the additional features required for an OO STDL should be available later this year. This specification will define upward-compatible extensions to the current STDL specification.

6.1 Inheritance

Inheritance is the ability of an object to inherit the interface of another object. This allows a programmer to specify common attributes and functionality for multiple objects in one place.
The STDL syntax for task group specifications can be extended to allow the use of one task group in the definition of another. For example, the following is possible STDL syntax for expressing inheritance:

```
TASK GROUP SPECIFICATION object1
    . . .
END TASK GROUP;

TASK GROUP SPECIFICATION object2 INHERITS object1
    . . .
END TASK GROUP;
```

Inheritance would be used when a programmer wanted to create a second group of tasks with a similar interface to the first group of tasks. Multiple task groups could therefore be defined from a single common source. This allows better specification of common functionality.

6.2 Full Polymorphism

Polymorphism is the ability to use the same method name with different argument lists. Full polymorphism can be added to STDL by allowing the use of the same task name more than once within a task group, and allowing the argument lists for those tasks to differ. For example, the following is possible syntax for expressing full polymorphism:

```
TASK GROUP SPECIFICATION object1
            . . .
    TASK task1 USING arg1 PASSED AS INPUT;
    TASK task1 USING arg2 PASSED AS INPUT;
END TASK GROUP:
```

The selection of which method (task) to execute can be based on the data type of the input data.

6.3 Dynamic Objects

As mentioned previously, STDL's shared workspace feature provides shared object context. Currently, STDL task groups implement static objects with a single instance, which is fine for many applications. Certain applications require more dynamic capabilities to keep multiple copies of some state (objects), for example managing the steps in a purchase order.
Dynamic objects can be added to STDL by specifying which shared workspaces are dynamic instead of static, by providing object construction and destruction mechanisms, and by adding object references to the specification.

7 Concluding Remarks

Due to the fact that STDL has been successfully layered on top of existing TP monitors, STDL represents a proven path forward to object-orientation, based on fundamental compatibilities such as the use of an IDL for network transparency and RPC protocols. STDL therefore represents the quickest possible way for users to begin to realize the benefits of object-oriented technology while retaining the essential

features of current procedure-oriented TP technology and moving towards object-oriented TP in a sensible series of steps.
Making STDL available for ACTranS offers a high-level, proven interoperability and portability concept for distributed TP, easily inserting new communications protocols underneath the application, for example CORBA's IIOP. This global view of STDL is independent of the underlying communication mechanism, so a user can describe a transactional application with one single concept by using only one description language.

List of Acronyms

ACTranS A Transaction Processing Toolkit for ACTS
ACTS Advanced Communication Technology and Service
CORBA Common Object Request Broker Architecture
CRM Communication Resource Manager
DCE Distributed Computing Environment
DTP Distributed Transaction Processing
IDL Interface Definition Language
IIOP Internet Inter-ORB Protocol
OMG Object Management Group
OTS Object Transaction Service
RPC Remote Procedure Call
STDL Structured Transaction Definition Language
TP Transaction Processing
TxRPC Transactional Remote Procedure Call

References

[ACT96] J. Liang, S. Sedillot, B. Traverson, H. Lejeune, G. Vandome, S. Thomas, EEC ACTS ACTranS: Interoperability OTS-TxRPC, specification, Deliverable D2e, 1996

[BCDW95] R. Baafe, J. Carrie, W. Drury, O. Wiesler: ACMSxp Open Distributed Transaction Processing, Digital Technical Journal, Vol. 7 No. 1, 1995

[BGW93] P. Bernstein, P. Gyllstrom, T. Wimberg: STDL - Portable Language for Transaction Processing, Proceedings of the Nineteenth International Conference on Very Large Databases, Dublin, Ireland, 1993.

[ISO92] International Organization for Standardization: OSI TP Model/Service, April 1992

[KVT96] T. Kunkelmann, H. Vogler, S. Thomas: Interoperability of Distributed Transaction Processing Systems, Proc. Int. Workshop on Trends in Distributed Systems, Aachen, Springer Verlag LNCS 1161, Oct. 1996

[OMG95] Object Management Group: The Object Request Broker: Architecture and Specification, Revision 2.0, 1995

[OMG94] Object Management Group: Object Transaction Service, 1994

[Sch93] A. Schill: DCE - Das OSF Distributed Computing Environment, Springer Verlag, 1993

[SUZ96] M. Steinder, A. Uszok, K. Zielinski: A Framework for Inter-ORB Request Level Bridge Construction; Proceedings of the ICDP'96 -IFIP/IEEE International Conference on Distributed Platforms, Dresden, Feb. 1996

[Tele95] Telecom ´95: http//www3.itu.ch/TELECOM/wt95/

[VoGr95] A. Vogel, B. Grey: Translating DCE IDL in OMG IDL and vice versa; Technical Report 22, CRC for Distributed Systems Technology, Brisbane, 1995

[Vog96] F. Vogt: Werkzeuge für die Transaktionsverarbeitung heute - morgen; Proceedings of the 19th European Congress Fair of Technical Communication - ONLINE'96, Congress VI, Hamburg, Feb. 1996

[XOP95] X/Open CAE Specification : Distributed Transaction Processing: The TxRPC Specification, X/Open Company Ltd., November 1995

[XOP96a] X/Open Guide: Distributed Transaction Processing: Reference Model, Version 3, X/Open Company Ltd., 1996

[XOP96b] X/Open CAE Specification: Structured Transaction Definition Language (STDL), X/Open Company Ltd., 1996

[YaVo96] Z. Yang, A. Vogel: Achieving Interoperability between CORBA and DCE Applications Using Bridges; Proceedings of the ICDP'96 - IFIP/IEEE International Conference on Distributed Platforms, Dresden, February 1996

JAE: A Multi-Agent System with Internet Services Access

Anthony Sang-Bum PARK, Axel KÜPPER, Stefan LEUKER
ap@i4.informatik.rwth-aachen.de
Department of Computer Science 4 (Communication Systems)
Aachen University of Technology • 52056 Aachen • Germany

Mobile agents offer unique opportunities for structuring and implementing open distributed systems. This new paradigm is an alternative to Remote Procedure Calls used in client/server systems. Consequently a wide range of applications are perfectly suitable to be implemented through mobile agent technology, e.g. electronic commerce. This paper introduces the Java Agent Environment - JAE, a multi-agent system comprised of mobile agents communicating with a fixed infrastructure of agent servers making services provided through the Internet accessible. An important constituent of this architecture is the exploitation of existing trader systems.

Key words: mobile agent, multi-agent system, trader system, electronic service markets

1 Introduction

Mobile agents offer unique opportunities for structuring and implementing open distributed systems. Consequently a wide range of applications are perfectly suitable to be implemented through mobile agent technology, including electronic commerce, group work and work flow management [13]. This paper introduces the Java Agent Environment (JAE), a multi-agent system comprised of mobile agents communicating with a fixed infrastructure of agent servers making services provided through the Internet accessible. Such an environment offers a higher degree of flexibility and efficiency to the user by performing much of the work on the server. In this context an agent is understood as a program that takes over some/most of the users', applications' or even other agents' tasks. If necessary, agents communicate with each other, and/or the environment.

First implementations of systems supporting mobile agents already exist and are partly commercially available, e.g. General Magic's Magic Cap [4] and Tabriz [3]. France Telecom is going to test a Magic Cap agents based network for the general public. Magic Cap is intended for the use as a platform for Personal Digital Assistant (PDA) and desktop communication systems respectively, whereas Tabriz is for creating electronic market place applications based on the World Wide Web (WWW). However, use of commercial products restricts the open access to the agent network and, particularly, interferes with the - highly desirable - use of services available on the Internet. The Aglets Workbench [5] is another research project at IBM labs Japan and also realizes agents that move from one browser to another by means of a special Agent Transfer Protocol (ATP). Using Java as the implementation language and SUN's Remote Method Invocation (RMI) library [6], it gains a much wider acceptance.

The agent environment JAE currently designed and implemented at RWTH uses Java as the implementation language and for the agent programming. It is system independent since no native code libraries are used and provides an easy **go** concept like Telescript [1,2]. A major feature however is the way services are offered in the engine and on the network. Long-standing research at our department on Trader concepts for distributed services access helps us with integrating mobile agents into distributed systems according to the *Reference Model of Open Distributed Processing* (OPD) or the *Common Object Request Broker Architecture* (CORBA). Today's electronic service markets are determined by an increasing number of services for storing, processing, and transferring of data. These markets depend on a common

technical platform that works mostly according to the client/server principle. The most popular platforms that follow this principle are the World Wide Web (WWW) and distributed systems according to the Reference Model of ODP or CORBA. Results of the research in this area and parts of the Java Agent Environment JAE will flow into the current Middleware of the ACTS project OnTheMove [7].
The following section presents a survey of electronic service markets and trading principles. Based on the assumption that the current WWW and distributed system architectures grow together and form an unlimited electronic service market with a variety of chances, the deployment of traders for finding suitable services within the arising infrastructure is described. The third chapter describes the Java agent environment giving a system overview with detailed description of the agent server architecture and the agent programming concepts.

2 Trading in Electronic Service Markets

The term service has a wide range of meanings depending on the context in which it is used. For commercial electronic markets, services offered in CORBA-based distributed systems and in the wide-spread WWW are becoming more and more significant. Distributed systems provide a type concept that arranges services into a type hierarchy. A service is an instance of a service type. Each service type is associated with a computational interface that defines the signature of a set of operations. This set again has to be offered by all services of that type, so services in distributed systems have a *type specific interface*. The operations are invoked by a client application at the interface of a server that processes the request and returns a result to the client after completion. The type specific interface makes the development of dedicated software and their distribution to all potential clients necessary. Consequently, this procedure is an enormous disadvantage of distributed systems due to the dynamic character of electronic service markets and the arising costs for software maintenance.
In contrast to distributed systems, conventional Internet applications like the WWW have no type concept. All services have a rather static character and are available in the format of the HyperText Markup Language (HTML). All HTML documents are arranged within the file system of a WWW server and can be accessed through one general interface called HyperText Transfer Protocol (HTTP), no matter of which kind or content the service is. That means no dedicated software, as it is necessary in distributed systems, is required for accessing WWW services. Each service can be invoked from a single application that is mostly given by a WWW browser and its plug-ins. However, services with general interfaces provide merely a straitened functionality.
To overcome disadvantages as mentioned above, new CORBA implementations for the Internet programming language Java were developed. They enable the offering of portable client software in the WWW according to the CORBA architecture. In this way, clients are able to load dedicated software of a chosen service type just in time and so benefit from powerful applications that exceed the possibilities of static HTML documents. As a consequence, distributed systems and well-established Internet applications grow together and form a powerful, nearly unlimited electronic service market with a wide range of available services.
To locate services within electronic service markets and to get a survey of the available variety of services, a user needs powerful mechanisms that exceed today's directory services and searching tools deployed in the WWW. Distributed systems

supply a more powerful solution for finding suitable services by using so-called traders. A trader acts as a kind of mediator that matches user requirements against the offered range of services. The matching bases upon service properties that describe essential features of a service. The set of properties associated with a service is determined by its service type. Service properties can be subdivided into static and dynamic properties. Static properties change rarely and infrequently during the life of service. The values of static properties are asserted by a server and stored within the trader's database. In contrast, dynamic properties are obtained on demand by the trader [9]. In an electronic service market, properties are very important, because service providers competing among themselves advertise qualitative features of their services using service properties.

The set of properties together with a service type and an interface reference form a service offer which represents an abstract description of a service. In [10] and [11] interfaces for the trading of services are standardised. These interfaces provide operations for registering and requesting service offers as well as joining several traders to a trader federation. Initially, traders have been planned for the deployment in distributed systems. However, our approach aims at the establishment of traders in distributed systems and the WWW simultaneously and so constitutes the consequent continuation of integrating CORBA technologies.

As conventional distributed systems categorize services according to their type specific interface, the missing of these interfaces in WWW services forces new criteria for classifying them. In our approach, a set of comparable WWW services forms a particular service type. Comparability is an important term in this context and means that common service property types must be supported. Two WWW services are of the same type if and only if all features of them are described by equal service property types. Therefore, each service type is allocated to a static set of service property types. By adding further property types to a known service type, a subtype is established. The arising subtype has a more specific meaning than its supertype. The hierarchy originating this way represents a service type hierarchy in the WWW. Each collection of HTML based documents that constitutes a service is arranged within such a hierarchy. Taking this concept into account, a service offer can be established by a Uniform Resource Locator (URL) which represents the interface reference, a service type and a set of service properties.

Figure 1 shows the relationship between the type InformationService and the type News. InformationService is a general service type which constitutes properties that have to be contained in every kind of service providing up-to-date information. News extends this set of properties with those needed to describe certain features of a news service. Furthermore, it is possible that a computational interface specification belongs to these type specifications, provided that ODP or CORBA service instances exist. Due to the context of the WWW, this specification is not necessarily required here. On the right side of Figure 1, a service offer based on News is shown. It is specified by the service provider using the trader's export operation. A client which is looking for a service of a chosen type may specify its requirements using the query operation, e.g.

query (News, Costs==0 **and** Language==english **and**
Location==World **and** Headline≈'Austrian Chancellor',
Max(LastUpdate));

This term describes the search for a free service of type news which provides the latest information about the Austrian Chancellor in English.

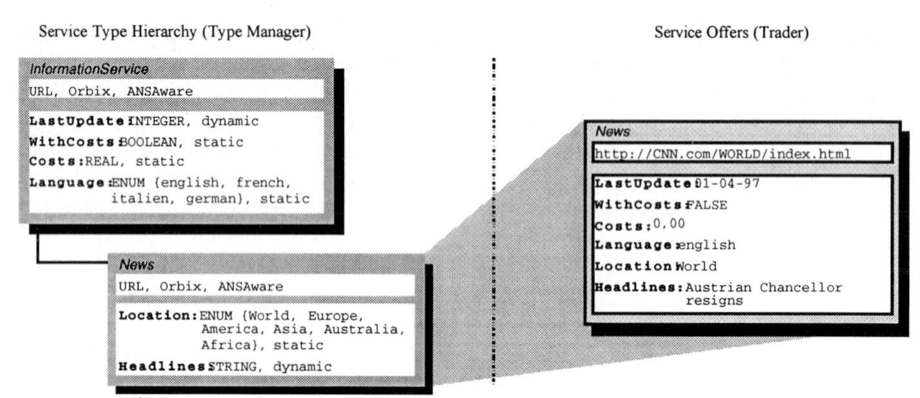

Figure 1: Service type and service offer specifications

Recently, we realized a trading approach for the WWW by providing the user with trading applets written in Java with support of the CORBA-implementation OrbixWeb [8]. Initially, the process of trading worked completely transparent. The invocation of query or export operations is handled by the type specific software without involving the user. However, our trading applets enable synchronous interactions between user and trading system. This system is not only limited to the trading of services, but also provides possibilities for keeping the user informed about various service types. Furthermore, policies for the parametrization of the trading process can be defined. Within this paper, we go a step further and present an architecture for asynchronous access to trading systems via mobile agents.

3 JAE - The Java Agent Environment

JAE is implemented as a Java library package, which consists of the following standard components: the engine, the agent pool, the *ServiceCenter*, security mechanisms and the agent transport facility. The functionality of these components is embedded in Java classes, and agents are programmed by subclassing the *Agent* class. The main part of the environment is the *Engine* class, which is started as a resident process and represents the agent system on a host. It takes care of the network communication, maintains a connection to the service providers on this host, and handles the incoming and outgoing agents.

Within JAE agents are not allowed to directly access the network or even remote hosts or services. Once an agent enters the agent pool, it can only process its internal data or call services controlled by the engine. However, fundamental to agents are the services that agents can rely on while they are migrating through the agent network. It is very important to mobile agents to know where to find services that help to complete their tasks. Therefore services are always looked up and called through the API of the *ServiceCenter* and *Engine* objects. The ServiceCenter takes care of any registration of agents and additional services connected to trader systems, while the engine allows for the actual call to a service.
The Java Agent Environment is a foundation for developing and deploying mobile agent based applications. It contains the core technology for security, services and mobility, and allows the easy implementation of user created agents. Before we will discuss the implementation and reusable parts of JAE however, here are some words on the concept of JAE.

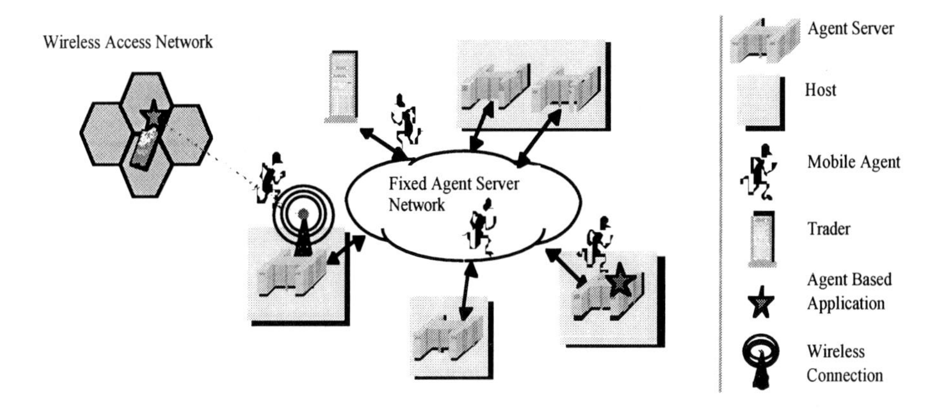

Figure 2: Components of the Multi-Agent Environment

Figure 2 shows the components that multi-agent environments are built of, both the wireless access network and the fixed network, and the physical machines hosting the agent environment. Agents are created and sent out by agent based applications to perform tasks, moving to agent servers on different machines and finding service providers that assist them. The actual deployment of the agent to an agent server by the application is an act of remote execution, whereas the migration of agents is code mobility. Remote execution of agents is particularly interesting in wireless access networks, where a permanent connection is not guaranteed and only short connections to the wired networks is desired due to high link costs. Agents that return to such mobile terminals not always connected to the fixed network must be able to send short messages e.g. via short message service (SMS) of GSM to inform the user to download the mobile agent. Therefore, hosts supporting SMS and the JAE agent server need so called local service agents as mediator of the external SMS service and the mobile agents. Then, mobile agents are able to use this external service through the ServiceCenter and the service agents.

3.1 Agent Server Architecture

The most central concept of an agent environment is the agent server, which is started as a resistant process and represents the agent environment on a host. As Figure 2 above depicts, there can be more than one agent server running on the same host simultaneously that is usually not necessary.

Figure 3 shows a single agent server that takes care of the network communication, maintains a connection to the services on the host, and handles the incoming and outgoing mobile agents. As mentioned before JAE differentiates two kinds of agents: the mobile agents which can enter the engine at any time and leave it after performing some operation, and service agents that stay resident on the engine to provide the mobile agents with services.

As the place where agents come to live and execute commands, the engine is the most critical part of the system. On the one hand, agents should not be deployed in a foreign system without a mechanism to protect the supposedly sensitive data (or program logic) it is carrying with it. On the other hand, a malicious agent may try to harm the agent server, other agents, or it's hosting system [12]. To prevent at least the latter, the JAE engine includes a security manager which controls all injurious operations as manipulating files or unrestricted network access. For a mobile agent, only actions that effect its own data and calls to special functions of the agent server and the service center are permitted. Non mobile service agents on the agent server may also access local properties of the host system, but the network access is not allowed.

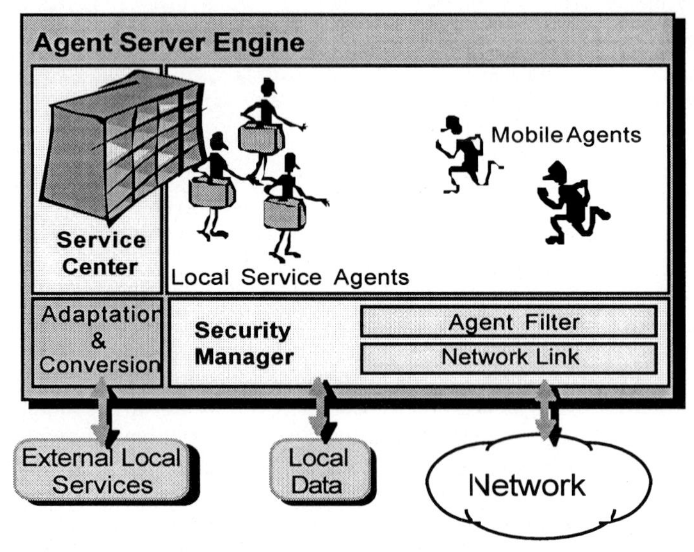

Figure 3: Agent Server Architecture

As for the security of the agent and its data, the agent server must be trusted, because there is no prevention for the access of the server to the agent in the current conception. However, agents cannot read or write each others' properties. In most cases, an agent does not even know who else is on the agent server. The inter agent communication is still very important. Thus, the concept we regard is to install local service agents offering exactly this kind of work, a mediator responsible only for the inter agent communication. One question still remains is, how these services can be obtained. Since mobile agents are not able to communicate with the network, nor can they access local properties of the host system, they totally rely on services that the agent server offers. A common way in agent environments is to provide service information — what is available and where to find it — in the agent program code. This approach is used by General Magic with Telescript [10, 11], but represents only a proprietary solution because Telescript applications are limited to local or metropolitan networks with restricted services only. It is far more desirable to have the agent look for a service when it is needed, so the latest available offers can be used.

Since trader systems can assist with the search for service information, JAE's agent server provides the *ServiceCenter* concept that communicates with local service agents and trading systems to supply a mobile agent the information it needs. A *ServiceCenter* instance can be accessed by an agent through the agent server, and it has methods to search and update the service directory. For every service offered by a service provider and available in the agent environment, the *ServiceCenter* has a record which lists the features and parameter requirements of that service. This list is similar to the signature of a service in a trader representation, including the engine controlling the service, the service type, the type of returned values, and the parameter specification, but goes beyond that in also being of help in creating the right kind of input for the service. An agent then can invoke the service with the help of this record and the service server that hosts the service.

3.2 Agent Programming Concepts

JAE is implemented as a Java library package, which consists of classes that represent the agent environment components. The table below lists the most important classes in the library and some of their methods.

Engine	Implementation of the agent server, including the security aspects. • **useService** invokes a *ServiceAgent* on the agent server.
Launcher	The basic class for every application that creates and remotely executes agents. It uses remote execution to deploy created agents on a designated agent server.
ServiceCenter	Facility for storing and retrieving *Service* objects. See chapter 3.3 for further information about *ServiceCenter* methods.
PropertiesStreamable	Java interface for the streaming mechanism provided by JAE.
Agent	Abstract class which contains most of the methods for mobility and persistence, but does not perform any task. • **go** lets an agent migrate to a remote agent server • **send** clones an agent and sends the clones to several agent servers
StreamableProperties	Key-value pairs that are streamable.
Ticket	Information about a destination for **go** or **send** invocations.
Service	A StreamableProperties subclass that stores the feature and parameter descriptions of a service. • **ticketForService** returns a Ticket with the destination of the agent server offering the required service
Provider	Generalized Java interface for service providers. The only method is • **use** which takes a *StreamableProperties* list of parameters to the service

Agents are actually programmed by subclassing the *Agent* class. Programmers only have to declare needed instance variables and implement two methods that are abstract in *Agent*, on the one hand the **initialize** method for initialization purposes and on the other hand the agent's program logic named **live**. These methods are written in Java code, making the creation of user agents easy for all programmers with some experience in C++ or other object oriented languages. However, some restrictions exist:

- Programmers must use the JAE storage classes for persistent data in agents, since JAE implements its own, system independent persistence concept. In general, every Class describing objects that are subject to migrate with an agent has to implement the *PropertiesStreamable* interface.
- In the actual prototype local variables are not permitted in the **live** method, because they are located on the Java Virtual Machine stack and could not be properly restored after the migration process. It is however possible to write a calculation method that uses local variables which are called from the **live** method.

- **go** may only be called from the user agent's **live** method. This is to ensure a save state when the agent migrates, allowing an exact code re-entry.
- **send** uses an unusual syntax for the code executed by the cloned agents. The call to the method is not terminated by a semicolon, but is followed by an execution block in curled braces; this is the code that is executed by the clones. A return statement is required in this execution block. **send** returns a list of all returned results.

Since JAE uses Java as its agent language, some security aspects come for free. In Java it is impossible to access memory directly, call methods in objects you permit or load program code that would crash the runtime system. In addition, Java already contains the concept of a security manager to overlook critical operations, which is used in the JAE agent server to prevent agents from direct network access and local file manipulation.

3.3 Agent Services

The JAE library contains a generic service exporter class called *ServiceAgent*, which implements the same underlying behaviour of agents, but does not directly relate to the *Agent* class since it is not mobile but stays resident. All *ServiceAgents* implement the service invocation interface *Provider* that is the common link between all service providing facilities and the engine.

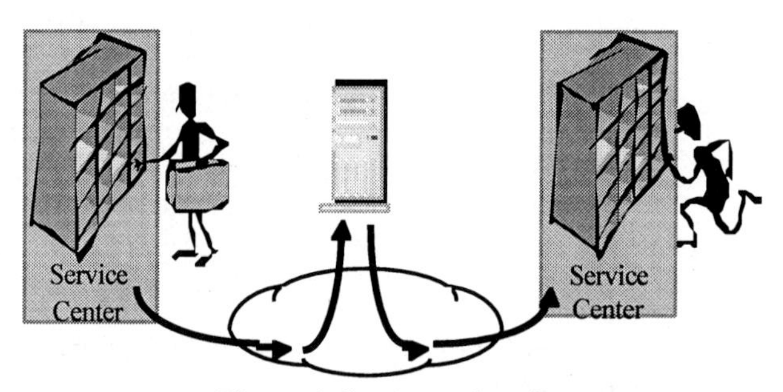

Figure 4: Service registration

When the agent server is started, service agents additional loaded, register their services through the service center by calling the **registerService** method. A Service object given to the service center with this method must be able to create a complete parameter list with default values that allow a successful service invocation. Service agents can either perform the offered task themselves or call an external, but local application, due to network communication is not allowed for services inside the agent server. Mobile agents are also able to register their own services through the same mechanism. Their registration, however, is only valid as long as the agent resists on the same agent server. To provide a registered service, the service agent or mobile agent has to implement the generalized Provider interface that has only one method called **use**. Since agents cannot see each other agents on an agent server however, the actual invocation of a service is done by calling the **useService** method of the *Engine*, which itself knows the right service agent and forwards the parameters to **use**.
Figure 4 shows a service agent registering its service through the local service center, which exports the information to the environment's trading system and on this way to all other service centers. A mobile agent then can access the service information on its current agent server through the local service center. Service providers can cancel the

support for a service by calling the **withdrawService** method signalling that this service will be no longer available.
Mobile agents use the *ServiceCenter* to search for a specified service with the **servicesFor** method, giving a property list of requirements as an argument. These requirements can be the return type or the type of service required. With the **firstServiceFor** method it is also possible to access only one service that matches the requirements. It returns the first service found by the service center. The service center however does not store all the service records itself, but relies on a trader community that stores the services for the whole agent environment.

Figure 5: **Service invocation**

A mobile agent on another agent server can access the *Service* record from its local service center using the lookup methods described above. With the *Service* records returned by **servicesFor** or **firstServiceFor**, the mobile agent can migrate to the engine that offers the required service, as depicted in Figure 5. At the new location the mobile agent creates a default invocation record, modifies it and calls the service with the generalized **useService** method of the engine. The result of the service can be stored and returned to the origin.

4 Conclusion

In this paper we showed how external services are linked into a multi-agent environment. The introduced agent system JAE is currently under development at University of Technology Aachen with demonstrations already running regarding the agent mobility, remote execution of agents and registering services. Recent work includes defining the interfaces for service invocation and agent server communication and the complete implementation of the JAE library. Future work will be done in implementing an agent development environment based on JAE including an agent compiler as well as creating an interface to trading systems that comply with the CORBA standards for distributed systems.

References

[1] James E. White: Telescript Technology: An Introduction to the Language. General Magic White Paper, 1995
[2] James E. White: Telescript Technology: Mobile Agents. General Magic White Paper, 1996
[3] General Magic: Tabriz White Paper. General Magic Inc., 1996
[4] General Magic: Magic Cap Developer Resources, Technical Documentation. General Magic Inc., 1996, http://www.genmagic.com/Develop/MagicCap/Docs/
[5] Danny B. Lange and Daniel T. Chang: IBM Aglets Workbench, Programming Mobile Agents in Java. IBM White Paper, 1996

[6] Java Soft: Java™ Remote Method Invocation Specification. Sun Microsystems Inc., Revision 1.0, Draft Oct. 2, 1996
[7] A. S. Park, J. Meggers: Mobile Middleware: Additional functionalities to cover wireless terminals. 3rd International Workshop on Mobile Multimedia Communications, Princeton, NJ, Sept. 1996
[8] Küpper, A.; Herzog, H.: Deploying Trading Services in WWW-based Electronic Service Markets. (in German). Accepted for KiVS '97, Braunschweig, 1997
[9] Küpper, A.; Popien, C.; Meyer, B.: Service Management using up-to-date quality properties. In: Proceedings of IFIP/IEEE International Conference on Distributed Platforms, Chapman & Hall, Dresden, February 1996
[10] Draft Rec. X.950 | ISO/IEC DIS 13235 - ODP Trading Function, 1996
[11] OMG Document orbos: OMG RFP5 Submission, Trading Object Service, 1996
[12] T. Magedanz, K. Rothermel, S. Krause: Intelligent Agents: An Emerging Technology for Next Generation Telecommunications?, INFOCOM '96, March 24-28, 1996, San Francisco, CA, USA
[13] C.G.Harrison, D.M. Chess, A. Kershenbaum: Mobile Agents: Are they a good idea? IBM T.J. Watson Research Center, NY, 1995

Motivation and Requirements for the AgentSpace: A Framework for Developing Agent Programming Systems

Alberto Silva
Alberto.Silva@inesc.pt
IST/INESC

M. Mira da Silva
mms@dmat.uevora.pt
University of Evora

José Delgado
Jose.Delgado@inesc.pt
IST/INESC

A framework for supporting agent programming systems, called *AgentSpace*, is proposed in this paper. Our goal is to define an abstract and generic world of agents that is a super-set of research and industrial proposals currently being developed. There are mainly two novelties in this paper. First, we argue that real-world complex agent-based applications span through three levels: framework, programming system and application level itself. The second novelty is that, unlike many other proposals, our programming model focus on heterogeneous multi-agent applications.

Keywords: AgentSpace, Agent Systems, Agent-based Applications, Internet

1 Introduction

This paper concentrates on the development of distributed applications based on agents and their support programming systems. A *software agent* [WJ95,Nwa96] is a program that acts on behalf of a user to perform a very specific—usually repetitive and laborious—task. The user is usually a person, such as a manager in a company, that needs to accomplish the task delegated to the agent. Central to the paper is the idea that one agent is already useful in a computer; however, it can only take advantage of all its potential when there are multiple agents co-existing in what we will call *an agent community*.

This community of agents, unimaginable even a few years ago, can now be supported by the Internet. Thus it comes as no surprise that the proliferation of proposals for *agent systems* help application programmers write agent-based applications. Possible targets for these applications can be found in the following areas: personal assistants, workflow and collaborative applications, electronic commerce, and network management systems [MRK96, HCK95]. Many other areas will surely appear if these initial areas see a number of successful agent-based applications being used to solve real-world problems.

However, despite the increasing popularity of agents and their support programming systems, there is a consistent shortage of useful agent applications. For example, in the ECOOP'96 Workshop on Mobile Object Systems [BTV96] there was not a single paper describing such an application. Except in the conclusion, there were also very few discussions, if any, of possible applications based on agents.

This paper starts with a concrete multi-agent application as a motivation exercise and then goes on to propose a complete *agent framework* as a basis to support independent *agent programming systems* that can support this kind of application, as well as to design other agent-based applications (see figure 1).

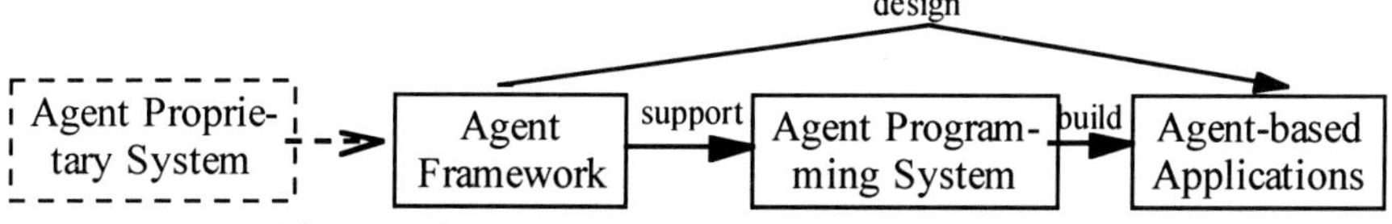

Figure 1: From Framework to Agent-based Applications

The motivation is followed by an overview of *AgentSpace*, our proposed framework for agent programming systems. The goal is to define an abstract and generic world of agents with the following characteristics.

- To provide an environment to discuss agents, agent systems, and applications.
- To propose an analysis and evaluation reference model as a first step towards a future integration and inter-operation between heterogeneous agent systems.
- To support the design of agent-based applications independently of their own proprietary agent systems.

There are mainly two novelties in the paper. First, the proposal goes through all three conceptual levels—framework, programming system and application—contributing towards a new global understanding of the agent research area. The second novelty is that we focus on *heterogeneous multi-agent applications*. Many agents that need to interact bring new issues such as agent communication, external services and resource access that in our opinion have not been sufficiently explored in other research work. Additionally, heterogeneity involves interoperability and integration issues between agents and between their corresponding (proprietary) agent systems.

2 Example Application

Although the current Web technology is already adequate to support the development of many distributed applications, it will not be adequate to support the kind of world-wide application that will be necessary in the future. This problem is illustrated below with a real large-scale distributed application that we are currently developing.

2.1 The Virtual Enterprise Network

The *Virtual Enterprise Network* (VEN) application is an Internet-based distributed information system that will support the cooperation between SME (small- and medium-size enterprises) in Portugal. Namely, VEN will maintain information about each company participating in the project, help their managers collaborate more and better, and in the future support remote training.

One of the services that VEN will provide is to support research projects being pursued by a number of these companies in consortium. They have to find the strengths and weaknesses of other companies, form alliances, write joint research proposals and then develop the project together.

A project is divided into a number of specific tasks. One of the companies is elected as the *coordinator* and becomes responsible for the whole project. The coordinator proposes who does what, negotiates with all *partners* the tasks to be accomplished, makes sure progress is being made, prepares the final demonstration and edits the technical reports. This implies a non-trivial set of tasks and coordination amongst disparate entities during the whole life cycle of the research project.

2.2 VEN Characteristics

The VEN application has a number of characteristics that by themselves have been dealt with independently in the past. It is their combination that poses problems.

- *Autonomous*—Each company creates and maintains their own applications using their own resources. (The last thing managers want is another manager from another company telling them what to do.)
- *Heterogeneous*—Each company has bought, got used to and uses different interfaces, machine architectures, programming languages, database systems,

communication packages, operating systems and so on. They also have different programmers with different backgrounds and levels of expertise.

- *Open*—Some services may depend on other applications and even external organizations, thus the VEN has to inter-operate with other (legacy) information systems: applications, databases and so on.
- *Dynamic*—Applications will be added, updated and removed at any time without previous notice. The VEN applications will have to cope with unavailability, new interfaces, oscillating bandwidths, etc.
- *Robust*—The VEN will have to tolerate different kinds of failures on machines, networks or at any level of software. For example, the application cannot stop executing just because a company is rebooting their gateway to the outside world.
- *Secure*—The system should provide different levels of security depending on each particular part of the whole application. There will be public, VEN-specific and administrative applications and data.

VEN is only an example of a large class of distributed applications that will be developed on the Internet. Cardelli even gave a name to them: *global applications* [Car96]. However, it should be noted that not all of these global applications will be global in the world-wide sense; an application geographically distributed over Portugal may have the same basic characteristics as truly global applications.

3 The Web Approach

The current state-of-the-art of Web information systems—based on server-centric combined with client-centric technologies—may already support the development of some global applications such as VEN. Examples of technologies that could be used for VEN include: HTTP, HTML, CGI, Java, relational databases, ODBC and CORBA—all well-known and tested in the-real world. However, they present some limitations and deficiencies that we have described elsewhere [SBD95,SMdSD97]. In the following section we show briefly why these technologies cannot be used to support VEN.

3.1 VEN as a Web Application

Figure 2 presents a (simplistic version of a) possible *decentralized solution* based on current Web technology. Each organization provides its own Web-based applications, eventually with connections to their own legacy applications.

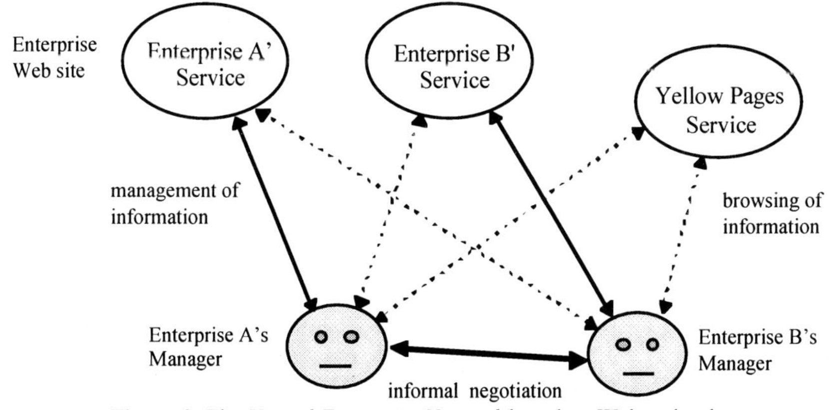

Figure 2: The *Virtual Enterprise Network* based on Web technology

Using the Web, managers are responsible for creating and maintaining their own Web sites with information about their companies' skills, products and services. In addition, if the manager is the coordinator of some project, then he or she has the responsibility of finding out the best set of available partners. In this case, the manager retrieves from the Web information on the best partners available (e.g., using a yellow pages service maintained by somebody else). When a good set of candidates is found, the manager then uses another technology (such as phone, fax, or e-mail) to negotiate directly with the other manager.

Another possible solution requires the existence of one special organization (or a restricted set of organizations) responsible by the development and maintenance of some *centralized service* in a single computer accessible to all managers. In addition to the yellow pages service—that tracks all members belonging to the VEN—this centralized service could provide all the desired activities (such as collaborative project establishment and management, search and retrieval of appropriate partners, and so on). However, a centralized solution restricts the autonomy and flexibility of each enterprise and cannot even use their applications if these already exist.

3.2 Current Web Technology Is Not Enough

Between the two solutions to implement the VEN described above, we still opt for a Web approach, although it presents some drawbacks. The manager must learn to use several applications, with different user-interfaces and eventually different interaction metaphors. Each application has to provide its own security and authentication mechanisms. The manager has to learn how to interact with all of them and keep a private list of authentication codes.

The manager in enterprise A, for example, has to maintain his/her own Web site, deal with B's Web site, search the yellow pages and talk to the manager of enterprise B. All these interactions have a different communication protocol. The problem grows quadratically to the number of companies in the application because all managers have to learn how to deal with all other companies! This is why the managers in figure 2 are not very happy.

Another issue concerns the synchronous mode of modern end-user interaction. Ideally, some tasks could be made asynchronously because of their isolated nature, low bandwidth, high-latency networks or just because they are too complex or too long. An example of such a task would be to match a set of requirements from a manager against a set of skills from all potential partners. A human interaction may take a long time (especially over lunch) and is prone to misunderstanding (especially after lunch).

On the other hand, if an automatic procedure was possible, then the answer could be found in the background by the VEN application itself. The current Web technology does not support this behaviour very well because it is based on human-computer interaction, not computer-computer interaction that is the realm of agents.

4 The Agent Approach

In order to support the development of Internet-based distributed applications like VEN, several proposals in the area of *agents programming systems* have been made. Examples of these systems include: Telescript [Whi94], PageSpace [CTV96], Cardelli's Obliq [Car95a], Agent TCL [Gra95], Sun's Java applets [AG96], Mole [SBH96], IBM's Aglets [IBM96] and Persistent Java [AJDS96,ADJ+96].

4.1 VEN as an Agent Application

Figure 3 depicts (again, a simplistic view of) the VEN application now based on agents. The managers (end-users) are now happy because they only need to understand and interact with a single application with a single interface: the agent. This is even more remarkable because the agent with which they have to interact is their own agent, meaning that it can be configured to a particular user or set of users. For example, we can envisage very patient agents with colorful interfaces for computer illiterate managers, agents with effective (shell-based) interfaces for knowledgeable managers, or even agents that adapt to their managers dynamically (it all depends on the effort put on developing these agents).

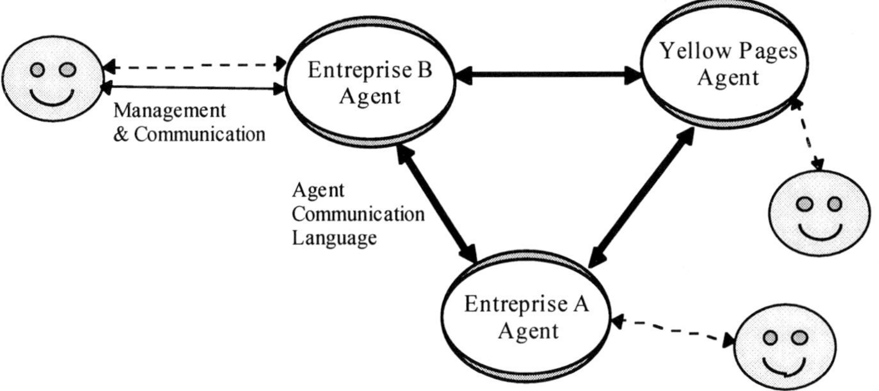

Figure 3: The *Virtual Enterprise Network* based on agents

This novel agent-based solution presents, when compared to the approaches described in the previous section, the following advantages:

- It is *decentralized* because the only centralized point is the yellow pages agent and *scalable* since this congestion point can always be attenuated or even eliminated by the incremental introduction of more yellow pages.
- It is *dynamic* because all agents can, with more or less flexibility, enter or leave the system. Additionally, the functionalities and complexity of the various agents may evolve independently.
- It promotes the *autonomy* and *flexibility* of each enterprise because they now become responsible for the development and maintenance of their own applications and particularly of their own agents.
- It offers a *single interface* because it hides the disparate complexities and user-interfaces of each application. As a consequence, the end-user only needs to interact directly with his/her own customized agent and eventually a few specific (management) agents.

4.2 Preliminary Issues

Although apparently an agent-based approach seems to offer an ideal development environment for VEN, there are some important drawbacks that will be described below.

In dynamic and open environments such as those we propose, *it is difficult to define and promote common agent-based application protocols and APIs.* All or the main involved organizations need to agree on a *common agent protocol* to allow agents to communicate. However, this protocol is specific to this application and so does not

need to comply with any existing standard—usually, a restricted number of pioneer organizations design and agree a common protocol, then all other organizations just accept and adopt it.
Also, due to the non-existence of experience, models, techniques, tools and environments there is nowadays a *great difficulty for designing and building* applications that are really based on agents. The opposite is the database world: their RAD systems make it extremely easy to develop client-server SQL-based applications from scratch. In the agent world we are still many years away from having such tools.
Finally, several research and commercial agent systems are emerging, each with its own *proprietary agent model.* In fact, the majority of proposed agent programming systems are still lacking some technological aspects. For instance, Telescript provides a complex, persistent and secure system. However, its object-oriented (proprietary) programming language is very difficult to learn and use. On the other hand, Tacoma and PageSpace provide high level scripting languages (Tcl and Java respectively) that are more "natural" to learn and to use. However, these two systems do not fully address security and robustness issues when compared with Telescript.
There are several other issues whose merits are not well understood yet. For instance, what really is an agent? How is it created? When and by whom? What behaviour can it present in generic applications? After being created and launched, agents then interact directly with each other on behalf of their managers. However, should the agent that interacts with its end-user be the same agent that also interacts with other agents? When the agent moves to another machine, can its end-user still interact with it?
All these difficulties and requirements would be even worse if the involved agents were based on distinct systems and technologies. Or, in other words, how should we address the heterogeneity issue? One elementary question should be, for instance: can a Telescript agent interact with an Aglet agent? If yes, how and to what extent?

5 Proposed Framework

This section, the core of the paper, gives an overview of *AgentSpace,* our proposed framework for supporting agent programming systems.

5.1 Agents, Nodes and Clusters

The *agent* is the basic entity of the framework. It executes some specific tasks on behalf of someone (a person) or something (an organization, another agent, etc).
The *AgentSpace* is the set of all agents and all *run-time environments*, that is, their corresponding computational (software) infra-structures where they execute (see below).
A *node* is a machine (hardware) infra-structure on which a computational infra-structure can be installed. These days a node is likely to be a computer, but in the near future it will probably include PDAs (now called "hand-held PCs"), TV sets, and mobile phones.
Nodes are logically organized in *clusters*. For example in the Internet context, all computers with an Internet name belonging to the domain "inesc.pt" belong to the same cluster. Clusters form a hierarchy in order to give some organizational and management functionality to the entire agent application. Usually a cluster suggests a geographical proximity, but it does not need to be so. An intranet, for example, can be implemented as a cluster. Depending on the application requirements, there may exist specialized agents for managing nodes and clusters called *system agents* (see below).

5.2 Agent Execution System

The agent run-time environment, provided in every node, is called AES—for *Agent Execution System*[1]. The AES can be built from scratch or, more likely, as a combination of existing hardware infrastructures (e.g., ordinary PCs), operating systems (e.g., Windows or JavaOS), communications packages (e.g., TCP/IP, HTTP or CORBA), and some kind of virtual machine (e.g., Java VM).

The AES provides a full computational environment to execute the agent, as well as other support mechanisms such as agent persistence, security and mobility. In order to support distributed applications and in particular agent mobility, different AES should communicate amongst themselves using some kind of low-level (read simple) protocols and agree in some common agent representation formats (e.g., using well-known marshalling techniques found in RPC systems). The AES should also provide or at least integrate specific APIs to allow access to external services and resources, such as databases, the file system and physical devices.

Furthermore, an AES should provide, in each node, one or more agent *execution places.*[2] These execution places are locations where agents execute and meet other agents. Every place is identified univocally by some electronic address. The implementation details of places vary and depend on each AES[3].

We agree with D. Chess et al. [CGH+95] in that, due to their complexity and inherent distributed characteristics, an AES should be better designed and built in terms of object services and supported by an existing distributed computing infra-structure such as DCOM or CORBA. In this way, several tasks of the AES can be delegated to its subjacent distributed infrastructure instead of "re-inventing the wheel" yet again.

5.3 Agent Communities

The basis for any *agent-based application* being built using our proposed framework is the notion of community. A *community* is formed by a set of agents that share their knowledge and communicate between them using a common language.

Knowledge is a description of some fact, some relationship between facts and/or other relationships in some restricted contexts. The knowledge maintained by one agent can be used by any other agent in the same community. (It is likely that at the implementation level the knowledge of all agents belonging to the same community will be maintained by a database, potentially distributed by a number of nodes.)

Communities are dynamic; new agents can enter the community at any time—if they are allowed by the community—or leave it. There should be protection against infiltration by foreign agents, perhaps based on real-world mechanisms: an identity card will be issued by the community, there will be friend communities, and so on.

A community is a logical concept that can be spread over a number of nodes or clusters. An agent belongs to one or several communities if it is allowed into those communities and can speak their languages. There will be *guest agents* that are allowed only limited access to the community knowledge, probably introduced by

---

1 The AES is usually called agent system (AS). We prefer AES because AS has a too general use and sometimes vague or prone to misunderstandings. Some authors also call the AES an "engine", "agent server", or "agent meeting point".

2 Some authors just call them "places", "locations" or "meeting points".

3 For instance, in Telescript a place is a process which can contain an arbitrary number of other places. This kind of places corresponds to an execution place and a stationary agent (that may provide several services) in our framework. In Mole, the location of a place is based on information of the IP address, port number associated with a Mole's engine, and a serial number.

another agent that is responsible for its behaviour. Other agents could be specialized as *translators* for different languages, *mediators* to resolve conflicts or *police agents* to stop or kill agents with bad behaviour.
Communities may be open or restricted. In the open case, all their agents (except the system agents) have similar characteristics and functionalities, and may work anonymously or not. In the restricted case, there should be different types of agents, identified (i.e., with some special tickets, permits or credentials) with different access levels and respective capabilities and available resources.

5.4 AgentSpace

As described in the previous section, a community is a set of agent-based applications sharing the same context and supported by a common AES. The inverse is not true. This means that the set of all agents supported by the same AES does not define, just by its own, a community.
The AgentSpace notion is an evolution relatively to the community concept in terms of desired capabilities and complexity. It is a dynamic set of agent-based applications related in the same restricted context, but supported by different AES. This means that an AgentSpace needs interaction and communication between heterogeneous agents.
Figure 4 shows the relationships between agent, community and AgentSpace. The agent presents basic capabilities, such as autonomy, persistence, mobility, and communication with its user. A group of homogeneous agents related in a common context (i.e., sharing the same AES) defines a community, which is the second level of the hierarchy.

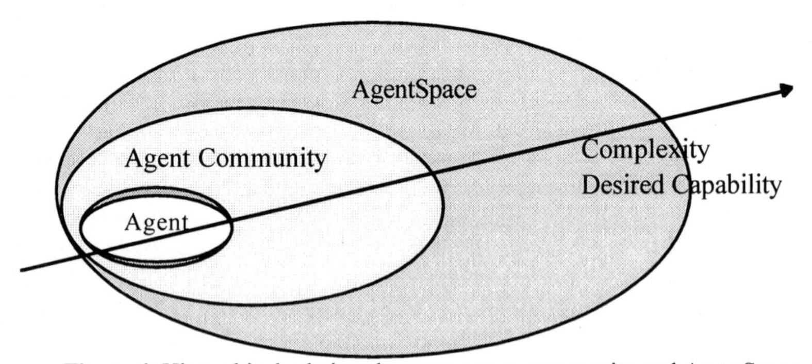

Figure 4: Hierarchical relations between agent, community and AgentSpace.

The community raises two new aspects. The first involves the need for a communication language between homogeneous agents. Basically there are currently two approaches: declarative (e.g., KQML) vs. procedural and/or object-oriented (e.g., Tcl, Telescript, and Java). The second aspect involves how to represent the specific knowledge of the community. There are also two basic approaches: knowledge representation languages (e.g., KIF and EDI) and specific APIs and protocols agreed amongst the principal entities responsible to the development and management of the involved community.
In the third level of the hierarchy, AgentSpace extends the community concept by allowing the communication and interaction between heterogeneous agents. Obviously, agent communication should be independent of any language or AES. In contrast to KQML (the declarative approach) we propose a *common interface language* for interacting between heterogeneous agents. The characteristics found on

CORBA IDL make this protocol a good starting point. A current effort on that direction can be found in [TF96].
Other issues that require in depth investigation concerns the notion of the AgentSpace delimitation. This issue raises several other questions, such as: does it make sense to define an AgentSpace of AgentSpaces? This means, does it make sense to define complex operations, such as aggregation and composition, around the AgentSpace concept? So, in the positive case, we may talk about an *open and universal AgentSpace*, where every agent may interact, subject to necessary restrictions, with any other.

5.5 More on Agents

The agent can exist outside a community but will only progress towards achieving its task after it is accepted in one of the existing communities (or starts its own!). Once accepted by a community, an agent can then interact with other agents; for example, to ask a question, negotiate a deal, provide a service, advertise a service, sell a product, buy raw materials, or simply help other agents achieve their tasks.
Although we don't intend, at this specification level, to detail agent internal implementation aspects, it is important to refer that, they should present a minimum set of well-defined characteristics: *identification* (i.e., name, electronic address and passport information relatively to their associated AgentSpace); *internal state representation* (i.e., code, specific data; common attributes, and execution image); and *external interfaces* (i.e., end-user interfaces, and agent knowledge- and services-based interfaces).
There are two basic kinds of agents: mobile and stationary (or static) agents. In general, *stationary agents* are created in the context of a very specific application at the user's initiative and become attached to that user during a long time period. Since these agents do not move, they do not present security problems to the system. However they should prevent other third agent attacks.
On the other hand, *mobile agents* are usually created by stationary agents and by other mobile agents. They are typically used to solve small and specific tasks and consequently have a short life. For example, the stationary agent of enterprise A, responsible for starting the project, may create a mobile agent to look for three potential partners for a task. This agent is launched by its parent to the yellow pages agent. There, it asks for electronic addresses of possible candidates, then it jumps to the execution places of the different stationary agents, meets and talks to them, and retrieves the information it needs. Finally, it comes back home and gives the information found to its owner (the stationary agent).
Finally, in applications like VEN mobile agents will make intensive use of sophisticated databases and carry data with them [MdSA96,MdS96]. Others have agreed that this integration between code and data will be crucial to the development of agent-based applications [Whi94,BC95,ADJ+96]. The reader is referred to the main issues involved with integrating persistence and mobility in a companion paper [MdSS97].

5.6 End-Users

All *users* of the agent application have at least one agent that executes some tasks on their behalf. Users give their agents specific tasks or brokerage and mediation capabilities. *User agents* are agents owned by end users that behave as "consumers" of services. The information consumed by user agents is "produced" by *application agents*, for example, those that manage databases of products on sale. Both of them are usually stationary agents but may occasionally create, at run-time, mobile agents to execute very specific tasks.

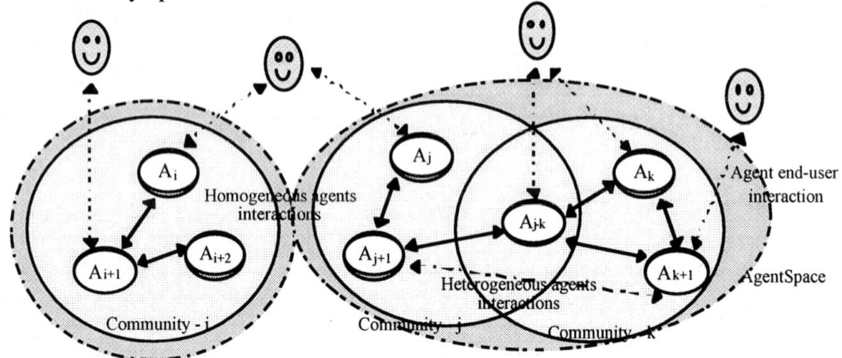

Figure 5: AgentSpaces and agent interaction patterns.

Users may interact with their agents in several ways depending on their interface characteristics. For instance, they may interact through e-mail messages, HTML forms, AWT-base Java applets, Active-X components, etc. (Ideally, this interaction should be voice-based in natural language!). They have all rights over their agents, namely to suspend, change their knowledge and goals, or even eliminate them. Nevertheless, these rights are restricted by the supported AES as well as by the political rules of the involved AgentSpace.

6 Conclusions and Future Work

The paper presented VEN, an example agent-based application, and how it can be implemented both by using the current Web technology and an agent programming system. We concluded that only agents can effectively support VEN, although current agent programming systems lack the functionalities needed to implement VEN.

We then introduced the main concepts and requirements of our framework for developing agent programming systems that will support better communication amongst agents and integration between code and data. These can then be used to implement the kind of application exemplified by VEN.

As part of our future research work we will develop a prototype VEN application using an existing agent programming system (probably the Aglet workbench from IBM or Persistent Java from the University of Glasgow). We will propose an independent agent development model, with which we will discuss two new emerging themes: how to design agent-based applications and what tools and components should agent programming systems provide in order to simplify their development?

References

[ADJ+96] M.P. Atkinson, L. Daynès, M.J. Jordan, T. Printezis and S. Spence. An orthogonally persistent Java. *SIGMOD Record.* 1996.

[AG96] K. Arnold, J. Gosling. *The Java Series - The Java Programming Language.* Addison-Wesley Publishing, 1996.

[AJDS96] M.P. Atkinson, M. Jordan, L. Daynès and S. Spence. Design issues for Persistent Java: A type-safe, object-oriented, orthogonally persistent system. In *Proceedings of The Seventh International Workshop on Persistent Object Systems* (Cape May, New Jersey, USA, May 29-31, 1996). Morgan Kaufmann Publishers, 1996.

[BC95] K. Bharat and L. Cardelli. Migratory applications. *Proceedings of the ACM Symposium on User Interface Software and Technology* 1995 (Pittsburgh, PA, Nov 1995). 1995.

[BTV96] J. Baumann, C. Tschudin and J. Vitek. *Proceedings of the 2nd ECOOP Workshop on Mobile Object Systems* (Linz, Austria, July 8-9, 1996). Dpunkt. 1996.

[Car95] Luca Cardelli. A language with distributed scope. *Computing Systems*. 8(1):27—59. Jan 1995. (A preliminary version appeared in Proceedings of the 22nd ACM Symposium on Principles of Programming Languages.)

[Car96] L. Cardelli. Global Computation. *Position Paper*. 1996.

[CGH+95] D. Chess, B. Grosof, C. Harrison, D. Levine, C. Parris. Itenerant Agents for Mobile Computing. *IEEE Personal Communications*. 2(5):34-49, Oct. 1995.

[CTV96] Paolo Ciancarini, Robert Tolksdorf, Fabio Vitali. PageSpace: An architecture to coordinate distribute applications on the web. *Computer Networks and ISDN Systems*. 28, 941-952, 1996.

[Gra95] R. Gray. *Agent Tcl: a transportable agent system*. Proceedings of the CIKM Workshop on Intelligent Information Agents, (CIKM'95), 1995.

[HCK95] C. Harrison, D. Chess, A. Kershenbaum. *Mobile Agents: Are they a good idea?*. IBM, 1995.

[IBM96] IBM Tokyo Research Laboratory. *Aglets workbench: Programming mobile agents in Java.* http://www.trl.ibm.co.jp/aglets. 1996.

[MdS96] M. Mira da Silva. *Models of higher-order, type-safe, distributed computation over autonomous persistent object stores*. PhD Thesis, University of Glasgow. (Submitted in December 1996.)

[MdSA96] M. Mira da Silva and M. Atkinson. Combining mobile agents with persistent systems: opportunities and challenges. In [*BTV96*].

[MdSS97] M. Mira da Silva and A. Silva Insisting on Persistent Mobile Agent Systems with an Example Application Area. Accepted to the *First International Workshop on Mobile Agents*, 1997.

[MRK96] T. Magedanz, K. Rothermel and S. Krause. Intelligent agents: an emerging technology for next generation telecommunications. *Proceedings of INFOCOM'96*, San Francisco, USA, 1996.

[Nwa96] H. Nwana. Software Agents: An Overview. *Knowledge Engineering Review*. 11(3), 1-40. Cambridge University Press, 1996.

[SBD95] A. Silva, J. Borbinha and J. Delgado. Organizational management system in a heterogeneous environment: a WWW case study. *Proceedings of the IFIP working conference on information systems development for decentralized organizations* (Trondheim, Norway, August 1995). Pages 84—99. 1995.

[SBH96] M. Strasser, J. Baumann and F. Hohl. Mole: A Java-based mobile object system. In [*BTV96*].

[SMdSD97] A. Silva, M. Mira da Silva, J. Delgado. A Survey of Web Information Systems. Submitted to the *WWW'97 Conference*. 1997.

[TF96] C. Tham, B. Friedman. Common Agent Platform Architecture. *General Magic*, Oct. 1996.

[Whi94] James White. Telescript technology: The foundation for the electronic marketplace. *General Magic White Paper*, General Magic, 1994.

[WJ95] M. Wooldridge, N. Jennings. Intelligent Agents: Theory and Practice. *Knowledge Engineering Review*. 10(2), 115-152. Cambridge University Press, 1995.

Intelligent Networks

Bert F. Koch
Siemens AG

This section deals with Intelligent Networks (IN) from various points of view. A lot of literature is available about the details of the basic concepts of IN, e.g. T. Magedanz, R. Popescu-Zeletin: *"Intelligent Networks - Basic Technology, Standards and Evolution"*, International Thomson Computer Press, 1996. For this section we included a couple of interesting papers that discuss different aspects of IN-based services and applications, models and even underlying architectures as well.

IN was first introduced for Narrowband ISDN in order to have powerful and flexible mechanisms to support services over the network. In this section, its role in the evolving new technologies, such as Broadband and Mobile Networks, will be shown from several points of view.

Within the Telecommunications Community there are mainly two "competing" architectures under discussion, which try to achieve the deployment of network services in a comfortable and "intelligent" way both for the user and the network operator or service provider: the so-called "classical" IN architecture and the Telecommunication Information Networking Architecture (TINA).

The paper of Giovanna De Zen et al. *"Proposal for an IN Switching State Model in an Integrated IN/B-ISDN Scenario"* is based on the classical IN architecture, as it is the basis for the ACTS INSIGNIA project. The paper presents a new functional model, designed for the integration of IN and Broadband ISDN (B-ISDN).

This integration is also the major topic of Jørn Johansen's paper *"Harmonisation/Integration of B-ISDN and IN"*, which presents an integrated and harmonized reference architecture for IN and B-ISDN as shown in the EURESCOM project P506.

The rapid changes within the telecommunication market, generated by the increased pressure of regulatory changes and by the rising demands of competition, as well as the requirements of individual users in terms of flexibility, Quality of Service or just availability of the service show the necessity of evolution of IN towards mobile networks. This demand is taken into account within the ACTS project EXODUS, which is represented here with the paper of Lucia Vezzoli et al. *"Intelligent Network Evolution for Supporting Mobility"* and which demonstrates the IN functionality needed to meet UMTS requirements. The paper presents the specified functional and physical architecture and the control model.

A totally different perspective is seen in the next paper from Young B. Choi and Adrian Tang *"A Generic Service Order Handling Interface for the Cooperative Service Providers in the Deregulating and Competitive Telecommunications Environment"*. In order to solve typical problems like customer friendly servicing and comfortable and efficient service handling and resource allocation they suggest an interface for the service order handling to meet the business needs of Service Providers, based on the TINA-C information modelling and computational modelling concepts.

The paper of Ulrich Herzog and Thomas Magedanz *"From IN toward TINA - Potential Migration Steps"* gives an interesting overview of the basic architectural migration from current IN-based service platforms towards TINA-based service environments. It includes also a description of the adaptation unit capabilities required for the most realistic evolution scenario, featuring an IN-based service access part and

a TINArized service control and management part, as well as some general considerations on the underlying SS7 / CORBA interworking.
Finally the paper of Subrata Mazumdar and Nilo Mitra *"ROS-to-CORBA Mappings: First Step towards Intelligent Networking using CORBA"* presents a way of mapping the principle concepts of Remote Operations Service (ROS) to the corresponding CORBA concept. Using the computational model of TINA-C the paper also shows techniques which may support interworking of existing implementations of data services with CORBA-based implementations.

Proposal for an IN Switching State Model in an Integrated IN/B-ISDN Scenario

G. De Zen*, L. Faglia*, H. Hussmann+, A. van der Vekens+
+SIEMENS AG, Public Communication Networks, Germany
*ITALTEL, Central Research Laboratories Settimo Milanese (MI), Italy
Tel: +39.2.4388.9087, Fax: +39.2.4388.7989
E-Mail: giovanna.dezen@italtel.it

1 Introduction

The provision of services that use Intelligent Network (IN) facilities upon the ISDN infrastructure has been successful from the point of view of both users and operators. This approach will also apply for the provisioning of multimedia services in the B-ISDN environment. Consequently, new IN features need to be investigated to handle more complex service scenarios.

Even if the B-ISDN signalling system will provide the user with the capabilities to establish very complex call configurations, it seems very unlikely that such enhanced services could be handled directly from the user application. A more realistic perspective foresees that services will be deployed in the network by dedicated service gateways handling simple service access requests coming from the users.

To achieve this goal two different architectures can be adopted. The first is the classical IN architecture while the second is the Telecommunication Information Networking Architecture (TINA). Ongoing studies in the ACTS (Advanced Communications Technologies and Services) INSIGNIA (IN and B-ISDN Signalling Integration on ATM Platforms) project has focused on the first solution because it does not dramatically change the typical service deployment architecture and therefore it can be realised as a slight evolution from the current IN to an enhanced one.

In order to allow a real and powerful IN control on service deployment, it has been decided to enrich traditional IN capabilities with the possibility for the SCP to manipulate the topology of the network configuration required by the invoked service. For example, this means that the service logic residing on the Service Control Point (SCP) can instruct the Service Switching Point (SSP) in order to set up new calls/connections as needed by the selected service.

A new functional model has been designed for the integration of IN and B-ISDN. It has been assigned a central role to the Service Switching Function (SSF), that links the Service Control Function (SCF) to the Call Control Function (CCF), and in particular to the IN Switching State Model (IN-SSM) offering the network resources view to the service logic.

In this paper the characteristics of this new IN Switching State Model are presented.

This model exploits the concept of session as co-ordination of different calls because it appears very general and powerful. Such a concept is very useful when interworking with a B-ISDN network implementing signalling CS-1 which allows only point-to-point single connection call. If more advanced features are available in the B-ISDN network (e.g. point-to-multipoint, multi-connection), the session concept can slightly reduce its role, as more complex call topologies will be handled at the B-ISDN level (and therefore in the CCF). However, the capability of co-ordinating different calls is always a powerful support for the provisioning of advanced services.

2 SSF model

One of the major differences of Broadband ISDN compared to Narrowband ISDN is the more complex notion of call. In B-ISDN, a call may comprise several connections, and calls and connections may involve more parties than two. In addition, a single service may comprise several B-ISDN calls.

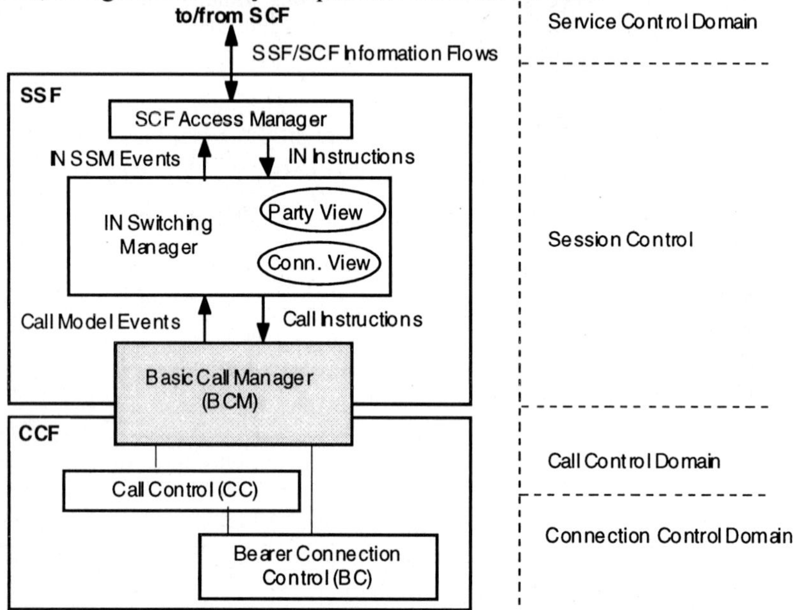

Figure 1: SSF/CCF Functional Model

Four control domains can be identified:

Service Control Domain

The overall control of a service is carried out by the IN Service Control Function (SCF).

Session Control Domain

The term session is used to denote an association of calls and connections for the realisation of a single service. This network view of a service is controlled by the IN Service Switching Function (SSF).

Call Control Domain

The realisation of a B-ISDN call is managed by a component of B-ISDN switching control which is traditionally named Call Control (CC).

Connection Control Domain

The physical switching resources are managed by a component of B-ISDN switching control which is traditionally named Bearer Control or Bearer Connection Control (BC).

Call Control and Bearer Connection Control functionality together form the so-called Call Control Function (CCF).

Figure 1 depicts the control domains and their mapping towards IN functional entities.

The focus of this paper is on the definition of an SSF model independent from the CCF one.

According to the [1], the *SSF* provides the set of functions required for interaction between the CCF and a SCF, see figure 1. In particular the SSF

- extends the logic of the CCF to include recognition of service control triggers and to interact with the SCF;
- manages signalling between the CCF and the SCF;
- modifies call/connection processing functions (in the CCF) as required to process requests for IN provided service usage under the control of the SCF.

This global definition can also be adopted for the Broadband SSF. In the Broadband environment, the need of enhancing the SSF with the capability of handling service sessions has been recognised. This new capability is offered by the IN Switching Manager (SM) who is also in charge of maintaining the view of parties and bearer connections involved in each active session. The SM keeps information on the status of an IN service by means of the IN Switching State Model (SSM) that allows the calls belonging to a session to be modelled using the defined objects. The SM co-ordinates several calls belonging to a session. It interprets the Call Model Events and translates them into SSM events (SSM state changes) to be communicated to the SCF. In the reverse direction, IN instructions coming from the SCF are translated into instructions to the BCM. Furthermore, it instructs the CCF to start call processing according to instructions from the SCF (SCP-initiated calls as a particular case of the capability "third party call set up").

3 IN-SSM model

Figure 2 depicts the graphical representation (using OMT notation) of the new SSM. The names of the associations have to be read from left to right or from top to bottom.

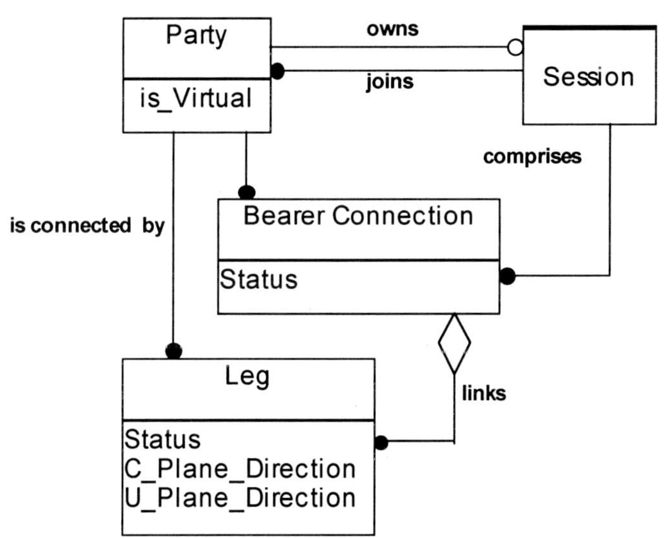

Figure 2: Object model of the IN-Switching State Model

The complete description of the SSM objects follows.

A "Session" is the representation of a complex call configuration, as it is seen by the IN functional entities.

A "Party" can either be an end user or a network component (e.g. the SCP), called a *virtual* party and distinguished by the attribute "Is_Virtual". This allows modelling of SCP-initiated actions, for example connection establishment, connection transfer or connection release. Several parties can join a session but only one party is the session *owner*. A session cannot exist without any party. During a session, new parties can be added to or joined parties can be removed from a session.
Between the parties of a session, a "Bearer Connection" can be established. If B-ISDN Signalling CS-1 is used, there is a one to one relationship between call and bearer connection. Therefore, each bearer connection can represent both a basic CS-1 call and a single connection of a multiconnection CS-2 call.
A bearer connection is composed of at least one "Leg". A leg represents the communication path to a party which is connected to other parties by a bearer connection. The multiplicity of the aggregation relation between leg and bearer connection determines the topology type of the connection. If a bearer connection contains exactly two legs, it is a point-to-point connection. For point-to-multipoint-connection each bearer connection has an association with more than two legs.
Table 1 summarises for each SSM object the attributes that have been identified and their corresponding values.

Table 1: Objects, attributes and values

Object name	Object identifier	Attribute	Attribute identifier	Possible values of attribute
Session	Session ID	None		
Party	Party ID	Virtual Party	Is_Virtual	False True
Bearer Connection	BC ID	Bearer Status	Status	Being setup Setup Being released
Leg	Leg ID	Leg Status	Status	Pending Destined Joined Abandoned Refused
		U-Plane Leg Direction	U_Plane_ Direction	Source Sink Bi-directional
		C-Plane Leg Direction	C_Plane_ Direction	Incoming Outgoing

The attribute "Is_Virtual" indicates whether the respective party is representing a network element (i.e. the SCP, when the attribute values is *True*) or a true party associated with an end system.
The "Status" of bearer connection indicates in which state is the bearer connection object in a given phase: establishment (*Being setup*), active (*Setup*) or releasing (*Being released*).
The "Status" of leg indicates in which state is the associated bearer connection in a given phase: establishment (*Pending*), route selected (*Destined*), active (*Joined*), abandon of calling party during establishment (*Abandon*) or refused by called party (*Refused*).
A relationship has been individuated between the "Status" attributes of the bearer connection and its legs: the "Status" of the bearer connection is *Being setup*, as long as the "Status" attributes of each associated leg is not *Joined*. A bearer connection is *Setup* only if all its legs are *Joined*.

The flow of user plane information can be indicated by the attribute "U_Plane_Direction" of the leg object, which can assume the following values: *Source*, *Sink*, and *Bi-directional*.
The attribute "C_Plane_Direction" is used to indicate the direction of the signalling relationship. The attribute value is *Incoming* for a leg connected to a calling party and *Outgoing* for a leg connected to a called party
Regarding the "Status" attributes of the objects some restrictions apply. The bearer connection "Status" can only change from *Being setup* to *Setup* and from *Setup* to *Being released*. The leg "Status" can only change from *Pending* to *Joined, Destined* or *Refused*, from *Destined* to *Joined* or *Refused* and from *Joined* to *Pending* or *Abandon*.
The SSM view, composed by the SSM objects and their relationship, is communicated from SSP to SCP (or vice versa) by means of Information Flows (IF).

4 CCF-SSF interface

An innovative aspect of this model is that the SCF and the Service Logic do ***not*** have a direct view of the CCF call states (Detection Points of the Basic Call State Model). The communication between SCF and SSF refers to the more abstract notions of sessions, parties, bearer connections and legs as they are represented in the SSM.
However, since the CCF is defined in terms of a BCSM, some translation is required between the abstract, SSM-related view offered to the SCF and the basic view of the CCF. In fact, this translation is carried out by the SM.

5 SSF-SCF interface

As indicated in the previous sections the SCF-SSF communication is based on the SSM knowledge.
In fact, the SCF handles only the objects the SSM model deals with. This means that all operations addressed by SCF to SSF act on these objects and not directly on the BCSM. The SSF will translate the commands received from the SCF in the actions on the appropriate BCSMs.

Table 2: SCF visibility on SSM object relationships

Relationship	Visibility
Join (session-parties)	Implicit
Comprise (session-bearer connections)	Implicit
link (leg - bearer connection)	Explicit
session ownership	No
bearer connection ownership	No
connect (party - leg)	Explicit

The SCF shares with the SSF the same view of SSM except for the ownership relations. The initial SSM state is typically provided by the SSF to the SCF with the IN service invocation by means of structures containing the existing objects and their "link" and "connect" relations (explicit visibility). The other relations can be simply derived by the SCF (implicit visibility). Table 2 summarises the SCF visibility respect to SSM object relations.
The SSM state in the SSF is aligned with the SSM state seen by the SCF. If an event changing the SSM state occurs, a communication flow able to keep the SSM state aligned in the two network elements is needed.

From the SSF point of view, SSM state changes and evolution are driven by events detected at CCF level and operations invoked by SCF. Consequently, IFs at the SSF/SCF interface have to be able to reflect all possible changes inside the SSM.
To ease the definition of the Information Flows, it is needed to identify a limited set of SSM transitions. This set has to contain the transitions meaningful from the service evolution point of view which usually involve operations on a group of related objects.
SSM transitions caused by actions requested by the SCF are in correspondence with SCF invoked operations, while IN-SSM state transitions due to CCF events are reported via SSF report operations.

Table 3: "Creation" Information Flows

Information Flow	Created objects and relationships
Join party to session and link leg to bearer	New party X New leg Y *Pending* leg Y is linked to an existing bearer connection in *Being setup*
Join party and bearer to session	New party X1 New leg Y1 *Pending* New bearer connection Z *Being setup* New leg Y2 (related to an existing party) Legs Y1 and Y2 linked to bearer connection Z
Add party and bearer to session	New party X1 and X2 New legs Y1 and Y2 *Pending* New bearer connection Z *Being setup* Legs Y1 and Y2 linked to bearer connection Z
Add bearer to session	New legs Y1 and Y2 (related to existing parties X1 and X2) *Pending* New bearer connection Z *Being setup* Legs Y1 and Y2 linked to bearer connection Z

Table 3 details the first identified set of "creation" transitions and corresponding IFs, while Table 4 provides the first set of "deletion" transitions. Moreover, Table 5 and 6 provide respectively the "event handling" and "activation" IFs. The terms creation (deletion) means that, whatever the starting SSM state is, the creation (deletion) transition adds (deletes) a fixed set of objects and relationships, reaching the target SSM state. The description of IFs given in the tables refers to a two party scenario, however, the same IFs can be adopted in a multiparty scenario. The introduction of enhanced signalling capabilities could require an enrichment of the SSM model in terms of new attributes or relationships and consequently of the associated Information Elements; it could eventually require an enlargement of the set of IFs without modifying the meaning of those already defined.
Three aspects are relevant:

- In every SSM state an instance of a party can exist if and only if it is connected to at least one leg; the only exception is a Party with Attribute "Is_Virtual" set equal to *True.*
- The identified transitions are all service independent.
- The name of the transition will be used as the name for the IF at SCF/SSF interface.

Table 4: "Deletion" Information Flows

Information Flow	Deleted objects and relationships
Release session	All the objects and related relationships
Drop party	Party X1 All legs Y1,...,Yn connected to X1 All bearer connections Z1,...Z2 that remain linked to only one leg All legs that are no more linked to a bearer connection All parties that are no more connected to legs
Release bearer	Bearer connection Z All legs linked to Z All parties that are no more connected to legs

The “event handling” IFs allow the SCF to request and to be informed about the selected state changes for the monitored SSM objects. For the time being, only the state changes of the “status” attribute of the “bearer” and “leg” objects have demonstrate to be useful in order to control the service evolution.

Table 5: "Event handling" Information Flows

Information Flow	Meaning
Request report SSM change	SSF is requested by the SCF to monitor in the specified monitor mode the indicated state transitions for the selected objects
Report SSM change	SSF reports to the SCF the monitored object and the current state

An IN service can start on the initiative either of an end user who sends a SETUP message containing an IN number or of the SCF, previously instructed to provide the service at a specific time or directly requested by means of the User Service Interaction mechanism. An “activation” Information Flow is exchanged between the two functional entities, SSF and SCF, co-operating in the IN service provision to share the initial SSM view and the other needed information.

Table 6: "Activation" Information Flow

Information Flow	Meaning
Service request	SSF invokes SCF in order to receive instructions on how to offer each specific service. The initial SSM state and the other needed information are provided within this IF
Create new session	SCF requests SSF to create a new service session and provides the initial SSM state

If the transitions from an SSM state are driven by SCF, the SSF creates/deletes objects according to what is contained in parameters of the appropriate IF that must contain the identifiers of the involved objects. In the case of creation transitions, this implies that the identifiers can be assigned either by SSF or SCF.

Uniqueness of identifiers is guaranteed by the following considerations:

- The IDs scope is meaningful only inside a session, that is the same Object ID can be assigned to different sessions.
- When the first trigger is detected, SSF generates the IDs and notifies SCF in the “Service Request” IF. For the remaining time of the session, only SCF creates new objects and assigns IDs.

Regarding the notification of SSM state changes, a request-report and report mechanism, is used, like in IN CS1. The main difference is that the report is related not to BCSM events but to changes of objects' attributes values.
In general, every action on the objects of the SSM model implies a corresponding action on the network resources of CCF. However, it can be useful for some services to terminate the relationship between SSF and SCF without releasing the associated calls/connections. As the "Release session" transition does not cover this case, an IF called "Drop session" has been introduced.

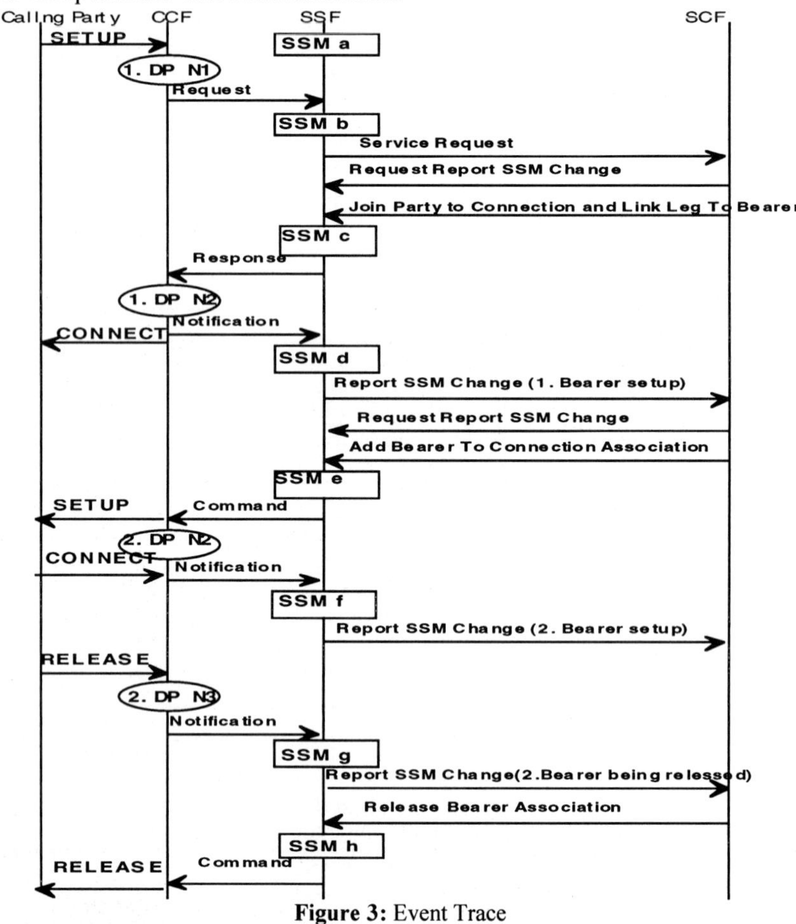

Figure 3: Event Trace

6 Example of dynamic behaviour of SSM

Consider a simple IN service involving a number translation, to connect the calling party to a specific called party, and a bearer connection addition, to establish a second connection.
Figure 3 depicts a possible event trace between the involved functional entities.
The SSF/CCF communication is represented with generic primitives (Request/ Response/Notification/Command) independently from the underlying BCSM.
Concerning the calling party, only the more relevant B-ISDN signalling messages are shown, while concerning the called party they are omitted.

A calling user sends a SETUP message, containing an IN-number. The message is processed by the CCF that, on encountering Trigger Detection Point N1 (the DP numbering is left intentionally generic), sends to the SSF the request to handle the new service session. The SSF consequently invokes the SCF service logic by sending a "Service Request" IF. To keep track of the call and to establish the second call only after the first call successful set up, the SCF requests for the notification of the bearer object state change to *Setup* and *Being released*. This request is transformed by the SSF so that the corresponding detection points N2 and N3 for this first basic call are armed. Then, the network address is provided within the "Join Party To Session And Link Leg To Bearer" IF to the SSF in order to complete the call set up.

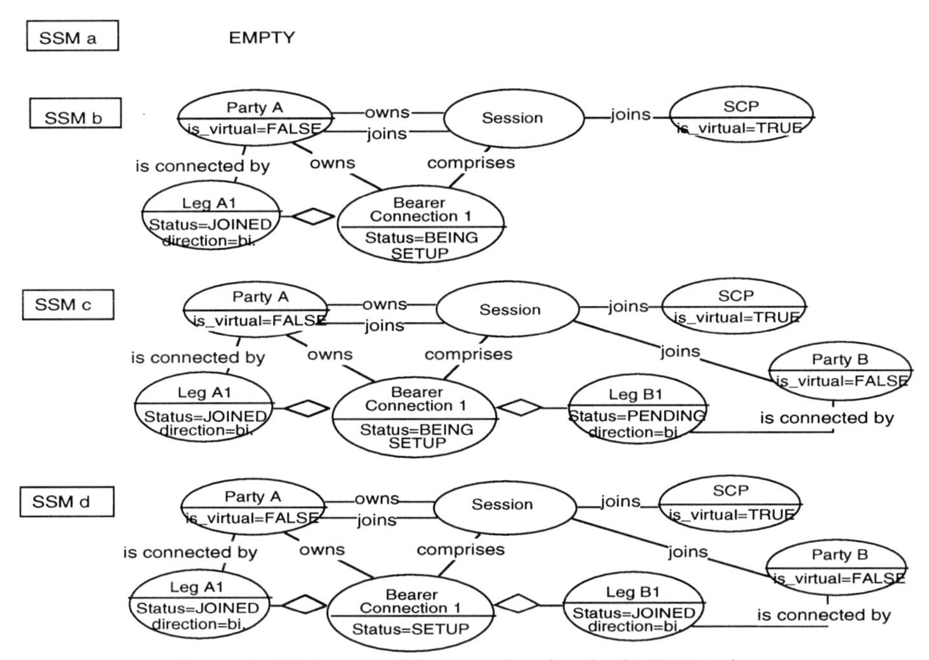

Figure 4: SSM states of the example using the OMT-Notation

The SSF provides to the CCF the information needed to complete the first call set up. The CCF will resume call processing until the call is active. This means that N2 detection point is reached and reported to the SSF within a notification. The SSF translates this event into a "Report SSM Change" IF to the SCF. After receiving the report the SCF can address the establishment of a second bearer. For this call a new BCSM instance is created in the CCF and a new instance of bearer connection with associated leg objects are created in the SSF. It is assumed that the second bearer connection is established successfully. If a call is released by a party, this is reported to the SCF via the SSF. Since, for the service, the second connection is useless without the released connection, the SCF initiates the release of the whole session.
The different SSM states visible in the SSF are influenced by CCF/SSF Primitives (Request and Notifications) and SCF/SSF-IFs. For the example, the different SSM states can be described with an Instance Diagram (provided by the OMT-Methodology) according to the Object Model for the SSM (see Figure 4 and 5).

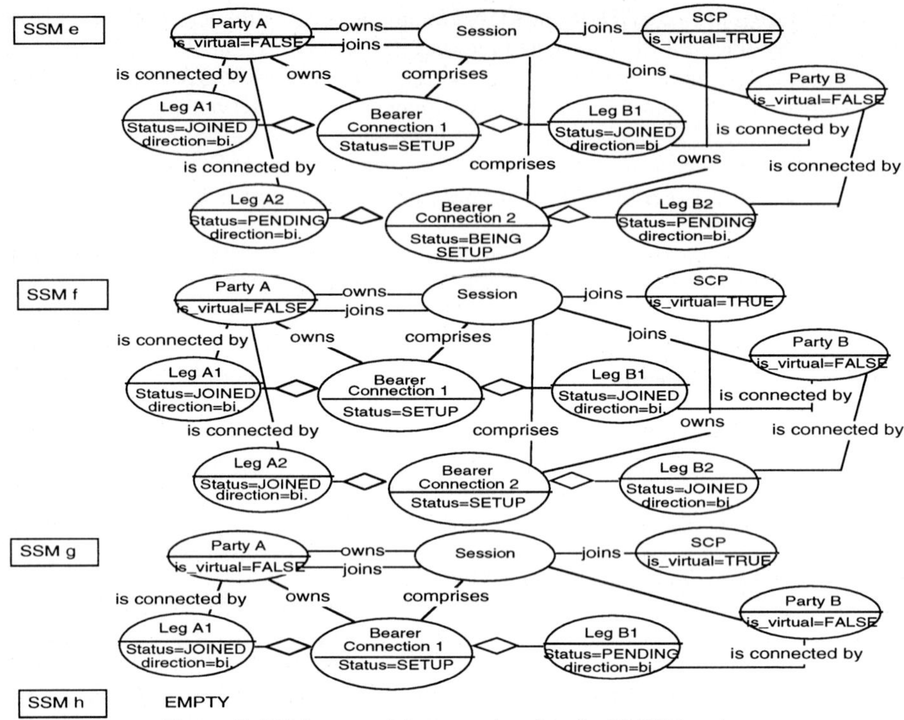

Figure 5: SSM states of the example using the OMT-Notation

7 Conclusions

The definition of a new IN SSM has demonstrated to be the key issue for successfully enhancing the set of traditional IN capabilities with those needed to support broadband multimedia services. In fact, the object orientation of the SSM has allowed the definition of generic Information Flows able to address operations on the objects. Moreover, the SSM design permits to model the new signalling capabilities available in a SSP by simply enriching the SSM with new attributes and relationships.
The identified Information Flows are only a subset of all the meaningful combination of low level SSM object transitions. In fact, the transitions needed to control the evolution of a generic IN service typically involve operations on a group of related objects. The identified Information Flows are service independent and hence allow to provide a large variety of services. Moreover, a unique service logic is needed whatever the signalling capability set is. This has been achieved by introducing at IN level the concept of service session. The session concept is very useful to correlate the different connections belonging to a single service instance.

Acknowledgements: This paper is derived from ongoing studies in the ACTS project AC068 INSIGNIA (IN and B-ISDN Signalling Integration on ATM Platforms). The following companies have participated to this project: Siemens AG, Italtel, CSELT, GPT, Telecom Italia, Deutsche Telekom, Telefonica, Fondazione Ugo Bordoni, CORITEL, GMD-FOKUS, Siemens Atea, Siemens Albis, National Technical University of Athens and University of Twente.

References

[1] ITU-T Recommendation Q.1214, "Distributed Functional Plane for Intelligent Network Capability Set 1"

[2] ITU-T Recommendation Q.1224, "Distributed Functional Plane for Intelligent Network Capability Set 2"

Harmonisation/Integration of B-ISDN and IN:

EURESCOM project P506

Jørn Johansen
Tele Danmark

1 Introduction

The EURESCOM project P506 "Harmonisation/Integration of B-ISDN and IN" was started in the beginning of 1995 and was running for 2 years. 13 Public Network Operators (PNOs) were participating in the project: Finnet Group (Finland), British Telecommunications, Swiss Telecom PTT, Tele Danmark, Deutsche Telekom, France Télécom, CSELT (Italy), Koninklijke PTT Nederland, Portugal Telecom, Telia (Sweden), Telefónica de España, Telecom Finland and Telecom Eireann. The purpose of the project was to study the issues involved in the integration and harmonisation of B-ISDN and IN needed for the IN Capability Set 3 (CS3). More specifically the project should develop a functional model of an integrated B-ISDN/IN architecture, study the possibility for harmonisation between the signalling protocols for B-ISDN and the INAP protocol and propose an extended Basic Call State Model (BCSM) able to exploit the new functionalities supported by B-ISDN and the demands coming from future sophisticated services (e.g. Multimedia services). Needed extensions to the IN functional entities should also be investigated.

In this paper the working method of the project is presented shortly. Then an integrated and harmonised reference architecture for B-ISDN and IN is presented. After this the main results of the project: The enhanced BCSM and the proposed changes to the Call Party Handling defined IN CS2 are presented. The enhancements to B-ISDN signalling and the INAP are not presented here.

2 Method

At first four benchmark services were selected to be representative for four different service categories: Multimedia conversational service (including multiparty features), multimedia distribution services, multimedia retrieval services and B-VPN facilities. It was also intended that mobility aspects should be taken into account, but mobility aspects were for various reasons only treated to a limited extent in the project.

For the selected services the ITU-T service description methodology described in ITU-T Recommendations I.130, I.140 and I.210 was used. The selected services were described in details, and introduction scenarios were developed for the same services. Then, based on these service descriptions and scenarios, the network requirements to support the selected services were identified. Requirements were identified for transport, management, signalling capabilities and various communication configurations. Furthermore the applicability of supplementary services to the selected services were identified.

After studying the identified requirements a reference architecture was proposed and verified against the chosen benchmark services. The functionalities needed in the different IN Functional Entities (FEs), SSF, SCF and SRF, were described. The existing IN BCSM had to be modified (extended) since it only supports simple call configurations with point-to-point connections that Public Switched Telephone Networks (PSTNs) and N-ISDN can provide. In contrast B-ISDN provides (or will provide) the support of general multiparty calls with the possibility of multiple connections between parties, and the individual connections can be unidirectional or bi-directional (symmetric or asymmetric) point-to-point connections or point-to-

multipoint connections. Furthermore B-ISDN supports modifications to connections and call configurations. B-ISDN has addressed these complications by introducing the concept of call and bearer separation. A similar concept was applied to the BCSM.
Information flows and functional entity actions were described for the benchmark services to be able to give a dynamic description of the Harmonised Functional Model. At the beginning of the project it was the intention to produce an SDL description of the different FEs. This was later dropped because of delay in standards, which threatened to make this work obsolete. Due to late decisions in ITU especially concerning Call Party Handling (CPH) modifications to already produced flows and the proposed BCSM were performed by the end of the project to maintain alignment and backward compatibility with standards.
Parallel to the development of a new reference architecture common for B-ISDN and IN and a new BCSM for IN CS3 an investigation was made of possible extensions for B-ISDN signalling protocols needed for B-ISDN/IN integration. Finally proposals were made for updating the CPH defined for CS2.

3 The Benchmark Services

The following benchmark services were selected:

- Broadband Video Conference Service
- Broadband TV Distribution Service
- Video on Demand

Two versions of the Broadband Video Conference Service were considered:

1. Add-on Broadband Video Conference Service
2. Meet-me Broadband Video Conference Service

B-VPN was considered but this topic is at a very early stage and it was not studied in the same depth that the other services were. So the influence from this service on the rest of the work in the project was limited. The same was true for the Mobility Aspects.

4 The reference IN/B-ISDN architecture

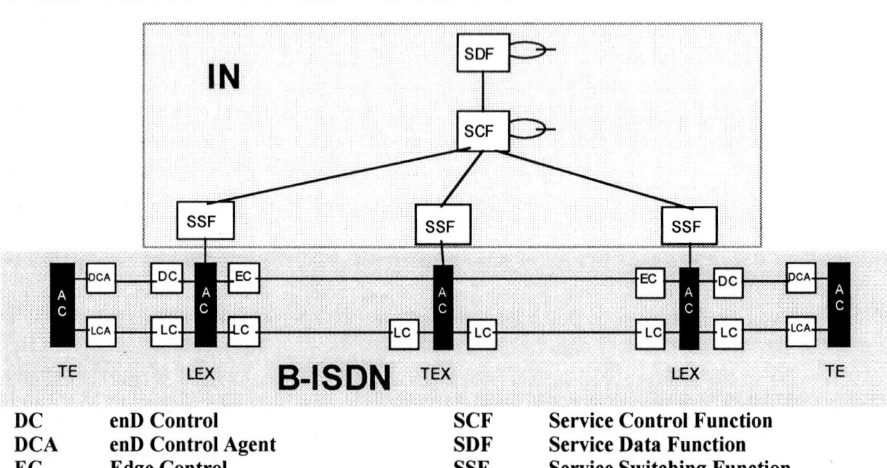

DC	**enD Control**	**SCF**	**Service Control Function**
DCA	**enD Control Agent**	**SDF**	**Service Data Function**
EC	**Edge Control**	**SSF**	**Service Switching Function**
LC	**Link Control**	**TE**	**Terminating Equipment**
LCA	**Link Control Agent**	**TEX**	**Transit Exchange**
LEX	**Local Exchange**		

Figure 1: B-ISDN and IN functional entities in one model.

The reference architecture shown in figure **1** is based on an integration of B-ISDN and IN concepts. Regarding the level of capabilities, the focus is on CS3 for both B-ISDN and IN. The integration of the existing models for B-ISDN and IN was very straight forward and no new functional entities needed to be introduced, but some of the functional entities will have to be enhanced. The main enhancements needed for SSF is treated later. Figure **1** is only showing the basic physical entities and the functional entities grouped into these. The analysis of the benchmark services showed that two different types of physical entities (not shown in figure **1**) containing the functional entity Specialised Resource Function (SRF) will be useful. These two entities, the Specialised Network Resource (SNR) and the Intelligent Peripheral (IP) are shown in figure 2.

Specialised Network Resource

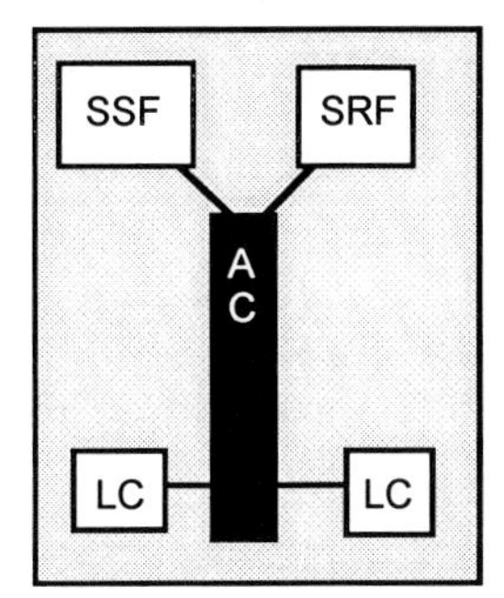

Intelligent Peripheral

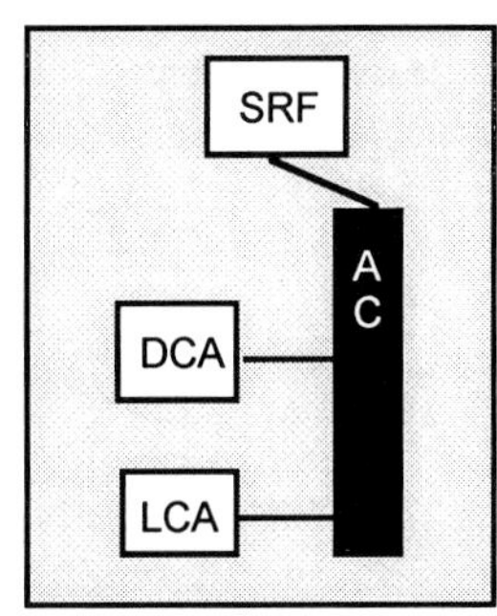

Figure 2: Two physical entities for resource modelling

The SNR is an "inside" network resource since no Edge Control functionality is present. Being an "inside" network resource means that the SNR communicates with other physical entities using the NNI protocol, B-ISUP. Two applications for an SNR were identified during the service case studies:

- Bridge - an entity responsible for setup, maintenance and release of multipoint-to-multipoint communication configurations for the Add-on Video Conference Service. A multipoint-to-multipoint communication is implemented by receiving, processing and merging information from many sources and sending the resulting information to many sinks. The bridge is capable of establishing and releasing connections to users.
- Code converter - an entity to be used in the TV Distribution Service. It converts the coding of the TV distribution program to a coding compatible with the user equipment.

The SSF in the SNR is needed to interact with the LC for controlling communication links.

The IP defined in the project is very similar to an IP in the traditional IN only enhanced with broadband capabilities. The broadband IP is intended for connection to the B-ISDN network through a UNI.

In the following are shown examples of the use of the reference architecture to provide the two types of Video Conferences.

4.1 Add-on Video Conference

In figure 3 an example is shown of an architecture for the add-on video conference service. Add-on implies that one user is the conference co-ordinator. He is responsible for activation and release of the service, and he has to authorise new participants.

In the architecture below, TE1 is the conference co-ordinator. It requests the service via LEX1 from the SCP. The SCP will then allocate the appropriate resource (SNR) and report this back to TE1. After this is done, TE1 can add new participants (TE2 and TE3). When a new participant is added to the conference, the SCP will instruct the SNR to add the required connections and to adjust the mixing of incoming signals (audio and video) into outgoing signals. Messages concerning the control of the involved connections will be sent over the SCF-to-SSF interface. Messages concerning control of the contents of the user plane connections will be sent over the SCF-to-SRF interface.
The SNR as used in this architecture is an entity usually referred to as a 'video conference bridge'.

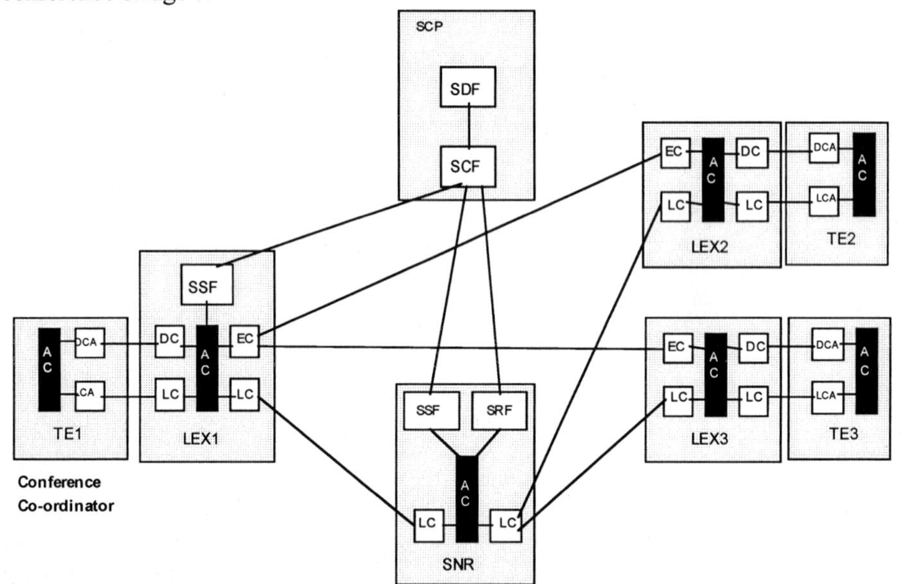

Figure 3: Architecture for an add-on Video conference service.

4.2 Meet-me Video Conference

The other variant of the Video Conference service is the meet-me conference, see figure 4. In this case, users can dial in to the conference, with of course the possibility to screen incoming calls for authentication etc. It is different from the add-on conference because in this case an Intelligent Peripheral is used.
In this architecture, the SCP performs an (optional) authentication check when a user requests to be admitted to a meet-me conference. It then instructs the Intelligent Peripheral that a new participant will be added and that the required changes with respect to the mixing of the incoming audio and video signals must be made. The Intelligent Peripheral is in this case connected to the SCP, which makes this Intelligent Peripheral a resource that will be owned by the network operator.
Another option would be to provide the service via an Intelligent Peripheral that does not have an SRF functionality. This implies that all the service control information that is exchanged over the SCF-to-SRF interface will then have to be exchanged over user plane connections, i.e. in-band. These connections must then be established for this purpose, for the time being.

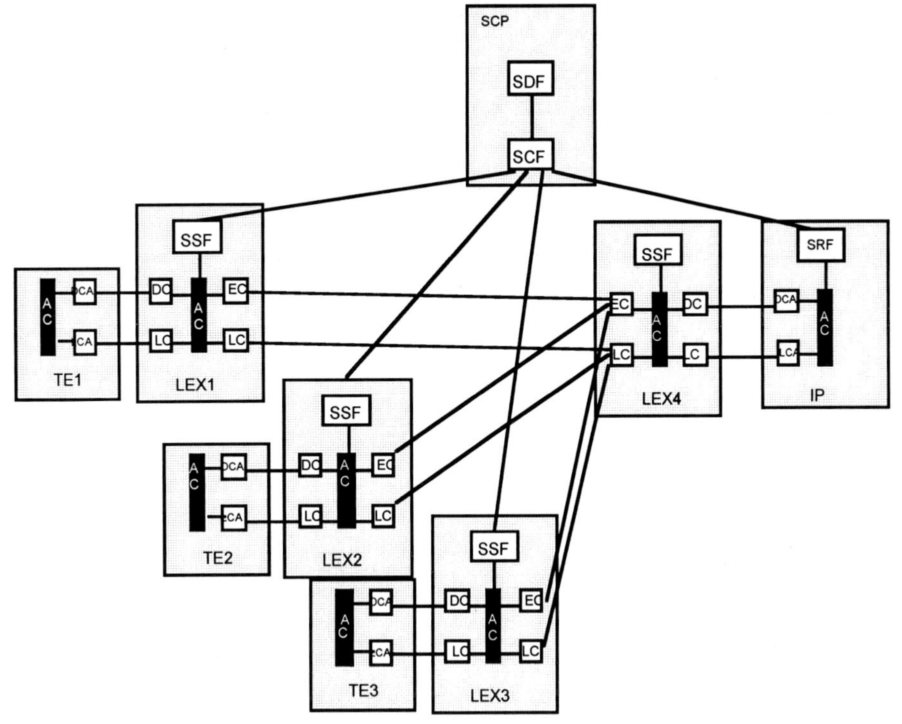

Figure 4: Architecture for a meet-me Video Conference service.

5 Enhancements to IN

As already referred to in the earlier sections, Introduction and Method, the CPH and the BCSM need to be enhanced to be able to cope with the extended capabilities offered by B-ISDN. The main reason for these enhancements are: In earlier systems a call was treated as a monolithic concept consisting of two parties connected with each other by a symmetrical bi-directional connection. In B-ISDN the concept of a call must be extended to describe a communication task involving more than two parties connected by an indefinite number of connection with different characteristics. As mentioned earlier these extensions in capabilities are handled in B-ISDN by decoupling of the call concept and the connection concept. It was decided to adopt a similar approach and work on a two level BCSM with multiple Finite State Machines at each level. The two levels were respectively called Edge Control (EC) level and Link Control (LC) level in accordance with the terminology used in the B-ISDN signalling. IN CS2 uses SSF Connection View States to represent situations for Call on Hold, Call Waiting, Transfer Forwarding etc. which are services involving more than two parties. Also in SSF it is proposed to use a two level approach and use Edge View States and Link View States[1] to represent the more complicated configurations to be supported in B-ISDN.

In figure 5 is shown the relationships between the Edge View objects, the Link View objects and the corresponding BCSMs for modelling a situation where a party is

[1] The terms "edge" and "link" is used here and in the rest of the paper to distinguish them from the IN CS2 terms "call" and "connection.

involved in four calls with in all six connections. The call a is a simple two party call with one point-to-point connection represented by Edge Segment (ESa), one BCSM_E, Link Segment (LSa) and one BCSM_L. The call b is a multiparty multiconnection call involving six parties and two connections: a point-to-multipoint connection from the represented party to the five others (LSb) and a point-to-point connection between the represented party and one of the other five. The call c resembles call b, and call d resembles call a. If a sharing of resources similar to the one used in IN CS2 is needed in CS3 then Link Segment Associations shown in dotted lines on the figure can be used.

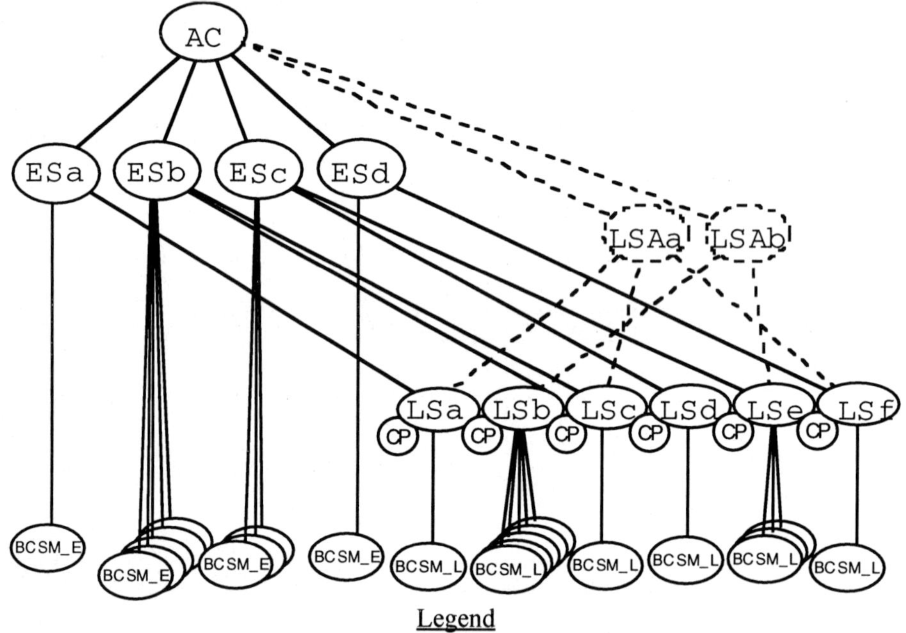

Legend

AC	Access Control	ESx	Edge Segment x
BCSM_E	BCSM for Edge Control	LSAx	Link Segment Association
BCSM_L	BCSM for Link Control	LSx	Link Segment x
CP	Connection Point	x= a,b,c to f	

Figure 5: A four-call, six-connection example

5.1 Enhanced BCSM

The project worked for some time on the possibility of introducing a token approach to the BCSM. This idea was abandoned by the majority of the project participants who judged the approach to be to revolutionary and demanding to much work in definition. It was also considered to have a unique BCSM both for the originating exchange and the terminating exchange, but it was decided to keep the discrimination between the two types of BCSMs as in IN CS1 and IN CS2.

The Originating BCSM for Edge Control is shown in figure 6. It is very similar to the BCSM in CS2. The extra connection bypassing "Collect_Information" reflects the B-ISDN case where overlap sending is not supported. The "Select_Route" PIC has been omitted because an Edge relationship is solely an end-to-end relationship; routes are necessary for bearer connections only. A new PIC "O_Look_Ahead" is inserted here to reflect the look ahead functionality of the signalling capability set 2. From this PIC, call setup may be aborted via one of the DPs "O_Reject", "O_Busy", or "O_No_Answer" through "O_Exception". Regular processing might indicate a forwarding address for which the "Authorise_Call_Setup" might need to be repeated.

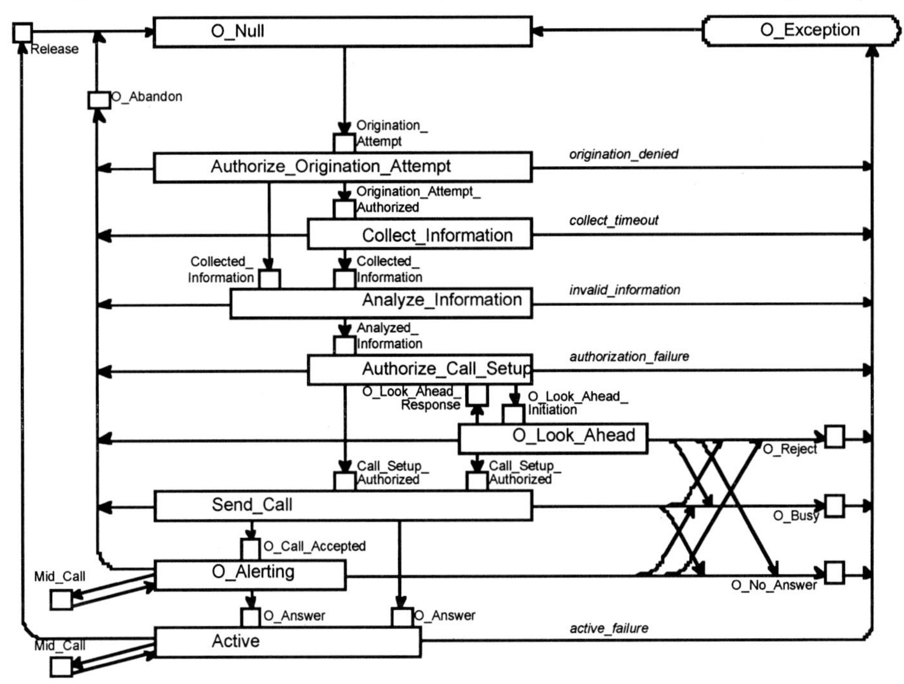

Figure 6: Originating BCSM for Edge Control (O_BCSM_E)

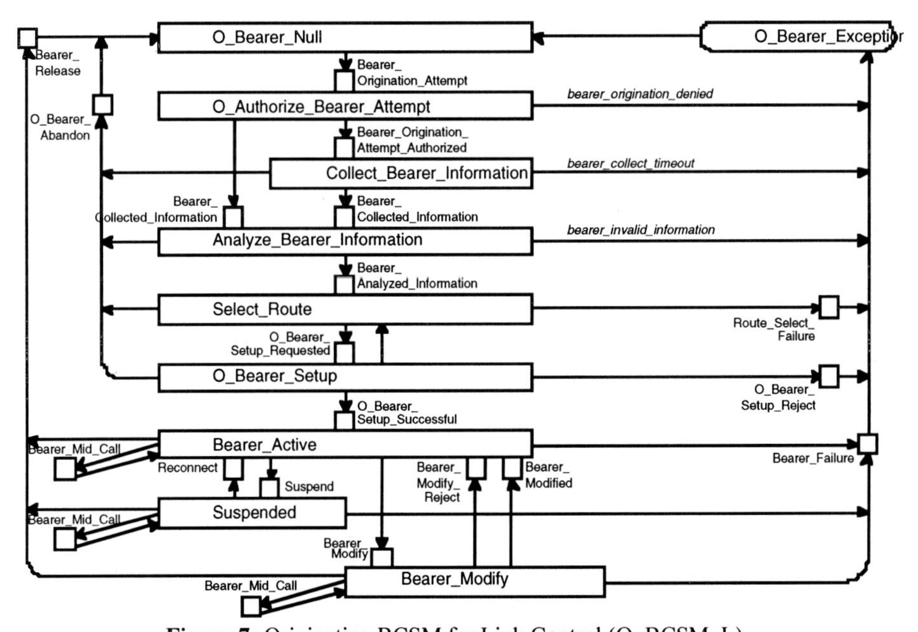

Figure 7: Originating BCSM for Link Control (O_BCSM_L)

Figure 7 shows the Originating BCSM for Link Control. In general, where similar PICs are proposed for Edge Control and Link Control, the term "bearer" has been introduced. For example, the "Null" PIC has been renamed "O_Bearer_Null" in order to distinguish it from the Edge Control BCSM PIC. The PICs "O_Bearer_Null", "Authorise_Bearer_Attempt", "Collect_Bearer_Information", and "Analyse_Bearer_Information" are copied directly from IN CS2. The extra connection bypassing "Collect_Bearer_Information" reflects the B-ISDN case where overlap sending is not supported. The PIC "Select_Route" which has been omitted in the O_BCSM_E is included in this O_BCSM_L. From this PIC, a bearer connection establishment progresses through PIC "O_Bearer_Setup" to "Bearer_Active". A PIC "O_Alerting" is not required as alerting is assumed to be an Edge Control activity only. Also the PIC "Suspended" can be found in this BCSM because - opposite to calls - connections may be suspended. Finally, a further PIC "Bearer_Modify" is proposed to be added to cater for the extended capabilities of B-ISDN. Terminating BCSMs for Edge Control and Link Control were also developed together with and BCSM for Access Control.

5.2 Enhancements to CPH

Figure 8 shows the proposed Edge View States (EVS) for Edge Segments. They represent:

N Null
This represents a condition where call and connection processing is not active. There is no controlling leg or passive leg present

A Add Parties to Call
This represents a situation where one or more remote parties are in the setup phase.

T Terminating Call Setup
This represents a terminating call in the setup phase.

S Stable Call
This represents a stable (or clearing) call. It is either an originating or a terminating two-party or multi-party call from the perspective of the controlling leg.

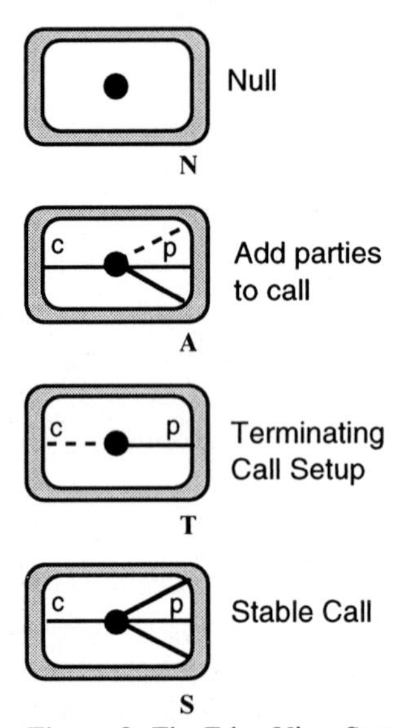

Figure 8: The Edge View States

The proposed Edge View States represent the possible states of a call. The legs in the Edge View States can be in one of the two states: Joined (solid line) or pending (dotted line).

Figure 9 shows the proposed Link View States. The states are described after the figure. The legs can be in the same states as legs in the Call View States of CS2: Pending, joined, surrogate and shared.

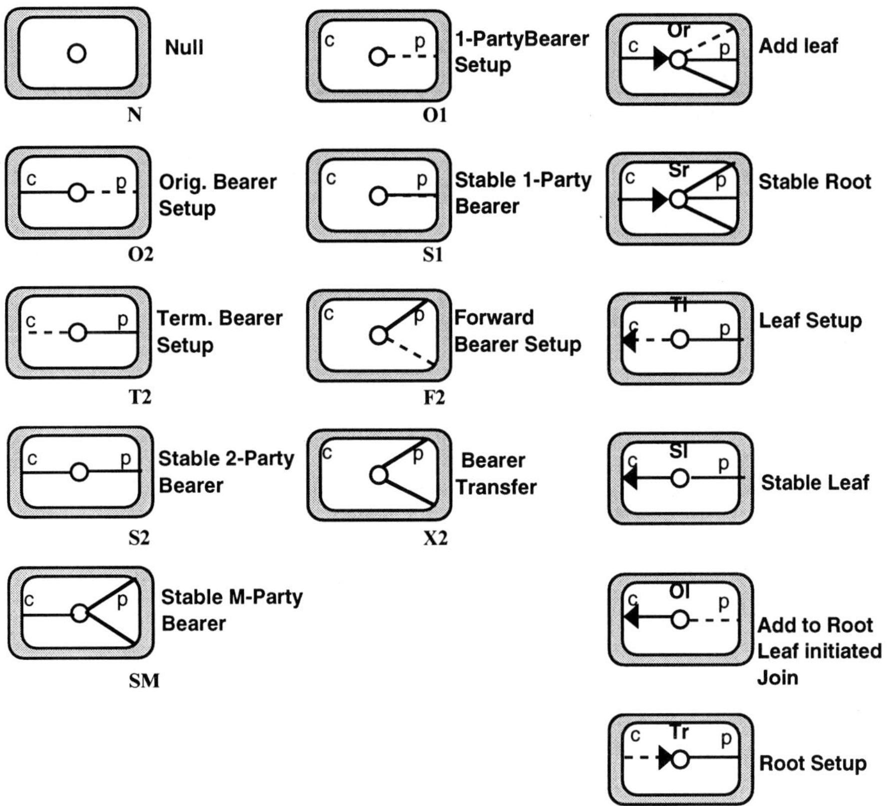

Figure 9: The Link View States

1. *Null*

   N **Null** (similar to IN CS2 state A: "Null")
   This LVS represents a condition where Link processing is not active. There is no controlling leg or passive leg connected to the connection point. Usually, this LVS is not shown.

2. *Bi-directional Bearer Connection*

   O2 **Originating Bearer Setup** (similar to IN CS2 state B: "Originating Setup")
   This LVS represents an originating two-party bi-directional bearer connection in the setup phase.

   T2 **Terminating Bearer Setup** (similar to IN CS2 state D: "Terminating Setup")
   This LVS represents a terminating two-party bi-directional bearer connection in the setup phase.

   S2 **Stable 2-Party Bearer** (similar to IN CS2 state C: "Stable 2-Party")
   This LVS represents a stable or clearing two-party bi-directional bearer connection, and is either an originating or a terminating connection from the perspective of the controlling user.

SM **Stable M-Party Bearer** (similar to IN CS2 state H: "Stable M-Party") This LVS represents a stable or clearing multi-party bi-directional bearer connection in one Link Segment. The connection point performs a "bridging" function between the controlling and both passive legs.

3. *Bi-directional Bearer Connection (Forward and Transfer)*

O1 **1-Party Bearer Setup** (similar to IN CS2 state M: "1-Party Setup") This LVS represents a 1-party bi-directional bearer connection being originated on behalf of the network (i.e., there exists no controlling leg). Note that there exists no controlling leg. However, a controlling leg exists within the Edge relationship and indicates the charging relationship for the originated passive leg (*leg* p1).

S1 **Stable 1-Party Bearer** (similar to IN CS2 state N: "Stable 1-Party") This LVS represents a 1-party bi-directional bearer connection originated on behalf of the network (i.e., there exists no controlling leg), that is in a stable or clearing phase. Note that there exists no controlling leg. However, a controlling leg exists within the Edge relationship and indicates the charging relationship for the passive leg (*leg* p1).

F2 **Forward Bearer Setup** (similar to IN CS2 state J: "Forward") This LVS represents a forwarded bi-directional bearer connection. Link processing for the first passive leg (*leg* p1) is in a stable or clearing phase, or a terminating connection setup phase, whereas Link processing for the second passive leg (*leg* p2) is in an originating call setup phase. Note that there exists no controlling leg However, a controlling leg exists within the Edge relationship and indicates the charging relationship for the forwarded passive leg (*leg* p2).

X2 Bearer Transfer (similar to IN CS2 state I: "Transfer") This LVS represents a transferred bi-directional bearer connection. The connection between the two passive legs is in the stable or clearing phase. Note that there exists no controlling leg. However, a controlling leg exists within the Edge relationship and indicates the charging relationship between the two passive legs after the bearer connection has been transferred.

4. *Unidirectional Point-to-Multipoint Bearer Connection*

Or **Add Leaf**
This LVS represents an originating two-party or M-party point-to-multipoint unidirectional bearer connection in the setup phase; it is originated at the root.

Sr **Stable Root**
This LVS represents a stable or clearing two-party or M-party point-to-multipoint unidirectional bearer connection at the root; the passive legs represent either an originating or a terminating connection from the perspective of the controlling user.

Tl **Leaf Setup** (terminating)
This LVS represents a terminating two-party or M-party point-to-multipoint unidirectional bearer connection at the leaf in the setup phase.

Sl **Stable Leaf**
This LVS represents a stable or clearing two-party or M-party point-to-multipoint unidirectional bearer connection at the leaf, and is either an originating or a terminating connection from the perspective of the controlling user.

Ol **Add to Root** (Leaf Initiated Join)
This LVS represents an originating two-party or M-party point-to-multipoint unidirectional bearer connection in the setup phase; it is originated at the leaf.

Tr **Root Setup** (terminating)
This LVS represents a terminating two-party or M-party point-to-multipoint unidirectional bearer connection at the leaf in the setup phase. Note that this LVS reflects the situation where a "Leaf Initiated Join" does not only add a leg but sets up the whole bearer connection.

In this paper it would be too detailed to go through the transitions between the shown Edge and Link View States, but in the project all the possible transitions have been investigated.

6 Conclusion

The paper has presented a functional model of an integrated harmonised B-ISDN/IN architecture validated through representative benchmark services. Proposals have been made for a two level BCSM and SSF. The enhancements made to the BCSM and SSF of IN CS2 follows the principles behind CS2 and will secure the support of future B-ISDN supported supplementary services and be backward compatible. The results from the project have been contributed to ITU-T and are included in the IN CS3 Baseline document.

Acknowledgements
Acknowledgement has to be given to the EURESCOM organisation and the numerous project participants from the organisation named in the introduction.

Abbreviations

B-	Broadband-
BCSM	Basic Call State Model
CPH	Call Party Handling
CS	Capability Set
DC	enD Control
DCA	enD Control Agent
EC	Edge Control
ECM	Edge Control Machine
ES	Edge Segment
EVS	Edge View States
FE	Functional Entities
IN	Intelligent Network
INAP	IN Application Part
IP	Intelligent Peripheral
ISDN	Integrated Services Digital Network
ISUP	ISDN User Part
LC	Link Control
LCA	Link Control Agent
LCM	Link Control Machine
LS	Link Segment
LVS	Link View States
N-	Narrowband-
NNI	Network Node Interface
PIC	Point In Call
PNO	Public Network Operator
PSTN	Public Switched Telephone Network
QoS	Quality of Service
SCF	Service Control Function
SDF	Service Data Function
SDL	Specification and Description Language
SNR	Specialised Network Resource
SRF	Specialised Resource Function
SSF	Service Switching Function
UNI	User Network Interface
VPN	Virtual Private Network

Intelligent Network Evolution for Supporting Mobility

L.Vezzoli (Italtel) [1], T.Bertchi (Ascom Tech), A.Markou (NTUA),
J.Nelson, C.Morris (Teltec Ireland, University of Limerick)

This paper analyses and investigates how Intelligent Network (IN) may evolve to support terminal and personal mobility for advanced B-ISDN services.
Taking as starting point current standard proposals like IN CS2 and Cordless Terminal Mobility [1], which is the first attempt to integrate mobility in the fixed network, requirements on Intelligent Network as well as on Broadband Network are identified focusing on the IN Distributed Functional Plane and IN Physical Plane.
The paper is based on the work done in the context of the ACTS Project AC013 *EXODUS* - EXperiments On the Deployment of UMTS, in which field trials to demonstrate the IN functionality needed to meet UMTS requirements are going to take place. The specified functional architecture, physical architecture and control model are presented.

1 Introduction

In recent years there has been an increasing demand for telecommunication services that satisfy individual users requirements in terms of availability, flexibility and Quality of Service.
Namely, the support of mobile terminals and mobile users in any kind of network is getting more and more important: the users are demanding advanced services, like B-ISDN ones, to be accessed anywhere and anytime, growing the need of introducing flexible and low-cost solutions.
During the 1990s telecommunications networks have become ever more diverse and complex. Fixed telephone network facilities have been extended by the introduction of first the ISDN and currently of Broadband capabilities. A number of different mobile networks, such as the GSM, have been developed to meet different market needs. Cordless access system such as DECT can be attached to both fixed and cellular networks.
Thus, at present there are different types of service network. Some operators are bringing about a convergence between fixed and mobile networks, with both network types providing similar kinds of network services.
The *Universal Mobile Telecommunication System (UMTS)* addresses the convergence of heterogeneous systems and networks, thus a number of different evolutionary paths can be envisaged in facing the road map from second generation technology.
In the context of the ACTS programme, the *EXODUS* (EXperiments On the Deployment of UMTS) project makes a clear choice facing the evolution towards UMTS on the basis of service interoperability between fixed and mobile network. The approach is based on adopting an Intelligent Network (IN) based service architecture that allows the integration of heterogeneous mobile and fixed systems by separating the service transport from the service control.
The application of IN to mobility is a key element in the integration of a mobile access to future advanced networks such as B-ISDN. In that integration, IN would take care of the user and mobility control with related data, while B-ISDN is responsible for the basic switching and transport mechanism.
The key element of IN that makes it suitable for handling mobility is its capability to provide services independently from the network implementation. This allows to isolate the services from the way the service-independent functions are actually

---

[1] Contact Person: Lucia Vezzoli, Italtel, 20019 Settimo Milanese (MI) - Italy
Tel. +39.2.43889085, Fax. +39.2.43887989, E_mail: lucia.vezzoli@italtel.it

implemented in various physical network infrastructures; in particular, independently from the fact that a network is fixed or mobile. IN allows to have a universal core network regardless of means of access: from a core network point of view a fixed/mobile Interworking Unit and a wireline terminal have the same behaviour.

2 Evolution Requirements

Following the evolution of IN Capability Sets, different steps for the modelling of mobility functions can be identified. They differ with respect to the integration degree of mobility functions into the IN service logic. Going through this evolutionary path, it can be seen that the role of intelligence as meant by IN (that is the service logic in the SCPs) increases gradually to embrace the system functions of user and terminal mobility.

In *CS1*, IN is utilised exclusively on the design of supplementary services at the top of mobile system. It is the underlying network that is fully responsible of handling mobility of the users.

To find the mobile specific IN services, we must consider the successive step: *CS2* standard. In this phase the scope of IN is widened to cover non-call related mobility functions themselves. IN is not only an additional intelligence to a basic call or a mobility function, but an elementary technology to implement those functions. New functional entities and new IN state models are needed to model signalling processes that triggers IN for requesting mobility services. A similar approach is proposed by the CTM standard where the mobile network is defined not as a separate overlay network, but as a system that allows true integration of mobile and fixed communications into a single narrowband telecommunications infrastructure.

The IN modelling of mobility functions can still be widened to call related mobility functions resulting in a situation where all mobility functions are defined in a IN way. The example of call related mobility functions are interrogation, paging and handover. This capability is partially covered by IN CS2 and will be finalised in *IN CS3*. Call related mobility functions imply also new state models in addition to the BCSM classical one.

Long Term IN is the final step for the IN view to mobility. It does not add any mobility function into the scope of IN, but it brings the object orientation, the new separation of services from underlying resources and the concept of distributed processing. It presumes renewing completely of IN modelling in a way that goes directly towards the TINA concepts.

The approach illustrated in this paper, which is the one defined by the ACTS EXODUS project, is in between CS2 and CS3: mobility services are clearly defined and specified as new services to be added to the service logic.

The concept of separation is used as a mechanism to address evolution and to derive a flexible functional architecture based on the current IN [2, 3], UMTS [4] and FPLMTS [5] functional models and architectures. The following requirements, applicable to IN evolution, are explicitly addressed:

1. separation of switching, service and data control;
2. separation of service and mobility control;
3. separation of terminal and network functions;
4. support of call related events, independent of the access technology;
5. support of call unrelated events independent of the access technology;
6. integration of terminal functions for both wired and wireless access.

The next section will review the impact of these requirements on the functional architecture.

3 IN Distributed Functional Plane

The proposed functional architecture that meets the listed requirements is illustrated in Figure 1. It is derived based on the ongoing, but not yet stable, work of standardisation with specific enhancements to support requirements 2 to 6. The first requirement had already been addressed within the existing functional architectures e.g. the separation of switching, service and data control is an inherent characteristic of IN based architectures.

The second requirement for the separation of service and mobility control is addressed by introducing two distinct FEs for the Mobile Broadband Service Control Function (MB-SCF), namely the SCFmm and the SCFsl, adopting the CTM approach. The SCFmm takes control of mobility management, handling both terminal and user mobility for a specific access technology, while the SCFsl controls the user services in the traditional IN sense, including the service logic support for management of multimedia services. The separation supports the execution of different types of service logic instances at the same time for the same call, if required.

The concept of service domain has been adopted. A "Service Domain" is the geographical area in which a certain set of services is provided to the user. In particular, in this approach, this corresponds to the area in which a single access technology is provided and *which is served by a single SCFmm*. The SCFmm is the basis of a domain, while the SCFsl is the basis of a network.

A corresponding SDF, i.e. an SDFmm and SDFsl, is associated with each control entity to store the related service data. The SDFsl stores the service related data such as the user's service profile, the security parameters, etc., while the SDFmm focuses on the terminal data, such as capabilities, and mobile related information including the temporary identifiers.

Integration of fixed and wireless terminal functionality has been performed by defining an single enhanced INAP protocol (called Mobile-INAP) from the SCFmm towards the Core Network Basic Switching Infrastructure.

The terminal related functional entities: MSF, MCF, CCAF and TACAF are adopted from the FPLMTS functional architecture. The M-CUSF and CRACF are introduced, in an IN CS-2 sense, to support call unrelated and call related events related to user and terminal mobility. Specifically, requirements are identified to support personal and terminal mobility in a B-ISDN environment and are extended to support the sixth requirement of integration of terminal functions. The M-CUSF is a merge of the IN CS-2 CUSF and NRACF functional entities, to handle call unrelated events for both wireless and fixed access in an integrated manner. It takes care of handling call unrelated events by invoking call unrelated procedures in the SCFmm e.g. for location management. A Basic Call Unrelated State Model (Mobile-BCUSM) has been developed to describe M-CUSF activities. The CRACF handles call related events such as paging and handover.

The evolved B-CCF and B-SSF have been introduced in the basic network switching infrastructure. The CCF is enhanced to support ATM based B-ISDN capabilities. In particular, as the core element of the CCF is the Basic Call State Model (BCSM), the development of an enhanced BCSM is needed to model B-ISDN calls. Accordingly, the SSF evolves to interface the new CCF and the enhanced SCFsl.

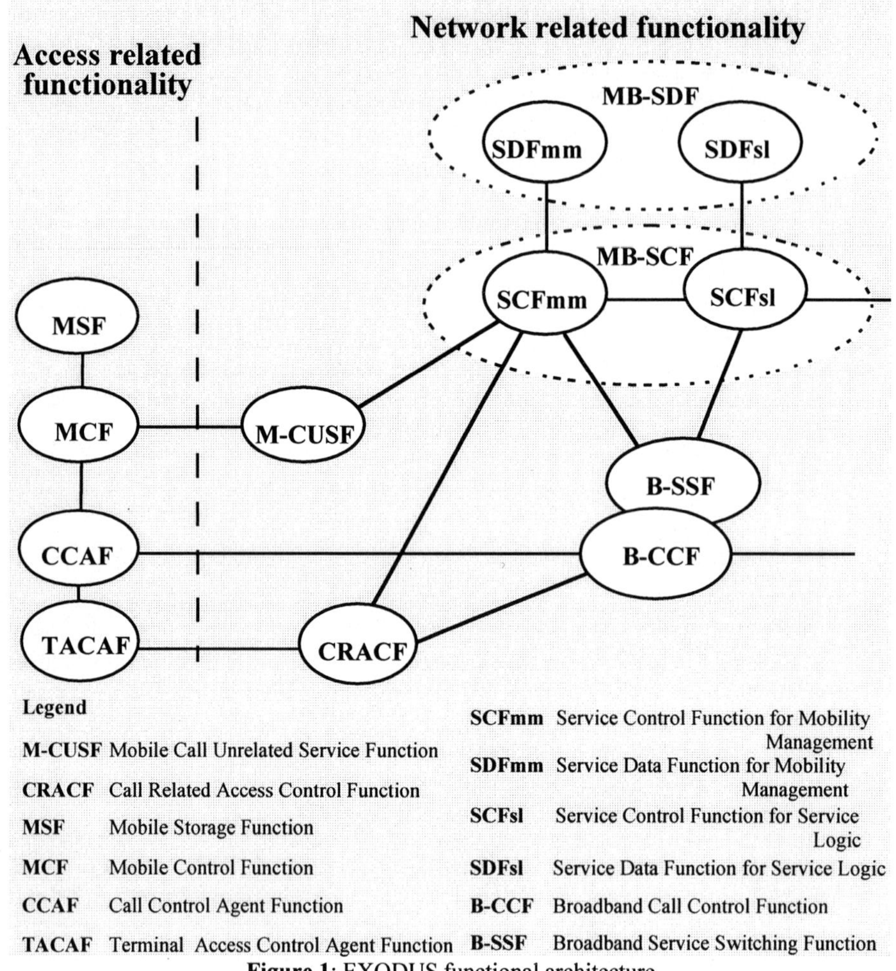

Figure 1: EXODUS functional architecture

3.1 Broadband Basic Call State Model

As the IN-CS1 BCSM provides an high-level model description of a basic two-party call (and connection), an evolved abstract model is needed to model B-ISDN basic call and connection activities to allow the interaction between IN service logic and a basic B-ISDN call. The new developed Broadband Basic Call State Model (B-BCSM) addresses the separation between call and connection control and provides a complete view of a broadband call to the service control logic.

3.2 Mobile Basic Call Unrelated State Model

To provide the MB-SCF with the possibility of handling the logical connection between the user/ terminal and the network, it is proposed to introduce a Mobile-Basic Call Unrelated State Model (M-BCUSM), to model M-CUSF activities.

The main concept of M-BCUSM is similar to the one of BCSM. It is a high level description of the M-CUSF activities required to establish and maintain an association between users and service processing, and to manage invoked operations.

The M-BCUSM identifies points in the basic call unrelated interaction where IN service logic instances are permitted to interact with basic call unrelated associated interaction processing. In particular, it provides a framework for describing basic call unrelated events, that can lead to the invocation of IN service logic instances and for describing those points in association and operation processing when transfer of control can occur.
Call Unrelated activities can be performed from wireless terminals as well as from wired terminals.
The M-BCUSM is described in terms of: Point In Associations (PIAs) that represent the states of an association for the call unrelated interaction, Detection Points (DPs) that indicate points in call unrelated logical connection at which IN is invoked, Transitions that indicate the normal flow of a call unrelated process from one PIA to another and Events that cause transitions into and out of PIAs.

4 IN Physical Plane

In Figure 2 the physical architecture and the correspondent mapping of functional entities into network elements are illustrated. The chosen wireless access is the DECT technology, enhanced with features (like multiple bearer) needed to provide multimedia services similar to those usually offered on fixed network.
The MB-SSP contains evolved B-CCF/B-SSF functionality together with mobility specific functionality, realised in the M-CUSF and CRACF functional entities. The MB-SCP contains the two groups of functional entities devoted to handle multimedia services and mobility management procedures, namely SCF/SDFmm and SCF/SDFsl.

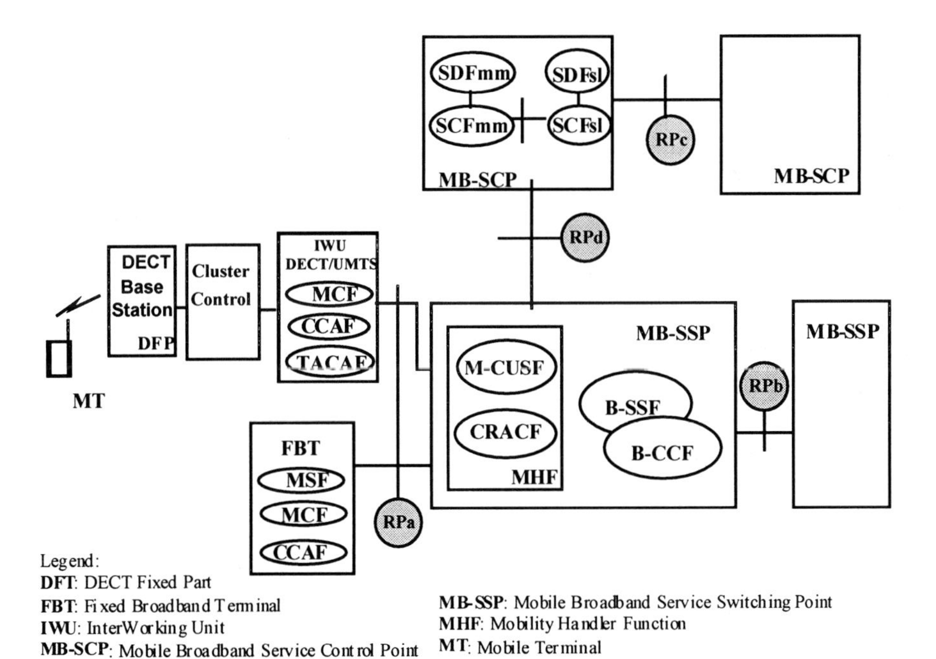

Figure 2: EXODUS Physical architecture

4.1 Protocols

The mapping between the physical interfaces and the signalling protocols is given in Table 1.

At the interface between MB-SSP and MB-SCP an evolution of the INAP protocol is identified to meet both the broadband and the mobility specific requirements. An extended INAP is defined also to support the dialogue between two different MB-SCPs.

The standard ATMF 3.1 signalling protocol is enhanced to support the UMTS UNI (RPa).

Physical Interface	Physical Entities	Signalling protocols
RPa	FBT/MB-SSP IWU/MB-SSP	ATM Forum 3.1 enhanced, parts of Q.2932
RP_b	MB-SSP/MB-SSP	B-ISUP
RP_c	MB-SCP/MB-SCP	M-INAP*
RP_d	MB-SSP/MB-SCP	MB-INAP

Table 1: Physical Interfaces and Signalling Protocols

The following sections present the enhancements on the existing signalling protocols needed in order to support IN-based terminal and personal mobility in a B-ISDN environment, focusing mainly on the evolution of protocol stacks at the IN-interfaces.

4.2 IN Interfaces

The signalling protocol stacks adopted at the IN-interfaces are depicted in Figure 3. Two kind of interfaces are under consideration: MB-SSP/MB-SCP and MB-SCP/MB-SCP. The standard NNI stack up to the TCAP layer is adopted for both the interfaces.

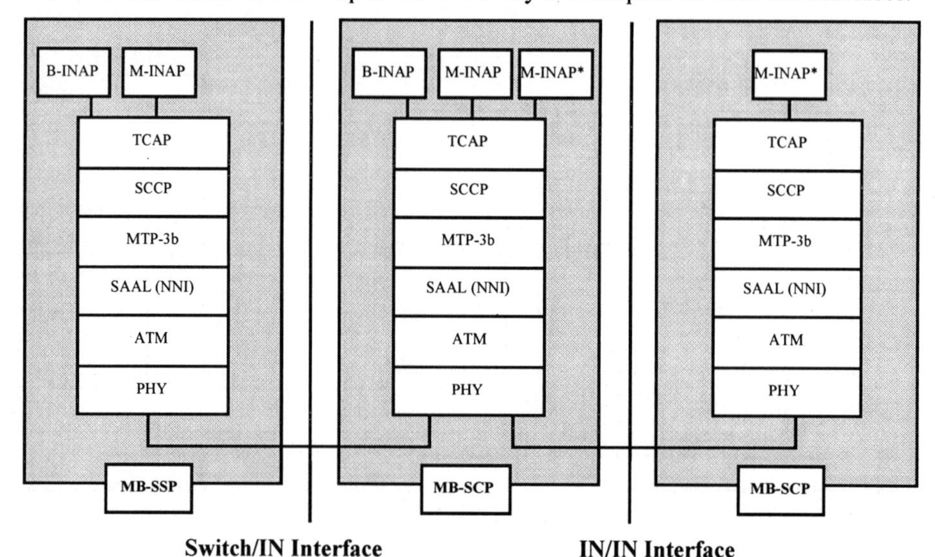

Figure 3: IN Interfaces Protocol Stacks

The Intelligent Network Application Protocol includes the following different protocols: the whole set of B-INAP for handling B-ISDN services in the context of B-SSF/SCF_{sl} interaction; the M-INAP for handling mobility management in the context of M-CUSF/SCF_{mm} interaction; the M-INAP* for handling communication between two SCF residing in two different MB-SCP.

M-INAP
Call-unrelated procedures for transport of mobility management operations between the MB-SSP and the MB-SCP are handled by the following messages (M-INAP):

- Initiate_Association
- Activation_Received_and_Authorized
- Association_Release_Requested
- Send_Component
- Component_Received
- ReleaseAssociation

This set of messages is partially defined in IN-CS-2 draft recommendation Q.1224 [3] for call-unrelated events which enable an association to be opened between the SSP and IN. Figure 4 shows how associations are established from the switch or from IN using these messages.

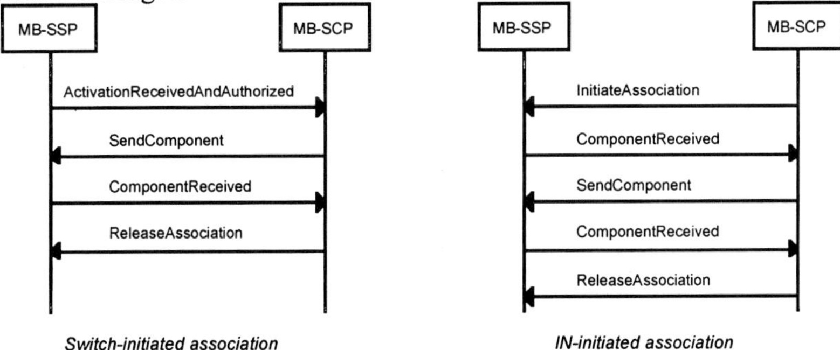

Figure : MB-SSP to MB-SCP Operations

M-INAP*
A relationship between two or more SCFs. is established when the service logic of one SCF needs to interact with the service logic of another SCF. The interaction can be call or non-call related. The following IN messages are used to handle such relationships:

- SCF Bind Request
- SCF Bind Result
- Handling Information Request
- Handling Information Result
- SCF Unbind Request

This set of messages is partly defined in IN-CS-2 draft recommendation Q.1224 [3] to handle call-related or call-unrelated events, by enabling a relationship to be opened between two SCFs. In principle, there are the following SCF-SCF relationships:

- SCF_{sl} to SCF_{sl}
- SCF_{mm} to SCF_{sl}
- SCF_{sl} to SCF_{mm}

The messages between an SCF_{mm} and an SCF_{sl} occur directly in the local network or indirectly between networks, being the SCFmm and SCFsl nodes organized in a hierarchy (where the SCFmm nodes communicate only with the local network SCFsl). Figure 5 shows how relationships are established between two SCFs using these messages.

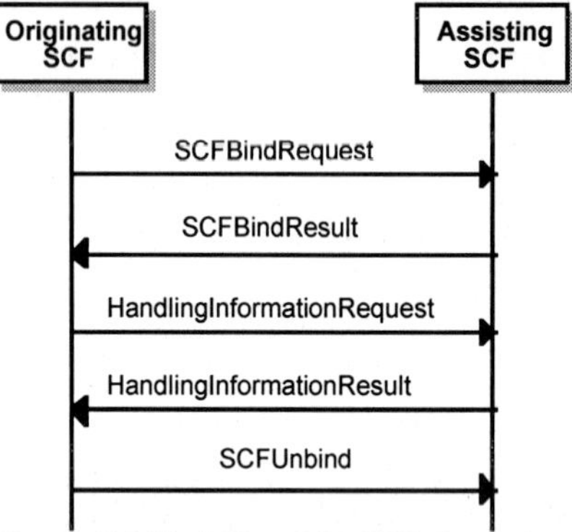

Figure 5: MB-SCP to MB-SCP Operations

4.3 UMTS UNI

The UMTS UNI signalling protocol stack assumes the following:

- Call unrelated messages are transported in a connectionless way by means of the Q.2932 CL-BI protocol.
- ATMF 3.1 has been chosen as call-related Layer 3 signalling protocol and enhanced in order to transport the required mobility identifiers.

4.4 Mobility Management Protocol Description

The Mobility Management protocol supports call unrelated operations for personal and terminal mobility in the UMTS network. The following mobility management procedures are defined:

- *Location Registration*: users have a unique identity contained in a movable smart card (the User Identity Module or UIM). However, a terminal has no identity as such and is assigned a temporary identity by the network when the first user registers by inserting the UIM. The procedure used to register the first user and terminal is called Location Registration. It is initiated by the MCF.
- *User Registration*: this procedure is used to register a user on a visited network.
- *User Deregistration*: this procedure is used to deregister a user. When the last user is removed, the terminal is also deregistered.
- *Location Deregistration*: this procedure is used to deregister all users from a terminal and to deregister the terminal.
- *Location Update*: this procedure is used to update the location area when a wireless terminal or wired terminal changes its point of attachment.
- *(Look Ahead) Paging*: this procedure is used before bearer setup to determine the point of attachment of a called user, and to find out whether the user is busy, the terminal is powered down or the user is otherwise not reachable. It is controversial the characterization of this procedure as call or non call related. The approach adopted by the Exodus project is to handle it by means of the non call related signalling even if it occurs during a call setup.

Moreover, the following procedures for user profile manipulation that allow the user to control the services offered by IN are defined:

- *User Profile Modification* initiated by the user to modify the user profile parameters
- *User Profile Interrogation* initiated by the user to read the profile parameters.

All the mobility management operations initiated by the user (terminal) are fully transparent to the SSP. The call-unrelated protocols (Q.2932 CL-BI and M-INAP) provide the transport layer at the UNI and IN interface for the mobility management operations between the access network and the MB-SCP.
On the contrary, the paging operation is initiated by the MB-SCP. The MB-SSP has to determine the location area where the paging message should be broadcast. The same call unrelated transport layer is used for this procedure, but this time it is not transparent to the MB-SSP. In figure 6 the Mobility Management Protocol stack is shown.

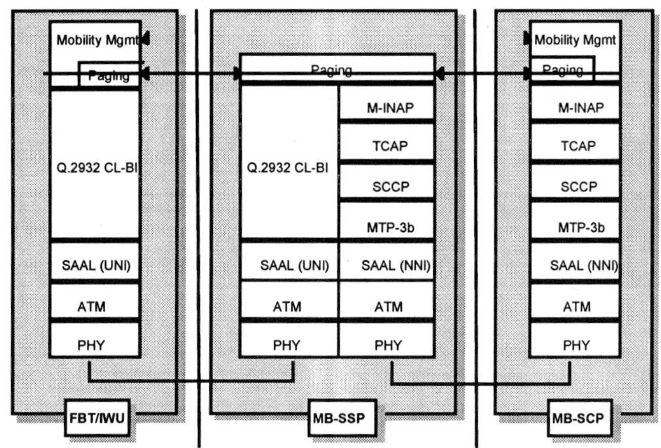

Figure 6: Mobility Management Protocol

5 A Service Example: the User Registration

Figure 7 illustrates the hierarchical structure of the network in case of a user registration operation. The main signalling links and direction of request messages are shown for the generic case of a user roaming from one visited network to another. The user is identified by the User Identity Module (UIM). In the new visited network registration is completed while de-registration, in the old network, is required. The associated database functions are DDF (SDFmm), VDF (SDFsl) and HDF (SCFsl) corresponding to the Domain, Visited and Home databases, respectively. The SCFmm in the New visited network contacts the home SCFsl (via the SCFsl of the New Visited Network), which then communicates with the SCFsl and (indirectly) with the SCFmm in the old network.

6 Conclusion

In this paper, an evolved IN platform that allows the integration of mobile and fixed networks has been presented. An enhanced IN CS-2 functional architecture, adopting CTM concepts, is specified. The adopting of this architecture allows to demonstrate the forceful of IN as a means to provide mobility (personal and terminal) to the user and to facilitate service/network implementation independent provisioning of services in a multi-vendor environment minimising the impact on existing underlying network.

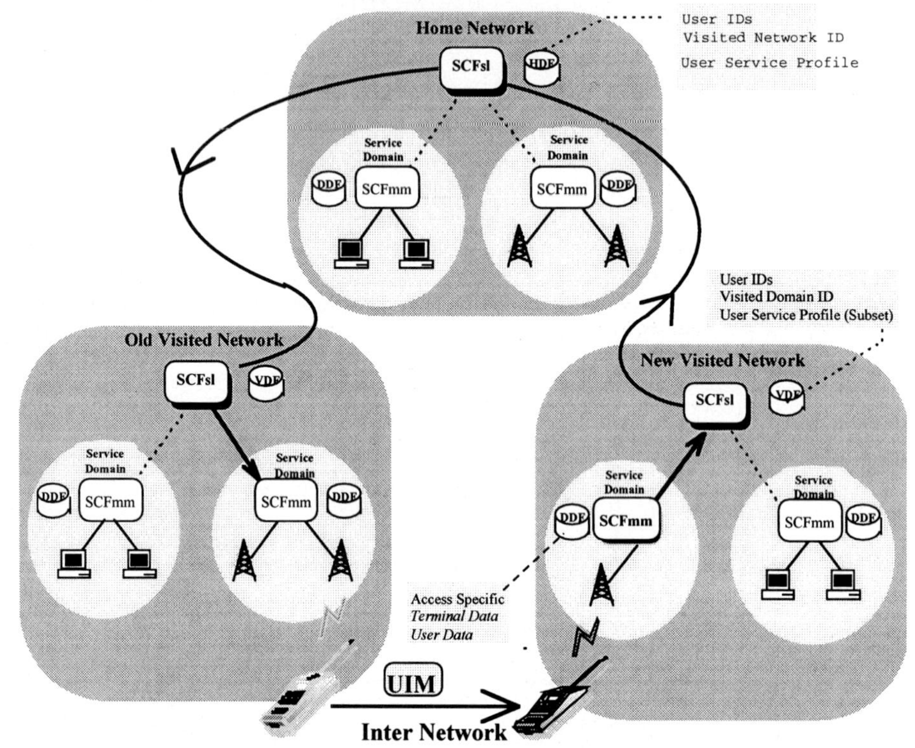

Figure 7: Inter-Network User Registration

Acknowledgement

This work has been carried out within the ACTS project EXODUS in which the authors are working. However, the views expressed are those of the authors, and do not necessarily represents those of the project as a whole. EXODUS is partially funded by the European Commission and the consortium consists of ASCOM Tech Ltd, Belgacom, CSELT, GPT Ltd, Universitatskliniken, Intracom, Italtel, Laboratoires D'Electronique Philips, NTUA, OTE SA, Swiss Telecom PTT R&D, Syndesis, TELTEC, Telecom Italia Spa., TMR.

References

[1] DTR/NA 61302 (ETSI): Cordless Terminal Mobility (CTM)-IN Architecture and Functionality for the support of CTM (Version 1.11), November 1995.

[2] Q.1214 (ITU-T Q.6/11): Distributed Functional Plane for Intelligent Network Capability Set 1.

[3] Draft 1224 (ITU-T Q.6/11): Distributed Functional Plane for Intelligent Network Capability Set 2, November 1995 version.

[4] DTR/NA 61301 (ETSI): IN/UMTS Framework Document (Version 8.1.1). October 1995.

[5] Q.FNA (ITU-T Q8/11): Network Functional Model for FPLMTS (Version 1.1.0), 4. - 15. September 1995.

A Generic Service Order Handling Interface for the Cooperative Service Providers in the Deregulating and Competitive Telecommunications Environment

Young B. Choi
Protocol Engineering Center (PEC)
Electronics and Telecommunications Research Institute (ETRI)
Taejon 305-350, Republic of Korea
ybchoi@pec.etri.re.kr

Adrian Tang
Computer Science Telecommunications
University of Missouri-Kansas City
Kansas City, MO 64110, U. S. A.
tang@cstp.umkc.edu

We propose a generic service ordering interface for the cooperative service providers in the deregulating and competitive telecommunications environment. In particular, we show the necessary operations and information objects used in the interface. As the speed of openness in the global telecommunications market is getting faster, the importance of mutually interoperable interface among the cooperative Service Providers (SPs) is growing day by day. Currently, the world wide effort to resolve this complex situation is going on in various ways. To achieve a smooth migration to the globally compatible interface from each SP's existing legacy service ordering interface, our model can be referenced as a good candidate.
Keywords: Service Order Handling, Service Provider (SP), Main Service Provider (MSP), Subcontracted Service Provider (SSP), Information Modelling, Computational Modelling, TINA-C.

1 Introduction

Current service order handling systems in the telecommunications industry have problems such as user-unfriendliness, long service waiting time, uneasy customization of services, limited capability of in-house services, and inefficient human resource allocation. Consequently, the current solutions for the ordering process cannot be used to meet the business needs of Service Providers (SPs). To solve this problem, a generic service order handling model is suggested. Based on the TINA-C information modelling and computational modelling concepts, service order handling information object types and interfaces were defined.

Three ordering interfaces in this model are concerned with service negotiation, service ordering, and order tracking for the Main Service Provider (MSP) and Subcontracted Service Provider (SSP). Five information object types defined are the rFP (Request For Proposal), pR (proposal), oP (option), oR (order) and sR (service).

Section 2 explains the service order handling problem and requirements. Section 3 introduces the order handling process model. Section 4 and 5 describe the SP-to-SP order handling information model and SP-to-SP order handling interfaces respectively. Finally, conclusions will be described.

2 Service Order Handling Problem and Requirements

2.1 Service Order Handling Problem

A fast and accurate service ordering process is very crucial for Service Providers (SPs) in an increasingly deregulating and competitive telecommunications market. This requires the necessity for efficient service order handling and tracking across many interrelated SPs. The current service ordering processes in most SPs have the following characteristics:

- Customer unfriendliness

- Delayed service delivery time
- Few service choices
- In-house services production
- Dependency on the manual service order handling
- Limited service order handling capacity.

Consequently, because of the above characteristics, the current ordering process cannot be used to satisfy the business needs of SPs. Needless to say, the ordering process is too slow, contains many steps, requires too much re-keying of data, is difficult to track the order to know what the status of order is, and is too inflexible. Also, the external interfaces such as phone, fax, and mail are inefficient, the accuracy is low and there is a risk for misunderstanding because of the terminology inconsistencies among the SPs. These problems with the current ordering processes are especially serious when several SPs are involved in providing end user service in the deregulating and competitive telecommunications environment.

2.2 Service Order Handling Requirements

As we reviewed in the Section 2.1, the traditional service order handling systems currently used by the SPs will become obsolete when they are applied to the new and changing global telecommunications market. In developing new solutions for the service ordering process, we must consider the evolution direction of telecommunications market besides the current problems. Some of the requirements to consider additionally are as follows:

- Service orders usually come out of a discussion, comparing different alternatives, with the customer in a competitive environment.
- A pre-order process is needed that can handle several design proposals in a short time, with low cost, even if several SPs and Network Providers (NPs) are involved in the proposals.
- The recommendation must be applicable in a real multi-SP market.
- Many different types of SP roles must be supported and roles may also change over time and between different markets.
- The entire ordering process must be taken into account when describing processes and information models, otherwise it will not be possible to get overall efficiency.
- During the ordering process, the best efficiency is obtained when different parties of customer and SP are involved and can address information to the right peer party.

3 Order Handling Process Model

The overall goal of the ordering process model is to meet customers need for new services as efficiently as possible. This means that a whole range of different type of cases covers from customers wanting a simple quote on a single service to customers going through a complete design and negotiation phase of complex offering. The ordering process can stop with a proposal to be considered by the customer or go the whole way to firm orders and implementation of the offer. This puts some extra demands on the process model which must be reflected in the model as a high degree of flexibility. The generic ordering process model [8] is given in the Fig. 1 below.

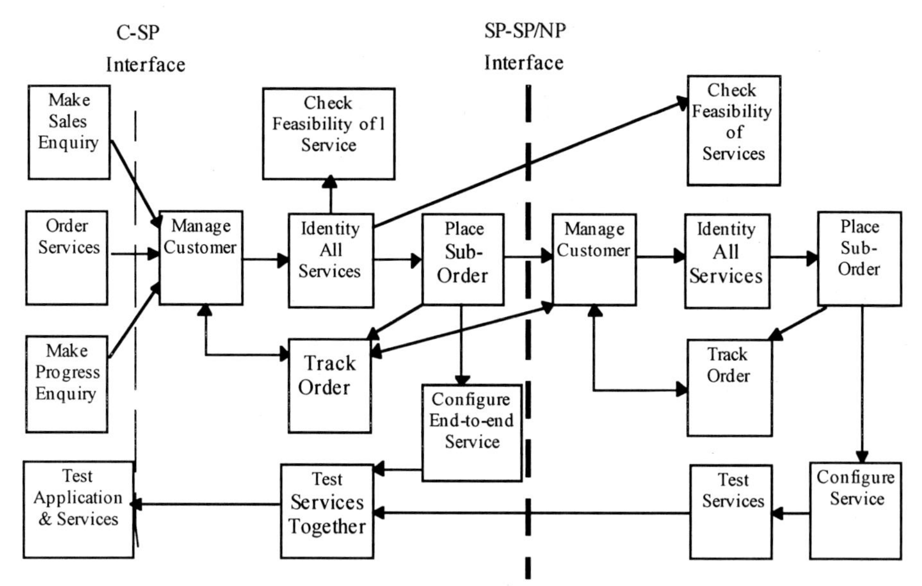

Figure 1: Generic Ordering Process Model

It is important to notice that the process model only focuses on the activities involved in ordering. There are other views of ordering that are important, like the information flow, that is not depicted in this model. There is a need to add more dimension to this model to show a more complete picture of the ordering area. The meaning of the signs and symbols in this figure can shortly be described as the following:

- Box: A process, i.e., a set of activities with a common goal, a well defined start and end point.
- Arrow: The hand over from one process to another, i.e., the trigger for another process to start. They are not showing the information flow.
- Broken line: Interfaces between the main actors involved in ordering. Besides these main interfaces there may be other interfaces depending on the actual market situation.

In the Fig. 1, we can find that there are two types of interfaces. The first type of interface is concerned with the interaction between a customer and Service Provider (SP). The second one is the interface between an SP and an SP (or a Network Provider (NP)). Here, we are interested in the second type of interface, i.e., the SP-to-SP interface.

Interactions between the Main Service Provider (MSP) and Subcontracted Service Providers (SSPs) are related to different processes in ordering. In a real environment, we can think of more than one instance of SP-to-SP interface and an SSP can be an MSP for the other SP(s) because the roles of an SP are relative. Three main interfaces needed to be defined in order to achieve efficiency throughout the ordering process are:

- Identify All Services -> Check Feasibility of Individual Services
- Place Sub-Order -> Manage Customer
- Track Order -> Manage Customer

- Place Sub-Order -> Manage Customer
- Track Order -> Manage Customer

4 SP-TO-SP Order Handling Information Model

4.1 Overview

The first step in the information modelling in the SP-to-SP order handling is identifying the information object types involved. The following information object types are relevant to the ordering handling interface:

- rFP (Request for Proposal)
- pR (Proposal)
- oP (Option)
- oR (Order) and
- sR (Service).

Here, the first three object types are used in the pre-ordering phase, and the last two object types are used in the ordering phase.

The rFP information object type represents the management view of Requests for Proposals (RFPs) which are entered by an MSP in the pre-ordering phase.

The pR information object type represents the management view of proposals offered by the SSP in the pre-ordering phase.

The oP information object type represents the management view of options which are created by the SSP in the pre-ordering phase.

The oR information object type represents the management view of orders which are placed by an SSP in the ordering phase.

Finally, the sR information object type represents the management view of service orders which are placed by an MSP in an order request during the ordering phase.

For all the above information object types, the detailed definitions on all the attributes for each information object type can be found at the document [14].

Next , we examine the state model of each information object type. We begin with the rFP information object type in the following Figure 2.

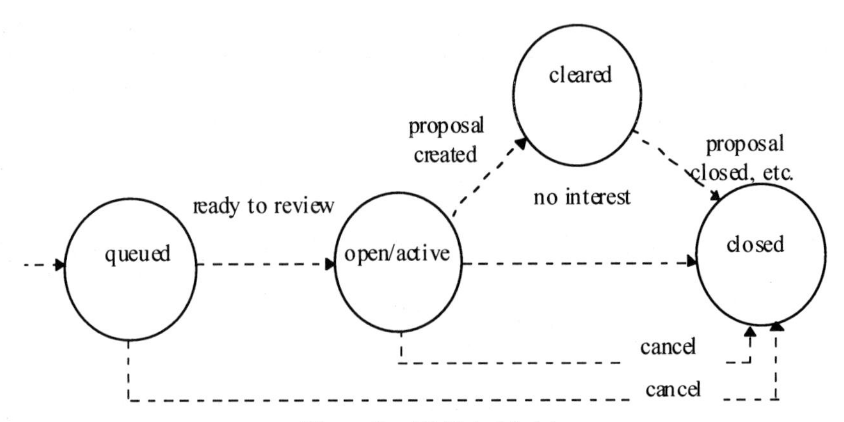

Figure 2: rFP State Model

The oP state model is shown in the following Figure 3.

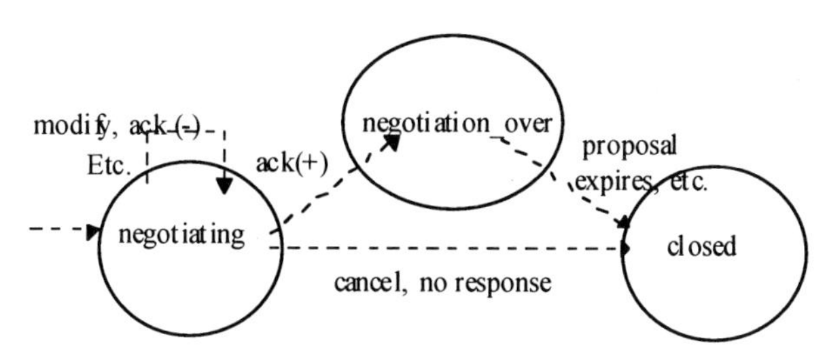

Figure 3: oP State Model

The state model of the sR information object type is shown in the following Fig. 4.

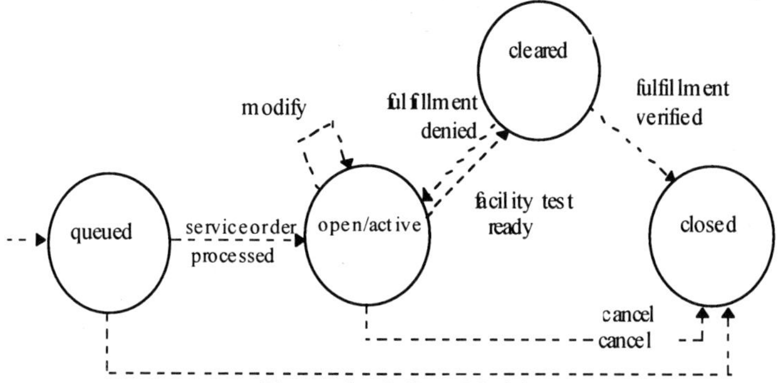

Figure 4: sR State Model

4.2 Technical Specification of Order Handling Information Object Types

In this section, we briefly describe a TINA-C specification of order handling information object types. The information specification of a distributed application provides the necessary knowledge to interact appropriately within the application. It describes such knowledge in the form of information object types and relationships among them.

TINA-C introduces an information specification language called quasi-GDMO+GRM (so called because of its resemblance to the GDMO [5] and [6]). In quasi-GDMO+GRM, an information object type is described in terms of states and actions. The state of an information object type is characterized by attributes. The actions of an information object type are the means to modify the state of an object. The state model does not associate events with an object type, although events are the causes to have actions performed on an object. Instead, events are associated with a computational interface, since the same information object type may be used for different computational interfaces. By associating events with a computational interface and then mapping events to actions of an information object type, one can give an operational semantics specification of a computational interface.

5 SP-TO-SP Order Handling Interfaces

5.1 Overview

Conceptually, there are three interfaces for the SP-to-SP order handling. They are Service Negotiation Interface, Service Ordering Interface, and Order Tracking Interface. These interfaces can be implemented as a single physical implementation according to the implementor's own idea. For these three interfaces, the full details of technical specifications can be found at the document [14]. Here, we summarize the framework of SP-to-SP order handling interfaces.

Service Negotiation Interface
Used in the pre-ordering phase, the two sub-interfaces of the Service Ordering Interface here provide capabilities to negotiate on an option (in a proposal) and track a request for proposal: SN_MSP and SN_SSP.
The operations in the SN_MSP sub-interface allow the MSP to;

- enter a request for proposal (rFPCreate),
- view an option or a proposal (oPView, pRView),
- modify an option or a proposal (oPModify, pRModify),
- acknowledge an SSP either positively or negatively (oPAcknowledge) and
- track the status of an RFP (rFPStatusTrack).

The operations in the SN_SSP sub-interface allow the SSP to inform an MSP of the following events:

- creation of a proposal (pRCreationNotify),
- modification of an option or a proposal (oPModificationNotify, pRModificationNotify),
- acknowledgement of an MSP either positively or negatively (oPAcknowledgeNotify),
- cancellation of an option or a proposal (oPCancelNotify, pRCancelNotify) and
- notification of a status update of the request for proposal (rFPStatusUpdate).

Service Ordering Interface
In an order, there are normally one or more service orders. The sub-interfaces, SO_MSP and SO_SSP here provide the capabilities to manage an active order or service order during the ordering phase.
The operations in the SO_MSP allow the MSP to:

- place an order (oRCreate),
- modify an order or a service order (oRModify, sRModify),
- view an order or a service order (oRView, sRView),
- cancel an order or a service order (oRCancel, sRCancel) and
- verify the fulfilment of an order or a service order (oRVerify, sRVerify).

The operations in the SO_SSP sub-interface allow the SSP to inform the MSP of the following events:

- modification of an order or a service order (oRModificationNotify, sRModificationNotify),
- creation of a history log record for an order (oRHistoryNotify) and
- deletion of an order (oRDeletionNotify).

Order Tracking Interface
Used in the ordering phase, the sub-interfaces here support tracking of an order or a service order in three different ways. First, they allow the MSP to poll the status. Second, they allow the SSP to inform the MSP of a status change. Finally, they allow the SSP to send a status progress report to the MSP at a pre-defined frequency.
The operations in the OT_MSP sub-interface allow the MSP to:

- track the status of an order or a service order (oRStatusTrack, sRStatusTrack) and

- set the status window size of an order or a service order (oRWindowSet, sRWindowSet).

The operations in the OT_SSP sub-interface allow the SSP to inform the MSP of the following events:

- status update of an order or a service order (oRStatusUpdate, sRStatusUpdate) and
- status update of an order or a service order at a pre-defined frequency (oRStatusWindowUpdate, sRStatusWindowUpdate).

5.2 Technical Specification of Order Handling Interface

TINA-C architecture [13] is decomposed into four main subsets: service architecture, network architecture, management architecture and computing architecture. Among these architectures, TINA-C computing architecture [10] defines concepts and principles for designing and building software and software support environment in a telecommunications information network (e.g., Bellcore's Advanced Intelligent Network). The concepts are based on the Open Distributed Processing (ODP) standard [3] which is a standard for the definition of generic, non-telecommunications specific distributed systems.

The major TINA computational modelling concepts [10] are:

- computational object: The components of a distributed application are represented as computational objects. Computational objects are the units of structure and distribution.
- computational interface: While computational objects are the units of structure and encapsulation of (application-specific) services, interfaces are the units of provision of services. They are the places at which computational objects can interact and obtain services.

ODL-95 [12] supports separation of computational object design from computational interface design. The separation offers freedom to application developers for independent declaration of objects and interfaces. An interface is not necessarily designed for a specific object. It can be used by any object where appropriate. Applying this separation principle to our context, it is also possible to develop customer-SP order handling interfaces independent of application developers designing customer or SP order handling objects.

6 CONCLUSIONS

Currently used service order handling systems used by the telecommunications arena cannot be used to successfully meet the diverse business needs of service providers.

To solve this problem, a generic service order handling model was suggested. Based on the TINA-C information modelling and computational modelling concepts, necessary information object types and attributes, and also interfaces were defined.

The five information objects, the RFP, order, service, proposal and option object types were defined and the operations and input/output parameters for three interfaces were defined for the service negotiation, service ordering and order tracking for the MSP and SSP respectively.

By using the service order handling model suggested, we can realize the provision of the automated and standardized service order handling between SPs in the deregulating and competitive telecommunications market. The possible positive effects are:

- By ensuring less manual interventions (e.g., re-typing) and errors, we can improve the QoS.
- We can achieve the capability for visibility of an ordering process. This is very important for SPs who sub-contract the elements of a customer order to multiple other SPs.
- We can minimize the service provisioning time of service to the end customer.
- We can reduce the order processing cost.
- We can provide an open interface to an SP's legacy systems and platforms.

The possible extensions to the proposed model would be the customer-SP service order handling interface, integration with the other customer care processes, and mapping to the real service order handling systems for the improvement of the model. Currently, some interested SPs are already in the process of implementing the service order handling system based on our proposed model and the NMF (Network Management Forum) SMART (Service Management Automation and Re-engineering Team) is in the process of reviewing our contribution [14] at the SMART Ordering Team now to adopt it as the international industry standard in the telecommunications arena.

REFERENCES

[1] A Service Management Business Process Model - Issue 1.0, Network Management Forum.
[2] Customer to Service Provider Trouble Management Interface Specification, The SMART TT Meeting, London, OSE Laboratory, Computer Networking Department, Computer Science Telecommunications, University of Missouri-Kansas City, October 30-31, 1995.
[3] ITU Draft Recommendation X.901, Information Technology - Open Systems Interconnection - Basic Reference Model of Open Distributed Processing, Part 1: Overview.
[4] ITU Recommendation M.3010, Principles for a Telecommunications Management Network, 1992.
[5] ITU Recommendation X.722 - Information Technology - Open Systems Interconnection - Structure of management Information: Guidelines for the Definition of Managed Object, Geneva, 1992.
[6] ITU Recommendation X.725 - Information Technology - Open Systems Interconnection - Structure of Management Information: General Relationship Model (GRM) Model, Part 7, 1991.
[7] Open Networking with OSI, Adrian Tang and Sophia Scoggins, Prentice Hall, 1992.
[8] SMART Ordering White Paper, NM Forum, Version 1.0, October 1995.
[9] Statement of User Requirements for Management of Networked Information Systems, Network Management Forum User Advisory Council, October 1992.
[10] TINA-C Computational Modelling Concepts, Version 2.0, Document No. TB_A2.HC.012_1.2_94, TINA-C, February 1995.
[11] TINA-C Information Modelling Concepts, Version 2.0, Document No. TB_EAC.001_1.2_94, TINA-C, April 3, 1995.
[12] TINA-C Object Definition Language (TINA-ODL) Manual, Version 1.3, TINA-C, June 20, 1995.
[13] TINA-C Overall Concepts and Principles of TINA, Version 1.0, 1995.
[14] "SP-to-SP Order Handling Information Modelling and Interface - Technical Specification", Version 2.0, Open Systems Environment Lab, University of Missouri-Kansas City, U. S. A., 1996.

From IN toward TINA - Potential Migration Steps

U. Herzog
Deutsche Telekom
Technologiezentrum Forschung, P.O.Box 100003,
D-64276 Darmstadt, Germany
e-mail: herzog@tzd.telekom.de

T. Magedanz
GMD FOKUS / TU Berlin
Hardenbergplatz 2, D-10623 Berlin,
Germany
Email: magedanz@fokus.gmd.de

This paper gives an overview of the basic architectural migration steps from current Intelligent Network (IN) based service platforms toward emerging Telecommunications Information Networking Architecture (TINA) based service environments. This evolution can be regarded as a stepwise replacement of IN functional entities by means of a set of appropriate TINA computational objects realised on top of a Distributed Processing Environment (DPE). Hence in the course of evolution, a service platform may comprise two parts; a still IN-conformant part and an already TINA-conformant part. This raises the need for the definition of appropriate adaptation units, allowing for example an IN Service Switching Function to interwork with TINA objects implementing the service control and management logic. Whereas the upper part of the adaptation concentrates on the mapping of concrete IN Application Protocol operations onto appropriate method invocations of TINA computational objects, on the bottom this adaptation has to deal with the interworking between the Signalling System No. 7 network protocols, and inter-DPE-node protocols.

Keywords: Adaptation Units, CORBA, DPE, IN, INAP, KTN, SS7, TINA

1 Introduction

Today *Intelligent Network (IN)* based [Q.12xx] service platforms provide the foundation for the provision of advanced telephony services on top of narrowband networks, such as PSTN and ISDN. Since these platforms represent a fundamental investment for network operators, there is no big motivation for operators to replace these IN platforms in the short term. However, with increasing demand for multimedia applications on top of emerging broadband networks, i.e. B-ISDN, it will become increasingly important for operators to study the enhancement and evolution of their IN service platforms in order to cope with the new customer requirements. Here the *Telecommunications Information Networking Architecture (TINA)* [TSA-96] is considered as a promising basis, since it is designed to support future multimedia multiparty services. In addition, *Distributed Processing Environment (DPE)* technology, such as OMG's *Common Object Request Broker Architecture (CORBA)* [OMG-95], the foundation for TINA, is gaining increased acceptance in the telecommunications environment. This raises the question for a possible migration path from the *function-oriented* IN architecture toward the *object-oriented* TINA architecture.

Although both IN and TINA represent service control (and management) architectures for bearer networks there are fundamental differences between both, which makes the migration quite complex [Bro-95, Dem-95]. Let us briefly take a look at both architectures.

The IN architecture is based on the idea of separating switching and service control by centralising service control in specific network nodes. These nodes remotely control the switches, which perform connectivity in accord with a *Basic Call State Model (BCSM),* using the *Signaling System No 7 (SS7)* network. Therefore the IN defines a set of dedicated functional entities, such as a *Service Control Function (SCF)* hosting service logic programs, a *Service Data Function (SDF)* hosting service related data, a *Service Switching Function (SSF)* enabling service triggering from the switches, and a

Service Management Function (SMF) supporting IN service management. These entities interact for the provision of IN service features by means of *Information Flows*, which are physically realised via the SS7 network by a dedicated *IN Application Protocol (INAP).*
TINA is based upon DPE technology, which has to be available in each TINA network node. Services in TINA are modelled by interacting *Computational Objects (COs),* where each object comprises logic and data and offers specific operations. TINA defines within the service architecture a set of generic COs. The DPE supports the arbitrary distribution and the communication of these objects to arbitrary nodes. A *Kernel Transport Network (KTN)* takes care for the communication within the DPE, i.e. between the DPE nodes. This means in TINA computational object interactions (supported by the DPE and the KTN) correspond to the information flows between IN functional entities (via INAP on top of the SS7 network). However, in contrast to the switch residing functions for controlling basic connectivity, TINA adopts a management oriented view of connectivity, known as *Connection Management* [TCMA-95], where connectivity control functions are performed in a top-down fashion. Finally it has to be stressed that most IN service capabilities are related to flexible screening, routing, and charging of calls. In TINA such functionalities are related mostly to the *generic Access Session* concept rather than to a particular *Service (Session).* In summary there is no one-to-one mapping of IN services to TINA services and IN functional entities and TINA objects.[1]
Migration or evolution of an IN-based service platform toward a TINA platform means a step by step 'TINArisation' of the IN platform elements, i.e. specific IN functional entities will be replaced by means of appropriate TINA service components, i.e. COs on top of a DPE. The resulting service platform comprises two parts, i.e. a legacy IN part and a new TINA(rised) part, which have to interwork for the provision of (IN) services. Consequently corresponding *Adaptation Units (AUs)* have to be defined for the mapping between INAP operations [Q.1218] transferred on top of the SS7 network [Mod-90] and interface operations of TINA COs [TSA-96] on top of a DPE [Kit-95].
In the following section we identify the basic migration steps from IN toward TINA. Section 3 provides a description of the adaptation unit capabilities required for the most realistic evolution scenario, featuring an IN-based service access part and a TINArised service control and management part. Section 4 provides some general considerations on the underlying SS7 / CORBA interworking. Section 5 concludes the paper.[2]

2 Migration of IN toward TINA

TINA concepts [TSA-96] and the inherent usage of DPE/CORBA technology [OMG-95] are regarded as enabling technology, to provide the necessary flexibility for the emerging telecommunication service market. The advantages of introducing CORBA / TINA based solutions within the IN are mainly related to the possible

---

[1] For more details on IN and TINA interested readers are referred to [Mag-96].
[2] The information presented in this paper has been primarily developed within the joint TINA-C auxiliary project / EURESCOM project P508 *„EVOLUTION, MIGRATION PATHS AND INTERWORKING TO TINA"* from 1995-96. However, this does not imply that the paper reflects necessarily the common technical position of all EURESCOM Shareholders/Parties.

rationalisation of the service aspects (e.g., the integration of service management and control), to a higher level of interoperability between applications running on DPEs, to the ability to extend service related capabilities (e.g., several points of control for a service), scalability of the service platform, vendor independence, etc.
In general, TINA concepts may be used in the following IN areas:

- *Service Management*: The introduction of TINA in the Service Management area seems to be promising because there is a lack of standardised IN management solutions and TINA offers the ability to *integrate* service management and control aspects by means of common objects and protocols.
- *Service Data*: TINA could be useful for supporting distributed incall and outcall signalling related personal profile access, in particular for personal and terminal mobility support.
- *Service Control*: Access Session and Service Session mechanisms could be usefully adopted in order to provide enhanced flexibility for supporting multi-party/multi-connection capabilities.

As outlined, the migration from IN toward TINA will be based on a step-by-step replacement of IN functional entities by means of an appropriate set of interacting TINA COs. This means that TINA COs are realising *quasi IN services.*[3] The replacement is believed to start with those IN elements that are deployed in small numbers in order to save the network operator's investments. Thus the SSFs already deployed on a global basis in the exchanges, representing the biggest amount of IN related investment, would most likely be the last elements to be replaced/TINArised.
From a technical perspective this migration is based on the extension of a DPE, forming the core part of a TINA platform, starting from a limited number of network nodes until all system parts (including switches and end-user systems) are covered. This results in a replacement of existing static interfaces and protocols for service control and management, such as INAP and CMIP, by more flexible DPE/CORBA mechanisms, i.e. the *Generic Inter-ORB Protocol (GIOP).*
The process of *TINArising* an IN step-by-step leads to the need of defining corresponding adaptation units, called *AUs*. These units should support the communication between the existing IN system part and the new TINArised part. In other words they should cope with all adaptation functionality (conversion of protocols, adaptation of models, etc.) needed. Therefore the selection of the IN entities to be TINArised in a particular step has to be carefully evaluated. Basically three major evolution steps seem most likely in this context, as illustrated in Fig. 1.

2.1 TINArisation of IN Service Data (and Management)

From an academic point of view the TINArisation of IN service data may be interesting. Thus functions envisaged for the IN SDF could be provided by TINA service components.[4] This approach allows one to take advantage of TINA data modelling concepts and data distribution transparency provided by the DPE. Consequently an AU would be required, enabling an IN SCF to access service operational data and service management data within the TINA platform. As an option within this evolution step the SMF may also be TINArised. However, notice

---

[3] Note that there are not any real *IN services,* but just a set of services which could be provided easily by an IN. The notion of a quasi In service is used here for a TINA service providing IN-like service capabilities!
[4] I.e. the personal profile, usage context, service profile, subscription agent, etc.

that this scenario is quite unrealistic because it decouples service logic and service data, which is in contradiction with an object-oriented approach promoted by TINA. Therefore the next step seems to be more realistic!

2.2 Harmonised TINArisation of IN Service Control, Data and Management

This scenario promotes the harmonised TINArisation of the SMF, SCF, SDF[5] in order to take full advantage of the TINA service architecture and the object-oriented approach. Only the SSF remains IN compliant, since it represents the biggest IN investment. This scenario allows network operators to keep the existing *Basic Call State Model (BCSM)* and the INAP interfaces in operation as long as possible/required. This means that IN service control and service management are TINArised, i.e. are implemented by means of appropriate TINA COs on top of a DPE. Note that this could be realised in a *TINA-based IN Service Node* or in a real distributed manner by a collection of distributed TINA nodes. Correspondingly dedicated AUs are required in order to enable the interworking between the TINA COs implementing IN SCF/SDF and IN SMF capabilities, and the IN SSF for the sake of service control and management. One basic rationale for an operator to adopt this migration step is to introduce TINA concepts for the realisation of IN services on top of an existing narrowband network, particularly for new operators without existing IN service infrastructure. Going straight for a TINA service infrastructure allows an operator to gain early experience with the TINA service architecture and to exploit this architecture in the long term, when the narrowband network evolves toward a broadband network.

2.3 TINA to the switch

This scenario goes one step beyond the previous one by also replacing the IN SSF. Therefore it represents the ultimate evolution step toward TINA. This scenario will be interesting with the introduction of new (broadband) switch generations, such as emerging programmable switches, which offer a very granular interface in order to control switching functions. This (proprietary) interface is described in terms of events (sent from the switch to the host controller) and commands (sent from the host computer to the switch). A programmable switch provides call control functions that can be controlled and augmented by functions located in a host computer. This organisation enables the control software for controlling the programmable switch to be organised according to TINA service architecture solutions and be deployed in host computers directly connected to the switch. The host computer itself can be equipped with the DPE so that it is integrated in a fully TINA compliant architecture. In this sense the “call processing” could be aligned to the TINA “call control” capabilities (i.e., the TINA service and connection management architecture).[6] By this approach the INAP interface is no longer required and the exchanges can directly invoke the operations provided by the TINA computational objects via the DPE. Furthermore, this approach may also enable the usage of *Mobile Agent technology* for downloading on demand service programs to the switches or the switch drivers [Kra-96].

---

[5] Note that the Specialized Resource Function is not considered here for simplification!

[6] For more information see [Lic-96].

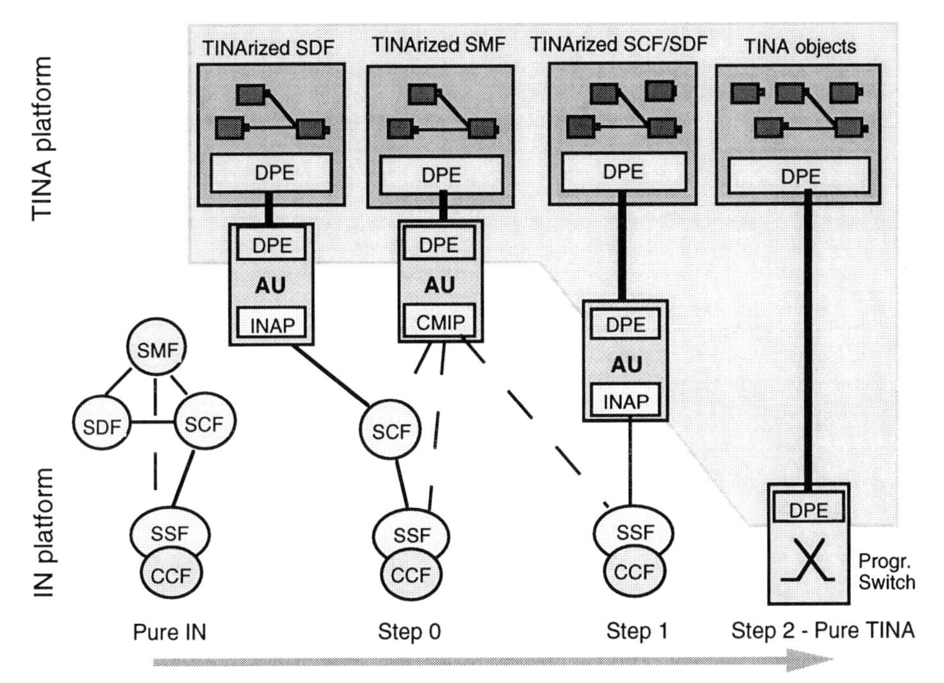

Figure 1: Step-by-Step Migration of IN toward TINA

Subsequently we concentrate on the harmonised TINArisation of the SMF, SCF, SDF presented in step 1 and the related interworking issues in the area of service control, since this scenario represents the most likely one in the near future. For more information on the other steps and the related adaptation issues interested readers are referred to [P508-97].

3 TINArised IN Service Control with IN-based Service Access

Below we assume that IN services, i.e. IN service logic and data, will be realised by appropriate COs in the TINA environment. In this context we concentrate on the AU between an IN SSF and the TINA part, also called *IN(-SCF)-AU*. The AU looks from the IN side as an SCF, whereas on the TINA side the AU behaves like a TINA (end user) system comprising multiple COs and maps SS7-based INAP operations invoked by the IN SSF into TINA CO method invocations via a DPE and vice versa. Thereby, the AU enables the establishment of a (service) control relationship between an SSF in the IN domain and a *Service Session Manager (SSM)* in the TINA domain, i.e. the TINA SSM controls the IN SSF and the related *Connection Control Function (CCF)* in the switch through the AU.

Obviously, the IN(-SCF)-AU definition depends strongly on the way of modelling IN services in a TINA environment:

- For a call which originates and terminates on the *IN side*, this means that both the calling and the called party don't use a TINA end system and are not modelled as TINA users. The call will be established entirely through the IN SSF and the CCF (under supervision of the TINA SSM via the AU).
- For a call which terminates on the *TINA side*, i.e. the called party is modelled as a TINA user and is using a TINA end system, things are much more complicated.

Here one part of the call establishment has to be handled on the IN side (i.e. through the SSF/CCF from the calling party terminal to an appropriate IN/TINA bridge) and the other part of the call establishment has to be handled on the TINA side via TINA Connection Management principles (i.e. through the *Communication Session Manager (CSM)* from the IN/TINA bridge to the called party end system).

Furthermore, looking at the realisation of IN services within TINA, a corresponding service session has to be implemented. However, it has to be recognised that most IN service capabilities are strongly related to the flexible routing and charging of calls (including authorisation, etc.). These capabilities are usually related to the *Access Session* within the TINA Service Architecture. Therefore two options exist for the modelling of IN service capabilities within TINA and consequently have impacts on the IN(-SCF)-AU design, which will be discussed in more detail in the following subsections:

1. The *wrapped IN* TINA service, where no call party is modelled as TINA user and the IN-like service capabilities are encapsulated within the TINA service session COs.
2. The *real IN-like* TINA service, where the call parties are modelled as TINA users and the IN-like service capabilities are provided jointly by means of a TINA access session and service session related COs.

3.1 The *wrapped IN* TINA Service

Within this approach the (quasi-IN) service capabilities are completely modelled within specific Service Session COs, i.e. a corresponding *User Application (UAP_{IN})* interrogating an SSM_{IN}[7], which makes use of service specific IN *Service Support Object (SSO),* e.g. a database, containing service subscriber information. This means that no call party (including the service subscriber) will be modelled as a specific TINA user. Consequently no user-specific access session can be performed, since there are not any call party specific access session related COs, such as a named *User Agent (UA).* Only an *anonymous user* is considered here.

From the TINA side the AU looks and behaves like an enduser system and thus comprises, besides the UAP_{IN}, a generic *Provider Agent (PA_{IN}),* and a *Generic Session Endpoint (GSEP).* Therefore the AU probably relates to the TINA *Retailer reference point (Ret-RP)* [TRP-96] as illustrated in Figure 2.

The function of the AU may be displayed by considering a Freephone service implementation. On reception of the first INAP request from the SSF, i.e. an **`Initial DP`** (1), the AU passes this information to the UAP_{IN}, which in turn initiates a corresponding TINA service session via the PA_{IN}. For that the PA_{IN} interacts with an *anonymous* UA_{IN} (2) within the provider domain in order to perform a generic access session for service session establishment.[8] Once the SSM_{IN} has been created and initialised (3) by a *Service Factory (SF),* a control relationship is established between the IN SSF and the TINA SSM_{IN}.

---

[7] Note that for simplification we omit the notion of an User Session Manager.

[8] The appropriate PA_{IN} may be determined by the given `Dialled Number` (or by the `IN Service Key`). Furthermore, the corresponding UA_{IN}, to be addressed by an *user name* (i.e. an object reference) has to be determined based on the dialled digits. This determination has to be performed by the UAP_{IN}.

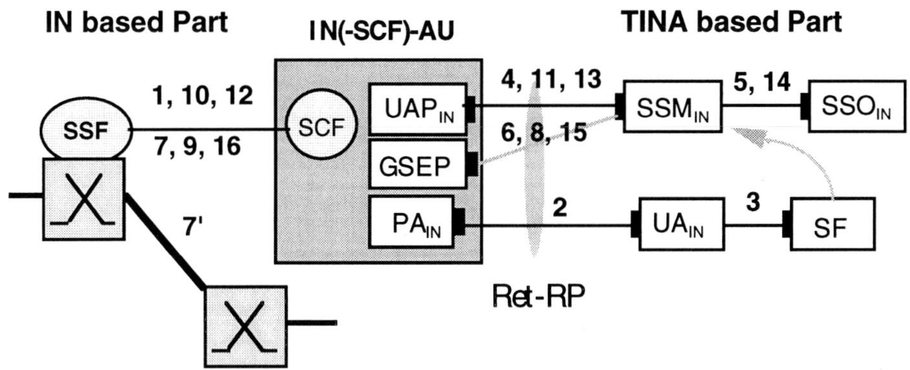

Figure 2: IN(-SCF)-AU Capabilities for *wrapped IN* TINA Service

The UAP$_{IN}$ can now start/use the service by passing the additional information given by the **`Initial DP`** INAP operation, i.e. the `Call ID`, `Miscellaneous Call Information` and the `Dialled Digits`, to the SSM$_{IN}$ (4), in order to obtain an appropriate `Destination Routing Address` required for the **`Connect`** INAP operation to be sent from the AU to the IN SSF (6, 7). Note that the SSM$_{IN}$ translates in accord with the implemented Freephone service logic and exclusive usage of an appropriate SSO$_{IN}$ the given dialled digits into the corresponding destination number (5). Note that the SSM$_{IN}$ does not create any logical connection graphs.
Subsequently, the SSM$_{IN}$ invokes another operation at the UAP$_{IN}$ in order to establish the monitoring of a specific call event for the reverse charging capability (8). Based on this operation the AU sends a corresponding **`Request Report BCSM Event`** INAP operation to the SSF (9) allowing the AU to be informed about the call duration. In turn the SSF will generate two **`Event Report BCSM`** INAP operations (one for the start and one for the termination of the call), which will be mapped by the AU onto appropriate notifications to the SSM$_{IN}$ (10-13). On invocation of the second operation the SSM$_{IN}$ determines the call charges and interacts with the SSO$_{IN}$ CO (14). Furthermore, the SSM$_{IN}$ issues a termination request to the AU in order to terminate the control relationship (15). The AU issues a **`TC_End`** INAP (i.e. TCAP) operation to the SSF for terminating the control relationship (16).
The above scenario only limits the required IN(-SCF)-AU capabilities to call/service control aspects, whereby the SSF being responsible for establishing the corresponding call connection to the called party's endsystem. No TINA *Communication Session* is established on the TINA side.
Note that in this scenario both UAP$_{IN}$ and SSM$_{IN}$ could be kept IN service independent, since the interactions between the UAP$_{IN}$ and the SSM$_{IN}$ are closely aligned with INAP operations. For example, operations related to specific SIBs or service features could be grouped at specific IN-SSM interfaces. Hence many IN services can be modelled uniformly within a TINA Service Session, without the need of separating IN service capabilities into Access Session and Service Session capabilities within TINA.

3.2 The *real IN-like* TINA Service

In the course of evolution, TINA conformant end user systems will be available on a global basis in order to enable advanced multimedia (TINA-based) services. Consequently each service subscriber and furthermore each call party could be modelled as TINA users and hence is represented by a corresponding *named UA* in

TINA, which interacts with other related Access Session objects. Since basic parts of an IN service are related to routing, charging, and authorisation capabilities, these may already be handled within the TINA Access Session instead of entirely within a specific (IN) Service Session (compare with the previous section). Particularly the user specific access session objects could be used for the realisation of some quasi IN capabilities. Only the additional *IN service intelligence*, which could not be accomplished within the access session, will be modelled within a specific SSM.
In this scenario we assume that the call will be initiated by a non-TINA user from a non-TINA terminal identical to the previous scenario (1, 2). However, in contrast to the *wrapped IN* service scenario, the SSM will receive from the UAP_{IN} a *service invitation* for the called party (3). As depicted in Fig. 3, the called party is represented here by the a corresponding set of access session objects, particularly a named UA. Hence subsequently the SSM contacts the corresponding named UA of the called party, instead of using a specific SSO (4). The UA of the called party interacts with the other access session objects, such as a personal profile, and returns the appropriate terminal information (in accord to time of day or call origin, etc.) to the SSM (5). This information is used by the SSM to deliver the invitation to the UAP of the called party (6). In case of an acceptance, the SSM generates a logical connection graph which will be passed (7) to the *Communication Session Manager (CSM)*. In addition (not illustrated in the figure) charging treatments may be established.

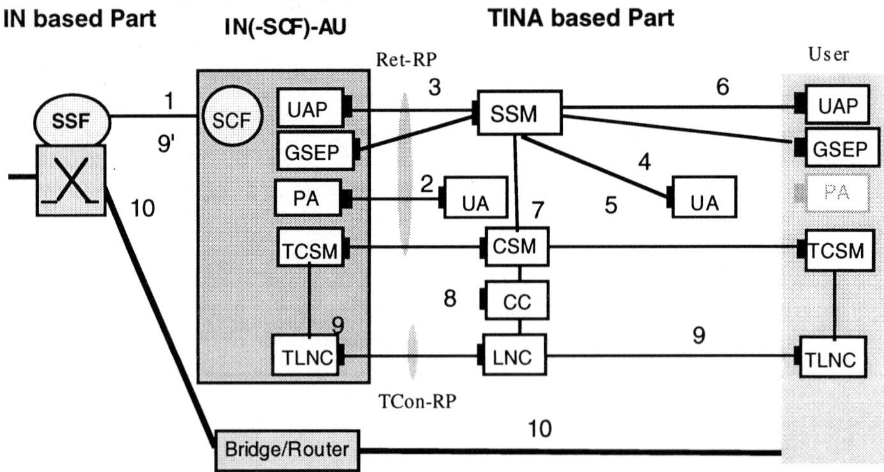

Figure 3: IN(-SCF)-AU Capabilities for *real IN-like* TINA Service

Note that in this scenario the call/connection establishment is separated! One part of the connection, i.e. from the calling party, is established via the SSF/CCF on the IN side via an INAP `Connect` operation (9). The other part on the TINA side to the called party is established by means of TINA *Connection Management* via the CSM (7, 8, 9) and the related connection management objects. Therefore the IN(-SCF)-AU or a dedicated node known by the AU has to provide some kind of bearer connection bridge, which can be regarded as a virtual endpoint for both the IN part and the TINA part of the entire call connection. Therefore the AU has to provide additional objects required for the connection management, such as a *Terminating Communication Session Manager (TCSM)* and a *Terminating Layer Network Coordinator (TLNC)*. Consequently, the AU also relates, in addition to the Retailer reference point, to the *Terminal Connection (Tcon)* reference point [TRP-96].

The basic advantage of this approach for TINA users is that there is no difference between IN triggered or TINA triggered services. The basic drawback of this approach is that one has to separate IN service capabilities into TINA access capabilities and pure IN service control capabilities, which requires a careful and comprehensive analysis.

4 SS7 / CORBA Interworking/Integration

We have illustrated that there are different scenarios for implementing adaptation units for IN-based service access with TINA based service control. Common to all AUs is the need to map INAP operations onto TINA CO interface operations, which on a lower level requires the interworking between the SS7 network, providing the basis for INAP, and the KTN, providing the basis for the communication *between Object Request Brokers (ORBs)* within the TINA DPE. This means that the interworking/integration of SS7 and CORBA represents the technical foundation and thus has to be studied in more detail.

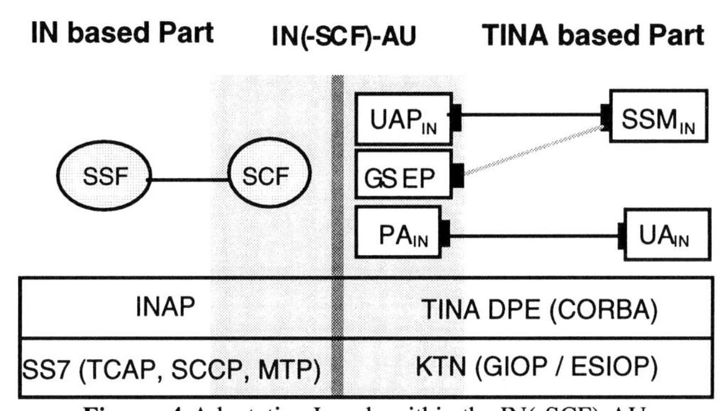

Figure 4: Adaptation Levels within the IN(-SCF)-AU

In respect to general IN-TINA migration steps, two prime aspects are of relevance:

1. Bridging/interworking between the SS7 / CORBA domains such that the external protocol interfaces to legacy systems could be maintained.[9] Note that the interworking could be related to different layers of the SS7 protocol suite, consisting of a *Transaction Capabilities Application Part (TCAP)*, the *Signalling Connection Control Part (SCCP)*, and the *Message Transfer Part (MTP)*, and systems of equivalent functionalities implemented using CORBA technology.
2. Implementation of a *Generic Inter-ORB Protocol (GIOP)* on top of SS7 or use of SS7 protocol suite as an *Environment-Specific Inter-ORB Protocol (ESIOP)* protocol for communications between CORBA-based implementations (of IN components). This means that the highly-reliable SS7 network, which is used for time-critical communications between signalling network elements, is used as a TINA DPE *Kernel Transport Network (KTN)*.

An analysis of these issues was started at the end of 1996 within the OMG Telecom Domain Task Force by issuing a *Request for Information (RFI)* on IN / CORBA Interworking.[10]

---

[9] The approach is more or the same as the one used for making CORBA DPEs interwork with existing TMN (CMIP) agents.

[10] Interested readers should contact the OMG anonymous FTP server. Path names are respectively: “ftp.omg.org” and “/pub/docs/telecom”.

5 Conclusions

Within this paper we have illustrated that TINA can play a significant role in IN evolution. On the road toward TINA, i.e. the step-by-step TINArisation of IN entities, adaptation between IN parts and new TINArised parts of a service platform becomes fundamental. This adaptation concentrates on the mapping of INAP operations onto TINA object interface operations at the top level and SS7 / CORBA integration on the bottom level. The later is now being considered as a major topic within the OMG Telecom Domain Task Force [P508-97b]. Hence in the near future CORBA technology could be introduced in the IN domain [Maz-97] paving also the way toward TINA.

References

[Bro-94] D.K. Brown: "Practical Issues Involved in the Architectural evolution from IN to TINA", pp. 270-275, Int. Conference on IN (ICIN), Bordeaux, France, Oct. 1994

[Dem-95] L. Demounem, H. Zuidweg: "On the Co-existence of IN and TINA", pp. 131-148, TINA 95, Melbourne, Australia, February 1995

[Her-97] U. Herzog, T. Magedanz: "IN and TINA - How to solve the Interworking" IEEE IN Workshop, Colorado Springs, USA, June 1997

[Kit-95] B. Kitson: "CORBA and TINA: The Architectural Relationship", in: Proceedings of the 5th TINA Workshop, pp. 371-386, Melbourne, Australia, February 1995

[Kra-96] S. Krause, T. Magedanz: "Mobile Service Agents enabling "Intelligence on Demand" in Telecommunications", IEEE Global Telecommunications Conference, pp.78-85, London, United Kingdom, November 1996

[Lic-96] C.A.Licciardi et.al.: "Would you use TINA in your IN based Network?" pp. 35-40, Int. Conference on IN (ICIN), Bordeaux, France, November 1996

[Mag-96] T. Magedanz , R. Popescu-Zeletin: "Intelligent Networks - Basic Technology, Standards and Evolution", International Thomson Computer Press, ISBN: 1-85032-293-7, 1996

[Maz-97] S. Mazumdar, N. Mitra: "ROS-to-CORBA Mappings: A First Step towards Intelligent Networking using CORBA", in: "Technology for Cooperative Competition - IS&N'97", Springer Publishers, May 1997

[Mod-90] A. Modaressi, R. Skoog: "Signalling System No. 7: A Tutorial", IEEE Communications Magazine, pp. 19-35, July 1990

[OMG-95] OMG: "Common Object Request Broker Architecture and Specification", Revision 2, August 1995

[OMG-96] OMG DTC Document: telecom/96-12-02 "Request for Information on issues concerning intelligent networking with CORBA", November 1996

[P508-97a] EURESCOM Project P508: "Evolution, Migration Paths and Interworking to TINA", Deliverable 2: "Migration Strategies and Interworking with legacy systems", Final Version, January 1997

[P508-97b] EURESCOM Project P508: "CORBA as an Enabling Factor for Migration from IN to TINA: A P508 Perspective", Final Version, January 1997

[TCMA-95] TINA-C: "TINA Connection Management Architecture", TB_JJB.005_1.5_94, March 1995

[TRP-96] TINA-C: "TINA Reference Points", Version 3.1, EN_RJ.030_3.1_96, June 96

[TSA-96] TINA-C; "TINA Service Architecture", Version: 4.0, TB_RM.001_4.0_96, October 1996

[Q.12xx] ITU-T Recommendation Q.12xx Series on Intelligent Network, Geneva 1995

[Q.1218] ITU-T Recommendation Q.1218 - Intelligent Network Interface Recommendation for CS-1

ROS-to-CORBA Mappings: First Step towards Intelligent Networking using CORBA

Subrata Mazumdar
Room 4G-634, Bell Laboratories
Holmdel, NJ 07733, USA
mazum@research.bell-labs.com

Nilo Mitra
Room 1L-301, AT&T Labs
Holmdel, NJ 07733, USA
dyons@mailnet.ho.att.com

The key to introducing CORBA-based object-oriented technologies in the real-time telecommunications signalling environment is to ensure interoperability by maintaining the existing, standard protocols using Signalling System No. 7 for the external communication interface between telecommunication equipment. Signalling applications such as Intelligent Networks (IN) are defined as Remote Operations Service (ROS) Application Service Elements (ASE). As shown in this paper, the first step to applying CORBA technology in the IN domain appears to be the ability to map the principal concepts of ROS, and the corresponding Abstract Syntax Notation One (ASN.1) information object classes that define them, to the corresponding CORBA concept and IDL-based constructs. Needless to say, the concepts of ROS and CORBA are mostly complementary but not necessarily always fully aligned. In the course of this work, it has been necessary to make use of constructs defined by the computational model of TINA-C which has adopted and expanded upon CORBA IDL. As the ROS concepts are more fully exploited by data services defined by the X.400-based Messaging and X.500-based Directory standards, the techniques described in this paper can also aid the interworking of existing implementations of these services with CORBA-based implementations.

1 Introduction

Protocols using the OSI Remote Operations Service [1] (ROS) are in widespread use in telecommunications signalling using Transaction Capabilities [2] (TC), such as Intelligent Networks (IN), as well as in key data applications such as the X.400-based Messaging, X.500-based Directory services, and OSI-based Network Management. ROS is a very general communications paradigm based on a simple request/reply model. The ROS model provides a number of constructs to describe the interactions between distributed objects, using the information object class construct of the Abstract Syntax Notation One (ASN.1) [3].

On the other hand, the Common Object Request Broker Architecture [4] (CORBA) defined by the Object Management Group (OMG) consortium is rapidly becoming the technology of choice for building distributed applications in many areas of information technology using, as a key component, the Object Request Broker (ORB) to enable the transparent communications between application objects. Such transparency is achieved through the definition of well-defined APIs between objects using CORBA's Interface Definition Language [5] (IDL).

Thus far, the application of CORBA technology to telecommunications has concentrated in the area of network management which essentially involve non-real-time communications. Specifications [6, 7] have been developed for interworking CORBA-based management applications with those that support the ISO/ITU-T Telecommunications Management Network (TMN) and/or the Internet's Simple Network Management Protocol (SNMP) interfaces. In these efforts, the interworking has concentrated on the mapping of the information models in these different domains with suitable protocol conversion at the domain boundaries.

This paper expands the application of CORBA technology to telecommunications and data services beyond the area of network management — principally to real-time applications based on the ROS paradigm and associated protocol. Such applications include the signalling for providing Intelligent Network (IN) services, ISDN and

Broadband ISDN supplementary services and services supporting Universal Personal Telecommunications (UPT). All these services use multiple, structured information models (ROS-User ASEs) and the ROS-based protocol, TC. *Instead of concentrating on individual services, we refer to all these ROS-based services by the generic term "Intelligent Networking".* As the ROS concepts are more fully exploited by data services such as X.400-based Messaging and X.500-based Directory, the techniques described in this paper will also aid the interworking of existing implementations of these services with CORBA-based implementations.

To this end, the principal concepts of ROS, namely operations, errors, operation packages, connection packages, contracts and ROS-objects, and the corresponding ASN.1 information object classes that define them are compared with and, where possible, mapped to the corresponding CORBA concept and IDL-based types/interfaces. It builds upon the already-completed work [7] on mapping the basic ASN.1 data types to the IDL data types. Needless to say, the concepts of ROS and CORBA are mostly complementary but not necessarily always fully aligned. In the course of this work, it has been necessary to make use of constructs defined by the computational model [8] of the Telecommunication Information Networking Architecture Consortium (TINA-C), which has adopted CORBA as its Distributed Processing Environment and expanded some CORBA concepts. As CORBA does not yet support object implementation specifications, we have found that the object template construct in the TINA-C Object Description Language [9] (ODL) for computational objects is very useful in representing ROS objects. Thus, this work also suggests constructs which can be adopted into the CORBA framework to better support interworking with applications in the telecommunications domain.

This paper is structured as follows: Section 2 describes interworking scenarios where the mappings addressed in this document can be employed. Section 3 provides the (informal) descriptions of various ROS constructs. Section 4 proposes the mappings of these constructs onto those in CORBA. This section defines the "specifications translation" aspects of the mapping, namely an algorithmic translation between specification templates in the ROS model using ASN.1 and those in CORBA based on IDL. Section 5 provides some conclusions and addresses our future work.

2 General Issues on Interworking

Facilitating the interworking of real-time telecommunications applications such as IN services which use existing SS7 protocols such as TC/SCCP/MTP (see [10] for a tutorial on the SS7 protocol architecture) with the same services implemented using ORB-based systems is essential if the CORBA technology is to succeed in the existing telecommunications environment.

If one considers the current architecture for a typical SS7-based "intelligent network", various entities such as switches, network servers/databases and adjunct processors are connected by various IN-standardized interfaces. Note that the ITU-T and regional telecommunications standards (in much the same way as the Internet standards) are based on defining only the external communications interface, i.e., standardized protocols, which define the syntactic and semantic rules for information exchange. On the other hand, the CORBA approach is that of transparent communication between distributed objects however implemented so long as the interaction interfaces (i.e., APIs) between such objects are defined in a standard way (i.e., using IDL). Such a difference in approach is naturally due to the emphasis each environment lays on what each considers most important: CORBA-based technology seeks to easily create

distributed applications using the advantages of object-oriented programming concepts and practices, and by provisioning services useful to all applications as a part of the ORB infrastructure. The ORB, among other things, shields the programmer from the intricacies of the communications infrastructure. On the other hand, signalling (and Internet) applications are built on top of a specific communications platform, and seek interoperability of multi-vendor equipment through standardized communications interfaces. This latter approach presumably spurs competition for telecommunications equipment by allowing vendors to implement their systems using any technology while maintaining interoperability with other implementations through strong on-the-wire conformance requirements. This is not to say, though, that CORBA is not concerned with interoperability. It is, just as much as the telecommunications industry; only the reference points at which it defines conformance to provide interoperability are different from those of the latter. Indeed, the chosen reference points — standardized programatic interfaces between distributed objects in one case, and information exchange via standard protocols in the other — need not be seen as conflicting but rather as complementary.
This suggests that the key to introducing CORBA-based object-oriented technologies in the real-time telecommunications signalling environment is to ensure that existing, standard protocols for communication between telecommunication equipment is maintained to ensure interoperability for as long as necessary to ensure interworking with existing implementations. The steps by which CORBA technology may be introduced and coexist in the signalling environment are illustrated in Figures 1.a—d. (While not explicitly shown, the same approach would also hold for introducing CORBA-based implementations in the Directory and Messaging services environment, where the SS7 protocol stack would be replaced by the corresponding OSI profile for these data services.)

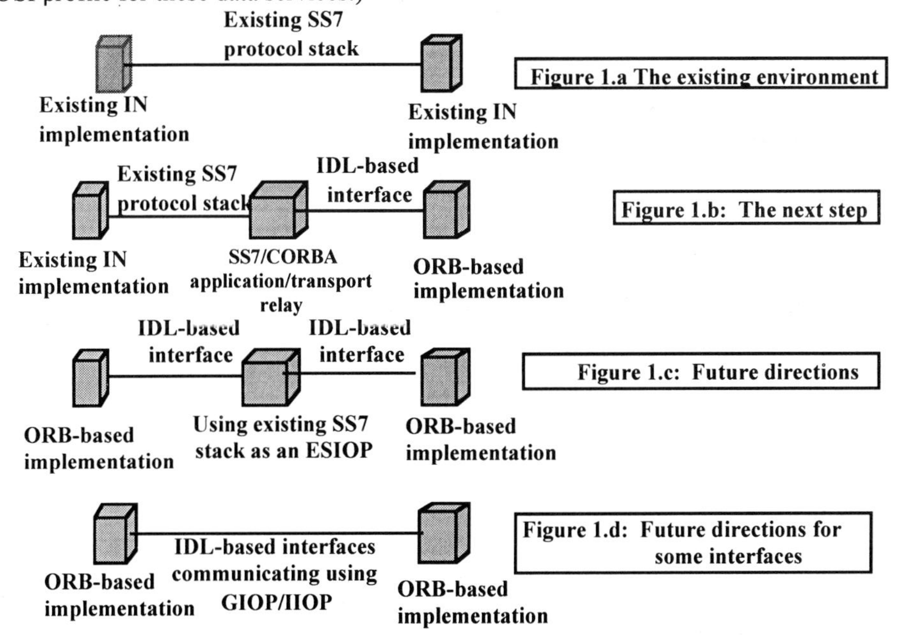

Figure 1.a The existing environment

Figure 1.b: The next step

Figure 1.c: Future directions

Figure 1.d: Future directions for some interfaces

Figure 1.a shows the current environment where two telecommunications equipment (a switch and a remote server/database, say, as in a typical IN scenario (see [11] for an excellent introduction to the IN standards, terminology, architecture and protocol)) communicate using the standardized SS7 protocol stack comprising INAP/TC/SCCP/MTP. A next step, as shown in Figure 1.b, might be to use an ORB-based implementation for either of the two equipments. This is not an unrealistic scenario considering various prototyping efforts [12] in research laboratories for providing CORBA-based distributed switching architectures. To be able to interoperate with the existing non-CORBA-based embedded base, such an ORB-based implementation would have to present an external *communications* interface that is identical to the currently standardized one. To be a CORBA object, however, its operational interface (i.e., API) would have to be specified in IDL. In our model, an application/transport bridge — which could be partially realized in the object adapter for the ORB-based implementation — would convert between the IDL-specified operation invocation or response and the standardized INAP/TCAP PDUs transported via the SS7 SCCP/MTP networking protocols. In other words, the relay would act as a static bridge converting invocations/responses from one domain into another. This paper defines the mappings that would comprise such a bridge. The approach is not very different from that for CORBA-based network management [6],[7]. Figure 1.c shows a scenario where there are many ORB-based implementations which may, for purposes of utilising the existing specially-engineered, highly reliable signalling network, use the SCCP and MTP protocols as an (potentially OMG-standardized) Environment-Specific Inter-Operability Protocol (ESIOP). And of course, as Figure 1.d shows, there may well be interfaces — for example between Customer Premises Equipment (CPE) such as workstations and the network — which may be based on ORB-based implementations utilising the already-standardized ORB-interoperability protocol, GIOP/IIOP. Interworking scenarios are discussed more fully in a paper [13] in this volume and elsewhere [14].

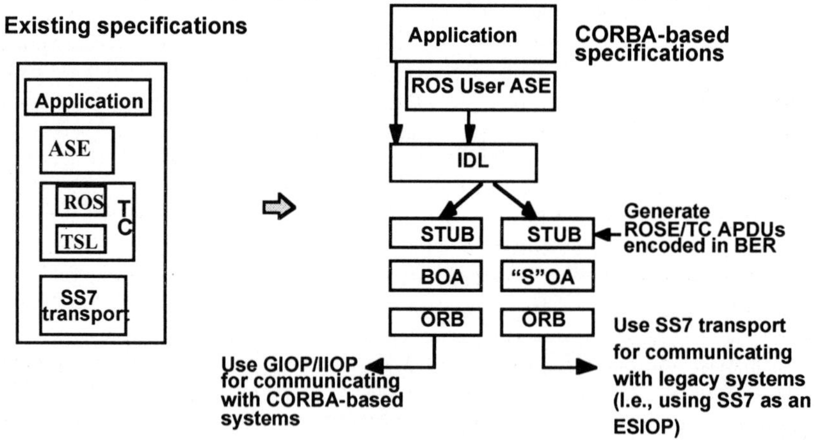

Figure 2: The nature of the interworking

The specific nature of the interworking to be performed at the bridge is shown in Figure 2. As shown on the left side of Figure 2, current signalling applications using TCAP, such as IN, are defined as ROSE-user Application Service Elements (ASE). These provide the syntactic and semantic aspects of the remote operations that describe the interaction between two signalling entities. The main aim of this paper is to outline how a ROSE ASE specification (written in ASN.1 and using various constructs/concepts of ROS) can be mapped to IDL to specify the interfaces of an object and implement ROS objects using CORBA object services. This is shown by the shaded portions of the figure on the right-hand portion of the figure. The IDL interface provides the ROS/ASN.1 constructs in IDL. Should the communication be with CORBA-based systems, the calls are transported using the OMG-standardized GIOP/IIOP protocols whereas, if communicating with systems that use the SS7 protocol stack, the stub generates TC/SS7 transport messages. The purpose of the "Signalling" Object Adapter would be, among other things, to map names in the two domains to the corresponding object references and map the dynamic requirements of a ROS-object (to be explained later in this paper) to those in the CORBA domain.

3 A Brief Overview of ROS Concepts and Constructs

ROS [1] defines a number of concepts and constructs to describe the interaction between objects that follow the request/reply interaction paradigm. Such objects are called **ROS-objects**, and the basic interaction is specified by the invocation of an **operation** by one ROS-object (the **invoker**) and its performance by another (the **performer**). Performance of an operation *may* lead to a return, from the performer to the invoker, of a report of the outcome — a **result**, to report a successful completion, or an **error** otherwise. During the performance of an operation, the performer may invoke **linked operations** to be performed by the invoker (of the original operation). The `OPERATION` information object class collects together all the syntactic aspects of what constitutes a remote operation, which is shared by the invoker and performer, and, for some aspects, the infrastructure through which they communicate.

The interaction between (pairs of) **ROS-objects** belonging to some **ROS-object-class** are defined in terms of sets of related operations, each set called an **operation package**. An operation package defines which operations each ROS-object in the pair may invoke of the other. Thus, unlike a traditional client-server model, which defines the operations which a client may invoke of the server, the ROS model simultaneously describes *both* the client *and* server aspects of a ROS-object. If both objects can only invoke the same set of operations of the other, then the package is said to be symmetrical. Otherwise, if there are some set of operations that one object can invoke, and a different set which the complementary object can invoke, then the package is said to be asymmetrical. In this case, based on some intuitive judgment of their roles, or arbitrarily, one of these objects is called the **consumer** (of the operation package) while the other is the **supplier**. The OPERATION-PACKAGE information object class defines this concept. (A given ROS-object could be playing the role of a consumer with respect to some operation packages, and that of a supplier with respect to some others.)

A pair of ROS-objects must have an association between them to serve as a context for the invocation and performance of operations. If the association is dynamically established, one of the ROS-objects plays the role of the **initiator**(of the association set-up), while the other is the **responder**. ROS defines a **connection package** as two special operations, called **bind** and **unbind**, that are available, as an option, to an application designer to dynamically establish and release, respectively, the association between two ROS-objects. An information object class, CONNECTION-PACKAGE, describes the `bind` and `unbind` operations and aspects of the binding used to establish/release the association. In addition to the means by which an association is established between two ROS-objects, the association is governed by an **association contract**, which is specified in terms of a set of packages which collectively determine the operations which can be invoked during the lifetime of the association.

The association contract is specified by the ASN.1 information object class CONTRACT.

Table 1 summarizes the different roles that have been described above for a ROS-object.

Role played by a ROS-object	With respect to which concept	Defined by which construct
invoker	remote operation	OPERATION
performer		
consumer	operation package	OPERATION-PACKAGE
supplier		
initiator	connection package	CONNECTION-PACKAGE
responder		
initiator	association contract	CONTRACT
responder		

Table 1: The roles played by a ROS-object and the corresponding ASN.1 constructs

Thus, *a ROS-object interacts with other ROS-objects only through the set of contracts that it supports*. Obviously, for each contract that it supports, there is another ROS-object which offers the complementary interface, i.e., if it supports a contract in which it plays the role of the initiator of the binding that enables the contract, then there is another ROS-object which supports the same contract in the role of a responder.

As our eventual goal is to determine the mappings from ROS concepts to the CORBA-based client/server concepts, it is necessary, for a given ROS-object, to identify the operations which it invokes (i.e., it plays the role of a client, in the traditional client/server terminology), and those which it performs (i.e, those for which it is a server). ROS provides several useful constructs that allow one to do this.. These concepts are illustrated through an example based on the X.500-based Directory services, and shown in Figure 3.

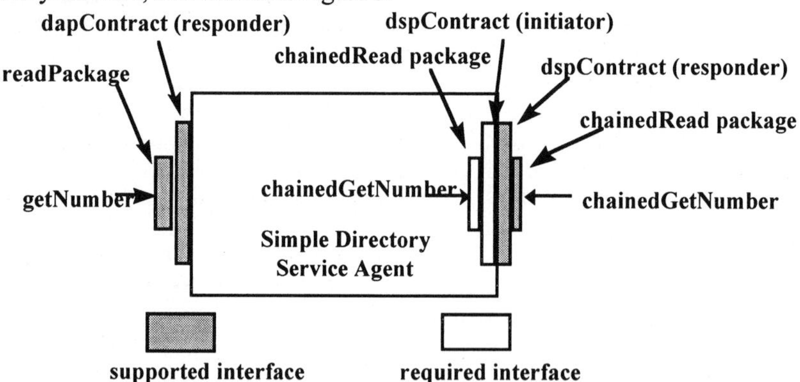

Figure 3: A simple Directory Service Agent (DSA) ROS-object

A simple Directory Service Agent (DSA), which is a ROS-object which represents a part of the distributed Directory, responds to bindings with objects representing users of the Directory (called Directory User Agents (DUA) ROS-objects) through a contract (called "dapContract" in the standards) which consists of an operation package, read, for which it plays the role of the "supplier". It also interacts with other DSA objects in interactions involving "chaining" of queries from a DUA for which it does not have the information. It does so through a symmetrical contract (called "dspContract") where it can play the role of the initiator or the responder. Figure 3 shows the complete set of operations supported by this simple ROS-object: its contracts, the packages associated with each contract and the operations it

invokes/performs. The operations “getNumber” and “chainedGetNumber” are performed by the DSA it invokes “chainedGetNumber” on another DSA.. Each operation is associated with a particular contract and a package within that contract.
In mapping the concept of a ROS-object to those of CORBA, we have found the constructs of **required** and **supported** interfaces in the TINA-C Computational Model [8] to be of value. A required interface is one that an object needs to support *as a client* in order to provide its services. A supported interface is an object’s server interface (in the client-server sense), i.e., the operations that it can perform. In Fig. 3, the DSA can be a client of another DSA and provides the dspContract (initiator) as its required interface. In turn it could play the server role over two supported interfaces, that of a dapContract (responder) in interactions with a DUA, and the dspContract (responder) in interactions with another DSA.

<table>
<tr><th>ASN.1</th><th>IDL</th></tr>
<tr><td>TelDirLookupProtocol {XXX.1}
DEFINITIONS ::= BEGIN
IMPORTS OPERATION, ERROR FROM Remote-Operations-Information-Objects
{joint-iso-itu-t remote-operations(4) informationObjects(5) version1(0)};

getNumber OPERATION ::=
- - timer = 5 seconds
{ARGUMENT Name
RESULT TelephoneNumber
ERRORS {unknown | noAccess}
LINKED {clarify}
CODE local:1}

clarify OPERATION ::=
- - timer = 60 seconds
{ARGUMENT Clarification
RESULT RequestedInformation
OPTIONAL TRUE
ERRORS {unknown}
CODE local:2}

unknown ERROR ::=
{ PARAMETER Cause
CODE local:1}
noAccess ERROR ::= {CODE local: 2}

Name ::= VisibleString
TelephoneNumber ::=
SEQUENCE SIZE(10) OF INTEGER (0..9)
Clarification ::=
ENUMERATED{provideZipCode(0),
provideMiddleInitial(1),...}
RequestedInformation ::= CHOICE {
zipCode[0] IMPLICIT SEQUENCE SIZE(5) OF INTEGER(0..9),
middleInitial [1] VisibleString}
Cause ::= ENUMERATED{noSuchName(0),
noSuchZipCode(1), ...}
END</td><td>Module TelDirLookupProtocol {

#pragma ID TelDirLookupProtocol XXX.1

// ASN.1 Types provided first, then execeptions, then
// interface with operations

typedef ASN1_VisibleString NameType;
typedef ASN1_Short TelephoneNumberItem;
typedef sequence <TelephoneNumberItem, 10>
TelephoneNumberType;
typedef enum CauseType { noSuchName,
noSuchZipCode };

exception unknown { CauseType error_param };

#pragma ID unknown “XXX.1:local:1:error”

exception noAccess { }

#pragma ID noAccess “XXX.1:local:2:error”

/* to be defined within the IDL interface
// timer = 5 seconds
TelephoneNumberType getNumber (in NameType name_param) raises (unknown, noAccess);

#pragma ID getNumber “XXX.1:local:1:operation”
#pragma ID Timer getNumber 5

//Behavior
//Linked operation: see signature of “clarify”
//operation in interface “....”
//timer = 60 seconds

// RequestedInformationType clarify(
// in ClarificationType clarification_param
//) raises { unknown };
#pragma Timer clarify 60
*/
}</td></tr>
</table>

Table 2: The ASN.1 to IDL mapping for the “Directory” example

4 Mapping of ROS Concepts to Those of CORBA

This section provides the outline for a specification translation, which is a static process of converting specifications written in ASN.1 to IDL. We define mappings for each ROS construct (operation, error, etc.) to those that are their equivalent in

CORBA (e.g., operation signature, exception declaration etc.). Needless to say, there are many cases where there are no direct mappings possible of the constructs.

4.1 Mapping of a ROS operation to a CORBA operation signature

With some exceptions, the fields of the ROS OPERATION information object class can be mapped directly onto the fields of the CORBA **operation signature**.
To illustrate the mapping by an example, consider the ASN.1 specification in the *left hand column* of Table 2 describing a simple query/response service for a telephone directory lookup. One ROS-object, representing the user, invokes the operation getNumber by providing the Name of the person whose TelephoneNumber is returned as a result of performing the operation. However there may be instances when, before completing the operation, the other ROS-object — the "directory" — may need further clarifying information (such as a person's middle initial or a zip code so as to be able to narrow down the choice of alternatives), which it does by invoking a linked operation clarify. The corresponding IDL mappings for this example are given in the right-hand column of Table 2.
In studying the example mapping in Table 3, it will be noted that certain design decisions have been taken. These are:

1. **ROS Argument and Result type:** ROS permits a *single* ASN.1 type to accompany an operation invocation (its argument) and its return (result). CORBA permits an operation signature to specify **in**, **out** and **inout** parameters indicating the "directionality" of the accompanying parameters. An identifier should be used as the "<simple_declarator> in IDL to identify the argument parameter.Note that the returned value, if defined, is mapped to the <op_type_spec> and is *not* returned as a distinguished "out" parameter as a part of the <parameter_dcls>.
2. **Handling of linked invocations:** As these are operations which the server *might* invoke on the client, there is no place in CORBA-based specifications for defining the signature of the linked operation as a part of the server's interface. The CORBA approach requires that it be defined elsewhere as a part of the client's interface. The fact that this operation is linked to the one being defined, and may need to be invoked under certain circumstances during the processing of the parent invocation, is a part of the *behaviour* specification of the performer of the parent invocation and we have chosen to express it as an IDL comment as shown in column 2. For every ROS operation invoked for which a linked operation is defined, the bridge must generate a **context_expr**(ession) containing a linked id value as a part of the mapping of the ROS operation to the corresponding operation signature. The context_expr can be defined as a sequence of name-value pairs, where both the name and value is of type string. The linked identifier of the ROS Invoke APDU can be passed as one of the items in the context_expr. The "name" for the linked identifier can be defined as LinkedId. If, as a part of the behaviour, the (server) CORBA object needs to invoke the linked operation on the (client) ROS object, this read-only context_expr must be reflected with a callback invoking the linked operation.
3. **Synchronous vs. asynchronous ROS operation invocations:** The `OPERATION` information object class has a field which permits defining whether an operation is synchronous or not , i.e., if defined as synchronous, the ROS-object may not invoke another *synchronous* operation before the return of the one being specified. CORBA currently supports only the synchronous or the deferred-synchronous mode of operation. If a ROS operation is asynchronous (as, indeed, are all IN operations standardized thus far), obviously the CORBA domain

requires support of a threaded environment. This has to be made known to the CORBA environment as an IDL comment. (Work is underway in the OMG on defining an asynchronous messaging service for CORBA).

4. **Operation codes:** We have used the "# pragme ID" construct to associate the operation code in the ROS specifications with the Repositoryld of the corresponding operation in IDL.
5. **Invocation/result priorities:** ROS permits (unlike CORBA) the application designer to quote a priority for sending an invocation/result to aid the infrastructure in choosing the order of sending for the case where there are several invocations/results waiting to be sent. The CORBA **context_expr** portion of the operation signature may provide a means for providing such extra information to the infrastructure regarding the conveyance of an operation. (It is likely that future work in the OMG, particularly on real-time ORBs will lead to well-defined contexts for such information, which are essentially Quality Of Service (QoS) issues pertaining to operation invocations and their returns.) At this time, the use of contexts in CORBA is implementation dependent; so, in the absence of a well-defined usage of this construct, any information on the priority, if specified, will be provided as an IDL comment and also via a new pragma called "Priority". The name for priority in the context_expr can be defined as Priority.
6. **Operation timers:** TC operation designers provide, as an ASN.1 comment, a timer value associated with an operation. This indicates a "time to respond" to an operation invocation, failing which the client presumably assumes the operation failed and proceeds to some other task. Again, CORBA IDL does not allow operation-specific timer specification. This information will have to be carried as a part of context_expr. The timer associated with an operation will have to be mapped to an IDL comment, as well as a new pragma, called Timer.

4.2 Mapping of ROS Errors

The CORBA operation signature template provides both implicit as well as explicit means for indicating exceptions, i.e., the inability to perform an operation. It does so explicitly through the use of the **raises** expression in the operation signature, which also allows *multiple* parameters to accompany the exception report. The ROS ERROR information object class maps almost exactly into the **raises** expression in the CORBA operation signature template. The ROS error report may be accompanied by a *single* ASN.1 data type clarifying the error. The handling of the error code and the priority for returning the error are handled in the same manner as for the corresponding invocation (see section 4.1).

All CORBA operation signatures *implicitly* (i.e., not listed via the **raises** expressions) include standard exceptions which provide an indication why an operation could not be performed. Standard exceptions also include a **completion_status** code indicating that either the server completed processing before the exception was raised, or it never started processing the invocation, or if the state of the operation performance is unknown. There are 26 standard CORBA exceptions, some to do with the server while others are related to problems with the ORB-based infrastructure. In any case, all these standard exceptions indicate problems at an *application* level.

ROS provides an exception reporting capability through the use of the Reject Application Protocol Data Unit (APDU) which reports the erroneous use of the other ROS APDUs. In other words, ROS exceptions indicate *protocol* errors, not errors at the *application* level. No completion status is provided. Clearly, the only case where a

<table>
<tr><th>ROS Object</th><th>TINA ODL-based computational object/interfaces</th></tr>
<tr><td>SimpleDirProtocol {XXX.2}
DEFINITIONS ::= BEGIN
IMPORTS ROS-OBJECT-CLASS, OPERATION-
PACKAGE, CONNECTION-PACKAGE FROM
Remote-Operations-Information-Objects {joint-iso-
itu-t remote-operations(4) informationObjects(5)
version1(0)}
getNumber, clarify, unknown, noAccess, Name,
TelephoneNumber, Cause, RequestedInformation,
Clarification, FROM TelDirLookupProtocol{};
simple-dsa ROS-OBJECT-CLASS ::={
BOTH {dspContract}
RESPONDS {dapContract}
ID global:XXX.2.3}
dspContract CONTRACT ::={
CONNECTION dsaConnectionPackage
OPERATIONS OF {chainedReadPackage}
ID global:XXX.2.4}
dapContract CONTRACT{
CONNECTION dapConnectionPackage
INITIATOR CONSUMER OF {readPackage}
ID global:XXX.2.5}
dsaConnectionPackage CONNECTION-PACKAGE
::={BIND dsaBind
UNBIND dsaUnbind
ID global:XXX.2.6}
dapConnectionPackage CONNECTION-PACKAGE
::={BIND directoryBind
UNBIND directoryUnbind
ID global:XXX.2.7}
readPackage OPERATION-PACKAGE ::={
CONSUMER INVOKES {getNumber}
ID global:XXX.2.8}
chainedReadPackage OPERATION-PACKAGE::={
OPERATIONS (chainedGetNumber}
ID global:XXX.2.9}
chainedGetNumber OPERATION ::= {
ARGUMENT Name
RESULT TelephoneNumber
ERRORS {unknown | noAccess}
LINKED {clarify}
CODE local:3}
END</td><td>#include
module SimpleDirProtcol
{
pragma ID SimpleDirProtocol “XXX.2”
interface dapContractResp
// operation interface template using TINA ODL
{
#pragma ID dapContractResp “XXX.2.5:Resp”
TelDir::TelephoneNumberType
getNumber (in TelDir::NameType name_param)
raises (TelDir::unknown, TelDir::noAccess);
#pragma ID getNumber “XXX.1:local:1:operation”
// clarify() is linked operation
};
interface dspContractRes
//operation interface template
{
#pragma ID dspContractRes“XXX.2.4:Resp”
TelDir::TelephoneNumberType
chainedGetNumber(
in TelDir::NameType name_param)
raises (TelDir::unknown, TelDir::noAccess);
#pragma ID
chainedGetNumber“XXX.2:local:3:operation”
// clarify() is linked operation
};
simple_dsa object template
{
#prgama ID simple_dsa XXX.2.3
supported interfaces
dapContractResp, dspContractResp;
required interfaces
dspContractResp, dspContractInv;
};
};
//END OF TINA-C ODL</td></tr>
</table>

Table 3: Example of mapping a ROS object class to a TINA computational object template

CORBA mapping for exceptions might possibly be suitable is for a Reject providing a problem code citing a GeneralProblem or an InvokeProblem. A server (respectively client) in a CORBA environment should not be expected to receive (respectively send) an exception mapping to a ROS Reject APDU with a ReturnResult or ReturnError problem code. A ROS GeneralProblem refers to badly formatted, mistyped or unrecognised PDUs, and is typically identified and generated by the infrastructure. Presumably such a cause is infrequent in an actual run-time environment, as problems with formatting of messages are detected and corrected during testing at compile time. The raising of such an exception in a run-time environment, therefore, is presumably a very infrequent occurrence arising from some bit error(s) introduced by the communications media which has escaped detection by the lower layer protocols. On the other hand, the ROS InvokeProblem notes that some aspect of the invocation was erroneous. There are 8 ROS InvokeProblems of which three refer to the unexpected use of linked invocations or their returns. There are no equivalent in CORBA for these. This leaves 5 codes which could possibly serve to funnel the richer set of CORBA exceptions. This is one area where there is great

flexibility of design, and a bridge mapping CORBA-generated exceptions to a ROS environment will obviously not be able to convey the true semantics of the exception. In the reverse direction a ROS-generated exception cannot obviously be mapped to a specific fine-grained CORBA exception, with similar loss of information. One must fall back on the argument that exceptions are, as the name suggests, exceptional occurrences and the effort to accurately handle infrequent situations is outweighed by the need to provide simple mappings and implementations of the bridge software.
Our solution, therefore, is as follows: For the case where a CORBA exception is raised by the CORBA-based server, the bridge aborts the association (for the OSI case) or aborts the transaction (for the SS7/TC case). Both the OSI user abort of an association as well as the user abort of a SS7/TC transaction permits the inclusion of a limited amount of user data, which could be used to carry the CORBA exception report and status indication if it was felt necessary to provide some information to the ROS-based client. If not, the information could be stored at the bridge for off-line analysis. For the case where the ROS-based server generates a `Reject` indicating a `GeneralProblem` or an `InvokeProblem`, the OSI association or TC transaction, as appropriate, is aborted and a CORBA standard exception of "unknown" with completion_status "no" is generated.

4.3 Mapping of the ROS object class to the TINA object template
As outlined in section 3, the ROS object class provides the complete description of the capabilities of a class of ROS objects by describing all the contracts that such objects can support — either through inheritance from their identified superclass(es) or explicitly defined. In short, a ROS object is the sum of the contracts it supports. The ROS object description, therefore, provides a richer description of the capabilities of distributed objects by providing both the client and the server aspects of an object, and, by implication, the server and client aspects, respectively, of the complementary object(s) which is(are) evidently necessary to complete the totality of the distributed application.
The description closest to that of ROS objects is that provided by the TINA-C computational object template [8] together with its support by the TINA-C ODL [9] , a super set of the CORBA IDL. A TINA computational object specifies two types of interfaces — called required and supported. **Supported interfaces** are the same as the CORBA object interfaces, namely the interface offered by an object (acting as a server) to its clients. **Required interfaces** describe those interfaces through which, as a part of the behaviour of an object, it invokes operations (i.e., plays the role of a client) to complete providing its services.
As before, we show the nature of the mapping through an example. Our example is a simplified Directory Service Agent (DSA) for the simple directory lookup protocol described in the ASN.1 module in Table 2 and illustrated in Figure 3. Figure 3 shows the interfaces for such a simplified DSA. We provide the DSA with an additional interface, one between two DSAs to permit chaining of queries, i.e., if a DSA does not have the TelephoneNumber related to a particular Name, it can forward the getNumber query to another DSA which may have the necessary information. To distinguish the operations on this interface, we append the word chained to the operation names, such as chainedGetNumber. Over the interface dapContract, the DSA interacts with a directory user and acts as a responder to any attempt to bind prior to receiving queries. Over the dspContract interface marked as a responder, the DSA receives chained queries from another DSA which initiates the binding. Both these interfaces for which the DSA plays the role of a responder are its **supported** interfaces. For the case where the DSA needs to query another DSA for information it is missing, it acts as an initiator of a binding with another DSA through the dspContract (initiator) **required** interface.The mapping of the simple DSA ROS object interfaces to those of the TINA computational object template is shown in Table 3.

5 Conclusions

We have provided an outline for the "specifications translation" of ROS to CORBA by showing the mapping of various ROS concepts and constructs defined in ASN.1 to those of CORBA using IDL. In a few instances, we have shown how some extensions to IDL, as proposed by TINA-C, can be used to convey the somewhat richer semantics of ROS-objects. Efforts are currently underway in the OMG to include these features in the CORBA object model. As the ROS constructs are the means by which a large number of SS7 services (what we have generically called Intelligent Networking) are specified, as well as data services such as OSI-based X.400 Messaging and X.500-based Directory, such a mapping can be used by a static bridge to implement a ROS-to-CORBA gateway. Such bridges would then permit the development of CORBA-based implementations which can interwork with current ROS-based implementations of these services. In a future paper, we shall further develop the interactions specifications for ROS-to-CORBA mappings by illustrating how contracts between a pair of ROS objects can be accommodated within the CORBA framework, and provide the mapping of the respective naming schema as well as the message formats in the two domains.

The ideas and figures described in this paper were presented at a meeting of the Telecom Domain Task Force of the Object Management Group in Washington, DC, in June 1996. The ideas described here have also been adopted in a draft OMG White Paper [15] on the subject of IN/CORBA interworking issues.

References

1. ITU-T Rec. X.880 (1994) | ISO/IEC 13712-1:1995, Information technology — Remote Operations: Concepts, model and notation.
2. ITU-T Rec. Q.771-774 Recommendations on Transaction Capabilities, Geneva, 1993.
3. ITU-T Rec. X.680 — 683(1994) | ISO/IEC 8824-1/2/3/4:1995, Information technology — Open Systems Interconnection — Abstract Syntax Notation One (ASN.1).
4. Object Management Group "Object Management Architecture Guide" Revision 3.0, 1995.
5. Object Management Group, "The Common Object Request Broker: Architecture and Specification," Revision 2.0, July 1995.
6. Subrata Mazumdar, "Mapping of Common Management Information Service (CMIS) to CORBA Object Services", submitted as contribution to Interaction Translation Specification of XoJIDM Taskforce, February 1996.
7. X/Open, "Inter-domain Management Specifications: Specification Translation", X/Open Preliminary Specifications, draft dated August 9, 1995.
8. TINA-C Stream Deliverable "Computational Modelling Concepts,version 3.2" May, 1996.
9. TINA-C Stream Deliverable, "TINA Object Definition Language (ODL) Manual, version 1.3," 20 June 1995.
10. Mitra and S. D. Usiskin, "Inter-relationship of the SS7 Protocol Architecture and the OSI Reference Model and Protocols," The Froehlich/Kent Encyclopedia of Telecommunications, Volume 9, Marcel Dekker, Inc., 1995.
11. Igor Faynberg *et al*, "Intelligent Network (IN) standards and their application to services," McGraw Hill, New York, 1996, ISBN 0-07-021422-0.
12. Center for Telecommunications Research, Columbia University, "Proceedings of the OPENSIG Workshop: Open Signalling for Middleware and Service Creation", April 29—30, 1996.
13. Herzog, T. Magedanz, "From IN towards TINA —Potential Migration Steps," IS&N Conference Proceedings, Como, Italy, May 1997.
14. Capellmann *et al*, "Migration Scenarios for Evolving to TINA," Proceedings of the TINA '96 Conference, Heidelberg, September 1996.
15. OMG Domain Technical Committee Document # telecom/96-12-03, "Intelligent Networking with CORBA", draft White Paper, December 1996

Communications Management

Vincent P. Wade, Trinity College Dublin, Ireland
Vincent.Wade@cs.tcd.ie

As the telecommunications industry moves towards a global 'service driven' market place, demand is increasing for management services which are capable of co-operation across multiple network operator administrations and inter-operation over heterogeneous network infrastructures. As a direct result of the deregulation and globalisation of the market place and the rising expectations of networked users for integrated trans-national multimedia information services, new challenges are confronting Network and Service management systems in the late 1990s.

With the emergence of such a 'co-operative yet competitive' telecommunications market place, the management of the required network connectivity and trans-national advanced information services has become increasingly complex. In the technical papers presented in the Communication Management section of this book, three inter-dependent types of challenges are investigated for the network and service management systems of the future:

1. the design, and implementation of network and service management applications which can co-operate across several service and network providers and the architectural frameworks within which such solutions execute
2. the development and management of broadband virtual private networks and network connectivity services which span heterogeneous network protocols and administrative authorities
3. the design and development of management platforms which are capable of inter-operation across different network management protocols and management domains.

The first category of papers addresses the issues of designing, specifying and implementing service and network management solutions which can achieve the required level of inter-management system co-operation (usually termed inter-domain management services since they span different service or network operators). This reflects the reality that a single service or network operator may be unable to satisfy the entire requirements of a corporate customer organisation and that service level agreements between operators must be supported by management solutions which can automatically co-operate to share management responsibility, functionality and information. Examples of such inter-domain management activities are accounting, subscription, fault, and configuration.

However, the design of service or network management applications are only part of the challenges facing the realisation of co-operative multi-domain management solutions. With such multi-domain services and management solutions being offered, there are significant problems in integrating and managing the underlying heterogeneous networks. Therefore the second set of challenges identify the need for management services capable of providing the necessary network connectivity and quality of service whilst hiding the underlying differences in communications technology and administration/ownership.

The third set of challenges involves the development and inter-operation of network management platforms themselves. Currently there are two principle management protocols in use. The Common Management Information Protocol [1] is the accepted protocol for telecommunications management systems, whereas the management of computer systems and the Internet is primarily based on the Simple Network Management Protocol [2]. The CMIP approach is generally accepted to be more powerful, but has been less widely implemented because of its complexity and suitability to large networks and systems of which there are fewer. However the SNMP approach has been implemented by most computer and equipment vendors and enjoys widespread usage but its alarm handling, security, and ability to scale to large network elements has been the subject of much heated debate and concern. It is no longer possible for the two approaches to pointedly ignore each other especially as asynchronous transfer model (ATM) is widely seen as being deployed in telecommunications networks as well as in the Internet [3]. Although the advocates of both approaches occupy opposite and well entrenched positions, the development of a coherent approach for management of both the telecommunications network and computer networks is a major challenge for network management in the late 1990s.

Overview of Communication Management papers

The first five papers describe the experience of several ACTS and Eurescom projects in designing, implementing and trialling co-operative management applications for both advanced service and network management. The first paper titled 'A Methodology for Developing integrated Multi-domain Service Management Systems' by Wade et al. outlines a full process development life cycle for service management systems which facilitates the requirements capture and specifications of a multi-service provider market place where management services must co-operate to assist the delivery of a tele-service across several provider organisations. This paper presents a methodology for designing and implementing Intra and Inter Domain (end-to-end) service management systems. In particular the methodology considers the integration & co-operation of management services from different service providers. To illustrate this methodology a case study is presented based on the development of subscription management services where the final delivered management service is based on the co-operation of different provider service management systems.
In Covaci et al's paper 'Towards Harmonised Pan European TMN Customer Care Solutions', the theme of developing co-operative multi-domain management systems is continued. This paper presents the design of a pan European Trouble Ticketing system whose goal is to harmonise the requirements of trouble ticketing systems and improve the performance of systems currently available. The envisaged system is developed within the TMN framework. Griffin et al. present the definition, design and implementation of TMN-like functions for resource configuration management. However this management system has been built within a TINA compliant framework and the paper identifies and addresses the issues of how TMN like services can be implemented using the TINA architectural framework.
The 'multi-domain inter-operation' theme is again discussed in Bleakley's paper titled 'TMN Specifications to support inter-domain exchange of accounting, billing and charging information'. The design of management functionality to share accounting information across multiple network operators is based on one of Network Management Forums ensembles and adheres strictly to the TMN management framework. Finally in Lewis et al's paper 'Inter domain Integration of services and service management' the theme of multi-domain management inter-operation is raised at the service level. This paper discusses the issues of integrating a range of advanced information services e.g. conferencing, hyper-media information services which are offered by different service providers and are required to be integrated to deliver the value added services supporting tele-education. The paper identifies the similarities of integrating services and service management systems and provides insight as to how

integrated managed telecommunication services can be implemented across multiple service and network providers. The paper identifies a suitable architectural approach based on TINA's service architecture and describes how Java/WWW and CORBA technology can be used to enable service and service management integration.
The next three papers in the Communications Management Section examine the issues of multi-domain inter-operability from the perspective of connectivity management. L.H. Bjerring et al in their paper 'Domain Interoperability for Federated Connectivity Management' identify key problems in achieving interoperability across administrative, technological, and service domains. The paper describes a generic model for interoperability between objects belonging to separate domains. The paper then illustrates how these models have been applied in the development and implementation of Broadband VPN service which is illustrative of the problems and solutions associated with inter-domain management. In Galis et al's 'Towards Integrated Network Management for ATM and SDH Networks supporting a Global Broadband Connectivity Management Service' the design and architecture of a connectivity management service is presented. The goal of this management service is to efficiently manage the network resources of SDH and ATM infrastructure, while meeting the quality of service requirements and the needs of a number of telecommunications actors: customers, value-added service providers and network providers. A more detailed treatment of the quality of service and routing issues in such an ATM-SDH network is presented in Verdier et al's 'QoS routing Solutions for Hybrid SDH-ATM network'. The paper presents the quality of service requirements for a global broadband connectivity service and presents a quality of service functional model. The routing requirements for inter and intra domain routing are also identified. A routing scheme is proposed based on a link-state algorithm tailored to allow for QoS constraints and simple policy routing. In addition, a centralised intra-domain routing algorithm for a pilot implementation is presented.
The third set of papers deals with the implementation of components of network management platforms. In particular, two papers discuss the issues of integrating TMN/CMIP and SNMP within such platforms. In Deri et al. 'Static vs Dynamic CMIP/SNMP network management using CORBA', the authors describe techniques that allow the management of CMIP/SNMP network resources using CORBA. Static techniques which map each managed object class into a corresponding CORBA interface are compared with dynamic techniques which rely on runtime information. The paper claims that CORBA-based network management applications are becoming attractive in terms of efficiency and application size, overcoming limitations of early solutions. In Dassow et al, a set of rules for the mapping between MIB (SNMP) and information model (GDMO/CMIP) are derived. Based on this mapping, principles for the implementation of a proxy are derived. The paper discusses runtime aspects related to the proxy implementation as well as describing how the different implementations fit into a hierarchical management architecture.
Rivalino at al's paper 'Managed Objects as Active Objects: A multithreaded approach' looks at the design of a different part of a network management platform, namely the managed objects in a network management agent. This paper presents a multithreaded infrastructure which facilitates the implementation of managed objects as active objects. The paper claims that the advantage of implementing agents with active objects is that they provide a higher degree of fault tolerance for agent processes and improve performance. The final paper in this section titled 'Internet - New inspiration for Telecommunications Management Network' observes that the

situation of carriers and carrier networks has changed significantly since work on TMN started in the 1980s and that traditional TMN solutions reflecting the centralized network structures of the past monopolistic era may no longer be viable in modern competitive multi-player networks. Perhaps TMN is in need of a new inspiration! The paper suggests that answers to some of the main (carrier) network management issues may now come from a direction that did not seem very promising a couple of years ago: the Internet. The paper asserts that recent developments in the management of enterprise networks will also set the trends for future management of public carrier networks. The paper suggests several possible ways in which TMN can benefit from Internet based technologies and in particular World Wide Web technologies.

Acknowledgement: As section editor I would like to thank all the authors for their valuable contribution to this very interesting set of papers and to gratefully acknowledge Patrick McLaughlin (Frontier Systems) for assisting me in reviewing and mentoring the papers in this section.

References

[1] ITU-T X700, OSI Systems Management, X700 Series Recommendations, OSI Systems Management

[2] Simple Network Management Protocol (SNMP), RFC 1157, May 1990

[3] A Gillespie, Management Models for Telecommunications, guest editorial published in IEEE Communications, March 1996, Vol 34 No.3 pp34-36

A Methodology for Developing Integrated Multi-domain Service Management Systems

Vincent Wade, Trinity College Dublin David Lewis, University College London
Mark Sheppard, Broadcom Eireann Research Ltd
Michael Tschichholz & Jane Hall, GMD Fokus,
Contact: Vincent.Wade@cs.tcd.ie

This paper presents a methodology for designing and implementing Intra and Inter Domain (end-to-end) service management systems. In particular the methodology considers the integration & cooperation of management services from different service providers. The methodology uses enterprise modelling techniques and object oriented design for requirements capture and description, system development, implementation and deployment. It also supports the generation of service management system specifications based on ODP viewpoints. The methodology facilitates the reuse pre-existing computational components e.g. TINA C service architecture components in developing and implementing management solutions. To illustrate this methodology a case study is presented based on the development of subscription management services where the final delivered management service is based on the co-operation of different provider service management systems. The methodology was developed and is being trialled within the PROSPECT ACTS project.

1 Scope

With the deregulation of telecommunication network services in Europe, there is increasing interest in telecommunication services being offered by third party service providers over different network providers. The benefit of such an open service environment would be 'one-stop-shopping' delivery of 'tailored' services to end customers without these customers having to deal with the multiplicity of underlying telecommunication services and network providers. The difficulty with such an environment is the complexity of managing the services across the different service provider organisations (administrative domains) i.e. integrating co-operative management services spanning individual service and network providers, value added service providers and final end users. The design facilitates both the design and development of individual management services and their co-operation and interoperation to support source-to-destination telecommunication services. The methodology facilitates the usage of Open Distributed Processing standards for the generation of system specification & the re-use of management component designs and implementations based on the TINA C standards. However the methodology itself is independent of these standards.

The paper first identifies the current trends in system analysis/design techniques and ODP based distributed systems development techniques. In accordance with these trends the paper describes the Prospect Design cycle which is at the center of the methodology. The paper also describes how ODP based specifications can be generated from stages in this design cycle. To illustrate the usage of the methodology the design of a multi domain subscription management service which spans several service and network providers is described. Finally conclusions are drawn as to the usage and benefits of the methodology.

2 Designing Multidomain Management Services

The late eighties and early nineties has seen the increased usage of 'second generation' object oriented analysis and design techniques. Principle among these methodologies are Rumbaugh's Object Modelling Technique (OMT) [Rumb-91], Ivar Jacobson's Use-Case driven OO software Engineering Model [Jacob-92] and Grady

Booch's Object Oriented Analysis and Design (OOAD) methodology. Another more recent object oriented methodology has been the FUSION methodology developed by HP Labs. The current trend in object oriented methodologies is to harmonise existing approaches rather than develop brand new modelling techniques. An example of this trend is the proposal of Unified Modelling Language. Also the Object Management Group (OMA) - the consortium responsible for CORBA standards, have recently called for standardisation of OO design techniques.
Since inter & intra domain Management Services inherently require distributed solutions, an open distributed approach must be adopted for multi domain *system description.* Currently several standards address the issue of specifying open systems. Principle among these is the Open Distributed Processing [ODP-94] standard which suggests the specification of five different viewpoints of the system (Enterprise, Information, Computation, Engineering and Technology). These viewpoints highlight, and provide a basis for the separate discussion of, different aspects of the design and implementation of the system. Several standards have taken these viewpoint description concepts and enhanced them for describing network and service management systems e.g. TINA [TINA-95] and ODMA.
In order to capture the idea that (management) services can be used independently of each other as well as being used in cooperation, the end-to-end management system must be viewed in two complementary (overlapping) models: (i) Inter Domain Management Model: This models the management services in co-operation to support end-to-end management services. (ii) Intra Domain Service Management Model: These models describe the management services (e.g. configuration, accounting etc.) of each of the tele-services in an Open Service Market e.g. management services for VPN, MultiMedia Conferencing, HyperMedia Services.
The reason for this decomposition is that each management service could exist on its own (i.e. is a usable management service regardless of its cooperation with other services management providers). This reflects the view of the Open Service Market where the management of a tele-service has its own objectives whether or not they are in fact integrated across provider organisations. Thus inter domain management captures the end-to-end management chain while the intra domain models capture the management services of each individual service in isolation. However, these models are not completely orthogonal as components used for inter domain management are also used in the intra domain management.

3 Service Management Design Cycle for Viewpoint Models

Rather than developing new modelling or systems specification techniques, the Prospect methodology harnesses existing modelling tools and concepts within a design cycle which facilitates the development of the inter and intra domain management systems. The methodology supports each stage of the design process lifecycle. The ODP viewpoints provide a well established approach to system model description. However, they do not provide a prescriptive methodology that can be followed in developing management systems [ITU-96]. Therefore unlike previous design methodologies which have attempted to specify design steps which allow the generation of one ODP viewpoint from a previous viewpoint, the PROSPECT design cycle sees the development of a full object oriented model as central to the overall design process. Figure 1 illustrates the Prospect design cycle, which at its centre is concerned with the design, development, implementation, testing & trialling of inter/intra domain management system.

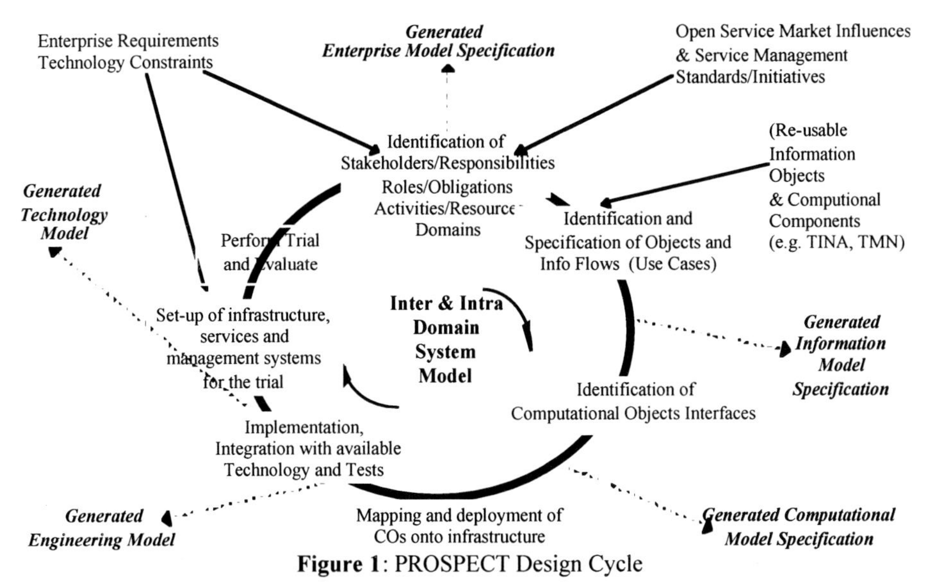

Figure 1: PROSPECT Design Cycle

The design cycle provides a structured way of developing, implementing and testing these object designs. It specifies the design steps from developing multi domain business models (which includes the representation of stakeholders, assignment of responsibilities, identification of obligations and activities etc), use case definition analysis, object identification and relation representation, definition of computational components, the integration and extension of pre-existing computational components (e.g. from TINA C Service Architecture), distributed placement of computational components, definition of platform architecture and platform services, generation of test sets and trial execution.

The Prospect design cycle also identifies key places in the development process from which specific viewpoint models can be generated and prescribes the contents of each viewpoint model specification and illustrates how these can be generated. The information needed to generate the specifications are a subset of the information needed to design and implement the actual systems. The design cycle ensures the consistency between each stage of the system model development and therefore provides a means of tracing the interrelationships between the ODP viewpoints. The benefit of generating the ODP viewpoint specifications is that a clear separation of the different aspects of the system can be captured. Also it provides a structured means of comparing different subsystems and services. The Prospect design cycle is iterative, allowing progressive deepening of the system models by iterating the cycle several times.

The modelling techniques integrated in the design cycle are *Organisational Requirements Definition for Information Technology* [ORDIT-93], Object Modelling Technique (OMT) [Rumb-91], OMG based interface definition language. The methodology has been influenced by the modelling work of previous ACTS projects PREPARE [Hall-96] and PRISM [Berq-96] and the TINA-C Service Architecture [TINAC-94] and modelling approaches.

4 Iterating the Inter-Domain System Development Methodology

This section illustrates the design decisions and experiences encountered when applying the Prospect design cycle. The example for this illustration is the design of multi-domain management services for a tele-educational service (which is itself composed of several teleservices e.g. Hypermedia information service, video conferencing service and makes use of connectivity services e.g. VPN service and ATM service). The example is based on the work of the Prospect ACTS project ACT 052 which is currently being trialled within this project.

4.1 Business Model and Use cases

The methodology for deriving the PROSPECT business model is based on that of the ESPRIT project ORDIT (Organisational Requirements Definition for Information Technology). ORDIT developed a model with roles, responsibilities and obligations that can be used for establishing relationships between the various parties involved in a socio-technical system which captures organisational requirements. It can be used at a number of different levels of abstraction and is iterative, allowing for revision and growth. Figure 2 illustrates the contractual relationships between the stakeholder organisations for the case study.

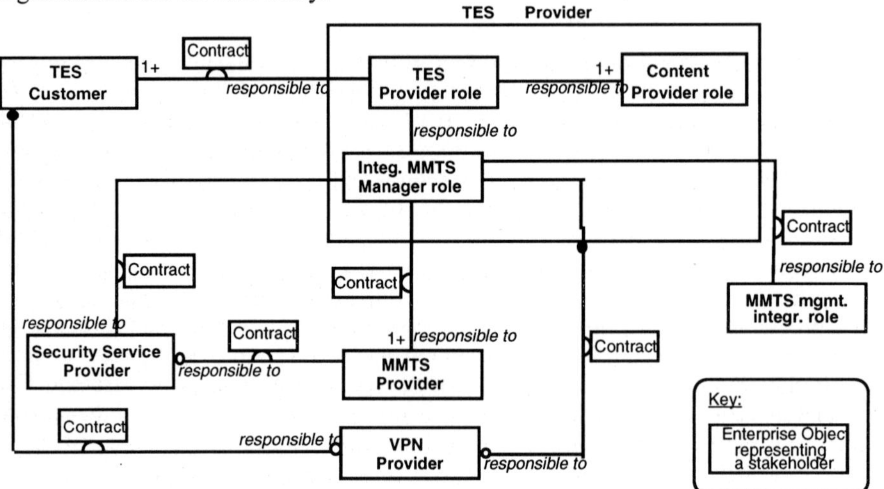

Figure 2: Contractual Relationships between Stakeholder Organisations

Once the model of stakeholder and roles had been established, the requirements on the stakeholder's systems were focused on through the definition of use case for the customer, provider and end user roles of the TES stakeholder. These use cases included; subscription to the TES, inclusion of a customer network site in a TES subscription, authorisation of a TES end user under a subscription and actual use of the service. To assess the inter-domain implications of these use-cases, i.e. the requirements they placed on the different stakeholders in the enterprise model, high level sequence diagrams were drawn up to help define the information that needs to flow between the different stakeholders and roles. An example of such a diagram for the "authorise a TES end user" use case is show in figure 3.

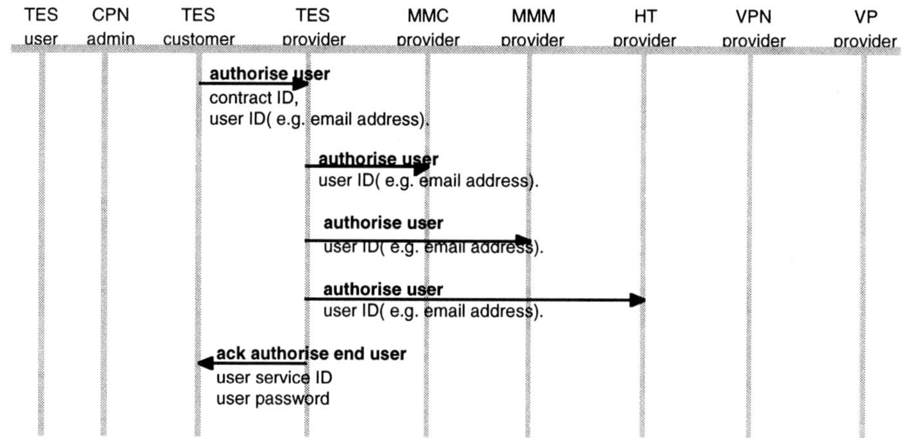

Figure 3: Use Case information flow between stakeholders

4.2 Reuse of Existing Models

At this stage functional requirements for the systems within each stakeholder that would be involved in the use cases had been outlined so other existing management system specifications were analysed to see if they could be reused in meeting the requirements. This is in line with current management system methodologies, e.g. TMN M.3020 and the NMFs Ensemble approach, which aim to make maximum reuse of existing functional components specifications and information model when designing new management system.

However, in the particular service management areas covered by the use cases, little was available in the way of existing specification, either in the TMN functional and information specification or in NMF solution sets. The TINA Consortium had however been examining areas of service management in detail in its service architecture (SA) [TINA-94]. In particular they have defined a generic service model for user telecommunications services that was closely integrated with management components for both subscription management and accounting management. This combined service and management model was seen as suitable for satisfying many of the requirements imposed by the use case for the Prospect systems and were therefore selected for reuse as the basis for the management systems of the TES and MMTS providers' systems. However, the TINA service architecture assumed only a single provider offering services to customers, whereas the use cases placed requirements on the TES system to integrate the MMTSs offered by other providers into a single service offering. This required the modification of the TINA architecture components used in order to suit the use case requirements.

The TINA service architecture is expressed in ODP viewpoints, and provided (i) an *Information Viewpoint* model in terms of information object (IO) descriptions and OMT object diagrams to express the relationships between the objects. (ii) a *Computational Viewpoint* model in terms of computational object (CO) descriptions and object block diagrams showing the client server relationships between objects.

Details of these viewpoints, expressed in TINA using Quasi-GDMO [TINA-95] and Object description Language [TINA-93] were not publicly available, however draft interpretations of some parts of the models by other projects were available.

The information and computational models were therefore required as the basis from which to extend the models to satisfy the requirements present in the use cases and from which to generate a detailed design specification sufficient to implement the components required. It was found however that the TINA design specification in the form of these two viewpoints was inadequate for this task. This was primarily due to the lack of an explicit linkage between the two viewpoint models, i.e. the mapping between IOs and COs was not present in any clear manner. This prevented both a clear understanding of the system from being made and hid the overall object model needed to actually implement this system. The first step to resolving this problem was to use the use cases as the basis for sequence flow diagrams showing and describing the flow of information between COs. This illuminated the dynamic aspects of the model and in the process clarified the relationships intended between the COs and IOs and their behaviour.

As COs are taken to be units of object distribution, some mechanism was required to map CO definitions to a form suitable for implementation on a distributed platform. The engineering viewpoint model for all TINA architectural components defines a Distributed Processing Environment, providing distribution transparencies to engineering computational object based on the COs of the computational viewpoints. However, no practical implementation of the DPE platform implementation was available to the project. Instead a commercial CORBA 2.0 [CORBA-95] implementation (Orbix from Iona) had been chosen as the platform for the components based on the TINA SA. This required mapping between the multiple interfaces of a TINA engineering computational object to the single interfaces of CORBA objects. This mapping exploited the similarity between ODL and CORBA's IDL, with ODL CO interfaces as mapped individual IDL interface definitions, grouped in a module mapped from the CO definition.

The combination of OMT IO definitions, IDL definitions of CO interfaces and sequence diagrams showing interactions between COs via the IDL interfaces provided enough detail for developers to understand the TINA SA components selected for implementation. In these cases a single object model was deemed too costly to synthesise, so extensions to these SA components were specified using the same combination of notations. The following section provides more details on how these notations were used in practice by examining the extension of the TINA SA subscription management component to satisfy the multi-domain requirements of the use cases.

4.3 Extending Reused Components

Since the design of the TINA SA subscription management component had been presented as a presumably consistent set of structures IOs and CO, and since the relationships between these sets of objects has been clarified through detailed sequence diagrams, the extension of the component was most readily performed by using the same notational structure.

Figure 4 shows a portion the OMT object diagram for subscription management from the TINA SA representing the parts that were actually implemented in Prospect. The shaded objects shown are those which were added to the model to extend it to handle the multi-domain requirements on subscription management. In this case information relating to subcontracted providers and the mapping of subscription-related IO in one domain to those in subcontractors domains was required. The intention of these extensions was to support the functionality required, while preserving the integrity of the existing information model.

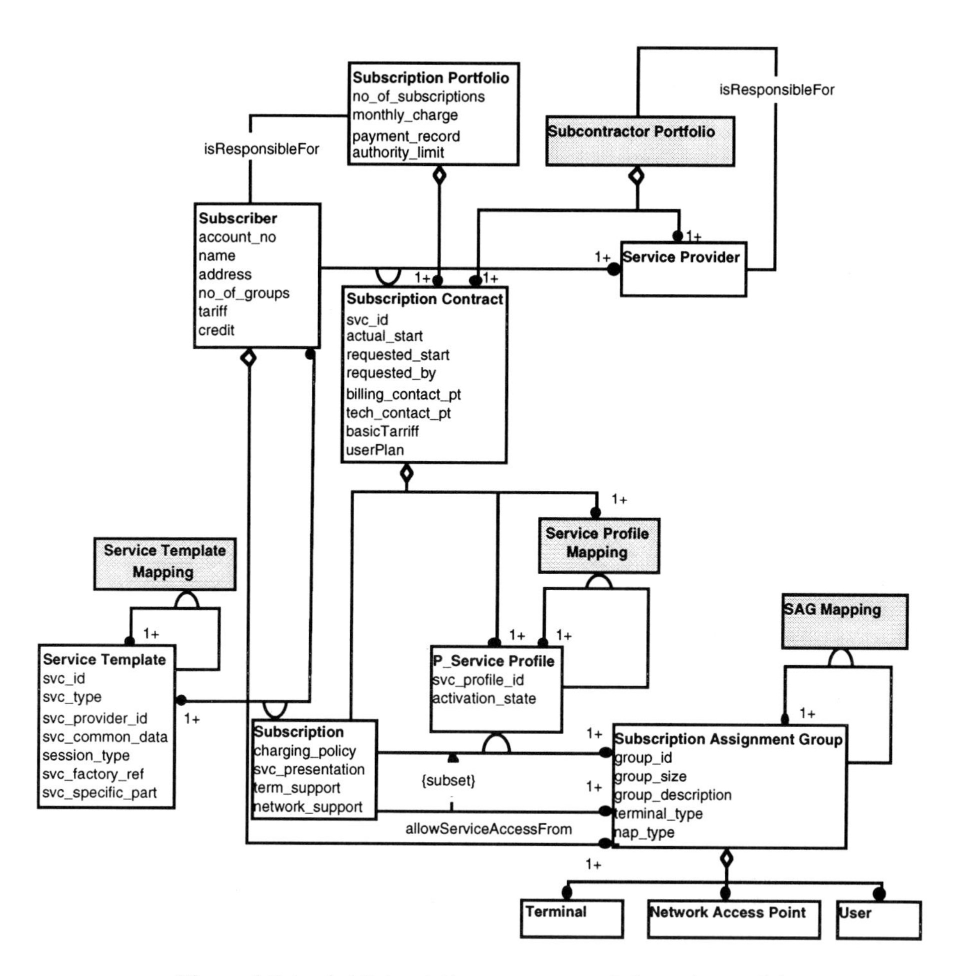

Figure 4:Extended Subscription management information model

A similar approach was taken when applying the extension to the computational model. It was deemed advantageous to retain as much as possible of the interface definition of the existing COs when developing the extensions required. In this way components designed to interact with an original CO interface of the component (SubMgmt in the figure) could also interact with an extended component (SubMgmt* in the figure) with minimum modification. This was performed simply by designing SubMgmt* as a wrapper for SubMgmt, with the new COs introduced to implement this wrapper (shaded in the figure) inheriting IDL interfaces from COs in SubMgmt. The SubMgmt* COs provide the functionality needed to interact with SubMgmt components as used in other domains, thus exploiting the same CO interfaces and minimising the complexity of information processing that needed to be performed.

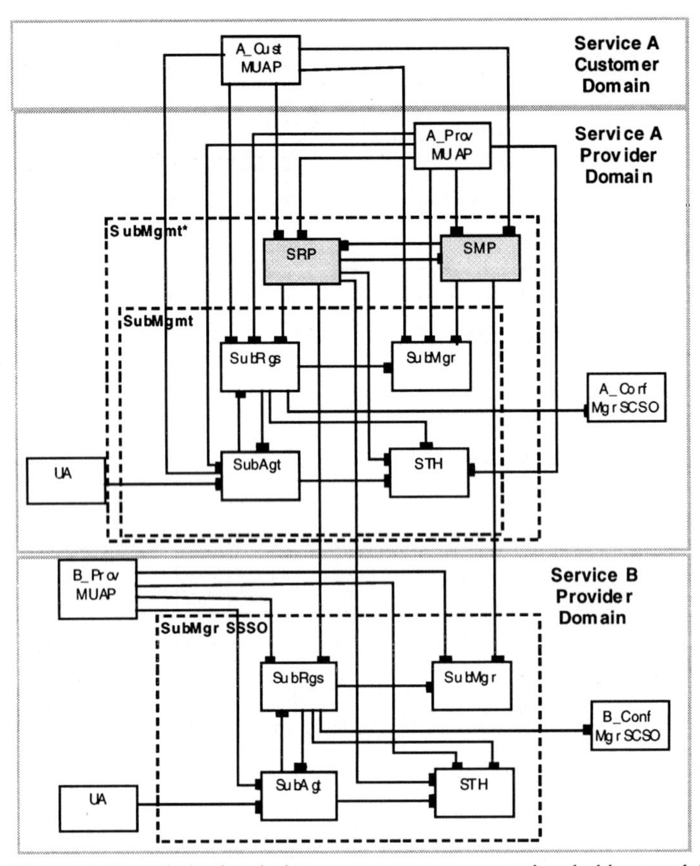

Figure 5: Extended subscription management computational object model

As had been performed with the SubMgmt object in the original TINA specification, the SubMgmt* COs were documented as a detailed block diagram identifying the specific server interfaces offered by the CO using the IDL interface names. In addition the other COs to which the CO was a client are also identified. An example of the notation used for this is given in Figure 6.

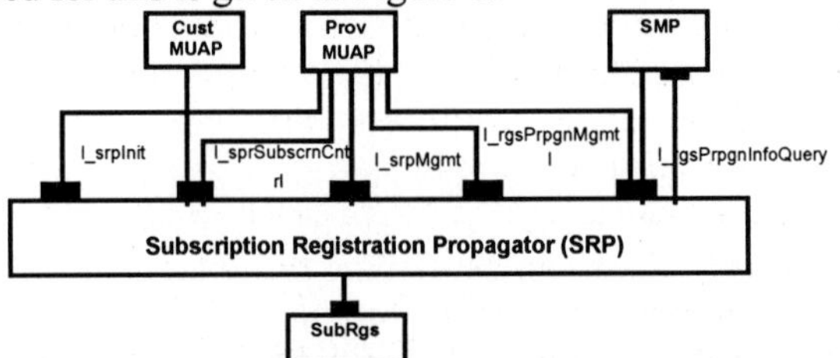

Figure 6: Sample of detailed CO block diagram

Such diagrams were accompanied with details of which IOs were handled by the CO and descriptions of the functionality provided by the different interfaces. As for the original TINA COs, sequence diagrams showing interface interactions between CO were used to develop these interface definitions and clarify which IOs are held in which COs. An example of such a sequence diagram is given in Figure 7.

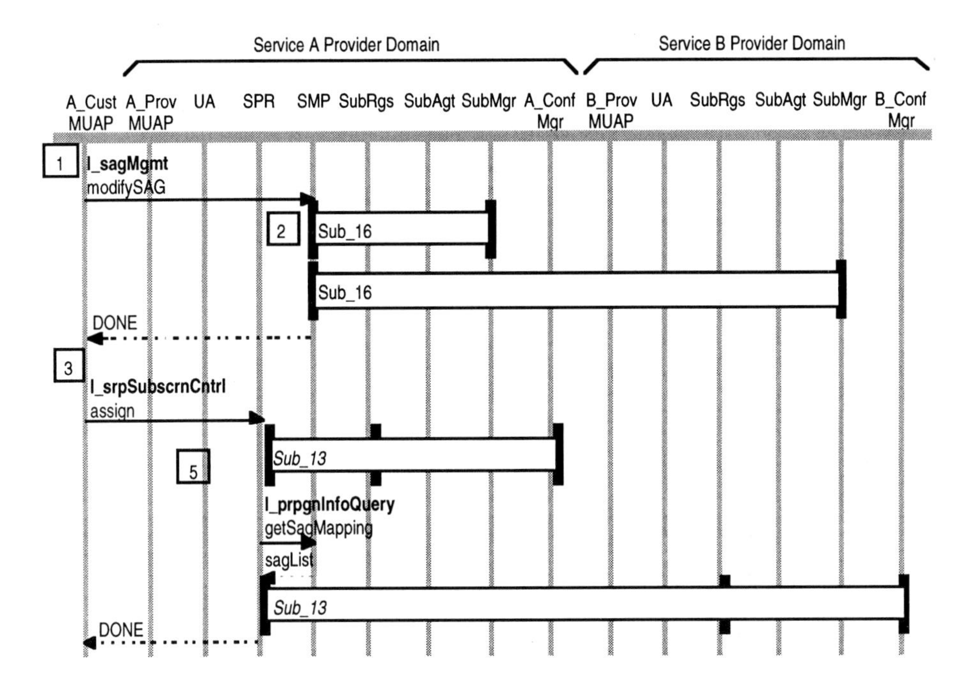

Figure 7: Example of sequence diagram showing interactions between COs

The functionality covered by these sequence diagrams was taken directly from the use case information flows used in the analysis, thus ensuring that the requirements were fully met by the design. Figure 7 shows the interactions that implement the use case information flows of Figure 3. As sequence diagrams showing interactions between multiple COs in multiple domains could easily become large and complex, a box notation was used to refer to sequences of interactions that were represented in other diagrams, e.g. the boxes marked Sub_16 and Sub_13 in Figure 7. This form of nesting sequence diagrams also simplified the drawing of situations where sequences of interactions were repeated. The boxed numbers referred to accompanying notes that explained each significant interaction in more detail, in particular referring to their effect on IOs contained within the COs shown.

As well as proving essential in clarifying the behaviour of CO interfaces and their internal operations on IOs, the sequence diagrams were also found to be ideal for producing test documentation. Integration tests performed between components implemented by different developers were specified by defining pre-and post condition values for IOs at the beginning and end of sets of interactions represented on a sequence diagram. Values could also be provided for the parameter of interface operations performed, so that appropriate test harness software could be developed and operated. This was especially important where interactions involved a chain of several COs, and these needed to be tested individually and in small groups before finally being able to test the complete end-to-end interaction.

Stage in Design Cycle from which Viewpoint is derived	*ODP Viewpoint*
ORDIT Method to describe (in text form) - Stakeholders, Relationships,Roles, Obligations, Activities OMT Object Notation diagrams for stakeholders,relationshp Jacobson USE CASES (text based descriptions of user interactions) describe what actions are required	Enterprise Model
Use OMT Object Model diagrams e.g. class, aggregation, etc	Information Model
Use Object Classes which satisfy the USE Cases. Use traces diagrams to describe and detail the interactions between object classes. Use IDL to specify CO interfaces	Computational Model

Table 1 Mapping from elements of inter & intra domain models to ODP viewpoint models

5 Conclusions and Further Work

The methodology has been used to develop multi-domain subscription, accounting and configuration management services. It has proved very useful both in supporting the full process development lifecycle and allowing the reuse, integration and extension of pre-defined (standard) computational components. Experience using the methodology clearly showed that a deeper understanding is required than just IO and CO models of pre-existing components, and the methodology provided techniques (use trace & interaction diagrams) to assist in the understanding within the context of the particular problem domain.

Acknowledgement: The work presented in this paper was conducted with partial funding of the European Commission under the Prospect project (AC052).

References

[Corba-95] *Object Request Broker 2.0.* Object Management Group, 1995.

[ITU-96] Text of draft recommendation G851-01, Study Group 15, ITU June 1996

[Jacob-92] *Object-oriented software engineering: a use case driven approach,* Jacobson Ivar ACM Press Wokingham Addison-Wesley 1992

[ODP-94] *Reference Model for Open Distributed Processing,* Part 1 Overview and Part 2 Foundations. ISO/IEC 10746-1 (DIS) & 10746-2 (IS) ITU-T X901 & X902. 1994

[ORDIT-93] *ORDIT* process manual, version 0.5 December 1993

[Hall-96] *'Modelling and Implementing TMN based multi domain management,* PREPARE Consortium, J Hall (ed) 1996

[Berq-96] *Succeeding in Managing Information Highways*, PRISM Consortium, Springer Verlang, Berquist, A. 1996

[Rumb-91] *Object Oriented Modelling and Design,* J Rumbaugh et al, Prentice Hall 1991.

[TINA-94] *TINA-C Service Architecture*, TINA Baseline document TB_MDC.018_1.0_94, Berndt H, Minerva R,

[TINA-93] *Computational Modelling Concepts*, TINA Baseline document TB_A2.NAT.002_3.0_93, December 1993

[TINA-95] *Information Modelling Concepts*, TINA Baseline document TB_EAC.001_1.2_94, H. Christensen, E. Colban, Version 2.0, April 1995

Towards Harmonised Pan-European TMN Customer Care Solutions: Interoperable Trouble Ticketing Management Service

Stefan Covaci, Dan Dragan
DeTeBerkom / GMD FOKUS
Hardenbergplatz 2, D-10623 Berlin, Germany
Phone: +49-30-25499200, Fax: +49-30-25499202
Email: covaci@fokus.gmd.de

The paper presents the current status in the development of a pan-European Trouble Ticketing management system, undertaken by a number of European Public Network Operators inside the EURESCOM1 P612 project. This system has the goal of harmonising the requirements for a customer/provider trouble management system at a European level without affecting the currently existing operational systems in this area. A subsidiary goal is to improve the performance of currently available trouble management systems using state-of-the-art TMN solutions. The paper describes the functional structure of such a management system, along with the chosen design options. The system specifications in the form of Ensembles are briefly described and issues related to the validation of the specifications via implementations and tests in a pan-European testbed are addressed.

1 Introduction

European Public Network Operators (PNOs), and Telecommunications and Information Service Providers (SPs) in general, are facing a number of challenges forcing them to undergo extensive Business Process Re-engineering programmes in a number of their activities. Some of these challenges are listed below:

- Increased globalisation of business activities of their customer base.
- European Commissions pressure to harmonise various telecommunication services across Europe.
- De-regulation in Europe.
- Increasing service complexity.
- Established Lead of the US Market.

One of the main customer care activities facing the above challenges is the management of trouble tickets [1], [2], [3].

The necessity for a TMN co-operative Trouble Ticketing (TT) service has been recognised by EURESCOM which started in march 1996 the project P612: "Trouble Ticketing and Provisioning for International Private Leased Data Circuit and International Freephone Service" with the following objectives:

- development of a generic, interoperable TT process model valid for at least 2 different telecommunication services, i.e. International Private Leased Data Circuit Service (IPLDC) and International Free Phone Service (IFS), different enough to stress that generic model;
- development of specifications for interoperable TT systems for these two services, i.e. the essential X interfaces involved in the TT management:
- X.user interface between a PNO-SP Management Domain and a Customer Network Management Domain, and
- X.coop interface between two peer PNO-SP Management Domains,

---

1 EURESCOM - European Institute for Research and Strategic Studies in Telecommunications - is a joint research institute of the main European Public Network Operators.

- definition of test suites required to validate any implementation of the above specifications,
- set-up of a pan-European testbed, and carry out interoperability tests for different implementations based on the specified test suites.

The rest of this paper has the following structure. Section 2 is presenting the overall organisational model of the trouble management system for the IPLDC service. Section 3 focuses on the functional decomposition of the trouble management system. Section 4 describes the actual specification of the trouble management system. Section 5 outlines the validation process of the specifications and implementations. Section 6 addresses issues related to the first implementations. Section 7 presents the early conclusions that can be drawn from this activity.

2 Enterprise Model for pan-European TT service

The TMN framework is based on a clear separation of the telecommunication plane and the management plane. In the telecommunication plane, an IPLDC can be made up from a number of different sections each of which being potentially owned/managed by a different administrative entity (PNO, alternative network providers). For the TMN plane, P612 has selected the "star" based organisational model [5].

This model is preferred in the existing European environment as it involves a reduced number of PNO-SPs and offers a better round-trip response time than an alternative "cascade" model. Additionally, it is more appropriate to the current reality of the telecommunication networks in the European countries, where there is either a single monopolist PNO-SP or a single dominant PNO-SP. Therefore, there is a strong habit of the average customer to deal with only one PNO-SP, and the respect of this habit has represented one of the project requirements. Consequently, the Customer sees only a single point of contact for all its trouble management operations concerning all its contracts (services) with a PNO-SP.

3 Functional architecture of the interoperable TT system

Three main sources have been the basis for the architectural design of the trouble management system: X.790 standard [1], Ensemble concept [6], and the current status of implementations of TT Management OSs inside the various European Network Operators.

Based on these sources, a number of TT operations (at the interoperable interfaces and within the specific management domains) to be implemented in a co-operative TT environment were identified. They briefly summarised below:

- Operations to allow a manager to request the creation of a trouble ticket or to be informed of such a creation by the agent.
- Operations required for the trouble ticket resolution.
- Operations to enables the manager to monitor (to require to be informed about) the progress of the trouble ticket resolution.
- Operations for various updates of the trouble tickets (including escalation).
- Operations to allow the clearing and closing of a trouble ticket and the archiving of (some of) its information.
- Operations to enable a PNO-SP to refer the trouble ticket resolution to a co-operating PNO-SP.

- Operations to enable a PNO-SP to refer the trouble ticket resolution to a co-operating PNO-SP.

It is quite obvious that not all these functions are relevant for the inter-jurisdictional interfaces. Moreover, the currently available OSs for trouble resolution - so called "legacy systems" have to be preserved.

Therefore a 2-layer approach was conceived for each implementation of a co-operative TT process . A first layer is represented by the "legacy system" product. The second layer is a generic TT Process Model Layer that maps (a part of) the existing trouble ticketing operations into those operations seen at the X interfaces. This layer enables therefore the passing of required information between the management domains. This layer interposes itself between the "legacy system" and the X.user/X.coop interface. It should acts as a common denominator of all interoperating "legacy systems".

A final characteristic of this generic TT layer was the effort to minimise the differences that may exits between X.user and X.coop interfaces. The rationale of this design requirement was two-fold. For the present time it was aimed to minimise the initial implementation/testing effort. On the long run it was felt that, in a rapidly evolving market, the customer/PNO-SP roles can change frequently, and such a change should be made possible, if necessary, with a minimal effort. As a consequence the management of "shared" trouble tickets between co-operative PNO-SPs (known as hand-off / referral in [1, 2, 7]) will not exists as a distinct component in our specifications: it be covered by creation of subsidiary level TTs between PNO-SPs (as stated in 4.1).

Therefore, there are 4 sets of operations (Management Service Components - MSCs) in the pan-European TT management system, as described in fig. 1.

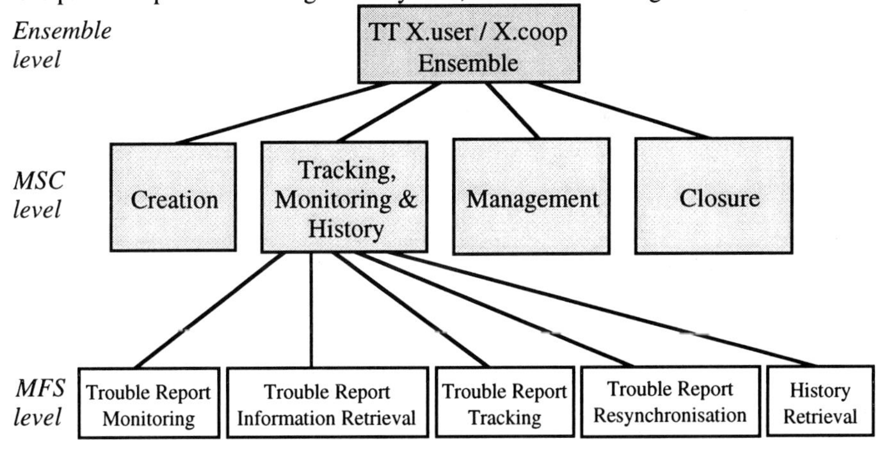

Figure 1: Decomposition of X.user / X.coop Ensemble

4 Specifications Status

The design phase of the generic interoperable TT process model is aiming to produce the specifications in the form of an (NMF) Ensemble. This type of specifications is further enhanced by the usage of Message Sequence Flow (MSF) diagrams - a form of simplified SDL description of the main usage (scenarios) of Management Functions (MFs) identified inside the Ensemble.

4.1 Management Service Components

If the contents of other MSC are quite straightforward, the TT Creation MSC requires some explanations. Here a number of distinct TT creation cases were identified depending on the specific interoperable X interface.

At the X.user reference point there are 3 cases of TT creation:

[u1] A customer requests the creation of a trouble ticket.

[u2] A PNO-SP creates a trouble ticket as result of an internal detected problem or as a result of a problem detected by a co-operating PNO-SP involved in the provisioning of the service to this particular user. The user is not and should not be aware of the source location of the trouble.

[u3] A PNO-SP creates a trouble ticket as a result of a planned maintenance (either its own planned maintenance or the planned maintenance of an involved co-operating PNO-SP). The user is not and should not be aware of the source location of the planned maintenance.

At X.coop reference point there are 4 cases of TT creation:

[c1] A requesting service provided requests the creation of a trouble ticket to a responding PNO-SP;

[c2] A PNO-SP creates a trouble ticket as result of an internal detected problem;

[c3] A PNO-SP creates a trouble ticket as result of a planned maintenance;

[c4] A requesting PNO-SP requests the creation of a trouble ticket to a responding PNO-SP as result of processing a request from one of its customers.

Obviously, for a given [u1] created TT there can be multiple [c4] subsidiary TTs but they will be processed sequentially in the same manner.

MF name	Purpose
TRAddTroubleInfo	The manager adds new data to an existing TT.
TRAttributeInfo	The manager obtains data from an existing TT .
TRCancel	The manager cancels and close a TT.
TRClearedVerify	The agent informs about the clearing of a TT and asks for verification.
TRCommimentTime	The agent updates the closure commitment time for a TT.
TRCreate	The manager requests the creation of a TT.
TRCreationRep	The agent informs about the creation of a new TT.
TREscalate	The manager requests TT escalation.
TREscalateRep	The agent informs about the escalation of a TT.
TRGrantAuthorisation	The manager informs the agent about granting a previous requested authorisation.
TRLog	The manager retrieves logged information about closed TTs.
TRMaintenanceRep	The agent informs about a planned maintenance.
TRModifyInfo	The manager modifies some data of an existing TT.
TRMonitor	The manger retrieves the current state of an TT.
TRRequestAuthorisation	The agent informs the manager of a need for access on manager premises.
TRSynchronisation	The manager recovers the status of existing TTs after a crash on its site or a communication failure.
TRTracking	The agent informs about a change in a TT state.
TRVerifyClear	The manager acknowledges or refuses the clearing of a TT.

Table 1 - List of Management Functions

4.2 Management Functions

Based on the above decomposition in MSC and using MSFs, a number of MFs were finally designed. The list is given in the table 1 containing the name of the MF, and its description. Note that in this table the acronym TR stands for Trouble Report, another commonly used name for a TT (trouble ticket).

4.3 Management Informational Model

The second part of an Ensemble describes the Information Model chosen for implementation of the previously identified functions. The managed object classes are those identified and defined in [1] i.e. *account*, *cnmService*, *troubleReport*, *telecommunicationTroubleReport*, *providerTroubleReport* and *troubleHistoryRecord*. To that we added a *M.3100-Network* managed class object to represent the main telecommunication service for which the trouble ticketing service is created and a *X.721-Log* managed object class as a container for history records.

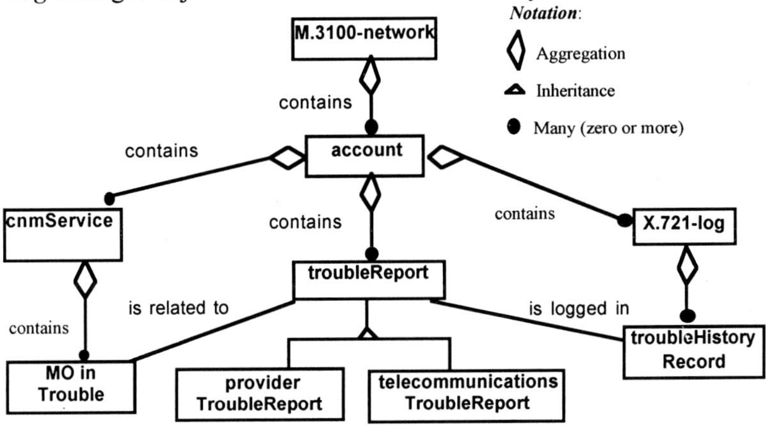

Figure 2 Information Model for Generic TT process

The inheritance tree for the above objects classes is that defined in X.790 [1]. The relationships between these objects classes are described in fig. 2 using OMT [8] object class diagram notation.

The model design proceeded with the profiling of the above objects classes, selecting the supported packages from those defined as optional in X.790 and ended with a formal description of the interface using the GDMO language. This represents the input for the implementation phase of the project.

5 Validation procedures

The validation process implies 3 tasks:

- definition of the testbed, i.e. selection of the involved laboratories;
- TMN platform interoperability tests: definition and execution;
- definition of the abstract test suite (ATS) for the Ensemble and main tests execution.

5.1 Laboratory Selection

The laboratory selection process was based on a number of carefully chosen evaluation criteria. Based on them, a comprehensive test questionnaire was defined in order to make the desired selection. This process ended up by choosing as laboratories for testing the TT X.coop the following participants: British Telecom (UK), Telecom

Eireann (Ireland), Telecom Italia (Italy), and Telia Sweden). For testing the X.user interface the laboratories of Deutsche Telekom (Germany) and Telecom Eireann (Ireland) have been chosen.

5.2 Definition of TMN platform interoperabilty tests

The TMN platform interoperability tests are defining a suite of tests having the goal of testing interoperability between the participants TMN protocol stacks. The TMN protocol stack is defined according to the standards described below:

- Physical Layer: ITU-T V.11/V.35, V.28/V24, X.21, X.21bis, X.27,
- Data Link Layer: X.25 LAPB, ISO 7776,
- Network Layer: X.25 PLP - ISO 8208;
- Transport Layer: ISO 8073;
- Session Layer: ITU-T X.215, X.225,
- Presentation Layer: ITU-T X.216, X.226 X.209 (ASN.1 Basic Encoding Rules),
- Application Layer: CMIP (X.710, X.711) based on ROSE (X.219, X.229) and ACSE (X.217, X.227).

Since the lower levels are implicitly tested by the operations on the highest level the interoperability tests are split up in the following phases:

0. definitions of tests including protocol parameters for all stack levels;
1. passing the interconnection tests verifying the A-ASSOCIATE, A_RELEASE and A_ABORT ACSE primitives;
2. passing a first level of CMIP tests covering the following CMISE primitives: M_GET (with single object selection), M_ACTION, M_EVENT-REPORT and M-SET;
3. passing the second level of CMIP tests covering the scoping and filtering mechanism for M_GET and M-SET primitives and M-CREATE and M-DELETE primitives.

5.3 Definition of the Abstract Test Suite (ATS) for the Ensemble

The test cases in an ATS are structured in a way which may be described by a tree with branches and leaves. The principle together with the naming convention has been derived from the output of a previous EURESCOM project [4]. The test tree is derived from the specification tree (as depicted in the fig. 1) extended with the following levels:

- an MF level under each MFS;
- a Manager /Agent Level under each MF;
- a Valid / Invalid Level under each Manager/Agent node;
- an Operation / Notification leaf under each Valid Behaviour node and an Invalid Syntax / Invalid Semantic under each Invalid Node.

Each leaf in the tree, together with some of the nodes of the tree, represents a situation which might be tested and these tests may be described as follows:

1. How the Manager reacts to a valid Operation message (i.e. a Response) received from the Agent; we want to check if it performs the expected actions, as defined in the specification which we are validating.
2. How the Manager reacts to a valid Notification received from the Agent.

3. How the Manager reacts to receiving an invalid message from the Agent; note that the PDU is syntactically correct from the CMIP point of view; the error could be, for example, the absence of a mandatory parameter.
4. How the Manager reacts to receiving an unexpected message from the Agent.
5. How the Agent reacts to a valid Operation request received from the Manager.
6. How the Agent reacts if the Management Information Base (MIB) is in a state in which it (the Agent) is supposed to issue a Notification.
7. How the Agent reacts upon receiving an invalid message from the Manager.
8. How the Agent reacts to an unexpected operation request from the Manager.

The above described leaf level tests, if correctly defined and passed, make useless the MF level tests. On the MFS Nodes tests should exist in order to verify, at least, the scenarios defined in the specifications phase. Since in our case the MSC do not have autonomous existence there will be no MSC level tests. For checking the system implementation of the TT ensemble tests have to be defined at the root level (Ensemble). These tests are based on real life scenarios defined with the assistance of people that are currently using the legacy systems for trouble report management.

6 Implementation issues

Even if the implementation is in its early stages, some conclusions became obvious, mainly using the existing experience gained during other previous EURESCOM TMN projects.

In the trouble ticketing system, the number of attributes that have to be transferred across the interoperable interface being larger than one hundred, it is practically impossible to be handled using "traditionally" XMP APIs (and XOM data structures), since implementation delays will be prohibitive. An automated tool to transform the GDMO source into some higher level, more user friendly entities to be used by the user code is strictly required. Since the number of such tools existing on the market is rather limited and their complexity is high, their price is an important part in the overall cost of the implementation.

On the other hand, even if above tools are often provided with proprietary APIs to implement the graphical user interface (GUI) requested by end users of the trouble ticket systems, those APIs are not well suited for GUI implementation purposes, for several reasons:

- These APIs are delivered only along with the tool itself making difficult the process of splitting the work (if necessary) between geographically distributed teams.
- They require that the GUI be implemented in the same language as the manager/agent pair of programs (usually C, C++) demanding the existence of skilled programmers for what is otherwise perceived as a secondary part of the implementation process, and limiting the fast-prototyping approach.
- There is at least a gap between a GUI written using such proprietary APIs and the world of WWW.

The solution we have chosen inside Deutsche Telecom was the development of GUIs using the Java(TM) language.

Although this approach implies the resolution of the problem of interfacing the GUI with the manager, a problem that has many solutions (see for example [9]), it presents from the point of view of a TMN/CMN service developer a number of advantages such as:

- excellent integration with WWW, allowing the downloading of GUIs if necessary;
- fast prototyping;

possibility of rapid integration of distinct GUIs for different services into a single client workstation function (GUI factory).

7 Conclusions

This paper has presented the current status of a complex distributed software (TMN) project. The main problems that this project - still under implementation process - has faced have revolved around the necessity to harmonise the requirements of pre-existent "legacy" operations systems, that were developed in different countries at different moments in time and reflecting sometimes even different cultural backgrounds. A difficult trade-off had to be reached between operations done automatically over the interoperable interface and some operations, less frequently used, that still would have to be made manually as at present. The list of allowable trouble error codes and trouble error causes codes had to be kept at a minimum in order to limit misunderstandings between operators having different cultural backgrounds.
The second problem was the fact that the X.790 [1] recommendation is in its early stages of existence and contains some errors and limitations (they are under a correcting process by the dedicated work groups of ITU-T.) For instance, a lot of time was allocated to the discussion of whether the history of state transitions of a given trouble ticket is to be seen at the X.user/X.coop interface. The actual version of X.790 precludes that, whereas a large number of participants felt that this operation is useful for security reasons. Finally, the requirement of compatibility with international standards has prevailed - at least for this first profile specification.

Acknowledgements: The active participation of all EURESCOM P612 team members in the critical starting phases of the project is hereby acknowledged. Special thanks to the GMD-FOKUS MILAN group for its GUI prototyping work.

References

[1] ITU-T X.790, Trouble Management Functions for ITU-T Applications, 11/1995
[2] SMART-NMForum, Technical Requirements - Interservice Provider Trouble Ticketing, 01/96
[3] SMART-NMForum, Customer to Service Provider Trouble Administration Interface - Requirement Specification, 12/1995
[4] EURESCOM P408, Pan European TMN Experiments and Field Trial Support, http://www.eurescom.de
[5] EURESCOM P223, TMN Guidelines and Information Model, Deliverable 2, Volume 1 - TMN Implementation Architecture for Joint/ Co-operative Scenarios, http://www.eurescom.de
[6] NMForum - Omnipoint, The "Ensemble" Concepts and Format NMF-025, 08/1992
[7] Elisabeth Ziesler, MITRE Corp., Customer Trouble Management of Circuits, 07/1996
[8] Rumbaugh, M. Blaha, W. Premerlani, F. Eddy, W. Lorensen, Object Oriented Modelling and Design, Prentice Hall International Editions, 1991
[9] Luca Derri, Network Management for the 90s, ECOOP'96 on OO Technology Workshop for Services and Network Management

Implementing TMN-like Management Services in a TINA Compliant Architecture: A Case Study on Resource Configuration Management

David Griffin, George Pavlou, Thurain Tin
University College London, UK

TINA aims to provide an architecture to enable telecommunications networks to support the flexible introduction of new, advanced services and to manage both the services and the network in an integrated fashion. While the specifications in the TINA Service Architecture are well advanced, network management aspects are less well defined. Resource Configuration Management is one of the most important management areas covering the management of static topology and dynamic connectivity resources; both of which are fundamental to the operation of TINA services. We present an analysis of RCM and a generic model for configuration management computational entities influenced by OSI/TMN design principles, but making use of the TINA ODP-based Distributed Processing Environment.

1 Introduction

The TINA (Telecommunications Information Networking Architecture) initiative aims at providing a framework for all telecommunications software encompassing components ranging from connection establishment through network and service management to service delivery and operation. One of the challenges of the TINA work is to bring together existing and established telecommunications software architectures, technologies, techniques and methodologies, such as the Intelligent Network (IN) [Q1200] and the Telecommunications Management Network (TMN) [M3010] in a future integrated framework.

In this paper we are concerned with the definition, design and implementation of TMN-like functions within the TINA Management Architecture. It is a fair criticism of TINA to state that its work on the Service Architecture is much more mature than that on the Management Architecture, and that the approach by TINA has been to assume that TMN functions and management services can be taken more-or-less en masse and incorporated into the TINA framework. This paper examines some of the issues behind this assumption through a case study covering Resource Configuration Management (RCM). We have taken the RCM specifications from TINA together with the requirements of a real prototype to create a design suitable for implementation.

The remainder of this paper describes the architectural issues related to the RCM management service, and demonstrates how the TINA specifications were enhanced to include Management Resource Configuration Management (responsible for the management of *management resources*), and how TMN and OSI systems management principles were applied to these functional areas to greatly simplify the design and implementation of the resource map - the heart of RCM systems.

2 The Telecommunications Information Networking Architecture

The main objective of the TINA consortium is to provide an architecture based on distributed computing technologies to enable telecommunications networks to support the rapid and flexible introduction of new services and the ability to manage both the services and the network in an integrated fashion.

One of the main motivations for the TINA initiative was the modernisation of the IN. IN operation is based on control plane functions with protocol based interactions between software embedded in local switches and centralised service logic. The IN techniques have been successful for implementing enhanced telephony services, but it is more difficult to introduce modern, advanced services such as multi-media, multi-party communications mechanisms to support applications such as joint document editing. Services such as these require advanced session management and control. Strictly speaking, more complex session control *could* be provided through signalling mechanisms, protocol based interactions, and centralised service logic, but traditional telecommunications engineering solutions such as IN are not as flexible as software engineering approaches based on object orientation and distributed systems.

TINA adopts the ODP (Open Distributed Processing) [X901] framework for specifying a ubiquitous software platform for service logic, covering both service operation and service delivery. In this way, service design and implementation can be achieved in a more flexible manner through re-usable software components. This is a revolutionary departure for the telecommunications industry and is characterised by a shift from protocol-based telecommunications engineering principles to software engineering techniques which are more closely related to the programming languages used to implement the service logic.

The TINA framework is decomposed into four architectures: Computing, Service, Network and Management [TINA-OVE]. The above discussion introduced the first two architectures: Service and Computing. The Network Architecture provides concepts for modelling the underlying network which implements the basic communications services required by the Service Architecture. TINA has based its modelling approach on the international Recommendations of the ITU, drawing on the Generic Network Information Model of M.3100 [M3100] and the SDH information model of G.803 [G803]. The resulting specification is the Network Resource Information Model [TINA-NRIM] which abstracts the communications resources forming the network infrastructure in a technology independent model.

TINA's Management Architecture draws heavily on the ITU's TMN architecture [M.3010]. The TINA specifications in the configuration management area are the most developed, especially those for connection management [TINA-CMA] [Bloem95] [DelaF95]. An interesting observation of the TINA results in this area is that they do not distinguish between the control and management planes in the same way that traditional telecommunications architectures do. Because of this, connection management is included in the Management Architecture as part of configuration management, rather than being part of the control plane of the Network Architecture supported by signalling mechanisms. This is perhaps the starkest example of the paradigm shift from telecommunications to software engineering principles.[1]

This paper concentrates on the Management Architecture of TINA, focusing on configuration management especially on the network topology configuration management aspects of RCM. This is dealt with in more detail in the next section.

3 Resource Configuration Management

3.1 Overview

The TINA Management Architecture is decomposed according to the five OSI functional areas, Fault, Configuration, Accounting, Performance and Security. Configuration Management is concerned with managing the configuration of

[1] However, there is an ongoing debate on whether DPE-based connection management services can perform as well as tightly engineered signalling mechanisms for time critical call set-up procedures.

resources in the TINA architectures. As such, it is more specifically called Resource Configuration Management. It was thought initially that RCM should only manage the resources of the Network, Service and Computing architectures [TINA-RCA]. We believe. though, that there is scope and necessity for managing the resources of the management architecture itself, as we explain next.

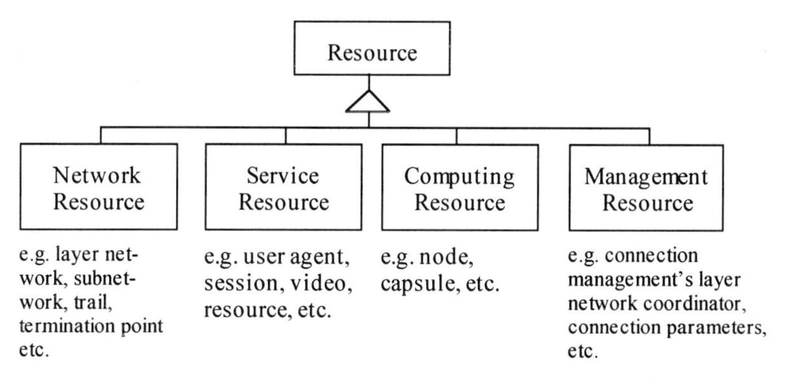

Figure 1: Classification of Resources

Layering is an important concept in TINA. Based on the TMN layering principles [M3010], TINA has defined the Service, Resource and Element layers [TINA-OVE] [TINA-REQ]. The concepts of layering and decomposition of the overall architecture are orthogonal: each of the architectures can be split into Service, Resource and Element layers. The management architecture supports management services which should be seen as specialisations of general telecommunications services; as such, they should conform to TINA principles. Because of this, the Management Architecture itself can be considered to be layered according to the three different layers, and therefore contains management resources i.e. the computational objects implementing and providing the management services. These resources need to be managed just like any other resource.
The dependency between the layering and decomposition concepts leads to the conclusion that RCM is applicable not only to network resources, but also to service, computing and *management* resources. The classification of resources addressed by RCM are depicted in Figure 1. TINA specifications recognise the fact that within the Configuration Management domain, there exist Network, Service and Computing Configuration aspects but the specifications do not address Management Configuration aspects, at least not explicitly. Despite this architectural omission, there is clearly a role for Management Configuration Management in TINA as demonstrated through the following example.
According to TINA, Network RCM (NRCM) consists of Network Topology Configuration Management (NTCM), which deals with static network resources, and Connection Management (CM), which deals with dynamic network resources. The CM part needs to be populated with computational components which are configured according to static topological information. In fact, CM needs to be \xd2 managed\xd3 and this is done through the Connection Management Configurator (CMC) [TINA-CMA]. The latter is in fact a Resource Configuration Manager for CM resources (Connection Performers, Layer Network Coordinators, etc.). In a similar

way, although not yet defined by TINA, there could be resource managers for the other functional areas e.g. fault, performance and accounting management. The above analysis leads to the conclusion that a new domain of Configuration Management is needed, namely Management RCM (MRCM), which we introduce to the TINA management architecture. The relevant set of managers are responsible for the *meta-management* of the TINA management architectural components. According to this view, it is now clear that the CMC belongs to the MRCM domain while it was previously thought (by TINA) to be part of Connection Management, and more recently as part of NTCM [TINA-NRA].

3.2 Resource Configuration Management Functions

The requirements specified in [TINA-RCA] [TINA-NRCM] [TINA-FMRCM] [TINA-NRA] and issues related to RCM in general, irrespective of whether it is Network, Service, Computing or Management RCM, can be summarised as follows:

- RCM should maintain an inventory of all resources under its influence showing relationships between the managed resources. RCM should ensure that the resource map is updated with newly installed or deleted resources.
- RCM should allow activation, deactivation, reservation and release of resources through queries and updates by other management components.
- RCM should support installation: Where physical installation is required RCM shall emit a notification

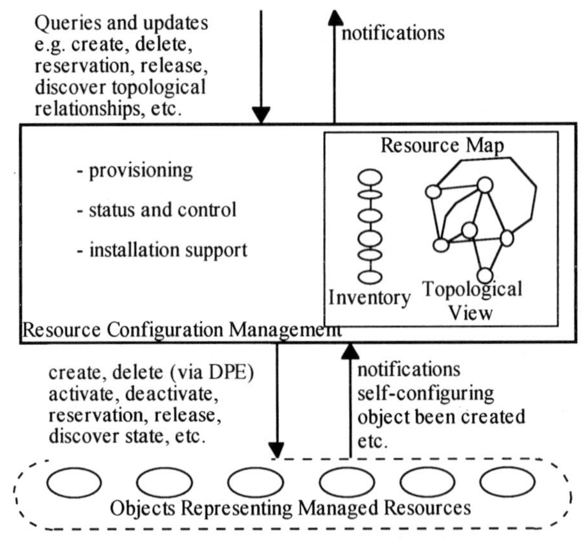

Figure 2: The scope of RCM

Figure 2 is a representation of the scope of RCM. The figure is equally applicable to Network, Service, Computing and Management RCM.

3.3 A Generic Model for Resource Configuration Management

A key aspect of RCM is the maintenance of resource related information, including relationships between resources. Since the resource information is object-oriented, a major requirement is an object-oriented database-like access mechanism. This should

provide access to information objects representing the resources and should allow their relationships to be navigated in a flexible fashion.
There are two main approaches to providing interfaces to support such services. The first approach is to define specific operations, in what we term a *task-oriented* interface, tailored to the particular resources managed by the RCM component in question. The definition of these operations will depend on the individual requirements of the clients of the RCM component as well as on the specific configurable resources being managed. The second approach is to define a generic set of operations applicable to all managed resources and providing the basic query, inventory, etc. services required by any client component. The advantage of the first approach is that the operations may be seen as simpler by specific clients as they are tailored for their exclusive use. On the other hand a significant disadvantage is that new interfaces and operations need to be specified for every new type of client and resource.
The first of the two approaches above is the one that seems most prevalent in the current TINA specifications, while the second approach is similar to that of TMN and OSI systems management. Our view is that both types of interface can co-exist, but the existence of the second approach is essential for generic RCM to allow re-use of specifications and software across all resource management areas.
A computational construct is required for this purpose, providing access to information objects specific to the nature of the resources and exhibiting behaviour that maintains consistency with respect to resource updates. We use the term *Resource Configuration Map* (RCMap) to name such a generic computational construct.
The computational interface offered by RCMap is general, offering maximum expressive power, but on the other hand this genericity may not be always desirable and clients of RCMap may prefer simpler, *task-oriented* interfaces. These are provided by Resource Configuration Manager (RCMan) computational objects which act as clients of RCMap and provide specific computational query and update interfaces, tailored to the nature of the particular resources held in the map and to the requirements of a particular client or group of clients of RCM.

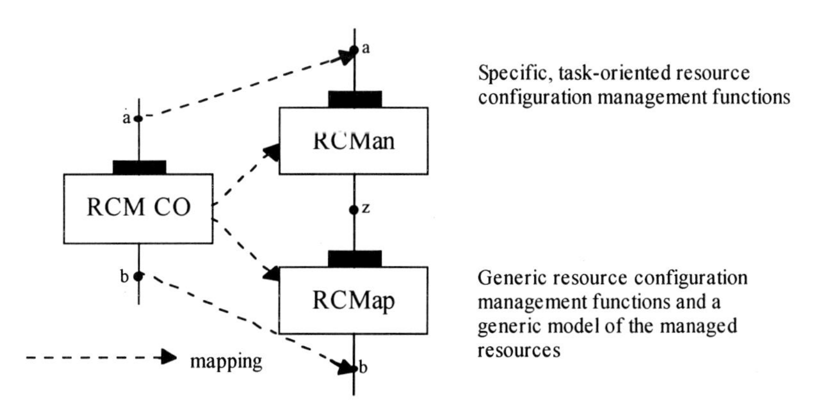

Figure 3: Specific and Generic RCM Interfaces

The RCMan/RCMap model is depicted in Figure 3. This approach, where computational objects hold internal information objects and provide access to them in a task-specific fashion, is not uncommon in TINA. The session graph in the Service Architecture and dynamic resource information in Connection Management are typical examples. However in TINA the computational interfaces offered by these computational objects are task-oriented rather than generic as defined above. This type of computational object, generically termed RCM CO, is depicted in the left part of Figure 3. The separation of specific functionality from generic database-like access as exemplified by the RCMan/RCMap model is depicted in the right part of that figure. Note that the applicability of this model to existing computational specifications, e.g. in connection management, is evolutionary rather than revolutionary as it retains the existing specific access interfaces (interface a in the figure) through the RCMan part of the model.

Because of the hierarchical modelling principles used in TINA, computational objects similar to the RCM CO may be layered hierarchically. In this case, relevant resource information is held in a distributed hierarchical fashion but there is no collective view of it through a single computational interface - for example, this is the case in Connection Management at present. When applying the RCMan/RCMap model to such hierarchical structures, there are two distinct possibilities, resulting in:

- a single RCMap, holding all the information previously held in a disjoint hierarchical fashion and accessed by hierarchically structured RCMan objects; or
- many hierarchical RCMan/RCMap pairs, with the topmost RCMap providing a global view of the overall resource space in a hierarchical federated fashion.

The choice between the two options depends mainly on scaleability issues since operations on managed resources should always be performed through the relevant representations in the resource map for maintaining consistency. Given the fact that OSI systems management-like principles are used for the computational interface of the resource map, a global federated view is feasible and similar approaches exist in today's OSI/TMN systems.

4 An RCM System Covering Network and Management Resources

Based on the generic RCM model presented in the previous section, we propose an architecture addressing the management of network and management resources, i.e. Network RCM (NRCM) and Management RCM (MRCM) domains. Network resources are further decomposed into static resources, representing topology information and covered by the Network Topology Configuration Management (NTCM) domain; and dynamic resources, representing connectivity information and covered by the Connection Management (CM) domain. The architectural decomposition of NTCM and MRCM domains has not yet been addressed in TINA, in fact the MRCM is a completely new domain. On the other hand, CM is relatively mature [TINA-CMA] and we are reusing its architectural decomposition as currently specified.

4.1 A General RCM Architecture

The overall RCM architecture covering the NTCM, CM and MRCM domains is shown in Figure 4. The initial approach taken regarding both the MRCM and NTCM is that of a single resource map. The use of a centralised resource map represents only a first approach to its architectural decomposition. We intend to investigate aspects of scaleability and possibly expand the current approach through hierarchical layering and federation in the future. Note that the CM decomposition is as currently proposed by TINA [TINA-CMA] and does not expose the generic RCMap interfaces to its clients - again, this is for future work.

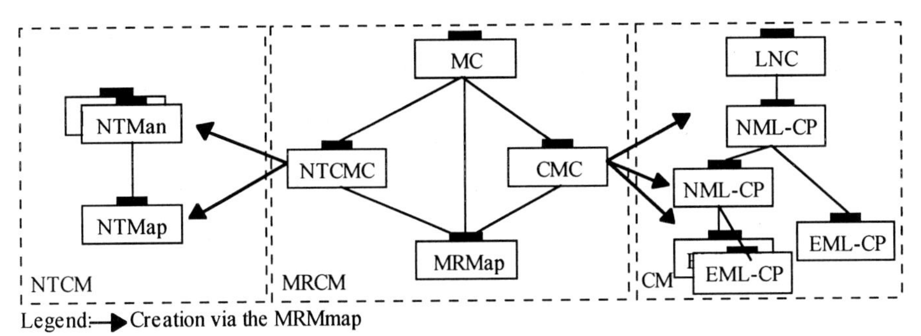

Figure 4: General RCM Architecture

The Management Resource Map (MRMap) contains a view of all the COs instantiated in the management architecture. In fact, this instantiation takes place *through* the MRMap, by creating the relevant representation of management resources. As such the MRMap is the very first CO, necessary for the bootstrapping of the whole management system. The highest level resource manager in the MRCM domain is the Management Configurator (MC). This is \xd2 launched\xd3 by creating its resource representation in MRMap and this operation initiates the instantiation of the whole system as described below. All subsequent CO instantiations take place through the relevant resource representations in MRMap.

By its design, the MC has *a priori* knowledge of what the management architecture should consist of, and in this case it triggers the instantiation of the NTCM and CM domains through relevant RCMan COs: the NTCM Configurator (NTCMC) and CM Configurator (CMC) respectively. It also knows that the CM domain depends on the existence of the NTCM domain since the configuration of dynamic network resources depends on knowledge of the relevant static topology. As such, the MC creates the NTCMC first. The latter is responsible for the NTCM domain and creates the NTMap (Network Topology Map) and NTMan objects. The NTMap has initially a predefined view of network resources, reflecting the underlying network topology. The CM domains are initialised next as the MC creates the relevant CMCs. There may be more than one CMC and corresponding CM object group, one for each *layer network* [G803][TINA-NRIM]. For example, in the case of ATM networks the NTMap contains static information about both the Virtual Path (VP) and Virtual Channel (VC) layer networks. As a consequence, there exist two CM domains, one addressing VP and the other VC connectivity. The MC has access to the network topology information and creates the corresponding CMCs according to the number of layer networks. Each CMC accesses topological information about its layer network and instantiates accordingly the relevant CM COs (LNC, NML-CPs and EML-CPs).

From the moment the whole system is operational, changes to static topological information can be made by (authorised) management applications. In addition, self-configuring resources may emit notifications which are received by the NTMap and result in automatic updates of the relevant resource information. In both cases, the NTMap emits notifications which are received by the CMC for that layer network. The latter may need to reconfigure existing CM COs, launch new ones or terminate others, according to the relevant topological changes.

4.2 A Prototype Implementation

We presented above a general RCM architecture that addresses the resource management needs of TINA systems. A subset of the proposed architecture was specified in detail, implemented and trialled in the context of a real TINA system. This prototype implementation served to validate and demonstrate the architectural concepts presented; as such, we describe it here.

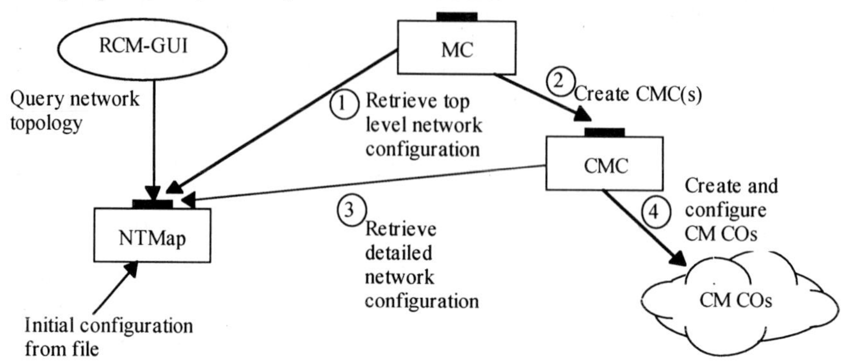

Figure 5: RCM Prototype

The prototype implementation did not address the following aspects of the RCM architecture as presented in Figure 4:

- there was no NTMan component but clients accessed directly the NTMap; in addition, there were no interactions between the NTMap and the real network elements;
- there was no NTCMC component since the NTCM domain was very simple; relevant functionality was embodied in the MC component for simplicity;
- there was no MRMap component; instantiation of the management COs took place directly and not through a relevant resource map; and
- the CM domain was as in current TINA specifications, without a federated RCMap that offers a collective view of dynamic connectivity resources.

The components of the implemented prototype are shown in Figure 5. The main components are the CMC, which configures the CM domain according to topological information, and the NTMap which provides access to the topology information. An application with a graphical user interface allows human network managers to access and manipulate the network topology e.g. in order to add, modify or delete topological information to reflect relevant network changes that took place in an off-line fashion.

The NTMap is the core of the RCM architecture since it maintains a central, consistent view of the resources in question. Given the fact that there is not yet a relevant computational interface in TINA, the NTMap computational specification is completely new. In addition, it is TMN-influenced in the sense of providing a generic CMIS-like [X710] interface that enables to access information objects with various relationships. Object discovery is supported through scoping and filtering constraints in order for clients of the NTMap (such as the CMC) to build up a picture of the network resources and their relationships. As such, it constitutes a cultural difference to the TINA approach to computational specifications, as described in the next section.

The information modelling approach for the network topology was based on the TINA NRIM [TINA-NRIM]. The fact that the NTMap computational interface is generic allows it to model different networks and instantiate the relevant system in a fashion independent of the particular network topology. The prototype system has been in fact exercised over two different ATM networks, with the NTMap initialised in a data-driven fashion. Future extensions will include interaction with the real network elements for on-line configuration purposes and the complete implementation of the RCM architecture.

5 Using a Management Broker to Provide Operations on Multiple Objects

Network resources are modelled as information objects in Quasi-GDMO (the NRIM \xd2 Network Fragment\xd3) and should be made accessible through computational interfaces in the computational viewpoint. These information objects have various relationships and the relevant computational constructs should allow for the navigation of those relationships in a flexible fashion e.g. to discover dynamically the network topology, provide inventory facilities etc.

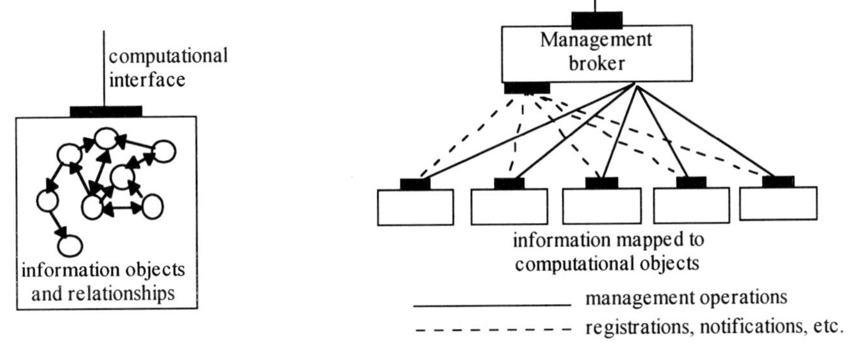

Figure 6: Mapping Information Objects to Computational Objects

The problem is related to a methodology for mapping an information model with various relationships to a set of computational specifications. There are two distinct approaches for this mapping: a specific approach according to the nature of the information model in question (following possibly some general guidelines); or a generic approach, applicable in all cases (the *RCMap* type of computational interface).

A generic mapping results in a computational interface that provides a generic style of access while it also allows for the reuse of the relevant infrastructure. Since OSI Systems Management [X701] exhibits exactly this paradigm, we have chosen to mirror its access facilities in a generic computational interface which we term a Management Broker (MB). This interface offers a CMIS-like [X710] style of access in IDL and provides TMN-like access services over the TINA DPE. We have used the MB approach to provide the RCMap computational interface in our case study.

The specific and generic approaches discussed above are depicted in Figure 6. The advantage of the generic approach is that it maps information to computational objects on a one-to-one basis and separates collective access facilities from behavioural aspects associated with the latter. These objects may be accessed through the MB, which acts as an object factory/naming server and provides multiple object access facilities based on scoping and filtering. They may be also accessed directly so that the relevant client benefits from strong typing with respect to the particular computational interface. The mapping of information objects in Q-GDMO onto equivalent computational interfaces in IDL is according to the guidelines of the NMF-X/Open Joint Inter-Domain Management (JIDM) group [JIDM].

The key advantage of the management broker approach is its genericity, which renders the MB as a server over the TINA DPE. Each MB groups together a cluster of information/computational objects while it is possible to organise those clusters hierarchically and provide a global federated view (federation issues in TMN-like object clusters have been solved through \xd2 chaining\xd3 of the relevant requests). In addition, MBs may also behave as notification servers, allowing for the fine grain control of notifications emanating from the relevant object cluster through event discriminators and filtering [X734]. An additional advantage is that the MB and the relevant administered objects may be distributed as they are separate computational entities.

The benefits of providing OSI Systems Management-like facilities over the TINA DPE are many, as described above. The current TINA approach to these facilities is to provide them in an ad-hoc manner as required. The obvious disadvantage to this method of designing and implementing management systems is that the same features have to be re-specified and re-implemented, as required for each computational interface. There is a distinct advantage in having a generic, \xd2 standard\xd3 way of providing these. Such a generic approach needs not necessarily to rely on OSI System Management and TMN methods and techniques. The advantage, though, of doing so is that we benefit from a host of research and standardisation in this area, we are able to reuse relevant methodologies and specifications and we lay the foundation for TMN and TINA coexistence and migration strategies.

6 Summary and Conclusions

In this paper we have proposed a generic model for Resource Configuration Management which may be applied to Network, Computing, Service and Management resources in a TINA compliant system. We have extended the scope of RCM to include Management resources, which was previously not considered by TINA.

We have shown that there are two approaches to defining configuration management interfaces to the resources being configured: generic and task-oriented interfaces, as demonstrated by our RCMap and RCMan model. The RCMan task-oriented interfaces provide high-level operations and queries on the resources which are tailored to the requirements of the managers performing the operations, and according to the specific types of resources being managed. This type of interface provides simpler operations (from the clients point of view) but loses some of the power and expressiveness of the generic approach. The alternative approach of a generic interface, based to a large extent on OSI systems management and CMIS-like operations, allows resources to be treated in the same way with common methods for object manipulation.

We discussed our approach to providing the RCMap interface through the use of a management broker allowing operations on many objects via a single interface. As well as providing an essential service to configuration management clients, this reproduces some of the important features of OSI agents, which not only aids interaction with existing management functions in coexisting TMN systems but also eases the migration path for deploying TMN management services in TINA compliant systems.

We have specified CORBA-based management brokers that mirror the facilities of OSI systems management. These may administer clusters of other CORBA objects; they may also act as generic adaptors between CORBA clients and TMN applications in agent roles. In the former case, the TMN methodologies for producing object clusters or ensembles may be fully re-used in TINA; in fact, this is how we

approached the NTMap. We have implemented a first version of such a management broker using the OSIMIS platform and the Orbix implementation of CORBA. While there is plenty of ongoing research regarding TMN to TINA migration and interworking, we believe our approach retains the relevant advantages of TMN for network management while it is fully complementary to the JIDM approach. In addition, this is a viable path for gradually migrating existing TMN systems over CORBA-based DPEs. We intend to propose our approach to standards bodies such as TINA, JIDM, OMG and the ITU-T Study Group 4.
Our view is that because the methods and techniques of the TMN have been demonstrated to be useful, and even essential to the design of complex management systems they should not be replaced without careful consideration. As we have shown in this paper, it is possible to retain many of the essential elements of the TMN, even though Q3 protocols have been replaced with IDL interfaces in the DPE. The logical architecture can be kept, meaning that existing information and computational specifications can be reused. This additionally provides a smooth migration path from current (and future) TMN-based management systems to those based on TINA-like DPEs.

Acknowledgements

We would like to acknowledge the development work undertaken by our colleagues in the VITAL project for implementing the first version of the RCM system described in this paper. In particular we would like to thank: Lisbeth J\xbf rgensen Veillat, Juan Carlos Yelmo Garcia and Jesus Pe\x96 a Martinez of UPM who worked on the CMC and the graphical user interface which formed the client of the NTM; Panagiotis Palavos of Infoline who developed the MC; and Juan Carlos Garcia Lopez of Telefonica who provided useful feedback on the design of the RCM system. This paper describes work undertaken in the context of the ACTS AC003 VITAL project. The ACTS programme is partially funded by the Commission of the European Union.

References

[Bloem95] Bloem, J., et al, The TINA-C Connection Management Architecture, TINA'95, Melbourne, Australia, Feb. 1995.

[DelaF95] de la Fuente, L., et al, Application of the TINA-C Management Architecture in Integrated Network Management IV, pp. 480-493, Chapman & Hall, 1995.

[G803] ITU, G.803, Architectures of Transport Networks Based on the SDH.

[Gri95] Griffin, D., Georgatsos, P., A TMN System for VPC and Routing Management in ATM Networks, in Integrated Network Management IV, pp. 356-369, Chapman & Hall, 1995.

[Gri97] Integrated Communications Management of Broadband Networks, pp 37-48, Chapter 3, The 1CM Methodology for TMN System Design, Griffin, D., editor, Crete University Press, ISBN 960 524 006 8, 1997.

[JIDM] X/Open/NMF Joint Interdomain Management specifications, GDMO/ASN. 1 to CORBA IDL translation, 1994.

[M3010] ITU-T M.3010, Principles for a Telecommunications Management Network.

[M3020] ITU-T M.3020, TMN Interface Specification Methodology.

[M3100] ITU-T M.3100, Generic Network Information Model.

[Pav95] Pavlou, G., Knight, G., McCarthy, K., Bhatti, S., The OSIMIS Platform: Making OSI Management Simple, in Integrated Network Management IV, pp. 480-493, Chapman & Hall, 1995.

[Q1200] ITU-T Q.1200 series, Intelligent Networks.

[TINA-CMA] Connection Management Architecture, Draft, Document label TB_JJB.005_ 1.5_94, TINA-C, March 1995.

[TINA-FMRCM] '94 Report on Fault Management and Resource Configuration Management, Version 1.0, Document label TR_MK.006_ 1.0_94, TINA-C. January 1995.

[TINA-MA] Management Architecture, Version 2.0, Document label TB_GN.010_2.094, TINA-C, December 1994.

[TINA-NRA] Network Resource Architecture, Version 2. 1.1, Document label NRA_v2.1.1 97_01_27, January 1997.

[TINA-NRCM] Network Resource Configuration Management, Version 2.0, Archiving label EN-DK.001_2.0_96, TINA-C, October 1996.

[TINA-NRIM] Network Resource Information Model Specification, Document label TB_LR.010_2.1_95, TINA-C, August 1995.

[TINA-OVE] Overall Concepts and Principles of TINA, Document label TB_MDC.018_1.0_94 TINA-C, February 1995.

[TINA-RCA] Resource Config. Architecture, Document labelTB_C.AMB.001_1.0_93, TINA-C, December 1993.

[TINA-REQ] Requirements upon TINA-C architecture, Document label TB_MH.002_2.0_94, TINA-C, February 1995.

[X701] ITU-T X.701, Information Technology - Open Systems Interconnection - Systems Management Overview, 1991

[X710] ITU-T X.710, Information Technology - Open Systems Interconnection Common Management Information Service Definitions, version 2, 1991.

[X734] ITU-T X.734, Information Technology - Open Systems Interconnection - Systems Management: Event Reporting Management Function, 1992.

[X901] ITU-T X.900, Information Processing - Open Distributed Processing - Basic Reference Model of ODP - Part 1: Overview and guide to use.

TMN Specifications to Support Inter-Domain Exchange of Accounting, Billing and Charging Information

Chris Bleakley
Broadcom Eireann Research Ltd., Kestrel House, Clanwilliam Place, Dublin 2, Ireland
Tel:+353-16046000
Fax:+353-16761532
email:cb@broadcom.ie

Willie Donnelly
Waterford Institute of Technology, Cork Rd. Waterford City, Co. Waterford, Ireland
Tel:+353-51302066
Fax:+353-51378292
email:wdonnelly@staffmail.rtc-waterford.ie

Arne Lindgren
Telia Research
S-136 80 Haninge
Sweden
Tel:+46 87075521
Fax:+46 87075480
email:Arne.S.Lindgren@telia.se

Harri Vuorela
Telia Engineering
S-12386 Farsta
Stockholm, Sweden
Tel:+46 87133186
Fax:+46 87134180
email:Harri.O.Vuorela@telia.se

This paper describes work carried out as part of the EURESCOM project P515 "European Switched VC-based ATM Network Studies for Advanced Service Features". The paper describes specifications for the automated inter-domain exchange of accounting, charging and billing information. The specifications define management functionality and information models which support the exchange of information, both between a customer Network Management System (NMS) and a Service Provider (SP) NMS and between two SP NMSs. The specifications follow the Telecommunication Management Network (TMN) framework and were developed according to the NMF ensemble concept. The paper presents an overview of the specifications, explains how they can be used to support the management of accounting in telecommunications networks and describes our experiences in using the ensemble technique.

1 Introduction

In an increasingly competitive telecommunications market, two of the major differentiators between Service Providers (SPs) are their successes in customer care and cost reduction [1,2]. In the area of accounting, improved customer care and cost reduction can be achieved by increasing the customer's access to accounting, charging and billing information. This paper provides an overview of TMN specifications developed for this purpose.

The work described herein was undertaken as part of the telecommunications management task of the EURESCOM project P515 "European Switched VC-based ATM Network Studies for Advanced Service Features" [3]. The goal of the task was to define a collection of solutions to aspects of the problems inherent in managing a Pan European ATM Network. One of these aspects was the problem of managing the exchange of accounting, charging and billing information, both between the customer and the SP (via the Xuser interface), and between Service Providers (via the Xcoop interface). The approach taken to the problem was to define the management solutions within the TMN framework [4] using the NMF ensemble concept [5] as recommended by EURESCOM project P414. This paper provides an overview of the ensembles produced by the project [6], it explains how the specifications can be used and it considers the effectiveness of the ensemble concept.

During the specification process, a three stage procedure was adopted with iterations as required to refine the work. First, the requirements of the customers and the SP for such a system were identified. The functional requirements were then mapped to management functions. These management functions precisely defined what inter-domain functionality is required by the actors. Once the management functions were satisfactorily defined, information models were developed to support them. The information models were defined using the Guidelines for the Definition of Managed

Objects (GDMO) [7]. Scenarios were then produced to relate the management functions to the actual CMIP [8] information flows.
This paper is structured as follows. First, the context of the work is explained. Second, the specifications to support billing and charging at the customer-SP (Xuser) interface are described. Third, the specifications defined to support accounting at the SP-SP (Xcoop) interface are detailed. In both cases, the sections are divided into sub-sections covering the functional requirements analysis, management functions and the information models. Fourth, some open issues and areas for further work are identified. Finally, a summary of our main conclusions is provided.

2 Management Context

In terms of the TMN Logical Layered Architecture [4], the Accounting Operations System Function (OSF) is located in the Service Management Layer. The Accounting OSF receives metering information, detailing service usage by customers, from the Network Management Layer. Based on this information, the Accounting OSF calculates customers bills. In addition, the Accounting OSF determines the amount of revenue to be exchanged with other providers for use of their services and vice versa. Historically, the invoicing and revenue settlement information so generated was passed to customers and other providers by post. The specifications described herein allow this accounting information to be passed to customers and other providers electronically via CMIP interfaces. The management architecture to support this is depicted in Fig. 1.

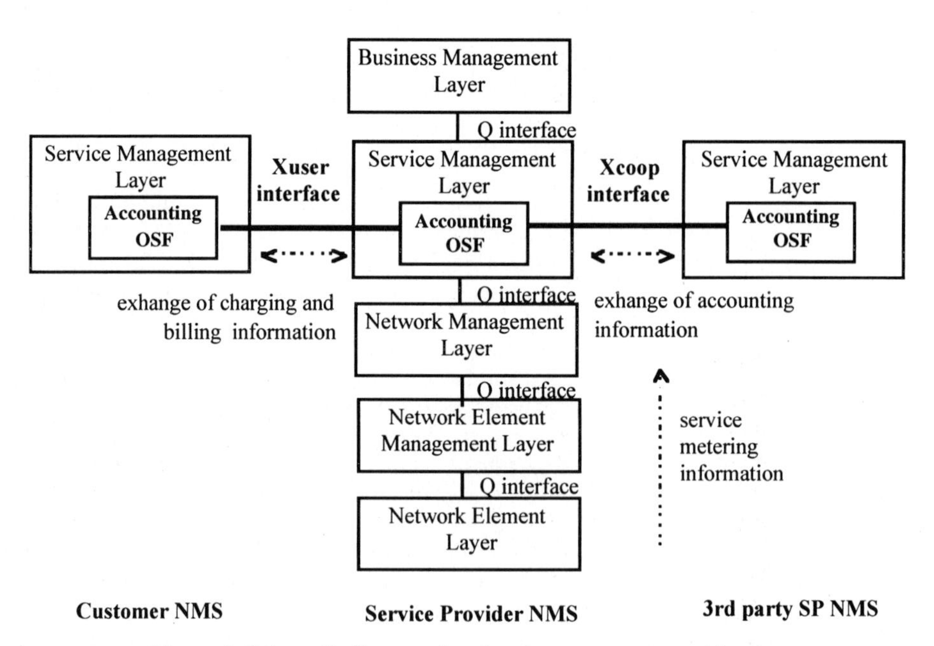

Figure 1: Schematic diagram showing the management architecture.

3 Billing and Charging at the Xuser Interface

3.1 Functional Requirements

The main functional requirements identified are as follows:

- The customer should have access to the tariff plan as defined by the SP. The tariff plan relates detailed service parameters to price.
- The customer should have access to its billing profile as stored by the SP. The billing profile contains information on how the customer is to be invoiced.
- The SP should be able to invoice the customer at the end of a billing period or when a subscription change occurs.
- The customer should have access to current and previous records of service usage by individual users in the customer organisation.
- The customer should have access to current and previous charging records on individual services. These charging records are generated as the result of the application of the tariff plan to a service usage record.
- The customer should have access to current and previous invoice records. These invoice records are generated from the charging records when an invoice is issued.
- The customer should be able to set, update and delete a billing threshold at any time.
- The customer should be notified by the SP when the customer-defined billing threshold is exceeded.
- The information model should support a variety of Service Level Agreements (SLA). For example, the customer should be allowed various payment options including the type of currency, the frequency of the bills, the bill design, etc.
- The information model should support a wide variety of services and billing schemes. For example, some services may have fixed monthly costs, while others may be traffic and time of day dependent.

3.2 Management Functions

Aside from customer queries of the management information held by the SP, the following management functions were identified:

- *Billing Process.* The billing process is triggered by a subscription change or the end of the billing period. The resulting invoice is sent to the customer.
- *Charging Query.* A customer can request a statement of current charging information in connection with a service. The customer may supply a time reference corresponding to a moment between the time of the last invoice and the time specified in the query. The contents of the statement is computed and a cancelled invoice is sent to the customer.
- *Set / Remove Billing Threshold.* A customer can set a new threshold or remove an existing threshold.
- *Threshold Warning Indication.* The customer receives a notification that the current total billing amount has surpassed the customer-defined threshold. Threshold checking is performed on a continuous basis by the SP.

3.3 Information Model

The information model defined to support the functionality discussed is depicted in Fig. 2. A set of tariffPlan objects are associated with each PNOServiceProvider. In turn, each customer associated with that PNOServiceProvider is associated with a customerProfile object, a set of user objects and a log object. Related to each user object is a userProfile object and a customerPNOusageRecord. Ultimately, the information held in the usage record is employed by the Service Provider to create chargingRecords and invoicingRecords in the customer's log.

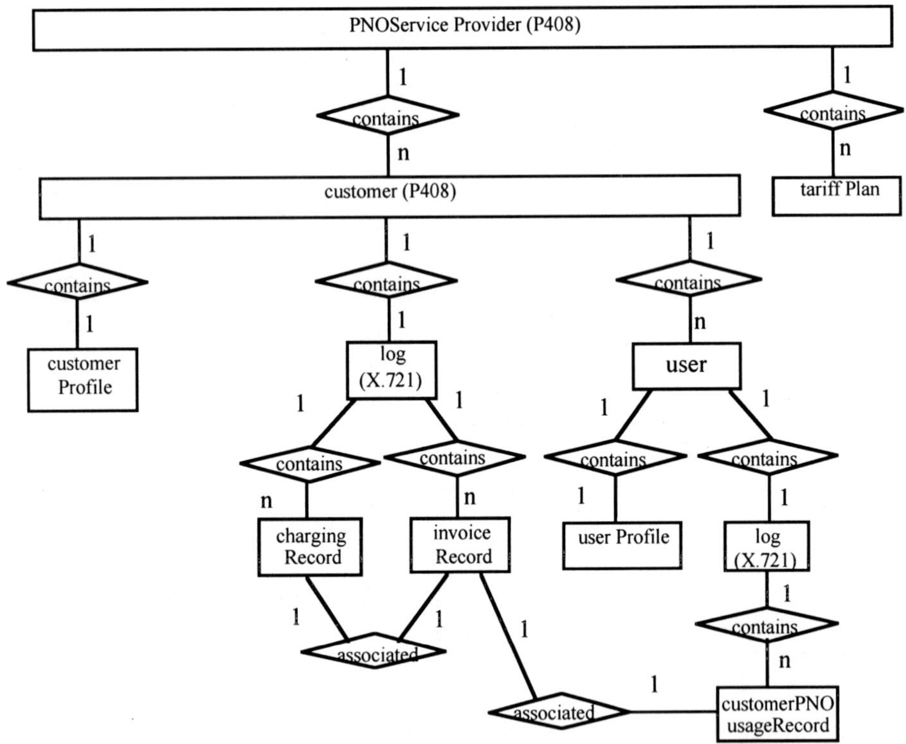

Figure 2: Entity-Relationship diagram of the information model specified to support the exchange of billing and charging information between the customer and the Service Provider [6].

The PNOServiceProvider object is that defined in EURESCOM project P408 [9], while the customer object is based on the P408 definition with the addition of a notification, issueInvoice, and an action, chargingQuery. The log objects are as defined in X.721 [10]. The chargingRecord provides the thresholdSurpassed and chargingRecordUpdated notifications. Similarly, the customerPNOusageRecord provides the customerPNOusageRecordUpdated notification.

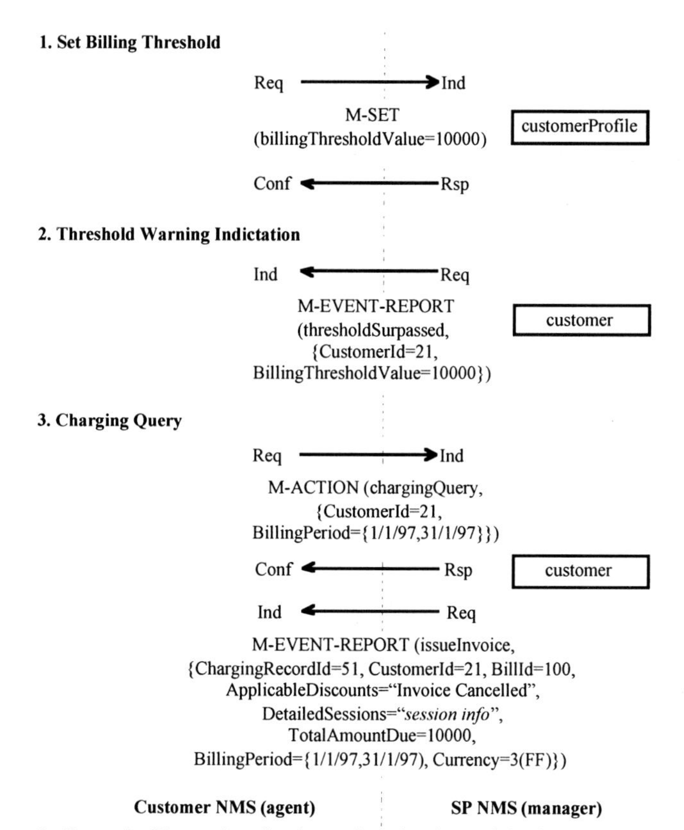

Figure 3: Scenario illustrating the dynamic behaviour of the information model.

4 Accounting at the Xcoop Interface

The basis for this ensemble was the view that, for any given interface, one Service Provider acts as a service user, while the other acts as a service supplier. In this way, the model mirrors that used at the Xuser interface.

4.1 Functional Requirements

The main functional requirements identified are as follows:

- The service user should have access to the revenue tariff plan defined by the service supplier.
- The service user should have access to it's billing profile as stored by the service supplier.
- The service supplier should be able to notify the service user when a revenue settlement is due.
- The service user should have access to current and previous records of its service usage.
- The service user should have access to current and previous charging records. These charging records are generated as the result of the application of the revenue tariff plan to a service usage record.
- The service user should have access to current and previous revenue settlement records. These records are generated from the charging records when an revenue settlement request is issued.

- The information model should support a variety of SLAs.
- The information model should support a wide variety of services and billing schemes.

4.2 Management Functions

Aside from service user initiated queries of the management information held by the service supplier, the following management functions were defined:

- *Revenue Settlement.* This function calculates resource usage based on data from the service supplier. The revenue settlement transfer method is decided by the SLA. If the SLA prescribes CMIP, a notification containing the resource usage data will be sent to the user SP. The periodicity/frequency of the notifications is also agreed upon in the SLA. If, on the other hand, the data transfer method chosen is file transfer, a notification of data availability is sent as prescribed in the SLA.
- *Record Transfer.* This function is employed when the data transfer method agreed upon in the SLA is file transfer. The service user retrieves resource usage data from the service supplier after a notification of data availability has been issued.

4.3 Information Model

The information model defined for the Xcoop interface is depicted in Fig. 4. The providerPNO object holds information on each service user via a set of customerPNO objects. The providerPNO object also contains an number of revenueTariffPlan objects. Each customer object contains a customerPNOprofile and a log object. The log object contains the customerPNOrecords, the accountingRecords and the revenueSettlementRecords.

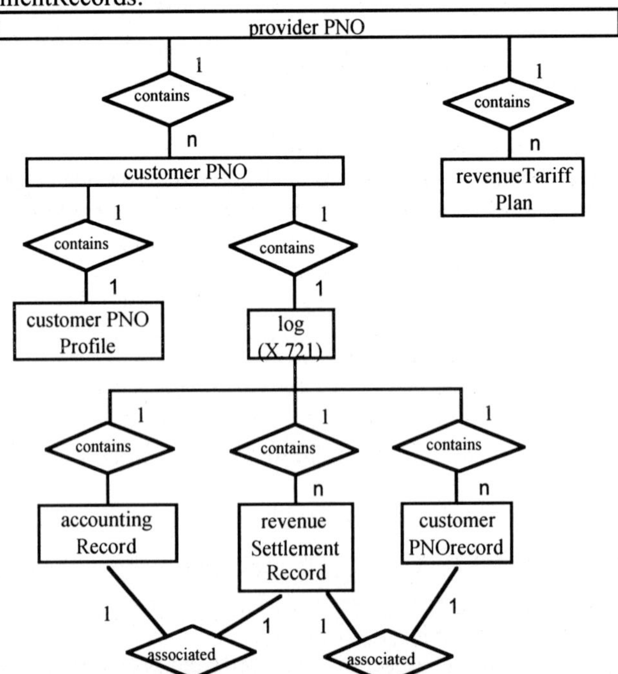

Figure 4: Entity-Relationship diagram of the information model specified to support the exchange of accounting information between the customer and the Service Provider [6].

4.4 Scenarios

The example in Fig. 5 shows how revenue settlement records are transferred between SPs.

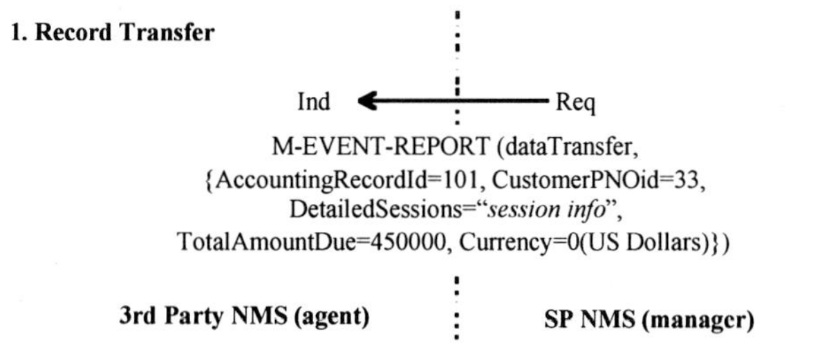

Figure 5: Scenario illustrating the dynamic behaviour of the information model.

5 Open issues and Further Work

The following issues remain open and are topics for further work:

- The work presented here gives a generic framework for the exchange of accounting information. However, customisation according to the SLA is required prior to implementation. For example, the charging parameters will vary between services. Also, revenue settlement might be done via a Clearing House, rather than directly between SPs.
- The invoicing process is a business issue and therefore cannot be addressed at a technical level. For example, the legal requirements on the system will vary from nation to nation. EURESCOM project P407 is currently working on more detailed specifications for accounting.
- Mechanisms for charging for an ATM service have not yet been fully agreed. The ACTS project CANCAN "Contract Negotiation and Charging in ATM Networks" (AC014) is currently studying schemes for ATM charging.

6 Conclusions

The main conclusions from the work are as follows.

The use of the ensemble approach has, in the view of the project, been very successful. Ensembles allow the designers to present advanced feature solutions in a focused and self-contained manner with the aid of scenarios. They impose rigor on the specification process, helping to identify errors and omissions, and they aid understanding by providing a standard format across projects. However, the addition of detailed use-cases and mappings between the management functions and CMIP primitives would make the technique more precise.

The detailed specification described in the ensembles provide generic solutions to the accounting, charging and billing management problem which need to be customised based on the commercial and technical implementation environment to provide the desired functionality.

References

[1] Adams, E.K., and Willets, K.J., *The Lean Communications Provider*, McGraw-Hill, 1996.

[2] Rainger, J. "Convergence billing: an operator's glimpse of the future", Telecommunications, May, 1996, pp. 90-97, 121.

[3] EURESCOM P515, "Summary specification for a VC based switched ATM Europe-wide network supporting B-ISDN capability set 2", Deliverable 1, vol. 1.

[4] ITU-T Rec. M.3010 "Principles For A Telecommunications Management Network", October 1992.
[5] NMF 025 "The Ensemble Concepts And Format", issue 1.0, August 1992.
[6] EURESCOM P515, "Management functionality", Deliverable 1, vol. 6.
[7] ITU-T Rec. X.722 "Information Technology – Open Systems Interconnection – Structure Of Management Information: Guidelines For The Definition Of Managed Objects", January 1992.
[8] ITU-T Rec. X.711 "Common Management Information Protocol Specification For CCITT Applications" March 1991.
[9] EURESCOM P408, "Xuser Interface: Subscription and VP Service Management", PIR 9.1 vol. 1, June 1995.
[10] ITU-T Rec. X.721 "Information Technology – Open Systems Interconnection – Structure Of Management Information: Definition Of Management Information", February 1992.

Acknowledgements

The work described in this paper was carried out as part of the EURESCOM funded project P515 "European Switched VC-based ATM Network Studies for Advanced Service Features". The authors wish to thank the EURESCOM Permanent Staff and the member of the P515 project for their assistance.

Inter-Domain Integration of Services and Service Management

David Lewis, Thanassis Tiropanis, Alistair McEwan
Department of Computer Science,
University College London
{D.Lewis,T.Tiropanis,A.McEwan}@cs.ucl.ac.uk

Cliff Redmond, Vincent Wade
Department of Computer Science,
Trinity College Dublin
Cliff.Redmond, Vincent.Wade@cs.tcd.ie

Ralf Bracht
IBM, European Networking Centre,
R.Bracht@heidelbg.ibm.com

The evolution of the global telecommunications industry into an open services market presents developers of telecommunication service and management systems with many new challenges. Increased competition, complex service provision chains and integrated service offerings require effective techniques for the rapid integration of service and management systems over multiple organisational domains. These integration issues have been examined in the ACTS project Prospect by developing a working set of integrated, managed telecommunications services for a user trial. This paper presents the initial results of this work detailing the technologies and standards used, the architectural approach taken and the application of this approach to specific services.

Keywords: Inter-domain service management, TINA Service Architecture, reusable service components

1 Introduction

The emergence of an open market in telecommunications services is occurring globally. Competition in this market-place will increase pressure on service providers to develop and deploy services of increasing functionality and diversity in shorter time-frames. This presents a major challenge for the telecommunications industry, i.e., to accelerate the service life-cycle so that new services can be developed and deployed at a faster rate. This will enable services to evolve fast enough to meet changing customer demands competitively. In such a competitive market, services will be difficult to offer as a vertically integrated system developed by one provider; instead services will have to integrate with other services and service components operated by different providers. This co-operation may be motivated by providers' different market positions, e.g. allowing specialised providers to use the greater geographical coverage or resource capacity of a larger provider, or allowing customised services to be offered by combinations of separate providers' offerings.

Combining service offerings in this way requires the integration of service software components as well as the integration of the management components that manage these services. One approach to such integration could be accessing service and management functionality via industry agreed interfaces. This provides opportunities for third party service software developers to provide "off the shelf" reusable components. Service management components may be suitable for such development, since they encompass relatively stable systems such as the underlying network management systems and legacy customer and billing databases. This approach may however prove problematic where competitive pressures force the pace of change beyond that at which industrial agreements can be readily made. In this case common management and service components intended for reuse in different service systems will require appropriate flexibility in order to suit the requirements for new services as they emerge.

This paper presents the practical results of the ACTS project Prospect in examining the integration of telecommunications services and management in the open service market. In such a market there is the possibility of a wide range of new service providers as well as complex relationships between them and their customers. The particular business situation adopted in Prospect (see figure 1) is intended to provide a reasonably complex business scenario to generate realistic service integration problems. This scenario is based on a provider of educational courses that offers a Tele-Education Service (TES) to its customers. This service is an integration of three examples of a Multimedia Tele-Service (MMTS, i.e. a distributed, multimedia telecommunications service); a Hyper-Media (HM) service, a Multimedia Conferencing (MMC) service and a Web-Store (WS) service, each provided to the TES by a separate provider organisation. In addition the TES provider subscribes to a general purpose Virtual Private Network (VPN) service that it uses to provide broadband network connectivity between itself, its customers and the MMTS providers. The VPN provider further extends the service chain by using the ATM Virtual Path management service offered by a public network provider, delivering international links between the connected customer and provider sites.

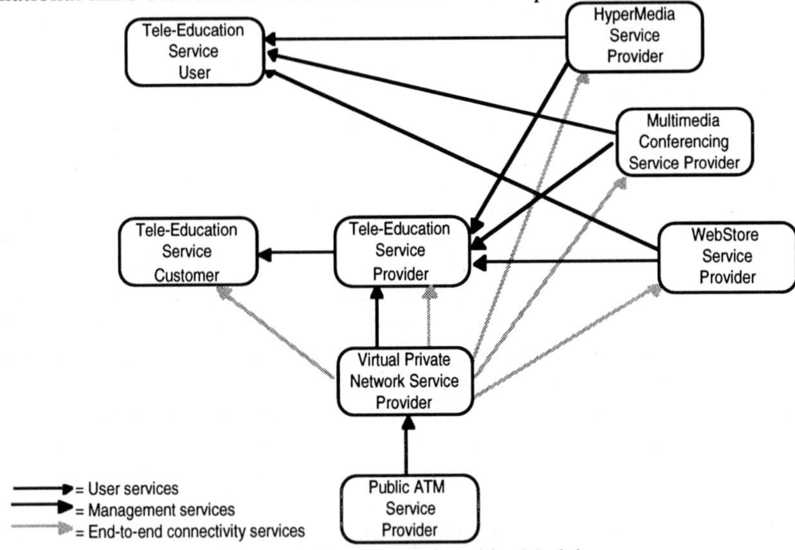

Figure 1: Prospect Stakeholder Model

The following section presents some of the contemporary technologies and architectures adopted for implementing this business model. Section 3 describes the overall approach to service and management integration in Prospect, and provides details on some specific design areas. The future work planned in this project is then outlined before concluding. The actual analysis and design methodology is not given here but is discussed in more detail in [Wade].

2 Architectures and Technologies

The aim of Prospect's investigations into the integration of telecommunication services and management was to develop a pragmatic approach, both to facilitate the development of a working trial and to ensure that the guidelines developed would be relevant to both existing and evolving technologies, architectures and services. The selection of the approaches adopted is summarised below.

2.1 Architectures: TMN and TINA

The Telecommunications Information Networking Architecture (TINA) Consortium has been specifically addressing the area of telecommunications services and their management. For this, TINA has developed a service architecture [TINA-SA], a TMN-based management architecture [TINA-MA] and a network architecture (influenced by ATM Forum and TMN network models). These architectures provide a consistent, detailed model for the integration of service, management and network components.

For Prospect, the integrated service and management model of the TINA Service Architecture (SA) was selected as the basis of the TES and MMTS providers' systems as described in the next section. TINA's integration of the network architecture with the service and management architectures was not, however, adopted, since it assumes a call-based model that cannot be straightforwardly resolved with the connectionless communication model used by the Internet applications used in Prospect (see section 2.3). The information technology approach taken in TINA is based on a distributed processing environment (DPE) defined in the TINA computing architecture [TINA-DPE]. However due to the lack of any DPE implementations, alternative approaches were adopted as discussed in the next section.

2.2 Software Integration Technologies: CORBA and Java

Recent development in distributed systems technologies have been led by the Object Management Group [OMA], and in particular their definition of the CORBA-2 standard. CORBA 2 allows for the specification of object interfaces and their abstract location in a (virtual) distributed environment. Object interface definition is separated from object implementation, thus allowing components of a CORBA 2 compliant object network to use different (possibly proprietary) implementation languages. The freedom this imparts to the developers of components, intended for integration into large systems, ideally places this technology for use in the open services market. In Prospect CORBA was used primarily for implementing computational objects derived from the TINA SA. These objects typically had multiple interfaces, which were mapped onto individual CORBA objects.

The user interface systems of the Prospect system has some specific requirements centred around the close integration of components and portability, i.e. the user interfaces have to integrate the interfaces offered by the different services being made available to users in an apparently seamless way on both UNIX and Win32 platforms. The definition of an IDL mapping for Java in the CORBA 2 standard allows for this language to be used to implement desktop service client components for users. The portability of Java and the well defined inter-applet communication interface meant that this was the most viable solution. In addition, Java provided a path to the integration of the user interface components with WWW browsers, either as Java applets or as Java implementations of Netscape plug-ins, and to the dynamic distribution of service client component software. The Java AWT library also was used to provide integrated, portable graphical user interfaces.

2.3 Network Technologies: Internet and ATM

The focus of Prospect is on an open market for digital broadband services. ATM is often assumed to be the primary technology to provide such services, yet the lack of commercial, switched wide area ATM services and the absence of any killer applications is delaying its widespread adoption. In comparison, Internet technology is the major solution for open digital communications today, with a huge growth in commercial Internet service providers and a wide range of applications and services being available. Two of the major developing strands of Internet services are represented in Prospect:

- The *M-BONE*, though still largely a non-commercial trial service, has shown the feasibility of real time, multiway, multimedia communications. This technology is easily scaleable from small to large groups of communicating individuals and makes optimum use of network resources.
- The *World Wide Web* has been largely responsible for the recent explosive growth in the use and extent of the Internet, as well as providing the motivation for the adoption of Intranet systems in many corporate networks. The ease of providing information, the free availability of browsers and the use of a simple communications protocols has enabled providers to develop and deploy new services in an open way, thus allowing a wide range of commercial services to be offered over telecommunications networks.

The Internet community is now starting to deploy mechanisms [RSVP] for providing the kind of range of quality of service guarantees that makes ATM attractive for integrated services. The resulting possibilities for convergence of IP and ATM technologies have already been commented on [Crowcroft], and interest in new technologies such as IP switching, point to a combined IP-ATM approach as being a good basis for the integration of services in the open service market. Prospect has therefore elected to use existing IP applications between hosts on mixed technology LANs, communicating via routers interconnected over a backbone of campus and wide area ATM networks. Such a configuration provides paths to both the parallel development of IP and ATM services and the convergence of these technologies. It is also typical of current ATM usage and therefore provides a realistic and practical basis for broadband user trials.

The required IP connectivity was provided in a generic way by the VPN service. This supports the assembly of a multi-domain Intranet (sometimes termed an "Extranet"), in this case connecting customer, TES provider and MMTS provider LANs. This was controlled in a TMN-structured management plane, separate to the control of the services themselves, i.e. the service sessions did not interact directly with communication sessions in the way prescribed in TINA. Issues of network resource and connection management in Prospect are not discussed further in this paper but are addressed in more detail in [Bjerring].

3 Integration Model

This section presents the approach taken for the integration of both service and management components in developing integrated digital services in Prospect. The development work reported here has attempted to simulate software development in the open services market, in particular by the development of common reusable components that can be readily integrated into the service and management systems of customers and providers for specific services.

3.1 Common Service and Management Components

The enterprise model contains several organisations providing different types of services, however many of the user and management activities that need to be performed by the systems used by customers and providers are similar, and therefore lend themselves to common solutions. This was appropriate for the TES and MMTS providers who all have to solve similar problems with the control and accounting of end user access to their services. The approach taken to these problems involved the development of common component interfaces and implementations that were reused in the different systems of the various service providers.

The TINA SA integrates the end user's access to and interaction with a telecommunications service with the management of service-related subscription and accounting information. This architecture identifies areas that are regarded as service specific, i.e. those that will be different for different service offerings, and those which are service independent, i.e. those that will remain constant over different service implementations. In this way, the SA model provides a core set of common inter-operable interfaces between its components, as well as a common semantic model around which services can be developed, both of which aid in the integration of separate services. The TINA SA was therefore deemed well suited as a basis for common service and management components in Prospect. The version of the SA publicly available when this work was conducted assumed, however, that all services were supplied by a single provider. This architecture therefore had to be augmented in order meet the multi-domain requirements of the Prospect business model.

The major common components of the Prospect design model derived from the TINA SA are; the integration of service control in the user and provider domains, termed the Desktop Service Integration (DSI) component; the Subscription Management component and the Accounting Management component.

The DSI component is based on the TINA separation of issues related to user access to services and issues related to usage and control of the service. These are modelled in TINA as separate access and service sessions, but augmented in Prospect to operate with existing desktop technologies and to use Internet protocols for service data flow, rather than an explicitly controlled communications session as prescribed in TINA. The aim was to provide a single model that each of the providers could use, but which could be easily integrated at the desktop in a way that would render the multi-provider composition of the services transparent to the end user. This required a clear distinction between the parts of the model which were service independent and could therefore be used directly by each service provider, and parts of the model that were service specific, and were thus expected to be augmented by the individual service providers to accommodate functionality particular to their service.

In the user domain the DSI model modifies the TINA SA to support the integration of existing applications, adopting a component-based approach using Java applets integrated into a WWW browser which together form the user application (UAP) component of the model. Also in the user domain a separate object, named the integrated session manager (ISM) mediated all the service independent interactions between the user and provider domain. This was based on the TINA notion of a Generic Session End Point but with additional support for hierarchical session relationships, thus supporting the composition of service sessions from different providers, into what appears to the user to be a single service session.

In the provider domain the TINA SA concepts of a User Agent (UA), Service Factory (SF) and User Service Session Manager (USM) have been adopted, providing the service independent components that allow the user domain components to interact correctly with the Subscription and Accounting Management components. The TINA service session concept has however been simplified by removing the requirements for group session functionality. Where needed, this functionality is provided by service specific components, e.g. multicast groups in the MMC service. The UA mediates between the user and the Subscription Management component of the provider's system, relaying the services the user may access and allowing the user to initiate a session in one of those services. The SF is then used by the UA to create a USM that allows the user to interact with a service session. The TINA SA service session interactions, e.g. initiate, terminate, suspend and resume, are handled by the service independent parts of the USM, including logging them with the usage metering parts of the Accounting Management component. Other service specific interactions need to be implemented by the service provider responsible, as extensions to the USM and corresponding UAP components.

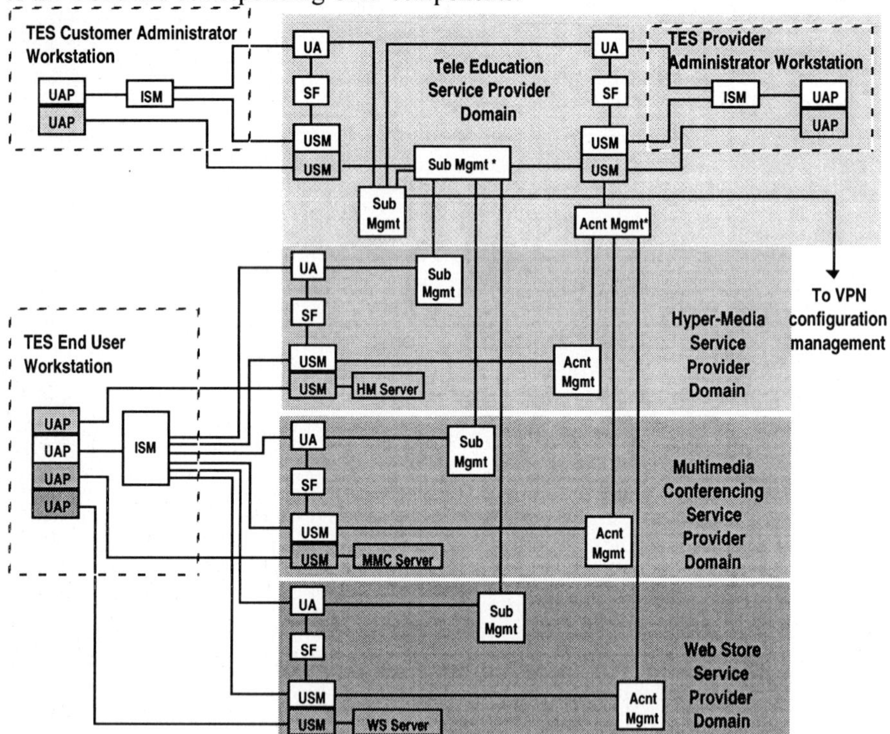

Figure 2: Integration of Service independent and service specific components

The Subscription Management component allows for the authorisation and barring of users to specific services, and the addition and removal of network sites supporting these users. Parts of the Subscription Management component are identified as areas for service specific specialisation; for instance the subscription registrar (SubRgs) component, which performs service specific operations on the basis of service independent ones it receives. The Accounting Management components allows service usage monitoring through interactions with a USM, charging based on tariffs agreed for a subscription, and bill generation based on these charges and a customer specific tariff plan which specifies discounts, payment schedules etc..

Figure 2 shows how the various service independent components (white boxes) and service specific components (shaded boxes) have been combined in integrating the service and management systems of the TES and MMTS providers. Within the project, these common components were implemented by one set of partners, and then reused by other partners in the implementation of the specific service provider systems they will operate. This simulated, to an extent, the development environment of the open service market which provided a means of gaining direct experience of some of the problems involved. All the objects shown in figure 2 were implemented on a single CORBA 2.0 compliant platform, i.e. Orbix, OrbixWeb and OrbixNames, to avoid any IIOP interoperability problems. Objects were actually implemented in either C++ or Java. The following sections describe in more detail how common components were used for specific services.

3.2 Integration of the Tele-education Service

The TES system is a good example of how a service can be constructed from an integration of existing services and common, reusable service and management software components. The integration for this service offering can be decomposed into the integration of the user services and the integration of management services.

The integration of user services was performed in the user application. This was based on a WWW browser that integrated Java applets which performed the service independent UAP functions of the DSI components as well as applets supplied by the MMTS providers to perform their service specific interactions. The problem of integrating the services was therefore reduced to that of defining inter-applet communications. This enabled the TES UAP developer to concentrate on the look and feel of the user interface, and the composition of the MMTS in delivering the TES to the user, rather than how the UAP interacted with the different MMTS providers' systems.

The development of the TES management system involved both the development of provider domain management systems and mechanisms for the TES administrator and TES customers to interact with these systems. The TES needs to co-ordinate its management with that of the subcontracted MMTS and VPN providers. For subscription management this required propagating operations such as the activation and deactivation of service profiles and the authorisation and barring of end-users, to the Subscription Management components of the MMTS provider. This was done by developing a wrapper (SubMgr* in figures 2) for the TES provider's common Subscription Management component, that forwarded the relevant operations to the identical common components in the MMTS providers' domains. In addition the SubRgs CO of the TES was modified to perform the correct configuration management operations when a customer site was added or removed, i.e. to use the VPN provider's configuration management services to connect to this site. For accounting management the interactions with sub-providers was limited to the collection of billing information from the common Accounting Management components used by each MMTS provider.

The customer and provider management UAPs were based the common DSI component. For the provider the UAP and the USM were extended to provide the user with access to subscription and accounting management information as well as management functions controlling how this information is propagated to specific sub-provider systems. For the customer the UAP and USM provided a subset of these operations, allowing the customer access only to specific operations related to their particular service subscriptions.

3.3 Integration of the Hyper-Media Service

The Hyper-Media service is a WWW service modified to facilitate accounting management, subscription management and service customisation management facilities for service customers.

The management service allows the generation, storage and access of per-user data on the service provider side. This can be used to provide functionality that is currently dependent on the client software implementation, which restricts the user to using the same client to avail of the bookmarks and history lists they generate.
Service specific enhancements to the USM component were needed to simulate a session context for WWW applications, which offer an inherently connectionless service. The overall requirements for a generic, unified, access session, an allowance for the provision of security services in the future, and the concept of a service session meant that the current access and security mechanisms the WWW were not sufficient for our needs. A revised definition of a service session, as a context in which HTTP service requests are made, was found more useful for a service which is not connection oriented. Simulating a session gives benefits, i.e. a known amount of users, user barring and user identification, and the solution developed here requires no modification of common client software, i.e. no service specific UAP is required. The TINA service session concept is implemented using a combination of browser state and a proxy mechanism. The proxy mechanism also ensures that each user accesses a document using a unique URL so organisational caching (which can occur in a organisational firewall application-level proxy cache) does not affect the per-user charge. The de facto support of a browser side state mechanism (Netscape 'Cookies') in current HTTP implementations lent itself for use as an application layer request context. This means that a conceptual session can exist in the application layer protocol interactions between a client and server.
The session mechanism works by initialising the client with a URL which, when accessed, creates a proxy for that user in the provider domain. Following a redirect, subsequent interactions to the service can only be made through this proxy. Usage information generated by the modified server is forwarded to the service independent metering component through a component that substitutes the user ID, which is originally passed to the service in the URL, for the 'Cookie' being used as the session identifier.

4 Further Work

The work described here was performed as an initial step in Prospect's on-going examination of key issues in multi-domain service and management integration. The TINA service architecture adopted in the model described here assumes a single service provider context, and therefore required extensions to handle the multi-domain situation assumed in Prospect. However TINA-C has recently refined the service architecture to provide support for multiple providers, and alignment of the current Prospect model with this will be attempted. In addition, a major feature of the TINA service architecture is its built-in support for mobility, the exercising of which will be an important theme of future Prospect work.
The introduction of RSVP based services in the Internet, and its integration with applications and management services, particularly accounting, is also a key component to placing this work in a relevant business context, and is therefore a subject for further work. The integration of security is also a key aspect of system development and integration in the open services market. Though this work does not aim specifically to implement security mechanisms, it will attempt to integrate existing and emerging mechanisms, e.g. SecuDE, to assess their impact on practical service development.

5 Conclusion

This work provides an example of how the definition of common, reusable components aids in the rapid development of integrated telecommunications services. The availability of implementations of such components eases the development of individual services by providing common, well understood functionality "off the shelf". Care must be taken however in clearly defining what parts of a reusable component are intended to be used as-is, i.e., the service independent parts, from those that must be modified by its users to integrate into the service, i.e. the service specific parts. The widespread use of common components could also ease the task of developing integrated services, since the common parts can be designed explicitly to support such integration as demonstrated by the Prospect management components.
Though this project does not aim to be TINA compliant, the TINA service architecture was found to provide a practical framework for controlling and managing multimedia service offerings based on IP technology. Not all aspects of the TINA SA were used however, with communication session management and group session functionality being omitted. CORBA was found to provide a suitable basis for implementing TINA-based components, though fuller use of CORBA Common Object Services, e.g. the life-cycle service, is required. From this work, it also appears that a combination of WWW, Java and CORBA provides a practical approach to presenting integrated services to the user. However a suitable security model for applets that need to communicate with multiple remote objects needs to be supported in WWW browsers.

Acknowledgements

The work presented in the paper was conducted with partial funding of the European Commission under the Prospect project (number AC052). The authors would also like to thank all members of the Prospect team for their collaboration in this work and Patrick McLoughlin of Broadcom for his comments.

References

[Crowcroft] Crowcroft, J., Wang, Z., Smith, A., Adams, J., A Rough Comparison of the IETF and ATM Service Models, IEEE Networks, Vol, 9, No. 6, July 1995

[Bjerring] Bjerring, L., Lewis, D., Thorarensen, I., An Inter-domain Virtual Private Network Management Service, IEEE/IFIP 1996 Network Operations and Management Symposium, Vol 1, pp115-123, IEEE, NJ, 1996

[OMA] Object Management Architecture Guide, ed. R.M. Soley, OMG Document Number 92.11.1, Revision 2.0 Second Edition, Object Management Group, 1992

[RSVP] Zhang, L., Deering, S., Estrin, D., Shenker, S., Zappala, D., RSVP, a new resource ReSerVation protocol, IEEE Network, vol. 7, no. 5, September/October 1993

[TINA-DPE] Graubmann, P., Mercouroff, N., Engineering modelling Concepts (DPE Architecture), TINA Baseline Document TB_NS.005_2.0_94, December 1994

[TINA-MA] de la Fuente, L., A., Walles, T., Management Architecture, TINA Baseline Document TB_GN.010_2.0_94, December 1994

[TINA-SA] Bernt, H., Kim, C., Kim, L., et al., Service Architecture, TINA Baseline Doc. No. TB_MDC.012_2.0-94, TINA-C, March 1994

[Wade] Wade, V., Sheppard, M., Tschichholz, M., Hall, J., Lewis, D., A Methodology for Developing Integrated Multi-domain Service Management Systems, Proceeding of IS&N '97, Como, Italy, April 1997

Domain Interoperability for Federated Connectivity Management

L. H. Bjerring, P. Vorm
L.M.Ericsson A/S
Sluseholmen 8, DK-1790 Copenhagen V, Denmark
E-mail: {lmdlhb, lmdvrm}@lmd.ericsson.se

Federated connectivity management represents many of the problems encountered in the development of multi-domain management systems for an open service market. This paper presents the key problems of interoperability across domain boundaries of diverse types. Three basic domain types are defined: administrative, technological, and service domains. Interoperability across the boundaries of such domain types are discussed and a simple model of interoperability is described. An example application of these principles is described through the Prospect VPN system architecture, which utilises the interoperability model as a structuring principle for separating between the system's core function and the required adaptations needed to accommodate differences of the supporting network management systems.

1 Introduction

1.1 Motivation

Traditionally, the interoperability requirements that have driven the standardisation process in ITU-T have been motivated by telecommunications equipment customers' interest in being able to construct open multi-vendor systems, from a technical point of view. These systems generally belonged to a single administration, even though technical standards for network interconnection have also been defined (e.g., recommendation X.75), and applied between administrations.

In recent years, a new dimension has been added through the introduction of an open service market. This implies that interoperability is no longer "just" a technical matter, but also an administrative problem when network and services have to be openly available across administrative domain boundaries. This problem is even more demanding in that the open service market is expected to become very dynamic and inter-organisational customer-provider relations, which in general will form the context for technical interoperability, will need to be negotiated, established, utilised and removed very frequently. This means that interoperability problems no longer are restricted to technical issues associated with multi-vendor systems but more generally can be characterised by the multi-domain nature of an open service market.

In practical systems development a relatively large amount of the work (and investment) is normally associated with integrating legacy systems (adaptation problems), rather than concentrating on the system's core functionality. In order to be able to plan the development process and to be able to control cost it is necessary to distinguish between the system's core function and the extensions, in terms of adaptations towards particular other systems/components. For that purpose an architectural framework which localises and isolates adaptations from core system functions would be beneficial. The approach suggested in this paper aims at providing such a framework for the most important adaptation problems.

An architectural approach with emphasis on modularity and reusability has for a long time been the desirable goal of projects and consortia like TINA-C. An important aspect of this is the interfaces of a system to be developed and the need for interaction with already existing systems.

1.2 Overview of paper

This paper outlines in section 2 the concept of domain and the basic domain types encountered in connectivity management in an open service market. In section 3 a generic model for interoperability between objects belonging to separate domains is defined. These models were developed and have been applied in the development of Prospect's Broadband VPN service, which is illustrative of the problems and solutions associated with inter-domain management. Section 4 describes the architecture of the VPN and its implementation. Finally, in section 5 we summarise the main points.

2 Domain types

2.1 Definitions

The concept of domain offers a useful approach to dealing with complexity of open distributed systems, such as the Prospect system. Domains provide a means for structuring the system into areas of homogeneity and for identifying heterogeneity problems needing to be solved.

The general concept of domain has been defined and discussed extensively in the context of systems management, and rules for domain membership and domain associations are defined in the literature [1] and standards [2-3].

Domains can be defined as "units of homogeneity", i.e., groups of entities which have something in common. This is sometimes expressed as groups of entities (for example, managed objects) to which a common policy applies, where the policy is defined by one administrative authority.

Actual domains will generally be different with respect to functional as well as non-functional details: functionally, in that the objects in the domains may not offer compatible functions (services); non-functionally, in that the conditions for service invocation in diverse domains may not be compatible (for example, quality of service characteristics may differ both in their description and also in their values). This explains why interoperability is problematic across domain boundaries.

The concept of federation is the enterprise language concept which provides the context for domain interoperability. It is defined in [4]: "...a federation... is a coming together of a number of groups answering to different authorities (and thus representable as distinct domains) in order that they may jointly cooperate to achieve some objective. ... The domains concerned in a federation may be administrative domains ... or technology domains...".

2.2 Domain types in a federated network infrastructure

Three basic grouping criteria have been identified as characteristic in the context of network infrastructure management. These criteria reflect characteristics of the open service market which imply that the global network infrastructure is basically heterogeneous in at least three respects:

- *administratively* in that different parts of the infrastructure will be owned and administered by different organisations;
- *technologically* (network as well as management technologies) in that the communications technology employed in different places will vary;
- *services*, in that the services and management services provided over different parts of the infrastructure will vary, both with respect to their capabilities and with respect to their representation (model, language, etc.). The latter point is considered here to be a technological aspect.

This leads to the definition of three corresponding types of domain:

- an *administrative domain* encapsulates resources belonging to one administrative authority; (or which can be said to be within the scope of the authorities' administrative responsibility);
- a *technology domain* encapsulates resources with common communications methods and interface types;
- a *service domain* encapsulates resources which offer the same service. (As an example CBDS is specified for both DQDB and ATM communication technologies.)

Each of these domain types are characterised below.

Administrative domains

We use the term administrative domain to represent infrastructure units of administrative homogeneity and autonomy. Such units are assumed to provide access to their services only within the scope of an agreement which is represented by a contract between each unit (in the role of provider) and some other unit (in the role of customer).

Administrative homogeneity means that the administrative domain has its own, individual administrative authority. This authority is autonomous and independent in that it is free to join and leave federations. Sometimes administrative homogeneity is described in terms of 'policy definition', in that each administrative authority has freedom to define its own policies irrespective of other administrative domains.

Technological domains

We use the term technology domain to represent infrastructure units of technological homogeneity. Technology may refer to many different aspects, such as, for example, management or distributed application platform architecture (e.g., DCE, CORBA), communications mechanisms or protocols (e.g., CMIP, IIOP, SNMP), and interface description languages (e.g., IDL, SNMP, GDMO). In the context of an open service market the primary aspect of diverse technologies is the means for communications between objects, i.e., communications technology.

Service domains

We use the term service domain to represent units of (management) model homogeneity. The service domain is defined by the configuration, semantics and naming of the model of the information entities in the domain. When objects are used for modelling the information entities, service refers to the services offered by an object (capabilities offered at object interfaces), and the ways object services are named (modelled).

3 Generic Domain Interoperability Model

3.1 Motivation and definition

The traditional understanding of interoperability is well described in this quotation: "Interoperability. The capability of two or more open systems which communicate using a specific mechanism, in a known environment, to achieve user objectives." [5]. Interoperability requires that the interoperable systems agree on the interface between them. However, interoperability, with the particular problems addressed here, always takes place in the context of federation. Interoperability problems occur in federations when the domains which are required to interwork across their boundaries are heterogeneous.

3.2 Model

In the RM-ODP computational model language (and elsewhere) object interworking is represented by "bindings" between objects in client and server roles [4].
As stated earlier, an open service market will be characterised by heterogeneity. This comprises technical heterogeneity of the objects, in terms of incompatibility of interfaces, as well as administrative heterogeneity. Therefore, in an open service market, we need a more detailed view of the client-binding-server model of object interoperability.
This model (**Fig. 1**), partly derived from the ANSA interoperability model [6] [7], shows the way objects interoperate in general, across domain boundaries, that is, the assumption that they both exist within the same domains is not valid. This implies that the binding of the client and server interfaces needs to be mediated by an interceptor, which is a concept from [4]: "Interceptors corresponds to the notions of "gateway", or "monitor" objects which stand between two domains and enable or permit interactions on the basis of a contract between the administrations that specifies the basis for their federation". The interceptor in Fig. 1 comprises three separate gateway (GW) functions:

- Administrative gateway functions which are functions associated with controlling service invocations across administrative domain boundaries. These functions include for example authentication of client and accounting for its use of service resources;
- Technology gateway functions which are functions associated with transforming between different protocols and data representations (syntax); such functions are also known as (network) interworking functions;
- Service gateway functions which are functions associated with transforming or adapting different service representations based on the same underlying universe of discourse, e.g., transformation of names of compatible resources.

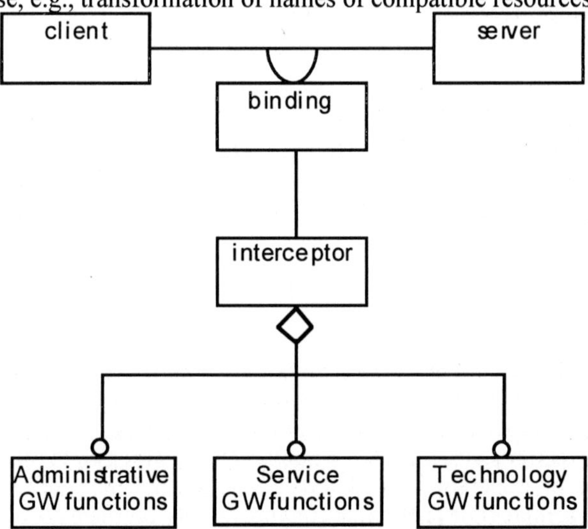

Figure 1: Interoperability model for heterogeneous systems

Depending on the domain types and the domain structure of the system under study the model in Fig. 1 will be instantiated in various ways. This is indicated by the cardinality of the aggregation association between the interceptor and the specific gateway functions. As a general rule it can be assumed that when the client and server are located in different domains of any of the defined types, the associated GW functions are needed, possibly in both domains.

Administrative gateways

Administrative gateway functions are functions associated with controlling access to and from other components, for the purpose of authorisation, accounting, etc. It should be noted that the TINA functional areas of subscription and accounting management [8] can be regarded as administrative gateway functions.

The effect of interworking across the boundaries of administrative domains is associated with non-functional concerns related to:

- *security and access control*: each domain will want to exercise control over who, when and from where service invocations are allowed towards objects in this domain;
- *monitoring and accounting*: each domain will want to monitor service invocations and resource consumption invoked from other domains, in order to exercise control and in order to determine the cost associated with invocations. Furthermore, each domain will need to be able to charge other domains for service consumption, to be able to send and receive payment and receipts for payment. Also, and in particular "client domains" will want to be able to initiate audits of "server domain" accounting, to be able to verify the validity of charging and bills;
- *autonomy*: each domain will want to control resource consumption and allocation at its own discretion.

Much of the ITU-T work on the TMN X interface is associated with administrative gateway functions [9].

Technology gateways

We distinguish between two types of technology gateway:

- *network technology gateway*: These gateways are typically interworking units (e.g., multi-protocol routers) or functions integrated in network elements (e.g., ethernet interfaces of MAN nodes);
- *management technology gateway*: This is basically the same type of gateway as a networking interworking function in that it transforms data representation and encoding from one protocol format to another (e.g., CMIP to SNMP). JIDM [10] and IIMC [11] specifications apply for CORBA, SNMP, OSI-SM gateways.

Service gateways

In the context of this paper, service gateways transform between different modelling representations of services (resources), for example between an ETSI model of ATM VP network and PREPARE's VPN information modelling constructs.

Service gateways are closely related with problems of type management [12-13]. In OSI standardisation some effort has gone into this area, for example use of X.500 directory for storing GDMO template information [14-15], and formalisation of managed object behaviour specification [16].

4 Example: Prospect's Broadband VPN Service

4.1 Overview

The generic domain interoperability model is representative of the problem and modelling approach to Prospect's VPN service system. This service aims at providing users with distribution transparent connectivity management services over the Prospect network infrastructure [17]. It is based on analysis and modelling work of PREPARE [18], and has a role similar to that of connection management in TINA [19]. The infrastructure is characterised by administrative, technological and service heterogeneity.

The core features conceptually are implemented by decomposition of user requests and forwarding derived requests to the various domains, and by aggregation of domain-specific resources into end-to-end (multi-domain) resources, and by monitoring the behaviour of each involved (contributing) domain, with respect to the VPN resources.

A VPN system architecture is designed to facilitate a modular development approach, and to facilitate later inclusion, if needed, of alternative network technologies, alternative model domains, and additional administrative domains into the network infrastructure, and to maintain the VPN system's control over these domains. The architecture therefore separates the various types of adaptation needed from each other, and from the core VPN functionality, see **Fig. 2**. These separations are in agreement with the domain interoperability model presented in section 3.

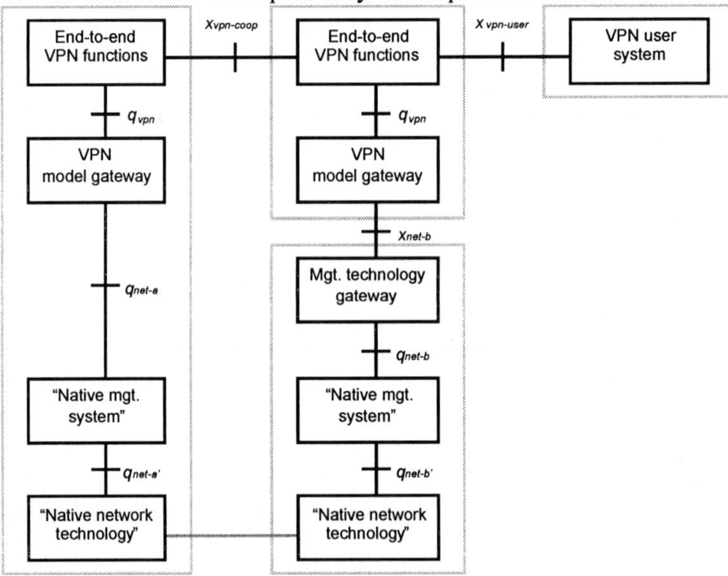

Figure 2: Illustration of VPN System Architecture

To manage the heterogeneity a set of VPN abstractions are defined as a common management model abstraction over each heterogeneous management model. Also, a particular management (communications) technology has been chosen. Finally, administrative gateways are defined to control interactions across administrative domain boundaries. Management technology gateways are developed to adapt between the various technologies that are present in the heterogeneous network infrastructure and the chosen communications technology. Likewise, management model gateways are developed to adapt between the management model domains and the VPN abstractions.

While it may be desirable from a user point of view, it is not feasible to expect in all cases each network owner to implement the VPN abstraction of the network domain himself - unless regulatory prescriptions impose such restrictions, or a clear business case proves it to be economical. Instead adaptations are isolated and the system is architected in such a way as to facilitate as much automation of interoperability provision as possible, irrespective of actual ownership problems.

Each network technology domain is assumed to be represented in its inter-domain interworking with exactly one management model and exactly one management technology. Moreover, it is assumed that each such network technology domain will appear as a separate administrative domain. Each domain will therefore, in its interworking with the VPN, have associated both administrative gateway, management model and management technology gateways. (Note, though, that these gateways are conceptual, not necessarily physical).

To solve these issues in the VPN we have defined inter-domain management "standard interfaces" and we have implemented gateways system which assists in overcoming differences.

This is illustrated in Fig. 2, which shows the VPN architecture in an example with a few domains. The architecture illustrates where we apply the two types of management gateway: model and technology, between some heterogeneous management domains. Administrative gateway functions can be located virtually anywhere, depending on where we define administrative domain boundaries. In Fig. 2 administrative gateway functions are located in components with X interfaces.

The architecture is basically layered. In the upper layer an end-to-end VPN function provides users with transparent end-to-end connectivity management services. This component utilises a common representation of the involved infrastructure domains, provided by a VPN model gateway component. This in turn uses the service offered by each domain directly, possibly via management technology gateways. In the bottom is shown the domain-specific management components and the infrastructure components. The end-to-end VPN functions are structured in such a way that administrative federation of VPN providers is supported.

Among the benefits of this architectural approach to the VPN service system is that it enables the possibility of replicating, rather than re-implementing, the end-to-end VPN functional component in several domains, provided, of course, they have deployed identical technology.

4.2 Implementation

The VPN service was implemented as part of the Prospect network and service management infrastructure. This section briefly describes aspects of the implementation.

The communications technology chosen is CORBA v2's IIOP, and the implementation technology for the VPN system is a CORBA v2-conformant ORB. The main reasons for that choice are:

- to minimise interoperability problems with VPN user systems (which are multimedia teleservice systems) the same management communications technology for this architecture (IIOP), has been chosen;

- distributed object technology eases the implementation of the VPN specification, and allows for distribution and migration of the VPN functionality into the individual infrastructure domains.

Fig. 3 illustrates a fragment of the VPN system implementation. A public network domain is managed with a TMN-structured system providing for external access via a TMN X interface for ATM VP service management by the customer (i.e., the VPN provider). A management technology gateway - TMN/CORBA interaction gateway - is located in the PN domain. The IDL representation of the TMN X interface was generated by a GDMO-to-IDL translation in accordance with JIDM's specification translation algorithm [10]. In the VPN domain a management model gateway is located which adapts (transforms) the ATM VP service model to the VPN domain abstraction. This in turn services the end to end VPN functions, which are accessible from the customer's system. Similar arrangements are made for the VPN to be able to manage the CPNs, which have SNMP management facilities.

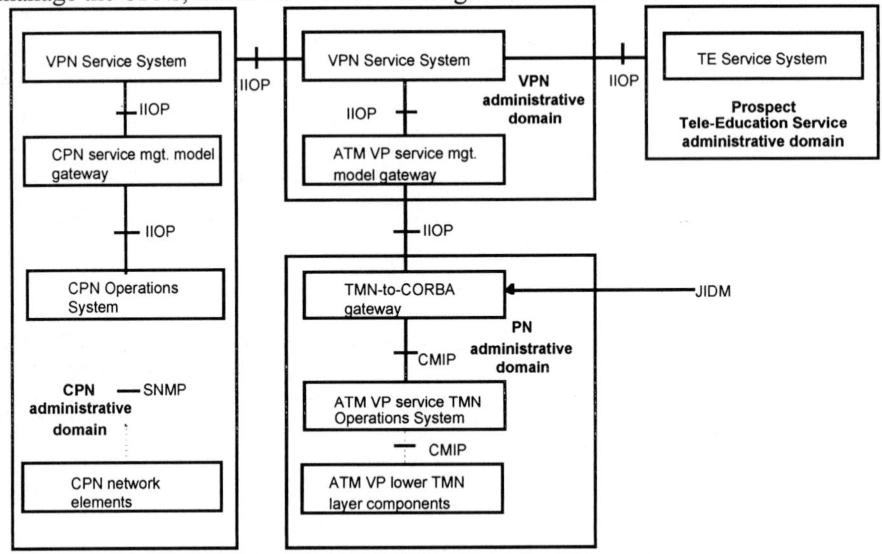

Figure 3: A fragment of the implementation architecture

5 Summary

In this paper we have discussed the key aspects of heterogeneity in infrastructure management and introduced the concept of domain as an appropriate modelling approach to manage the development work. A simple and comprehensive domain interoperability model has been presented which models the identified domain types and their interoperability: domains encapsulate resources which need to interwork in a controlled fashion when resources are to be part of a federated system. Interworking is required across the boundaries of all of the domain types described. In order to facilitate, monitor and, when required, prevent interworking (that is, interoperability) between the systems belonging to the domains, we assume that appropriate administrative, technology and service gateways are in place to control interaction at domain boundaries. Since we basically can assume that the network domains in an open service market are autonomous and that, over a long period of time they are likely to act as both clients and servers in their mutual relationships, we assume that

gateways in principle should exist in all domains allowing them to interoperate with other (peer) domains.
To enable inter-domain management in a heterogeneous system the following may be needed at or across domain boundaries:

- translation of service invocation, both syntactically and semantically, because we cannot assume homogeneity of client and server technology and service model;
- logging of service invocation requests, monitoring of resource consumption, controlling of identity and authority, and verification of request against contract.

The way in which these models have been applied in the development of the Prospect VPN service has been described. The philosophy behind the VPN service is to allow users to request and manage connectivity transparently. What needs to be masked is heterogeneity of the network infrastructure. To achieve this means that the VPN must provide an abstract connection management interface to users. To achieve that in turn means that heterogeneities must be adapted to a common abstraction, which is represented by a consistent set of service interfaces.
For the various aspects of heterogeneity we make the basic assumption that no single solution (in terms of network technology, management model, management technology or administration/contract interface) is ideal or superior to all other solutions. The choice of a particular solution will in virtually all cases be based on politics, rather than on technicalities. Therefore, understanding the heterogeneity problems and attacking them in an architectural framework is essential to successful integration of management systems, such as the Prospect VPN system.

Acknowledgements

This paper is based on work carried out in the Prospect project (project number AC052), under the European ACTS-programme. The views expressed are those of the authors. The development of the VPN is the result of the collaborative effort of the members of the project consortium. The contributions by H. Khayat, R. S. Lund and C. Grabowski are especially acknowledged.

References

[1] M. S. Slomann, J. D. Moffett, Domain management for distributed systems. In: B. Meandzija, J. Westcott (Eds), *Integrated Network Management, I.* Elsevier Science Publishers B.V. (North-Holland), 1989.
[2] ITU-T Recommendation X.701, Systems Management Overview, and ISO/IEC JTC1/SC21 N10352, Amendment 2 (Management Domains Architecture)
[3] ITU-T Recommendation X.749, Management Domain and Management Policy Management Function.
[4] ITU-T Recommendations X.900-series, Reference Model for Open Distributed Processing.
[5] Discovering OMNIPoint. PTR Prentice Hall, Inc. New Jersey, 1992.
[6] Y. Hoffner, Inter-operability and distributed application platform design. In: A. Schill et al (Eds), *Distributed Platforms*. Chapman & Hall, London, 1996.
[7] Y. Hoffner, B. Crawford, Federation and Interoperability. ANSA Architecture Report, 6th October 1995.
[8] H. Berndt, R. Minerva (Eds), Service Architecture. TINA-C, 1995.
[9] ITU-T draft recommendation M.Xreq, X interface requirements.
[10] N. Soukouti, Inter-Working between CMIP, SNMP and CORBA Technologies. In: Proceedings of ICDP'96.
[11] ISO/CCITT and Internet Management: Coexistence and Interworking Strategy, Forum TR107. Network Management Forum, Morristown, NJ, 1992.

[12] D. Arnold, A. Bond, An Interaction Framework for Open Distributed Systems. In: Proceedings of ICDP'96.
[13] P. Kähkipuro, et al, Reaching Interoperability through ODP type framework. In: TINA'96 Conference proceedings, 1996.
[14] ITU-T Rec. X.750, Management Knowledge Management Function.
[15] RACE Common Functional Specification CFS H430, Inter-Domain Management Information Service (IDMIS).
[16] GDMO+. ISO 10165-4 Amendment 4, 1996.
[17] L. H. Bjerring, I. H. Thorarensen, TMN and TINA: Two Approaches to Broadband VPN Services. In: TINA'96 Conference proceedings, 1996.
[18] L. Bjerring, J. Hall, M. Louis, P. Vorm, Virtual Private Networks for Integrated Broadband Communications. In: *Speakers' papers, 7th World Telecommunication Forum, Technology Summit*. ITU, Geneva, 1995.
[19] J. Pavón, J. Tomás, Broadband VPN Service Prototype: Application of TINA to Service and Network Management. In: TINA'96 Conference proceedings, 1996.

Towards Integrated Network Management forATM and SDH Networks Supporting a Global Broadband Connectivity Management Service

Alex Galis University College London, United Kingdom; Email: a.galis@eleceng.ucl.ac.uk
Carlo Brianza Italtel, Milan, Italy, Email: brianzac@settimo.italtel.it
Crescenzo Leone CSELT, Turin, Italy, Email: Crescenzo.Leone@cselt.stet.it
Christiam Salvatori SIRTI, Milan, Italy, Email: C.Salvatori@sirti.it
Dieter Gantenbein IBM Zurich Research Laboratory, Switzerland ; Email : dga@zurich.ibm.com
Stefan Covaci GMD Fokus, Berlin, Germany, Email : covaci@fokus.gmd.de
George Mykoniatis NTUA, Athens, Greece; Email : giorgos.mykoniatis@telecom.ece.ntua.gr
Fotis Karayannis NTUA., Athens, Greece; Email : fotis.karayannis@telecom.ntua.gr

This paper reflects the initial work of the ACTS Project AC080 MISA, a three year project, started at the beginning of 1996, whose main task is to accomplish and validate an integrated end-to-end management of hybrid SDH and ATM networks in the framework of open network provision (ONP) via European field trials.
This paper provides the initial description of an integrated network management of ATM and SDH networks and the definition of a new Service called Global Broadband Connectivity Management (GBCM), enabled through TMN. The enterprise model supporting both topics is described.
The GBCM service offers the required functionality to establish and manage end-to-end broadband connections in a multi-domain business environment. Its implementation by the MISA project, within trials across Europe, aims for an efficient management of the network resources of SDH and ATM infrastructure, while meeting the quality of service and the needs of a number of telecommunications actors: customers, value-added service providers and network providers.
Keywords: ATM, ACTS MISA Project, Co-operative Network Management, Global Broadband Connectivity Management, Integrated Broadband Communications, Open Network Provision, SDH, and TMN.

1 Introduction

The end-to-end management of broadband connections has been found to be very complex, especially in a multi-provider, multi-domain environment.
Typically, today, the setting up and re-configuration of such connections are performed through manual actions using faxes and telephone calls.
One of the main goals of the ACTS AC080 MISA project is to automate the operator configuration and maintenance functionality, so as to satisfy end-user requirements. This will be achieved by provisioning of open interfaces to the management centers for the necessary co-operation between management systems. This goal is vital to continue the development of the Integrated Broadband Communication (IBC) Infrastructure.
Currently, ATM and SDH are the most popular technologies for supporting broadband telecommunication. While ATM is designated by most standardisation organisations such as ITU and ETSI as the B-ISDN interface for end-user applications and is already used in LAN/CPN domains, the SDH transmission technology is more ubiquitous among the national carrier and bearer transmission service provider.
General efforts towards SDH network management (e.g. ITU-T G.784, G.774, G.774-01, G.774-02, G.774-03, G.774-04 and G.774-05) and ATM network management (e.g. ITU-T I.751, ETSI NA5-2210, ATM Forum M2 and M4 interface specifications) are largely uncorrelated. The availability and provisioning of Broadband Services on

top of hybrid SDH/ATM networks require a uniform and integrated network management system to control such different network resources.
The paper is structured as follows: first the enterprise model supporting the provision of the GBCM service is described highlighting the requirements for ATM and SDH integration from the different actors' points of view, then the reference architecture for the management of broadband connections spanning hybrid ATM & SDH domain is presented, identifying the reference points and the interactions among the system components and system interfaces; finally, a more detailed description of the components responsible for integrating the management of ATM and SDH networks at the network management level of the TMN is provided.

2 The Global Broadband Connectivity Management (GBCM) Service

The development of the Global Broadband Connectivity Management (GBCM) service is one of the main themes of the work of the MISA project. MISA assumes that Integrated Broadband Communication Networks (IBCNs) will consist primarily of ATM and SDH equipment providing basic ATM bearer-services. The MISA GBCM service will allow these network resources to be used in a co-operative and efficient manner, addressing the needs of:

1. customers requiring connectivity services,
2. network providers needing to make optimal use of their own resources and requiring timely identification and notification of performance or quality of service (QoS) changes and faults,
3. network providers who need to negotiate co-operative services with their peers.

3 GBCM Enterprise Model

In the MISA project, a number of actors and roles have been identified as relevant in the context of provision of the GBCM service [2]. The following subsections are devoted to presenting the enterprise model.

3.1 Actors

The actors composing the GBCM Enterprise model are Public Network Operators (PNOs), Business Customers and Value Added Service Providers (VASPs).

Public Network Operators

PNOs are the owners of the network infrastructure and thus are able to provide broadband connections through their ATM and SDH networks.
They need to manage their network resources and the bearer and tele-services (basic services) supported by the network. The interactions with their customers (VASPs or Business Customers) and with other PNOs, should be performed through management activities covering the standardised general functional areas of Fault, Configuration, Accounting, Performance and Security.
In the case of GBCM, the management differences of ATM and SDH networks are hidden, providing the service management layer with a common, homogeneous view of the underlying management systems and network resources This helps to satisfy the PNO's requirements concerning the integrated management of broadband networks. Below a non exhaustive list of such requirements is provided.

- In many PNO's domain an SDH network exists or is planned to exist in the near future. It is expected to be used as the transport network for carrying the ATM traffic originated by the customer's applications.In such a scenario, where ATM equipment is connected to SDH equipment, integrating ATM & SDH configuration management is necessary for a PNO.

- In order to satisfy the QoS requested by the customer, PNOs may adopt SDH resources to protect ATM connections (i.e. exploring the protection capability defined for the SDH layer; this should happen transparently to the customer).
- On the basis of a request for bandwidth modification for an ATM connection, the PNO may have to reconfigure the route between two end-points using a different SDH path as server trail.
- In the case of connections spanning both ATM and SDH networks, the integration of ATM and SDH fault management would not give any additional benefit in determining that a fault has occurred (someplace), but might help to say more precisely why or where a particular fault has occurred (single SDH faults may cause multiple ATM failures).
- The PNO is required to identify the layer (ATM and SDH) at which failures occur to activate the appropriate restoration mechanisms.
- An operator needs to gather and correlate performance information from its underlying ATM and SDH networks, as input to its network planning activities. This can be useful for determining whether their SDH and ATM networks are over/under utilised, how to plan for future growth of these networks, and what types of physical connections to their customers will give the greatest return for money invested.

Business Customers

A business customer is a collection of users, using various advanced network services mainly for the purpose of their work activities (e.g. tele-conferencing, advanced voice and data transfer services).

Communications within large organisations (e.g. companies, enterprises) with various location spread world wide, have become crucial to their effective operation.

Instead of a customer organisation itself establishing and maintaining its own private network, the responsibility for network and service provisioning has passed to third parties, the VASPs, who may provide: network facilities, end-to-end services as well as service management facilities to potential customers.

In the MISA framework, it is assumed that the services offered to the customers are provided either by the network providers directly or through VASPs.

As far as the integration of ATM and SDH is concerned, the following requirements can be identified from the point of view of a business customer:

- The lower layers at the customer access point in terms of protocols and line speed should be provided according to the preference of the customer: the access point can be both ATM and SDH.
- The customer by means of the same management interface should be able to request both SDH leased line and semi-permanent ATM connections.
- The customer should be able to specify the traffic and quality of service parameters for all layers that he/she is using. In the case of using ATM on top of SDH he might want to choose a protected underlying link, for example.
- The GBCM service should provide management information (status, performance data etc.) about all connections of the customer, covering the whole distance from

customer access point to customer access point via the management interface of the GBCM service provider the customer is registered with.

- Notification of reconfigurations due to faults should be sent to all customers affected by the connection failure.
- The GBCM service should try to avoid multiple notifications due to the propagation of faults from lower layers within a transmission network to the upper layers or due to multiple transmission systems involved in the case of a broken link. This implies the capability of a network provider to correlate multiple redundant alarms.

Value Added Service Providers (VASPs)
As the telecommunication market increasingly follows an open network provision policy, Value Added Service Providers (VASPs) are likely to appear, satisfying the user needs for one-stop shopping, one-stop accounting and one-stop complaining.
VASPs rent physical resources from one or more network providers, services from one or more service providers, add value to those services by guaranteeing better performance, by lowering subscription and operation costs and by offering user-management facilities and providing these value-added services upon a customer's request, as a homogeneous service package. The customers of a VASP (VASP users) are business or residential users.
The general objectives of a VASP are itemised below:

- The service provided to the customer should appear homogeneous, even if several parties are involved in the provisioning of the particular service.
- Provide one-stop shopping, accounting and complaining facilities to the customer. The multi-provider, multi-operator nature of today's networks and the consequent negotiation between customers and several providers, is difficult and un-satisfactory. That is why one-stop shopping for configuration control, billing etc. is demanded.
- Provide the customers with the ability to manage, to some extent, their own service.

3.2 Roles:

The GBCM Enterprise Model identifies and defines different roles in which the above actors may be recognised. The described roles are represented in Figure 1.

GBCM/GBC Service Provider
This role is played largely by the PNOs. They provide a GBC (Global Broadband Connectivity) service through the network infrastructure. GBC Management provision is through the associated GBCM service. Thus this entity provides both GBC and GBCM services to users.

GBCM Users:
Four different users of the GBCM service have been identified.

GBC-SP (GBC Service Provider)
This role is played by the PNOs. The establishment of a GBC service, when the broadband connection involves more than one administrative domain, requires that a GBC-SP co-operates with one or more other GBC-SPs, using the GBCM, through a co-operating reference point. The GBC-SP requesting the co-operation of the others

will play the GBCM User role, while the GBC-SPs replying to the co-operation requests will play the GBCM Provider role.

VASP, Non GBC-SP (Value Added Service Provider, Non-GBC - SP)
This role provides value added services using the GBC (e.g. Video Conference Service, VPN Service). It does not, by itself, provide the GBC, but uses one or more GBC-SP to do this. This type of VASP is likely to offer the selection of the most economical GBC-SP for its users or one stop shopping for this and other services.

VASP, GBC-SP
Such role is played by a PNO who offers to its end-users Value Added Services on top of its network (e.g. a PNO offering a VPN Service). This role is clearly different from the one presented above since, in this case, the VASP is co-located with the GBC-SP.

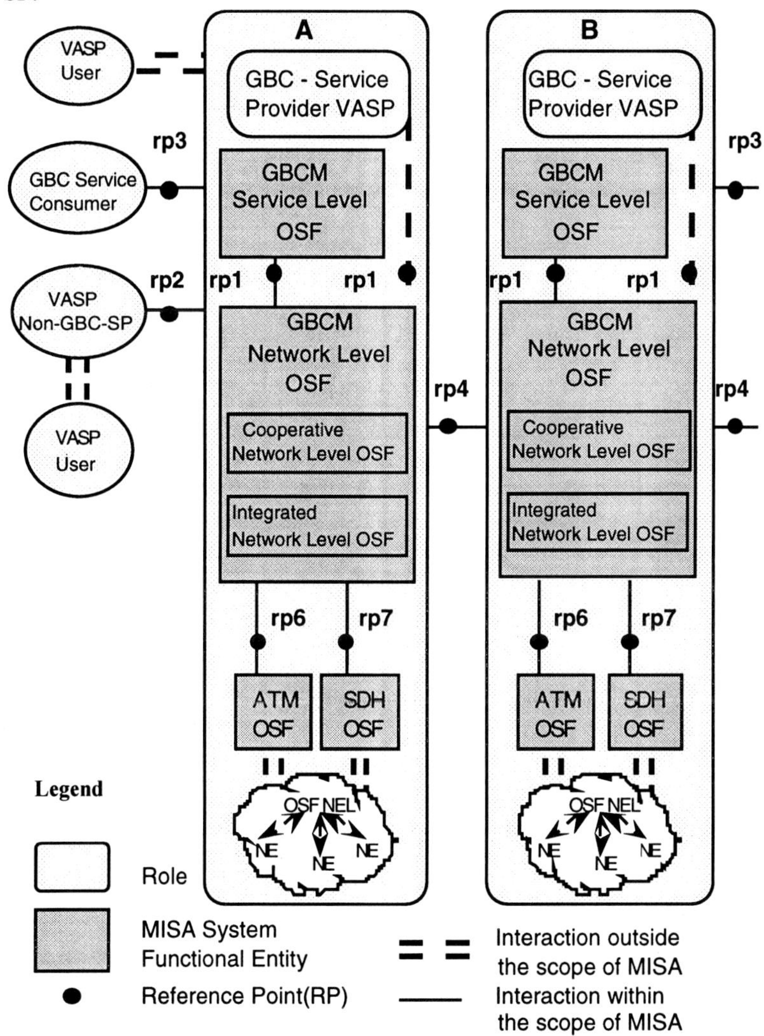

Figure 1: MISA reference functional architecture

GBC Consumer
The entities which connect to the GBC network service are consumers of the GBC. They may put information on the network, take information off the network, or both. The GBC consumer interacts directly with a GBC-/GBCM-SP to whom they have subscribed and who is responsible for the establishment, maintenance and billing of the GBC service. Such entities are expected to be, for example, big companies offering to their members services (e.g. video-conference) that require the establishment of broadband connections between far located premises, i.e. require the interaction with a GBC-SP.
Thus this role is expected to be played by business customers.

4 MISA Reference Functional Architecture

On the basis of the Enterprise Model presented in the previous section, the following reference model can be derived, showing the identified roles and the interactions among them [3]. The system is decomposed into Operation System Functions (OSF) with reference points (rp) among them compliant with the TMN architecture [4].
At the reference points rp2 and rp3 the interaction between the customer (VASP and business customer respectively) and the GBCM service provider takes place at the service management level of the TMN. Here the customer subscribes to the GBCM service, requests GBC connections, gathers information and receives notifications about them.
The rp1 lies between the service level and the network level of the TMN. The connection requests coming from the customer and their responses pass through this reference point to/from the GBCM Network Level OSF.
As evidenced in Figure 2, the network management layer has been split into two sub-layers: one technology independent, the other one technology dependent. The separation between these layers is at the rp5 reference point. This enables satisfaction the customers connection requests without being bound to a particular technology (ATM or SDH): the technology independent layer is able to employ, according to the network operator policy, the resources and functionality made available by both, the ATM and the SDH networks. As a consequence, GBC connections can be established spanning only the ATM network, only the SDH network or both.

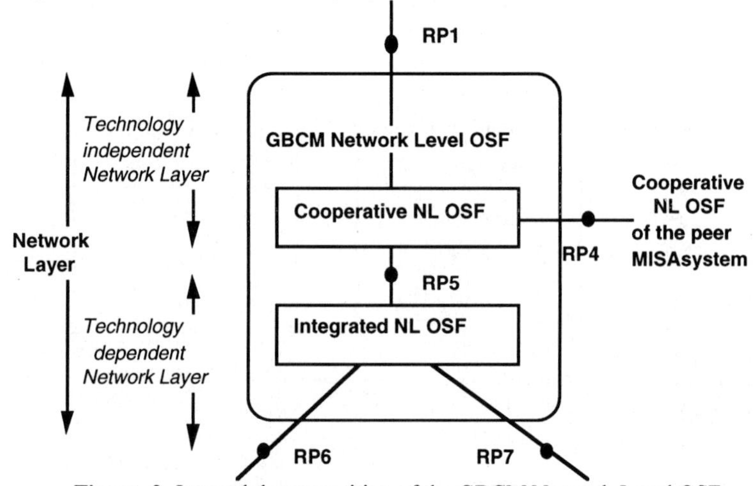

Figure 2: Internal decomposition of the GBCM Network Level OSF

The Co-operative Network Level OSF is in charge of interacting with other peer entities when the destination of the GBC connection requests belongs to other administrative domains. The interactions between peer GBCM Co-operative NL-OSFs applies at the reference point rp4.
The Integrated Network Level OSF performs the integration of the management of ATM and SDH and interacts with the underlying ATM OSF and SDH OSF through the reference points rp6 and rp7 respectively.
The ATM OSF implements the network level management of the ATM network resources, while the SDH OSF implements the network level management of the SDH network resources.

5 MISA Interface Identification

On the basis of the functional model presented in the previous section, the GBCM management system architecture can be derived.
Figure 3 shows the interfaces identified in the MISA project and the mapping of reference points into interfaces.

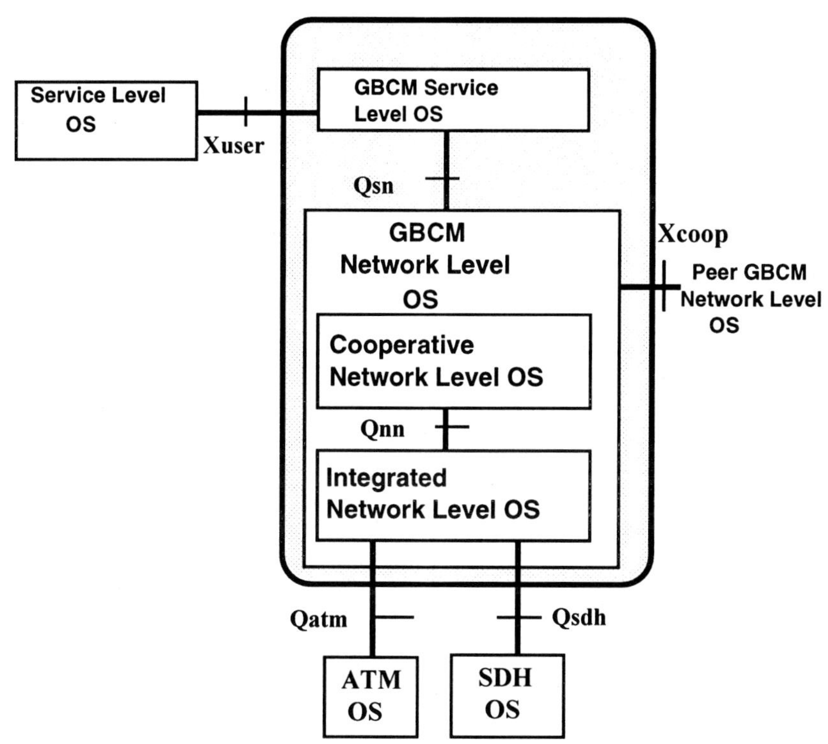

Figure 3: MISA system Interfaces.

In Figure 3 the following components can be identified:

Service Level OSs perform the Service Level OSFs in the management systems of the following GBCM user categories: VASP GBC-SP, VASP Non-GBC-SP and GBC-Consumer.

GBCM Service OS performs the GBCM Service OSF, i.e. it offers the GBCM service management level functionality.

GBCM Network Level OS performs the GBCM Network Level OSF, i.e. it implements the network management level functionality for the provision of the GBC (global broadband connectivity) by managing the local network domain and by co-operating with other peer entities belonging to different domains.

SDH Network OS performs the SDH Network OSF, i.e. it provides the functions for the management of the SDH intra-domain network.

ATM Network OS performs the ATM Network OSF, i.e. it provides the functions for the management of the ATM intra-domain network.

Between these blocks, the following interfaces can be identified:

Xuser is the realization of reference point rp3 and lies between the GBC-Customer Service Level OS and the GBCM Service/Network Level OS; it supports customer access to the GBCM service.

Xcoop is the realization of reference point rp4 and lies between two GBCM Network Level OSs. It allows the co-operations between the GBCM Network Level OSs to establish and manage the global connectivity.

Qsn is the realization of reference point rp1 and lies between the GBCM-Service Level OS and the GBCM Network Level OS.

Qnn is the realization of reference point rp5 and lies between the Cooperative NL OSF and Integrated Network Level OSF.

Qsdh is the realization of reference point rp7 and lies between the GBCM Network Level OS and the SDH OS.

Qatm is the realization of reference point rp6 and lies between the GBCM Network Level OS and the ATM OS.

For the implementation of the MISA system, the GBCM Service OS and GBCM Network Level OS can be realised as a single physical entity (MISA OS). In this case the Qnn interface becomes an internal interface.

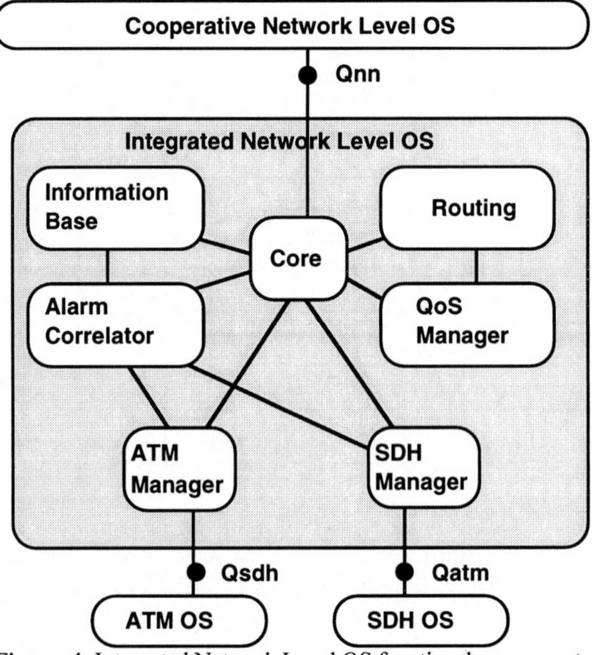

Figure 4: Integrated Network Level OS functional components

6 Integration of ATM and SDH network management

Key topics for the integrated management of ATM and SDH networks are the correlation of the alarms emitted by the ATM and SDH networks and the definition of appropriate routing algorithms to find paths satisfying the requested QoS. These aspects have been taken into account in the design of the Integrated Network Level OS. Figure 4 depicts its functional components.

A brief description of the functionality that each functional block supports is provided below:

Core functional component: it interacts with the Co-operative Network Level OS (through the Qnn interface) by receiving requests and by providing their results. Such requests concern operations on the ATM and SDH resources of the network. Moreover it reports to the overlying OS the relevant notifications affecting the connections requested by users. In order to accomplish its job, the Core exploits the functionality provided by the other functional components and co-ordinates the execution of the tasks within the Integrated Network Level OS. The Core is also able to activate a procedure which enables to retrieving all the MIB information from the underlying ATM and SDH agents within the ATM and SDH OSs respectively.

Information Base functional component: it permits the access to all information necessary for the other functional components to perform fault, configuration and QoS management operations under their responsibility. The Information Base contains in particular the end-user access points to the network, the end-points of the links connecting two adjoining administrative domains (inter-domain access points), the configuration of the intra-domain network in terms of access points interconnecting the ATM and the SDH networks (inter-technology access points), the status of the connections established in the network, the event notifications sent by the underlying ATM and SDH OSs and the measured or estimated values of the relevant QoS parameters related to the connections between the access points viewed by the Integrated Network Level OS.

Routing functional component: it is invoked when hybrid ATM/SDH intra-domain connections have to be established. This happens when an ATM connection or part of it has to be supported by an SDH path or when the Co-operative Network Level OS requests a connection between an ATM end-user access points and an SDH inter-domain access point. The routing functional component is in charge of identifying the set of routes between the intra-domain connection end-points satisfying the network provider's routing schema. It interacts with the Information Base to take the needed topology information about the interconnection points between the ATM and SDH networks. The result of the routing algorithm is the input of the processing activity of the QoS Manager functional component.

QoS Manager functional component: it is in charge of selecting, among the routes calculated by the Routing component, those routes that satisfy the QoS requested by the user. In order to perform this job, it interacts with the Information Base in order to retrieve all available information concerning the QoS values guaranteed by the network.

Alarm Correlator functional component: aims at the correlation of the alarm notifications coming from the SDH-OS and ATM-OS the Qsdh and Qatm interfaces. A set of rules have been defined to identify the alarm root cause, taking into consideration the network configuration (i.e. the relationship between SDH and ATM network layers) and alarm notification types. The objective is to emit towards the Co-operative Network Level OS (through the Qnn interface) the minimum possible

number of notifications providing the most useful and fruitful set of information about the original causes of the faults.
Atm Manager functional component: it interacts in the manager role with the ATM OS and adapts the commands received from the Core (e.g., the establishment of a connection over the ATM network) to the specific Qatm interface. The results of the operations requested by the Core are reported to the latter; the alarms coming from the ATM OS are transferred to the Alarm Correlator block, while the other event reports are transferred to the Core.
SDH Manager functional component: it interacts in the manager role with the SDH OS and adapts the commands received from the Core (e.g., the creation of a protected SDH connection) to the Qsdh interface. The results of the operations requested by the Core are reported to the latter; the alarms coming from the SDH OS are transferred to the Alarm Correlator block, while the other event reports are transferred to the Core.

7 MISA System Validation

The validation of the MISA architecture specified for the integrated management of ATM and SDH networks is being realised using and adapting the resources (physical equipment and management systems) available in some European experimental ATM and/or SDH networks called National Hosts (namely Germany, Greece, Italy, Portugal, Spain, United Kingdom). The realisation of the MISA system is being performed by using two different commercial management platforms.
At the end of 1996, ACTS MISA milestone CRISTINA [5] established a partial (test-lab) demonstration of the MISA components and testing systems. It shows our architectural design concepts and reflects the status of the implementation work, that is geared towards achieving the ultimate MISA project goals of a trial of integrated end-to-end management of hybrid ATM and SDH networks.
In particular, CRISTINA features implementations of X.700 agent/managers running on full Q3 stacks and is based on the implementation of the interfaces mentioned in this paper. It includes snapshots of the management information models specified in the interface ensembles for configuration and fault management, and will emulate the real resources of the eventual trial environment. The prototype is able to demonstrate features related to the integrated management of ATM and SDH networks and the co-operative management of the network infrastructure.
Web-based access to the ME15 CRISTINA demonstration allows a high degree of location and platform independence. The translation between the full Q3 stack access to managed objects and the Web is achieved with management servers offering high-level symbolic CMIS interfaces for languages including HTML and JAVA.
In the CRISTINA demonstration of the MISA system (see Figure 5), the Webbin CMIP and Java language/environment are used for the integration of the demo components with the graphical user interfaces. Webbin CMIP is a gateway that converts http requests to CMIP requests and it can be used for communication with the designed agents/managers.
The development is progressing in the MISA project towards a full implementation of specified components and interfaces.

8 Conclusion

We have described in this paper an approach to integrate the network management of ATM and SDH networks to support the provision of the GBCM service. It will be validated and demonstrated by the MISA project, in a multi-domain environment, through Pan-European trials.

The proposed architecture and enterprise model make possible the development and provision of the GBCM and integrated network management as distributed system. This will allow many PNOs to co-operate in providing end-to-end manageability of their broadband networks.

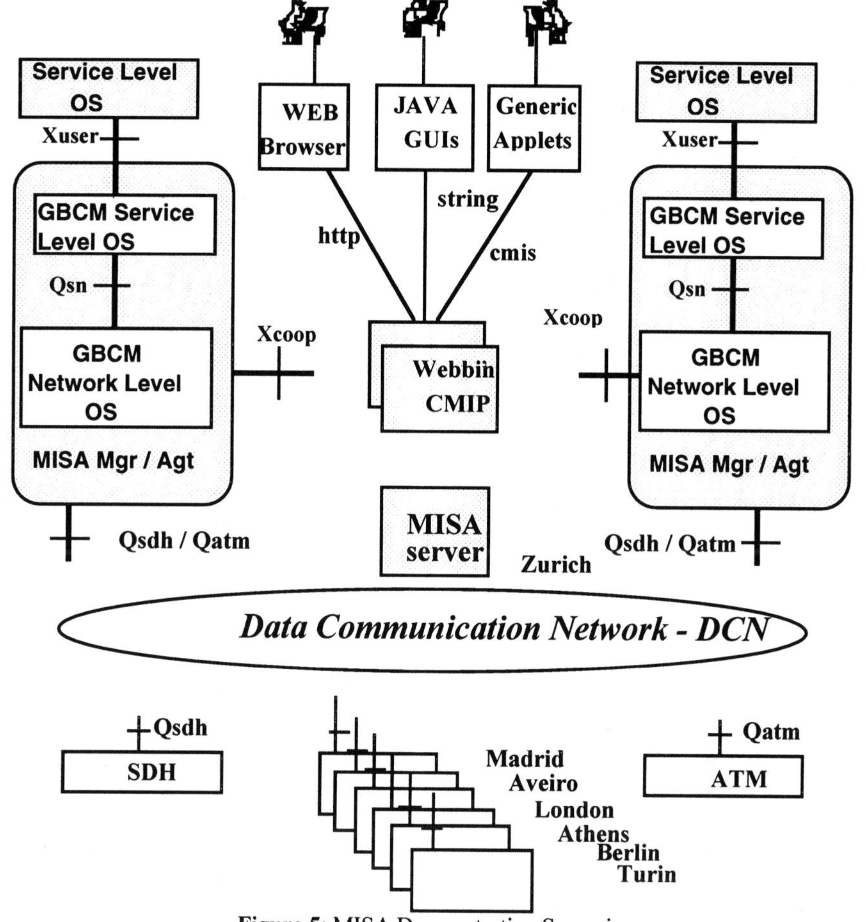

Figure 5: MISA Demonstration Scenario

References

[1] ACTS MISA Consortium - Management of Integrated SDH and ATM Networks, http://misa.zurich.ibm.com/World/, December 1, 1995

[2] ACTS MISA Consortium- AC080 Deliverable 1 - MISA System Specification, September 1996

[3] ACTS MISA Consortium- AC080 Deliverable 3 - Initial MISA High Level Design, September 1996

[4] ITU-T Recommendation M.3010 - 1995 - Principles for a Telecommunication Management Network

[5] ACTS MISA CRISTINA - Demonstration, http://misa.zurich.ibm.com/cristina/, November 30, 1996

[6] Integrated Network Management for ATM and SDH Networks: Alex Galis - UCL, United Kingdom, Dieter Gantenbein - IBM Zurich Research Laboratory, Switzerland,

Stefan Covaci -GMD Fokus, Berlin, Germany, Carlo Bianza CSELT- Italy, Fotis Karayannis- NTUA Telecom. Lab. Greece, George Mykoniatis - NTUA, Greece,- Workshop on State of the Art Technologies in Inter-Domain Management, Brussels, June 3, 1996

Abbreviations

CPN	Customer Premises Network	PNO	Public Network Operator
GBC	Global Broadband Connectivity	QoS	Quality of Service
GBCM	GBC Managemen	RP	Reference Point
IBC	Integrated Broadband Communication	SP	Service Provider
OS	Operation System	TMN	Telecommunication Management Network
OSF	Operation System Function	VASP	Value Added Service Provider

Acknowledgements:The authors gratefully acknowledge the support and funding of the European Commission, EU ACTS project AC080 MISA (Management of Integrated SDH and ATM Networks). This work was also supported by grants from the Swiss Federal Office for Education and Science. This paper represents the view of the authors, while reporting on the work of MISA Consortium team, whose project contributions we recognise and appreciate.

QoS-based Routing Solutions for Hybrid SDH-ATM Networks

Céline Verdier, Magda Chatzaki,
Graham Knight, Rong Shi

Many efforts have been made to manage homogeneous B-ISDN networks across multiple domains. However, less effort has been made for hybrid (i.e. ATM and SDH) B-ISDN network management. The ACTS **MISA** (**M**anagement of **I**ntegrated **S**DH and **A**TM networks) project addresses the issue of hybrid ATM and SDH management, aiming to automate the end-to-end **GBC** (**G**lobal **B**roadband **C**onnectivity) service in the hybrid ATM and SDH network environment by realisation of the **G**lobal **B**roadband **C**onnection **M**anagement (**GBCM**) service according to the TMN management architecture. The routing issue is fundamental to automating GBC in order to choose an optimal or nearly optimal route for the purpose of meeting QoS constraints imposed by customers and, at the same time, achieving good network performance. In this paper we discuss the QoS and routing issues, within a domain and across domains. By analysing the characteristics of the MISA network and MISA management system, the MISA routing requirements are identified. Through reference to ideas from the existing datagram network routing approaches and an understanding of the distinction between MISA routing and datagram network routing, a MISA inter-domain routing scheme is proposed for use in the pan-European demonstration. The scheme is based on a link-state algorithm tailored to allow for QoS constraints and simple policy routing. In addition, a centralised intra-domain routing algorithm for pilot MISA implementation is presented.

1 Introduction

A great deal of effort has been made towards achieving global homogeneous ATM or SDH network management and considerable progress has been made. However, in today's increasingly deregulated telecommunications world, it may be that not all domains support the same B-ISDN technology, furthermore in certain domains, instead of having a homogeneous ATM or SDH network, a hybrid network with ATM over SDH fulfils B-ISDN tasks. The management of such a global hybrid network is understood to be very complex regarding integrated management of ATM and SDH networks as well as co-operation between administrative domains. Particularly, end-to-end **G**lobal **B**roadband **C**onnections(**GBC**) are set up and configured today by manual actions using faxes and telephone calls across different telecom operators. A network management system which enables automation of **GBC** formation is highly desirable. One benefit of such a management system is that the heavy burden of manual actions for setting up and configuring the end-to-end **GBC** is removed. Another benefit is that it pioneers in the direction of managing different inter-connected networking technologies in a unified way. The ACTS **MISA** (**M**anagement of **I**ntegrated **S**DH and **A**TM networks) project aims to address the above issues by building a TMN compliant **G**lobal **B**roadband **C**onnection **M**anagement (**GBCM**) service in a multi-provider and multiple domain environment. GBCM allows the network resources provided by multiple network providers to be used in a co-operative and efficient manner in order to:

- satisfy users' end-to-end connectivity requirements
- allow network providers to make efficient use of their own network resources by on-time identification and notification of performance or Quality of Service (QoS) changes and faults
- allow negotiation of co-operative services between network providers and their peers.

It is important that the issue of QoS for Broadband Communications will be considered since it is a prerequisite for the development of Broadband Services through various network providers. The QoS management model for GBCM we propose, provides a mechanism to propagate users' QoS requirements through the network management levels and enables the GBCM system to verify the delivered service.

A routing mechanism is essential in the GBC formation procedure and this mechanism must make efficient use of network resources as well as satisfying customer QoS and policy requirements. The GBC is perceived as being able to support 'a wide range of audio, video and data applications' with different traffic and QoS requirements. Besides these, there may be policy-related requirements regarding the GBC customer's preference, for example, in choosing a network. Therefore MISA routing can be seen as a process of selecting the best end-to-end paths for the purpose of satisfying various GBC requirements, minimising the cost, whilst at the same time meeting network provider's various constraints and achieving good network performance in the multi-provider and multi-domain environment.

In this paper, we discuss MISA QoS and routing issues including QoS and policy-based inter-domain routing and QoS-based intra-domain routing. The remainder of the paper is organised as follows. In Section 2, the MISA system architecture is briefly introduced and the nature of MISA routing is analysed by comparing with the strategies employed in datagram networks. Section 3 introduces the QoS requirements for GBCM services. In Sections 4 and 5 the MISA routing requirements for inter-domain and intra-domain routing are identified and corresponding schemes are proposed. The QoS model is then described in Section 6. Finally, some unsolved problems and open issues are discussed in Section 7.

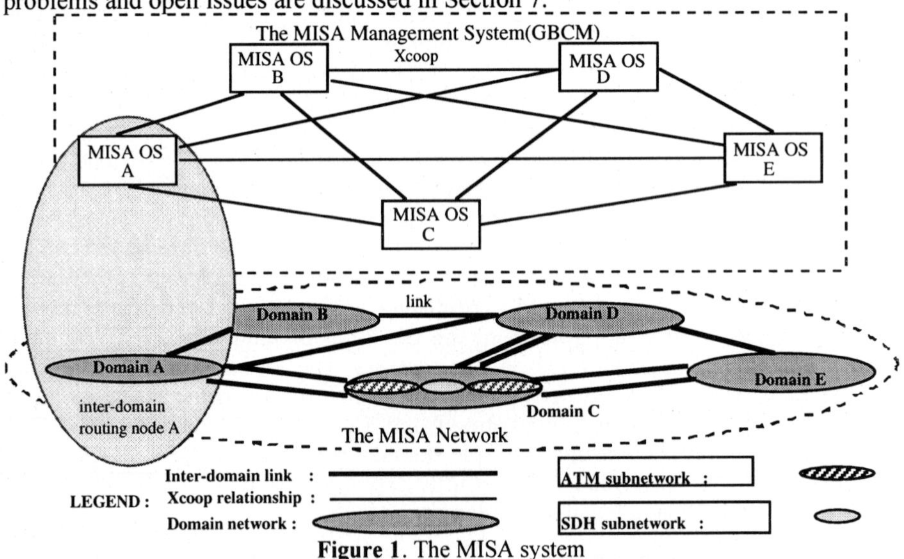

Figure 1. The MISA system

2 The MISA Architecture

2.1 The MISA System

The MISA system can be subdivided into two separate but tightly related systems - the *MISA network* and the *MISA management system* (see Figure 1). Obviously, the

MISA network is a global broadband ATM and SDH hybrid network on which the GBC is established. The MISA management system performs management actions on the MISA network through configuration, performance, fault, accounting and security management functions [9]. The MISA management system is designed according to the TMN architecture and contains as many co-operating MISA management systems as the number of administrative domains. Here, the administrative domain (**AD**) is defined in terms of the PNOs' administrative authority.

2.2 Inter-Domain Routing and Intra-Domain Routing

In the MISA system, GBC set-up is realised by the GBCM in which the routing mechanism is embedded. As described above, each domain is only visible to the other domains as a black box across Xcoop and does not expose its internal topology or ATM (SDH) sub-network capabilities. **Inter-domain routing** makes routing decisions based on the domain network capacity and domain inter-connection information for the purpose of meeting the various constraints given by both customer and network provider. The route achieved at this level is the best or near best set of ADs with ingress (on-ramp) and egress (off-ramp) access points. Thus each GBC request can be broken into a series of GBC sub-network connection requests each of which is passed to the MISA OS residing in the selected AD across Xcoop. The overall inter-domain routing procedure involves only Xcoop, therefore the inter-domain routing algorithm should be embedded in the Xcoop OSF of the MISA OS.
During the whole process of inter-domain routing a MISA OS plays the role of an inter-domain routing node (as shown in Figure 1), which is responsible for collecting and propagating network state information via *Xcoop*, maintaining the routing and QoS information in the MIB and calculating routes for a GBC request.
In the case of a domain with two or more ATM islands connected by an SDH sub-network (See Figure 1, Domain C) or accessing ATM via an SDH circuit, the GBC sub-network connection cannot be supported by a single ATM or SDH sub-network. Therefore, an **intra-domain routing** mechanism is essential at the level of GBC sub-network connection to choose a route. This route is defined by a set of access points which includes the termination access points of the GBC sub-network connection as well as the access points inter-connecting the ATM and SDH sub-networks. A GBC sub-network connection request is again broken into ATM sub-network connection requests (which are passed to the ATM OS via Qatm) and SDH sub-network connection requests (which are passed to the SDH OS via Qsdh) [5, 9]. Intra-domain routing is invoked by inter-domain routing to form the domain network capacity information. This is based on the ATM and SDH sub-networks capacity information as well as the ATM and SDH inter-network topology information. In the case of a domain with only a single ATM or SDH network, the **intra-domain routing** will perform a direct mapping function between the GBC sub-network connection and the ATM or SDH sub-network connection. The intra-domain routing still involves operations across Qatm and Qsdh, therefore the intra-domain routing algorithm fits into the INL (Integrated Network Level) OSF of the MISA OS.
The inter-domain routing and intra-domain routing are tightly coupled because inter-domain routing relies on the domain network capacity information that intra-domain routing provides. Further, inter-domain routing resolves each hop on the route to a pair of end-points (APs at the boundary of a domain) which can then be presented to the intra-domain algorithm.

3 QoS Requirements for MISA

In this section we specify the QoS requirements of MISA. The GBCMUser [5] is a logical entity of the actual consumers of the GBCM Service. It represents the entity that interacts with the GBCM Service Level OS in order to obtain the GBCM Service. The GBCMServiceProvider [5] is the entity that owns the network infrastructure providing the GBCs and owns the GBCM management system, thus providing the GBCM Service.

The GBCMService is the entity that represents the different kinds of GBC connections a GBCMUser could ask for. These are the APPS (ATM Path Provisioning Service) and the SPPS (SDH Path Provisioning Service) [5]. The APPS provides the establishment and management of connections through semi-permanent ATM VPs and/or ATM VCs. SPPS provides the establishment and management of leased lines offering SDH connectivity. The nature of GBCs supported by the APPS and SPPS is inherently different in terms of properties and characteristics.

The GBC accommodates both an ATM end-to-end semi-permanent path provisioning service (APPS), as well as an SDH end-to-end leased line path provisioning service (SPPS). The GBCs supported by the APPS (atm-GBC) includes permanent (or semi-permanent) VPCs and VCCs which are maintained by the MISA management system in the sense that no Inter-Carrier-Interface ATM signalling is supported. The atm-GBCs can be pure ATM connections or a mixture of ATM and SDH connections, that appear to the GBCUser as ATM connections only. The GBCs supported by the SPPS (sdh-GBC) can only be pure SDH connections.

The QoS management system supports an efficient way to serve various users according to their connectivity requirements. According to the traffic characteristics and QoS requirements, it is able to find the optimal provisioning service (APPS or SPPS) to support the connection. For example, suppose a GBCMUser subscribes both APPS and SPPS service and requests a Constant Bit rate end-to-end connection with specific traffic and QoS parameters. The QoS management system has the intelligence to investigate APPS and SPPS solutions in order to serve this connection request. In addition, optimisation of the selected route can be achieved, based on routing and network performance information.

3.1 The GBCM Traffic Contract

The GBCM Traffic Contract specifies the characteristics of a GBC connection. Motivated by the background information in [6, 7], the GBCM Traffic Contract consists of the following parts:

- Kind of connectivity (APPS, SPPS, both) and (source address, destination address pair)
- Connection guarantees: best effort, compulsory, guaranteed
- the QoS class: Constant Bit Rate (CBR), real-time Variable Bit Rate (rt-VBR), non-real time VBR (nrt-VBR), Unspecified Bit Rate (UBR), Available Bit Rate (ABR)
- Traffic specification
- Quality of Service parameters

Due to the different nature of SPPS and APPS, their QoS characteristics are different. Therefore, for each type of connection, a different set of QoS and traffic parameters are used to describe the connection.

3.2 QoS Requirements of the ATM Path Provisioning Service

APPS is able to accommodate a variety of GBCs. ATM VP connections are usually requested by a GBCMUser that represents for example a broker selling ATM services, or even the manager of a local CPN (Customer Premises Network). They accommodate end-user traffic. We assume that they are permanent in the sense that they are requested for long- medium-term usage (months or weeks). They are modifiable in the sense that originating GBCMUser could request the modification of certain parameters upon demand. They could be scheduled in time or periodically scheduled.

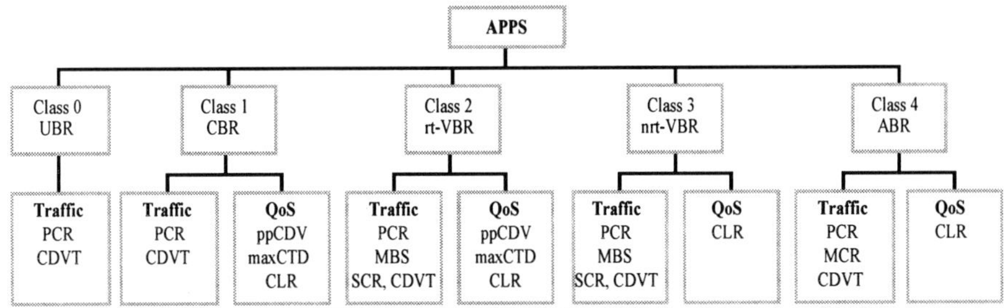

Figure 2: APPS QoS data structure

The GBC customer expresses its connectivity requirements according to the *ATM Service Architecture* proposed by the *ATM Forum* [6, 7]. Each service is specified by its traffic descriptor: Peak Cell Rate (PCR), Sustainable Cell Rate (SCR), Cell Delay Variation Tolerance (CDVT), Maximum Burst Size (MBS); and its QoS class defined by the parameters: Cell Loss Ratio (CLR), maximum Cell Transfer Delay (maxCTD) and peak-to-peak Cell Delay Variation (ppCDV).
The parameters characterising the connection depend on the service chosen. The traffic and QoS parameters are summarised for the APPS in Figure 2.

3.3 QoS Requirements of the SDH Path Provisioning Service

SPPS accommodates end-to-end SDH leased paths. End-to-end leased line paths are usually requested by a GBCMUser who, for example, represents the manager of a VPN (Virtual Private Network), that connects remote sites of the same company. They are permanent, in the sense that they are requested for a long period of time (months or weeks) and are not permitted to be modified in the mean time. Modification of such connections is interpreted as release and creation of another connection with different characteristics than the first one.
APPS and SPPS connections can not obviously support the same services. An SDH connection can not support Variable Bit Rate services. The SPPS supports Constant Bit Rate services, as well as any service with unspecified QoS. The rest of the services (rt-VBR, nrt-VBR, ABR) are treated as CBR when served by an SDH underlying network. The QoS parameters that characterise such kind of connections are allocated bandwidth, Bit Error Rate, and end-to-end delay (propagation delay + transmission delay). The QoS data structure and the parameters associated are described in Figure 3.

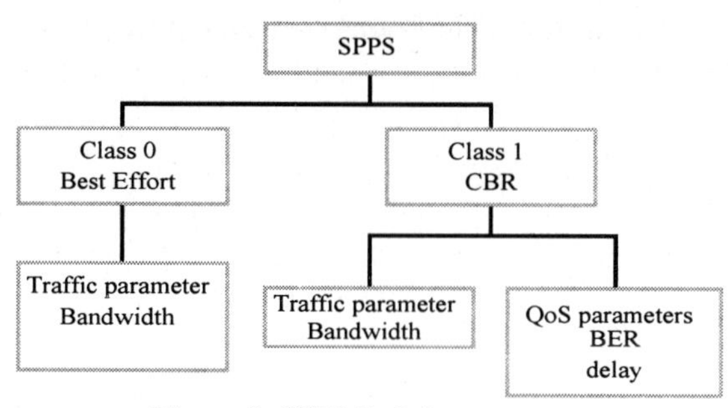

Figure 3: SPPS QoS data structure

4 Inter-Domain Routing

4.1 QoS and Policy Requirements

The negotiation and set-up of connections across independent domains takes place at the inter-domain network management level. QoS is a major factor in the choice of the path for the GBC service. As the result of the routing algorithm, a set of possible inter-domain paths between the origin and the destination is defined. Information on the performance of the networks on the paths between the access points involved are collected based on the GBCM specified QoS and traffic parameters. At this level, the negotiation between domains takes place. The QoS system requires QoS values sustainable by the domains crossed and calculates the end-to-end value of each QoS parameter. If the "best" path with the required QoS is available then the connection can be reserved. The achieved performance and QoS parameters are then stored for monitoring and management.

Policy factors have two aspects. On the one hand, ADs may have different administrative policies concerning which traffic may be transported on their network or which customer's traffic is allowed to terminate etc. On the other hand, GBC customers may express their preferred domains in a GBC request. MISA inter-domain routing is expected to support simple policy routing which only concerns policy factors that are relatively stable. Both *stub control* (which specifies who can send traffic to my internal resources) and *transit control* (which specifies who can send transit traffic through my domain) should be supported by the MISA inter-domain routing mechanism.

4.2 Inter-Domain Routing Mechanisms

Routing mechanisms can be classified in a number of ways:

Explicit or Implicit Paths

In the classic datagram-based network, although the network holds sufficient information to be able to route datagrams from A to B it does not retain knowledge of "connections" or "paths" as such. In a virtual-circuit based network, connections and paths are explicitly part of the state information maintained by the network and there is a clear separation between the connection set-up phase (during which route selection takes place) and the data transfer phase. The networks used in MISA are of the latter type and we expect that, normally, once an inter-domain GBC has been set up between A and B it will remain in place for a significant time and will continue to use the same path for the lifetime of the GBC. Thus we do not expect to see the highly dynamic re-routing associated with classic datagram networks. MISA routing will adapt, but only to relatively long-term load trends or major topology changes.

Centralised or Distributed Route Calculation

In a *fully centralised* scheme a single entity is responsible for calculating all routes and distributes results to the constituent nodes. There is an assumption that all information relevant to the calculation of a route is simultaneously available in one place. In such a situation it is possible to guarantee loop-free routing and that the routing information distributed to nodes is consistent. It is difficult to see how such a scheme could work in the MISA context. Firstly, there is an administrative problem of multiple independent domain administrations agreeing on the setting up of such a scheme and then jointly managing it. Secondly, QoS and policy based routing means that there are many "best" routes between a particular source/destination pair depending on the particular values of QoS and policy parameters in the customer request. It is not feasible to pre-compute and distribute multiple routes just in case they may be needed later.

In a *fully distributed* scheme each node performs routing calculations for its local environment and exchanges routing information with other nodes. No single entity has knowledge of any entire route. This means that inconsistencies and routing loops may arise unless precautions are taken. Although it is mainly associated with datagram networks there is no reason why this "hop by hop" approach should not be considered for MISA. Essentially each domain would choose the next domain along the route according to some criteria and then pass the responsibility for the calculation of the remainder of the route to this new domain.

Mid-way between the fully centralised and fully distributed schemes we have schemes in which a single node or domain is responsible for negotiating a route through the network. The classic example of this is "source routing" in which the route is selected by the source node. In datagram networks the source route has to be explicitly stated in all datagrams. In a virtual-circuit network the initiating node (which need not necessarily be the source node) would select the route and then negotiate with all the nodes along the route in order to set up the connection. Knowledge of the connection is then held in the per-node state information and does not have to be explicit in the data packets. We call this scheme "*initiator controlled*" routing. Since the whole of the route selection is performed by a single entity it is possible to guarantee loop-free routing within an initiator controlled scheme.

Initiator controlled routing is attractive in a multi-domain environment as it keeps the route selection exclusively in the control of the domain to which the customer belongs. This allows that domain great flexibility in the sort of contract and guarantees it offers to its customers. Further, as stated above, it is straightforward to avoid loops in an initiator controlled scheme.

On the other hand initiator controlled routing requires the initiating domain to have full knowledge of the entire inter-domain network all in one place. Further, it takes no advantage of the parallelism which is present in a fully distributed scheme. It seems likely that a pure initiator controlled scheme may have scaling problems and that a hybrid scheme may eventually prove to be the best approach.

4.3 Routing Information Collection and Calculation

A number of existing inter network routing approaches such as EGP (Exterior Gateway Protocol), BGP (Border Gateway Protocol), IDRP (Inter Domain Routing Protocol) and IDPR (Inter Domain Policy Routing protocol) etc. have been extensively researched in the context of *datagram networks* such as the Internet, DECNET and OSI networks [10]. Each of these approaches addresses the problem of distributing routing information and it is worth considering what relevance this work has for MISA. It is evident that MISA routing and datagram network routing have many common requirements; both must have the functionality of collecting and maintaining network state information, both need to calculate and re-calculate routes based on network state information. However there are important differences. In particular, datagram networks maintain no connection state and try to adapt to load changes on a comparatively short timescale. A more subtle distinction is that in datagram networks the links used for exchange of routing information are precisely those which carry the data. This means it is relatively easy to exchange routing information with neighbours, relatively hard to do so with distant nodes. In the MISA context routing information is exchanged between OSs which are on a conceptually independent control network; it may be as simple for an OS to consult the OS of a distant domain as it is for it to consult the OS of a neighbour.

Dynamic routing algorithms may be classified as *distance-vector* or *link-state. A distance-vector* algorithm requires that each node maintains the "distance" from itself to each possible destination and sends this information to its neighbours at intervals. New distance estimates are computed using the information in neighbour's distance vectors. In a *link-state* algorithm, each node has complete information on the network topology so that each node can carry out route computation independently. Nodes send out updates to all other nodes by broadcasting either regularly or when a node detects any changes in local link distances,.

The chief problems of *distance-vector* algorithms can be categorised as:

- Slow convergence of network state information: A node cannot pass routing information on until it has re-computed its distance vector. If the maximum diameter of a network is N hops, N routing updates are needed to propagate changes across the network.
- Routing information inconsistency: Due to the slow convergence it is possible that not all domains will receive network state updating information in a timely way. This leads to inconsistent views of the networks state and may result in routing loops.
- QoS and policy routing: Analysis of distance-vector algorithms shows that it is not suitable for QoS and policy routing as each routing node only maintains distance information to each potential destination. Network policy and QoS tolerance are hard to express.

Link-state algorithms avoid the defects above. They also have the advantage that each node accumulates complete information about the network. This is required if we adopt the pure initiator controlled routing scheme. All this suggests strongly that a link-state-like algorithm would be the best choice for MISA inter-domain routing. However, we must consider the fact that conventional link-state algorithms employ a flooding strategy to propagate routing information and ask how this fits into the MISA/TMN world.

The straightforward approach would be to set up semi-permanent CMIP associations between "neighbour" OSs (via Xcoop) for the purpose of forwarding routing update messages. For simplicity the topology of these associations would mirror that of the physical inter-domain links - though this is not essential. Each domain would have to undertake to provide routing updates at prescribed intervals or after defined events. Each would also have to undertake to forward routing updates from other domains in order to implement the flooding. The disadvantage of this scheme is that its success depends on all (administratively independent) domains playing the game correctly. Are possibly competing organisations likely to be willing to trust each other to do this?

A modified scheme would remove the "forwarding" duty by having each domain notify all others directly (as noted above, the OSs of *all* other domains are directly accessible via TMN). In its raw form this idea would scale very badly generating a large number of messages each of which incurs a connection set-up overhead and delivers very little information.

A final alternative would be to have the updates driven by polling rather then notification. This leaves each domain more in control of its own destiny at the expense of generating management traffic when there has actually been no change of state.

One objective of MISA is to build a large-scale demonstration system which will allow practical evaluation of the various approaches described above.

4.4 Frequency Requirement of Routing Information Update

Routing information and route computation are the essence of routing in general. In order to avoid GBC QoS degradation and network congestion, MISA inter-domain routing must be able to automatically adapt to inter-network topology changes as well as domain network capacity and policy changes. Ideally, optimal routing decisions rely on up-to-date routing information which is obtained at the moment a GBC request is made. However, given that our objective is to adapt routing to long-term trends and major topology changes it is not worth paying too heavy a price to have timely information. In fact a quasi-static model can be used in the MISA inter-domain scheme, in which the routing tables remain unchanged until significant changes occur, such as inter-network topology change, domain policy change and domain network capacity change. It must be remembered that there will always be a route set-up phase following the route calculation. Occasionally we will find that a domain cannot satisfy a request to provide a GBC segment at the required QoS because its circumstances have changed since it last provided routing information. When this happens an alternative route must be found. Our aim in setting our update strategy must be to keep the frequency with which these recalculations are required within acceptable bounds.

4.5 The Inter-Domain Routing Functional Component

Having analysed the MISA inter-domain routing requirements, we propose the following inter-domain routing scheme for use in the pilot MISA implementation:

- initiator-controlled routing with link-state based routing information collection by direct notification of state changes;
- quasi-static approach for routing information updating;
- supporting QoS and policy routing with stub control and transit control policy routing mechanism.

This means that each domain in the MISA system, (i.e. the MISA OSs), should maintain full knowledge of the inter-domain network topology in its MIB by querying all the other domain OS that are installed. Further, each domain should notify all other domains of major state changes and should maintain local domain network state information in its MIB. This includes the local network capacity, inter-domain link capacities and domain policy. It should calculate the route based on full knowledge of the inter-domain network topology by using the QoS routing calculation algorithm.

Accordingly, five fundamental functional elements (shown in Figure **4**) are essential to MISA inter-domain routing for the purpose of keeping routing information consistent and computing the inter-domain route for a GBC.

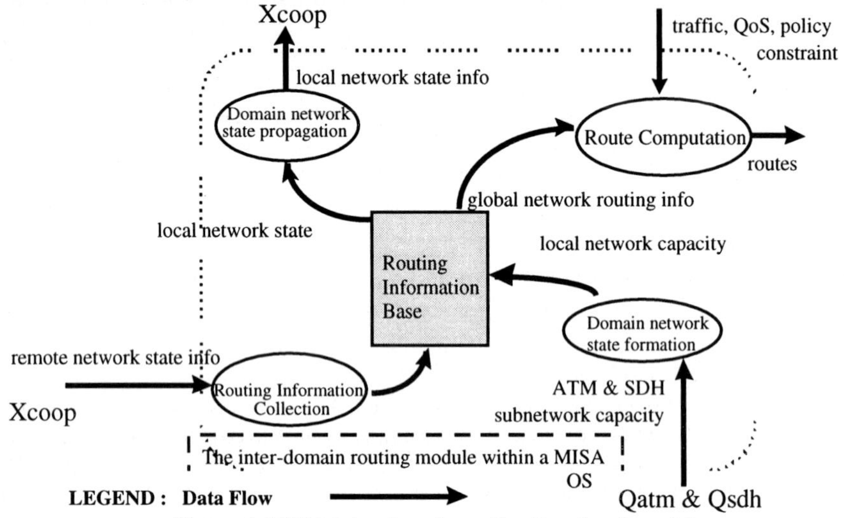

Figure 4. MISA inter-domain routing functional model

Domain Network State Information

This function is responsible for measuring local network capacity in the form of domain network traffic load allowance and QoS tolerance by querying the intra-domain routing module.

Due to the quasi-static character of inter-domain and intra-domain routing, a **domain network state formation** action occurs only when there is an ATM or SDH sub-network state change caused by

- ATM and SDH inter-network topology changes;
- ATM and SDH sub-network capacity changes.

In the QoS-based routing case, local network capacity is measured as potential domain network connections with the perceived traffic allowance and QoS tolerance synthesised from ATM and SDH sub-network capacities.

After obtaining the local domain network state this must be stored in the local routing information base while waiting for the **Domain Network State Propagation** module to send it to all the other inter-domain routing nodes.

Domain Network State Propagation

As the routing is link-state based, every MISA OS will have full knowledge of the MISA network topology and capacity. Each OS is obliged to propagate any changes regarding its local network capacity, inter-domain link status and domain policy to all other domains by using a CMIS notification primitive via Xcoop. Here we can use different notifications to represent local network capacity change after a domain network capacity change, domain policy change, inter-domain link status change(topology change) and inter-domain link capacity change. Here the inter-domain link capacity is measured in the same way as the domain network capacity, except that one inter-domain link is from one end-point located at the boundary of the domain to the end-point located at the boundary of an adjacent domain.

Route Computation
The best routes for a GBC is determined based on the full network topology information stored in the **routing information base** under the restriction of various requirements given in the GBC traffic contract which contains traffic, QoS and policy constraints. Due to the fact that the GBC is end-to-end based, the traffic and QoS parameters specified in the GBC traffic contract are end-to-end. Therefore the traffic and QoS required must be jointly fulfilled among the domain networks crossed.
The details of the route computation algorithm to be used in the first MISA system are being finalised and will be presented in a subsequent paper.

Routing Information base
In addition to the full network topology information for inter-domain routing, the **Routing Information Base** maintains the local network capacity, inter-domain link status and capacity, and the domain policy information. The **Routing Information Base** is part of the MISA OS MIB.

Routing Information Collection
This function is mainly responsible for receiving the remote network state change information issued by the **Domain Network State Propagation** module of inter-domain routing residing in other domains and updating the corresponding information in the **routing information base** accordingly.

5 Intra-Domain Routing

- QoS requirements: The QoS parameters considered in the intra-domain routing metric must be the same as those used in inter-domain routing. It is the intra-domain routing's responsibility to map from the GBC QoS parameters to the technology-dependent QoS parameters.
- Requirements on routing information collection and calculation: Considering that the intra-domain routing decision is made centrally in the INL (Integrated Network Level) OSF of the MISA OS (see Figure 5) and routing information is only collected from two sources (ATM OS and SDH OS), routing information collection can be simply done by transferring a message from the ATM OS or SDH OS to the MISA OS. All we need is an algorithm for calculating a route based on QoS.
- Frequency requirement of routing information update: To be aligned with the inter-domain routing, the same quasi-static approach is used in intra-domain routing.

5.1 The Intra-Domain Routing Functional Model

By analysing the MISA intra-domain routing requirements, we also can conclude that the intra-domain routing scheme should have the following centralised routing decision based on a centralised routing information store [4];

- routing information collection is a one-step information transfer via two interfaces Qatm and Qsdh;
- quasi-static approach for routing information updating;
- support for QoS routing.

6 QoS Functional Model

The QoS model is based on the architecture shown in Figure 5 [5].

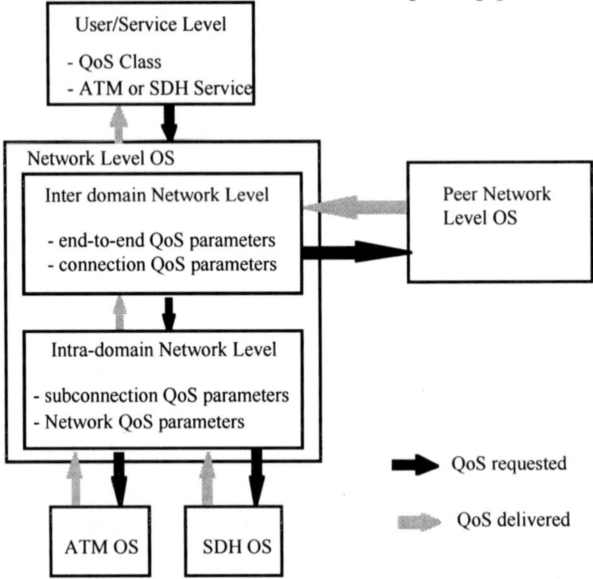

Figure 5: QoS Management architecture

The management levels considered are:

- User and Service Level
- Inter-domain Network Level
- Intra-domain Network Level

6.1 User and Service Management Level

The QoS management system supports an efficient way to serve various users according to their connectivity requirements and the availability of resources without affecting already existing connections. The service classification we propose is based on the ATM Forum recommendation [7] as described in section 3.

In the QoS negotiation process, the GBCMuser requests a service with a specific QoS. This QoS is negotiated by the GBCM on behalf of the user with all the domains that the connection involves. The negotiation process follows the following steps:

1. A QoS class is chosen and requested by the user and is mapped onto a set of QoS parameters at the Network Level.
2. At the Network Level, QoS parameters are collected from the networks involved in the connection.
3. The sets of QoS parameters negotiated with the networks are re-integrated into a set of end-to-end QoS parameters.
4. The end-to-end QoS negotiated will be accepted or rejected by the User Level depending on whether it complies or not with the requested QoS.

6.2 The Inter-Domain Network Management Level

The Inter-Domain Network Level is responsible for the negotiation and set-up of connections across independent domains. QoS is a major factor in the choice of the path for the GBC service. According to the GBCM Traffic Contract (traffic type and

QoS class), information on the performance of the networks on the paths between the access points involved are collected as described in section 4.5.

6.3 The intra-domain Network Management Level

Each domain involved in a connection has to define its own intra-domain subconnection through ATM and SDH subnetworks. The inter-domain Network Level sends an intra-domain subconnection request with a specific set of QoS parameters for a certain type of service (i.e. ATM or SDH service). The QoS block evaluates the possibility of fulfilling the requirements.
Subconnections can be pure ATM, pure SDH or hybrid ATM/SDH connections. The QoS block collects the network-dependent parameters through from ATM or SDH Network Agents and maps them onto connection-dependent parameters, i.e. it calculates the equivalent value of the parameter for the connection. This is particularly important in order to harmonise the calculation of the end-to-end subconnection parameters. Indeed, if an ATM connection is requested that has to be supported by an SDH network, then the QoS parameters delivered by the SDH network (e.g. Frame Error Rate) have to be transformed onto ATM like connection parameters (e.g. Bit Error Rate).
The QoS block then computes, for each complete subconnection, the value of the end-to-end parameters. The best path that fulfils the requested QoS is chosen, the values of the corresponding Network QoS parameters are reserved in the ATM and SDH subnetworks, and the QoS parameters are stored in the database.

6.4 QoS Computation

The equations to compute the end-to-end values of the QoS parameters as defined in section 3 are similar at the inter-domain and the intra-domain level [8]:

$$\text{PBR[end-to-end]} = \text{Min}\left(\text{PBR}_{\text{atm}}, \text{Bandwidth}_{\text{sdh}}\right)$$

$$\text{CDVT[end-to-end]} = \sum_{\text{atm}} \text{CDVT}$$

$$\text{Delay[end-to-end]} = \sum_{\text{atm}} \text{CTD} + \sum_{\text{sdh}} \text{delay}$$

$$\log(1 - \text{BER[end-to-end]}) = \sum\left(\log(1 - \text{BER}_{\text{atm}}) + \log(1 - \text{BER}_{\text{sdh}})\right)$$

$$\text{MBR[end-to-end]} = \text{Min}\left(\text{MBR}_{\text{atm}}, \text{MBR}_{\text{sdh}}\right)$$

At the intra-domain network level, network dependent parameters (i.e. ATM or SDH specific parameters) are translated into connection dependent parameters (i.e. SPPS or APPS specific parameters), so as to be able to integrate them in the calculation of the end-to-end value of the parameters for the connection.

7 Conclusion

An architectural level intra-domain and inter-domain routing scheme for use in MISA has been proposed by studying various routing approaches in datagram networks and adapting these to the rather different MISA environment. From an understanding of the MISA network and its routing requirements an information model for the routing scenario can now be defined. This paper also presents a model for a QoS management system for Broadband connections in a multi-domain environment. This model decomposes the problem of QoS from the user level down to the network level. In an heterogeneous environment, it provides a mechanism to translate technology independent user requirements into technology dependent network performance

characteristics and vice versa. We describe these mechanisms for ATM and SDH services and we show how the mapping of QoS parameters propagates through the management levels. This enables the network operator to optimise its resources according to the specific service required and provides a way to verify the delivered service.
Further work is required in some areas including:

- details of the algorithm for calculating the inter and intra domain routes based on a set of QoS metrics
- scaling issues in the routing update mechanism;
- studies of the trade-offs between initiator controlled and fully distributed approaches.

Pan-European experiments involving several European National Hosts will take place by the end of the MISA project. Furthermore, these field trials are intended to provide inputs to standardisation bodies as ITU-T and ATM Forum.
In order to have an inter-domain routing algorithm for the MISA robust enough for the real world, with optimal functionality, further research work in this area within the various MISA trials should be carried out since:

1. route computation for the QoS routing is very complex. Issues regarding,

- how to combine the multiple QoS graphs to form a single tree as basis of the final route selection for the purpose of QoS and policy routing;
- how to divide or recalculate the end-to-end QoS parameters such as end-to-end delay and the error bit ratio etc. This will require a great deal of effort to be spent in the next stage of the project.

2. routing information in the MIB should be organised in a way which is scaleable and being best able to support route computation.

References

[1] Brijesh Kumar, “Policy Routing in Internetworks”, Ph.D thesis, Dept. of Computer Science UCL; 1993.
[2] Miguel Angel Ruz, “Intra-domain routing in MISA”, MISA Consortium internal document; 1996
[3] Graham Knight and Rong Shi, “Inter-domain routing in the MISA”, MISA Consortium internal document; August 1996.
[4] Rong Shi and Christian Bitard, “The specification of intra-domain routing”, MISA Consortium internal document; Oct, 96.
[5] MISA Deliverable 3, “Initial MISA High Level Design”, MISA Consortium, Sept. 1996.
[6] ATM Forum User Network Interface ver.3.1, September 1994.
[7] ATM Forum Traffic Management Specification ver.4.0, April 1996.
[8] G.V. Bochmann, A. Hafid, "Some Principles for Quality of Service Management", Université de Montréal.
[9] A.Galis et. al, “Towards Multi-domain Integrated Network Management for ATM and SDH Networks”, Conf. On Broadband Strategies and technologies for Wide Area and Local Access Networks, BSTW’96, Berlin, Germany, Oct. 96.
[10] Martha Steenstiup, “Routing in Communication Networks”, Prentice Hall, 1995.

Static vs. Dynamic CMIP/SNMP Network Management Using CORBA

Luca Deri
IBM Zurich Research Laboratory [1]
& University of Berne [2]

Bela Ban
IBM Zurich Research Laboratory [1]
& University of Zürich [3]

The increasing complexity and heterogeneity of modern networks is pushing industry and research to look for a single and consistent way of managing networks. With the advent of open object-oriented distributed computing models such as CORBA, there are efforts to make the operational and management models the same, i.e. to manage and operate the network using CORBA.
The aim of this paper is to show some techniques that allow to manage CMIP/SNMP network resources using CORBA. Static techniques which map each managed object class into a corresponding CORBA interface are compared with dynamic techniques which rely on runtime information. Finally this paper demonstrates that CORBA-based network management applications are becoming attractive in terms of efficiency and application size, overcoming limitations of early solutions.
Keywords Network Management, CORBA, Scripting Language.

1 Introduction

With the growing impact of CORBA [13] in the telecommunications sector, the need has risen to employ CORBA to manage CMIP/SNMP agents. Since the CORBA object model (using IDL as specification language) is easier to learn than CMIP [3] and SNMP [14], anyone who is able to create CORBA applications can immediately use services offered by CMIP/SNMP agents, given a CORBA interface to them, without having to have a specific knowledge about CMIP or SNMP. It is the authors' opinion that this asset will become a widespread need as more applications in the telecommunications business will be programmed using CORBA. Given the large investment of carriers in CMIP/SNMP, however, there will still be a need to manage CMIP/SNMP agents in the future. If CORBA can be employed to transparently manage those, then a smooth transition/cooperation between the two worlds can be achieved. If a strategy of a company is to move towards CORBA, then they can still access their legacy agents, maybe gradually phasing them out and replacing them by CORBA applications.
There are several, partly orthogonal, partly overlapping, approaches that use CORBA for CMIP/SNMP management which will be outlined and compared in this paper. We present two schemes that use CORBA in a dynamic manner for network management and contrast them to static approaches. The structure of the paper is as follows: first, we will give a short overview of CORBA. Then, a static and two dynamic approaches to use CORBA for network management will be presented and compared with each other, with the focus being on the dynamic approaches. Finally, some conclusion will be drawn.

---

[1] IBM Research Division, Zurich Research Laboratory, Säumerstrasse 4, 8803 Rüschlikon, Switzerland. Email: {lde, bba}@zurich.ibm.com, WWW: http://www.zurich.ibm.com/~{lde, bba}/.
[2] Universität Bern, Institut für Informatik und angewandte Mathematik, Software Composition Group, Neubrückstrasse 10, 3012 Bern, Switzerland. Email: deri@iam.unibe.ch, WWW: http://iamwww.unibe.ch/~deri/.
[3] Institut für Informatik, Universität Zürich, Winterthurerstr. 190, 8057 Zürich, Switzerland. Email: ban@ifi.unizh.ch.

2 CORBA Overview

CORBA allows instances to be created locally or remotely either in local host or on a remote one. If the instance is remote it can be accessed from every CORBA compliant client, whereas local instances can be accessed only by the application that created it. On each host on which CORBA instances are to be created there is a daemon called *CORBA server* which is responsible for handling object creation/deletion and other requests sent by client applications in collaboration with the Basic Object Adapter. In order to serve such requests, the server accesses two databases: interface repository and implementation repository. The *interface repository* is a persistent type repository of objects representing the elements of interface definitions and is created and maintained based on information supplied in the IDL source file. The *implementation repository* is a database which contains the implementation definitions of CORBA objects, i.e. the shared information about the location of the libraries which implement the CORBA objects and the classes which are offered by a server. Whenever a client application creates a remote object, the server uses the implementation repository in order to have access to the code that implements the requested CORBA object and starts a server application where the CORBA object gets instantiated. The client application receives back from the object creation an object reference, allocated in the client's address space, which identifies the real instance created in the server application. Due to this mechanism, clients are not allowed to manipulate instances directly but they do access them transparently through the proxy instances. Instead, if the instances are created locally, i.e. in the client address space, there is no interaction with the server and the instance behaves like a normal non-CORBA instance since it is "private" to the applications.

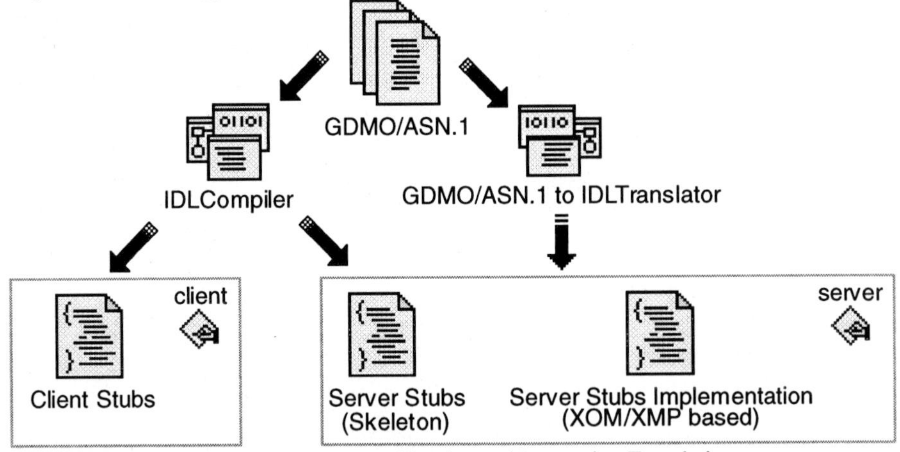

Figure 1: XoJIDM Specification and Interaction Translation

3 Using CORBA for Network Management

3.1 Static Approach: XoJIDM

X/Open's Joint Inter-Domain Management task force (XoJIDM) is working on the mapping between GDMO/ASN.1 [16][1] and IDL and vice-versa (only the first mapping is of interest to us here) [8]. Their approach consists of two parts: the first *is Specification Translation* [15] and defines the static translation of GDMO/ASN.1 to IDL while the second is called *Interaction Translation* [10] and deals with how the

mapping is used at runtime. The goal of this effort is to translate the MIB of an agent (GDMO/ASN.1) to CORBA IDL which can subsequently be used to manage the agent using CORBA. The approach is shown in figure 1.
The GDMO/ASN.1 documents describing the agent's MIB are translated to IDL and then to a server implementation. GDMO templates are mapped to IDL interfaces and actions and attributes to operations. For each GDMO attribute, a potential GET-, REPLACE-, ADD-, REMOVE-, and SET-TO-DEFAULT-operation is defined. Each ASN.1 type is translated to a corresponding IDL type, e.g. SEQUENCE is mapped to an IDL struct, OCTET STRING to string etc. The generated IDL interfaces representing GDMO class templates will include IDL attributes generated from ASN.1 types. IDL is then compiled to produce the client and server stubs in the desired language binding (e.g. C++). The server stub and the implementation code generated by the GDMO/ASN.1-IDL compiler are compiled and linked to produce a CORBA implementation. The client stubs are compiled and linked with the user application. Information about available CORBA interfaces (which represent CMIP instances) is contained in the generated client stubs and therefore known to the client at compile time. At runtime, the client proxies will forward any request they receive to their corresponding objects in the CORBA server. These will use the implementation code generated by the GDMO/ASN.1-IDL compiler to communicate with the managed objects in an agent using CMIP/SNMP (e.g. by using X/Open's XOM/XMP API).
It is worth to mention that the Telecom Special Interest Group of OMG defines an approach for CORBA-based management based on the results of XoJIDM [17].

3.2 Dynamic Approaches

The limitations of the static CMIP/SNMP to CORBA mappings and their extreme complexity has been the main reason that pushed the authors towards a more dynamic approach to the problem. In this section we will present two dynamic models defined by the authors which use CORBA for network management.

GOM

The Generic Object Model (GOM) [2] is a framework for management of instances of multiple object models such as CORBA, CMIP or SNMP.

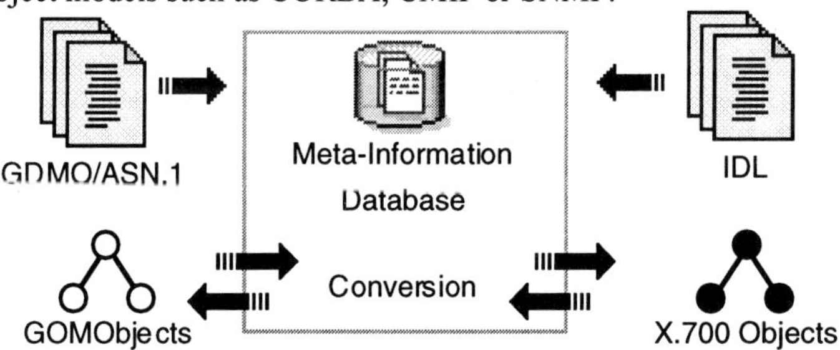

Figure 2: GOM Architecture

It uses the concept of reification, modelling elements of object-oriented models as objects themselves [11]. Thus, all CORBA interfaces or GDMO class templates that will ever be encountered are mapped to an instance of the generic GOM class `GenObj`. All attributes are mapped to instances of `Attribute` and all values to instances of `Val`. It has a list of attributes and operations which are instances of the classes `Attribute` and `Operation` respectively. Since client applications do not have to include type information about available classes at compile-time, they typically have a very small size. Also, when a specification is modified, clients do not have to be recompiled, but it is only the modified specification that has to be parsed and fed into the Metadata Information Database (MID) (c.f. below). Once it is there, adapters (see below) that use that specific type information just have to flush their metadata cache in order to access the modified type information, which is a runtime operation. Having no tight binding between client and server and providing a reified object model has a number of advantages especially in X.700 which is more complex than CORBA. It is for example possible to accommodate conditional packages on an 'as needed' basis. This means that no elements of conditional packages will be initially available when an instance is created. When a request is sent to the instance referring to an attribute residing in a conditional package, the attribute will be created ad hoc using the MID and will be added to the instance's attribute list. This scheme of on-demand loading allows for memory-sensitive applications to be created. Another example is the use of the X.700 ANY DEFINED BY type which is a type that can be determined only at runtime using metadata. It is simply not possible to map this type at compile time to an IDL interface or C++ class in the static approach. As shown above, there is one proxy instance of `GenObj` in the client's address space for each underlying object, be it a CMIP, SNMP or CORBA object. Requests sent to those proxy instances will be forwarded to an adapter which knows how to translate between the generic and exactly one specific model.

Since, contrary to the static approaches described above, there is no compiled-in knowledge of any classes in the system, adapters rely entirely on metadata about the CORBA, SNMP or CMIP classes available to perform their work. GOM includes a MID which is a repository of type information about the various specific models. It is populated by compilers, e.g. in the case of CORBA, an IDL compiler will parse the IDL specifications and add metadata about CORBA interfaces, their operations and attributes to the MID. In the case of X.700, a GDMO compiler will add metadata about GDMO templates, their operations and attributes and an ASN.1 compiler will provide information about the ASN.1 types. Conceptually, the MID is a single logical database, whereas it uses a separate database for each object model internally. The MID can also be used for other purposes such as lookup of type information for interpreters or debuggers, documentation of class specifications etc.

In order to handle instances of GOM, a C++-like interpreter has been written which lets users create, access and delete instances either interactively or by running scripts. *GOMscript*[4] has simple values such as numbers, boolean, strings and aggregate values such as struct, union and list. It has the usual control statements for repetition (for, while) and conditional branching (if, else) and is object-oriented in the sense that it implements (single) inheritance, polymorphism and encapsulation. Its core is very small and can be extended (additional functions and classes) through user-written components [4] which are located in shared libraries. GOMscript can be started in daemon mode in which it waits for (potentially remote) clients to send scripts to be executed. This allows for implementation of simple roaming agents [12] which are essentially scripts moving from machine to machine and taking their state with them.

---

[4] A prototype for IBM AIX and Linux is available for public download at http://www.zurich.ibm.com/~bba/gomscript.html.

CORBA-Liaison

Liaison[5] [5] is a software application which allows the management of CMIP/SNMP resources through the Web. Liaison is based on a special type of software component called *droplets* [4]. Droplets have the ability to be replaced and added at runtime allowing to dynamically modify and extend the behaviour of the application that contains them. Among the droplet part of Liaison's standard configuration, there are some which allow Java/C++ applications to do network management through Liaison. Basically Liaison provides some Java/C++ classes, called *external bindings* [6], which are linked to the C++ application or Java applet and which allows management operations to be performed. The management application uses the external bindings as normal classes, invoking methods, creating/deleting instances. Transparently, external bindings communicate with the Liaison server using the HTTP protocol, which is used extensively in the Internet by the World-Wide Web.

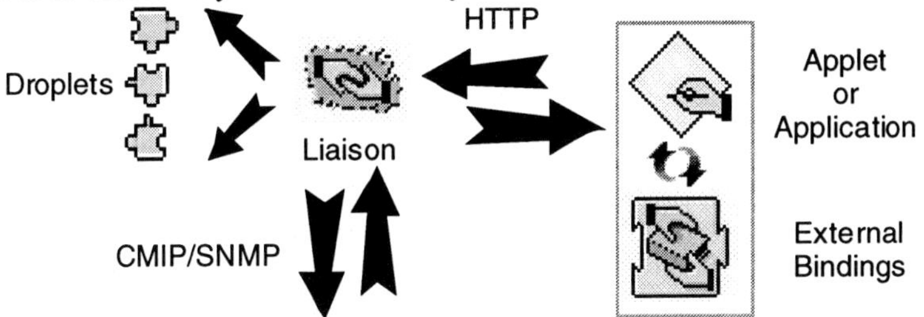

Figure 3: Liaison's Java/C++ Bindings

Whenever a method of the external bindings that performs a management operation (e.g. CMIP M-SET) is called, an HTTP request is sent to Liaison. Such an HTTP request contains all the parameters necessary to issue the management request. Liaison issues the management request, receives the response(s) and handles all the possible errors. Once the operation has been completed, an HTTP response containing the result of the operation is sent back to the application. This mechanism allows external applications to do network management in a simple and effective way by exploiting Liaison and without having to handle the complexity of management protocols. In fact, external bindings deal only with object oriented concepts shielding completely the underlying management protocol. Based on external bindings some CORBA interfaces to the CMIP/SNMP protocols have been defined. An important design choice has been not to map each CMIP/SNMP object to a CORBA object as seen in other translation methods, but to map the CMIP/SNMP protocols to CORBA in a very generic way. This choice has been motivated by the following reasons: a) ability to fully support CMIP/SNMP from CORBA, b) low complexity and flexibility since there is no need to generate new CORBA classes for new CMIP/SNMP objects that need to be supported, and c) user-defined abstraction level: users can define further CORBA classes based on the basic ones depending on their needs without the complexity of having to pay a CORBA class for each CMIP/SNMP object even if not all of them are currently used.

CORBA-Liaison interfaces (CL) for CMIP/SNMP, defined using the IDL (Interface Definition Language) language, have been implemented using DSOM [7], IBM's CORBA compliant ORB (Object Request Broker). Since we do not rely on any

[5] An on-line demo and a version available for public download (supported platforms: AIX, OS/2, MacOS, Linux, Win95/NT) can be found at http://misa.zurich.ibm.com/Webbin/.

specific characteristic of DSOM, similar considerations can be done for other CORBA implementations. ASN.1 datatypes, like external bindings datatypes, have been mapped to strings, hence to the native `string` IDL datatype. CL interfaces representing CMIP and SNMP objects, defined in a way very similar to Liaison's external binding, are depicted below:

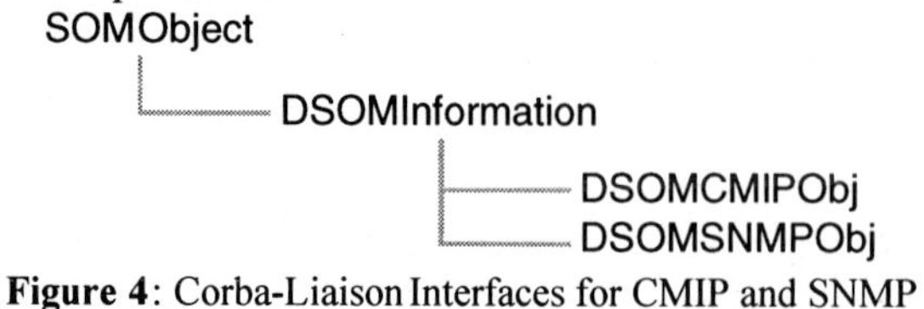

Figure 4: Corba-Liaison Interfaces for CMIP and SNMP

The interface `DSOMInformation` contains the information relative to the request and to the response(s). Internally the values are stores in a hash table where the attributeId constitutes the key and the attributeValue the value of each table entry. The use of hash table associated with the mapping of values into strings allows objects independently from their class and complexity to be handled. In the case of CMIP, the presentation layer or a thin layer on top of the stack, converts attribute values into strings and vice–versa, whereas in the case of SNMP it is Liaison that handles the conversion. Because `DSOMInformation` is built upon a hashtable, it is possible both to retrieve and store elements efficiently and to have only a few methods that handle all situations. In order to further simplify attribute manipulation, the classes `DSOMSNMPObj` and `DSOMCMIPObj` have been defined. These classes simplify the access to `DSOMInformation` by defining macros, for example

```
DSOMCMIPObj::GetObjectInstance()
```

mapped into calls to `DSOMInformation` methods to give

```
(DSOMInformation::GetAttribute("objectInstance")).
```

C++ methods of CL are almost identical to the ones defined in the corresponding class part of the external bindings hence the stub implementation is very straightforward. The similarity between these interfaces and the corresponding classes part of the external bindings has the advantage that developers can use both DSOM and the external bindings, and thus need only learn one object model. Additionally, code can be written once and them slightly modified to use either the C++/Java external bindings or the C++/Java language bindings of the DSOM interfaces. This is because methods and classes have the same names and parameters. Basically the only code that has to be added is related to a) the DSOM initialisation/termination, b) the `Environment` parameter needed in each DSOM method call, and c) exception handling that cannot catch all the DSOM exceptions using the try/catch mechanism since DSOM may use the Environment parameter to report about error conditions. The design choice to implement the DSOM stubs using the external bindings instead of wrapping the whole Liaison into a DSOM object has the following advantages:

- since the external bindings are quite light, the DSOM interfaces implementation is very light (less than 70Kb);
- DSOM has to be installed only by users that need to access Liaison using DSOM, i.e. applications based on external bindings do not need to have DSOM installed in order to run;
- DSOM allows the creation of objects on hosts where Liaison is not installed exploiting a remote Liaison, without the need to have DSOM installed on the host

where such Liaison runs (the communication DSOM server/Liaison is HTTP based);

- depending on the situation, users can decide to access services provided by Liaison using HTTP, DSOM or both (if Liaison had been wrapped in a CORBA object then users would need DSOM to access Liaison);
- it is possible to manage hosts outside firewalls using a local Liaison and DSOM interfaces since they are based on HTTP (DSOM cannot cross firewalls, HTTP can).

The drawback of this solution is that every time a management operation needs to be issued, there is a communication between the DSOM client, the DSOM server and Liaison instead of having Liaison contained inside the DSOM server. In the tests we have performed, the slowdown of the proposed solution is not more than 10-20% with respect to a full integration of Liaison inside a DSOM object. Considering the many advantages of this solution with respect to a total DSOM integration, we believe that this overhead is acceptable and almost negligible if client applications can perform multiple operations concurrently (multithread) without active wait.

4 Static vs. Dynamic Mapping

The work done on using CORBA for network management as described above can be broadly divided into two categories: approaches which statically translate GDMO/ASN.1 to generate code which is included by management applications that therefore know at compile time the extent of classes that they can handle, and dynamic approaches without dependency on compile time knowledge because they are either string- or metadata-based. Table 1 lists some of the major differences between the various approaches:

	CORBA-Liaison	GOM [2]	XoJIDM [15][10]
Mapping Type	Dynamic	Semi-dynamic	Static
Typing	Untyped	Runtime-type checked	Strong
Type Checking	Runtime (by Liaison and the OSI Stack)	Runtime (using metadata)	Compile Time
Implementation size	Small (<70Bk)	Medium (it needs MID)	Large
ASN.1/CORBA type mapping	All datatypes are mapped to a string	Datatypes are mapped to GOM types (15)	Every datatype is mapped to one (or more) CORBA type
#CMIP/SNMP vs. #Corba Classes	N:1	N:15	N:M (N <= M)
CMIP Support	Yes	Partial (CMIP M-EVENT-REPORT and M-ACTION are not supported)	Yes
SNMP Support	Yes	Not Yet	Yes

Table 1: Approaches compared

The CORBA Liaison (CL) interface approach is completely untyped since all types are mapped to strings. Conversion between strings and the desired data type of the

host language (e.g. C++) has to be done by the programmer. This may be easy for simple types such as strings or numbers, but increases complexity for the programmer considerably for aggregate types such as struct or sequence. Also, the probability of errors being introduced in user-written conversion functions is increased. This trade-off, however, was accepted by CL because its main goal was the creation of a light-weight model for network management that should be flexible (no compiled-in knowledge) and that should be integrated with the World Wide Web which uses strings as the major data type anyway. Also, network management is nowadays still predominantly based on SNMP which uses mainly atomic data types such as strings or integers. The programmer specifies the types in a string-based syntax which is checked at runtime by Liaison in the case of SNMP, or by the OSI stack in the case of CMIP. Compared to the static approaches with their strong typing enforced at compile time, GOM enforces typing at runtime using metadata. Contrary to CL, which knows only the string type, it has types for representation of classes (`GenObj`), attributes (`Attribute`) and values (`Val, Integer, String, Struct, Sequence` etc.). Whereas CL maps all types to strings, GOM maps them to an instance of this set of fixed types, and the static approach maps each type to a corresponding IDL type.
Whereas the static approaches fully integrate the translated code into the target type system using the target's native types (e.g. C++), GOM offers an abstraction of the target's type system (ca. 15 types) as API to the user whereas the API of CL is the single type string. In the case of the static approaches, the client of the API may mix types of the target system and the generated code since they are the same whereas using GOM, native (C++) types have to be converted to/from GOM types (e.g. `int` to instance of `Integer`), which is easy since all predefined GOM types offer conversion operators to/from C++. Using CL, it is the responsibility of the client to convert the string types to/from the native type system (C++, Java etc.). The dynamic approaches have two major advantages over the static ones: they typically produce smaller client applications and they are much more flexible. Since clients do not possess a-priori knowledge about classes available in the system, but rather use strings or metadata, they are independent of class modifications and can continue working while clients using the static approach may need to be recompiled. This is an essential asset in areas such as topology browsers and roaming agents that do not know all the classes they will encounter when compiled. Including at compile time a fixed set of classes may yield potentially large client applications that have to pay (in terms of size) for all the classes they carry with them even if only a few are actually used. In the dynamic approaches, when a client needs to handle a new/modified class, either the latter's metadata is dynamically loaded (GOM) or it need not be loaded since a string type represents all types (CL). Areas where metadata is useful or needed are X.700's conditional packages, attributes and the ANY/ANY DEFINED BY ASN.1 types which can only be resolved to a correct type at runtime.

5 Conclusion

This article shows that management of OSI and SNMP network resources through CORBA is possible and feasible. The two main directions of research currently done on using CORBA with network management have been analysed and compared. Besides known techniques, this paper described a novel technique named CORBA-Liaison. Relevant characteristics of this technique are: efficient, full CMIP/SNMP protocol support, light and simple object model independent from a particular CORBA implementation.

The goal of this paper was not to demonstrate that one approach is better than another but to understand the benefits and limitations and then to identify the solution that better fits user requirements. Finally this paper has shown that CORBA-based SNMP/CMIP management is now becoming a mature technique which overcomes most of the limitations of early solutions.

6 References

[1] ISO/IEC, CCITT, Specification of Abstract Syntax Notation One (ASN.1), ISO/IEC 8824, CCITT Recommendation X.208, 1988.

[2] B. Ban, Towards a Generic Object-Oriented Model for Multi-Domain Management, Proceedings of ECOOP '96 Workshop on Systems and Network Management, Linz, Austria, July 1996

[3] ISO/IEC, CCITT, Information Technology-OSI, Common Management Information Protocol (CMIP)-Part 1: Specification ISO/IEC 9596-1, CCITT Recommendation X.711, 1991.

[4] L. Deri, Droplets: Breaking Monolithic Applications Apart, IBM Research Report RZ 2799, September 1995.

[5] L. Deri, Surfin' Network Management Resources Across the Web, Proc. of 2nd IEEE Workshop on Systems and Network Management, Toronto, June 1996.

[6] L. Deri, Network Management for the 90s, Proceedings of ECOOP '96 Workshop on Systems and Network Management, Linz, Austria, July 1996.

[7] IBM Corporation, DSOM Development Toolkit, October 1994.

[8] J. Hierro, Architectural Issues For Using CORBA Technology in OSI Systems Management, Append of draft to XoJIDM forum, August 1994.

[10] Joint Inter-Domain Working Group, X/Open and Network Management Forum, Inter-Domain Management Specifications: Preliminary CORBA/CMISE Interaction Translation Architecture, April 1995.

[11] P. Maes, Concepts and Experiments in Computational Reflection, Proceedings of the 2nd OOPSLA Conference, 1987, pp. 147-155.

[12] T. Magedanz and T. Eckardt, Mobile Software Agents: A New Paradigm for Telecommunications Management, Proc. of 2nd Intl. IEEE Workshop on Systems Management. Toronto, Ontario, 1996.

[13] Object Management Group, The Common Object Request Broker: Architecture and Specification, Revision 2.0, July 1995.

[14] J. Case, M. Fedor, M. Schoffstall and C. Davin, The Simple Network Management Protocol (SNMP), RFC 1157, May 1990.

[15] Joint Inter-Domain Working Group, X/Open and NMF, Inter-Domain Management Specifications: Specification Translation, April 1995.

[16] ISO/IEC, CCITT, Information Technology - OSI - Management Information Services - Structure of Management Information - Part 4: Guidelines for the Definition of Managed Objects, CCITT Recommendation X.722, ISO/IEC 10165-4, 1992.

[17] Object Management Group, CORBA-based Telecommunication Network Management System, OMG White Paper, Draft 2, Telecom Special Interest Group, January 1996.

SNMP and TMN: Aspects of Migration and Integration

H. Dassow and G. Lehr
Deutsche Telekom AG · P.O. Box: 10 00 03 · D-64276 Darmstadt
dassow@tzd.telekom.de, lehr@tzd.telekom.de

1 Introduction

For the purpose of network management two different management protocols exist which have evolved from the management of different types of networks. For the management of small private Local Area Networks the Simple Network Management Protocol (SNMP) [1] is developed. Regarding the management of large public networks, a special architecture, called Telecommunication Management Network (TMN) has been standardised, which uses the Common Management Information Protocol (CMIP) [2] for the transmission of management data.
The demand for integrating these different kinds of networks on the level of network management is strongly increasing at present in the context of Customer Network Management (CNM). Due to the importance of this problem, extensive research is being carried out, but there are still open questions. The difficulties to be tackled result from the different origin of the two protocols which arise from totally different management philosophies. These different management philosophies lead to a different structure and meaning of the information model which in turn has a strong impact on the implementation philosophy.
In addition to the problem of integrating different management protocols, the difference in management philosophy also imposes severe restrictions on the choice of an adequate management scheme. These restrictions are mainly related to the fact that the polling approach used in SNMP causes management overhead which is much larger in volume as compared to CMIP. At present, several approaches to minimise the management overhead in SNMP are being considered.
The approach described by Siegel and Trausmuth [3] is based on the concept of an intelligent agent. In this approach the management program is downloaded during runtime into a local manager, which performs a local polling algorithm. One disadvantage of this solution becomes visible as soon as the network has to be reconfigured. In this case a modified program has to be downloaded to the manager. In addition, an overall view of the network is missing, because all processes are performed locally in the managers. Furthermore, the non-standardized functionality of these local managers represents a serious drawback.
Another approach might be the use of the Common Object Request Broker Architecture (CORBA) instead of SNMP for management purposes, as suggested by the Object Management Group [4]. CORBA has been developed to provide a distributed platform for all kind of applications. However performance aspects, such as arise when managing a large number of network elements, have not been considered yet when using CORBA.
It is obvious that the rebuilding of large SNMP managed networks according to the TMN architecture is another solution to avoid the SNMP management overhead.
This article addresses both aspects, the integration of SNMP- and CMIP-based management environments as well as the migration from SNMP to CMIP. It is organised as follows: In section 2, rules for the mapping between MIB (SNMP) and information model (CMIP) are defined. Based on this mapping, principles for the implementation of a proxy are derived. Section 3 discusses runtime aspects related to

the proxy implementation. Sections 4 and 5 describe, how the different implementations fit into a hierarchical management architecture.

2 Mapping rules for the definition of the information model

The syntax used for the definition of a Management Information Base (MIB) in SNMP is termed Structure of Management Information (SMI) [5]. Additional changes to this syntax are made for version 2 of the SNMP [6]. In CMIP, however, a SNMP MIB is called an information model and is defined according to the Guidelines for the Definition of Managed Objects (GDMO) [7]. Both definition languages use the Abstract Syntax Notation (ASN.1) to define basic structures.
For the translation of a MIB to an information model two alternative approaches were identified: the syntactic and the semantic approach. Due to their totally different functionality these approaches are explained separately in the subsequent subsections.

2.1 Syntactic translation

The idea of syntactic translation is to redefine an existing information model using a different definition language. In this context, the (re)definition of an SMI based MIB into a GDMO information model is discussed. Due to the fact that the SMI language for the definition of an MIB contains more restrictions and limitations than GDMO, a mapping from SMI to GDMO represents no insuperable obstacle.

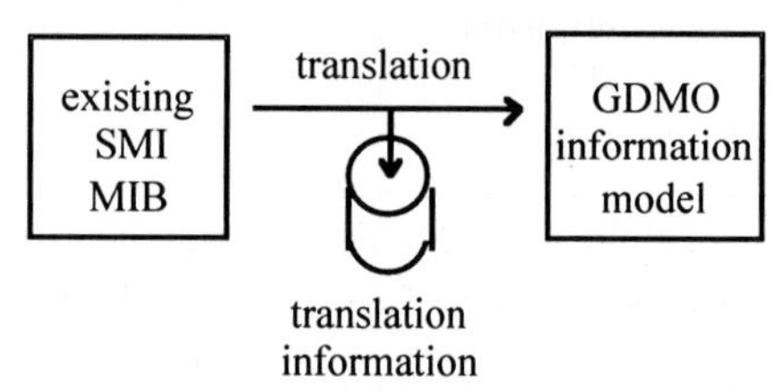

Figure 1: Basic principle of a syntactic translation

The basic principle of the syntactic MIB translation is shown in Figure 1. An existing SNMP MIB has to be translated into GDMO templates. Information about the translation process should be stored in a local database to enable a proxy realisation which is presented in section 3. This information can be, for example, the mapping between registered SMI and GDMO object identifiers.
An algorithm for the syntactic translation from SMI to GDMO has already been defined by the Network Management Forum (NMF) [8]. This approach offers a good starting point, but further modifications and extensions are required. Furthermore, the latest version of the draft SNMPv2 standard is not taken into account in the NMF approach. Hence a verification of the corresponding mapping rules is required to guarantee a uniform approach for both versions of the SMI. In this section the NMF algorithm will be illustrated and proposed changes and enhancements will be explained.
Figure 2 illustrates the basic mapping rules for the translation of an MIB into an information model. Obviously an ASN.1 type of the SMI definition corresponds to an ASN.1 type in the GDMO definition. As an SNMP object contains only a simple value without internal structure, it is mapped to a GDMO attribute. The GDMO propertylist specifying access mechanisms of an attribute corresponds to the access type of the SMI object definition. The behaviour of the attribute definition is equal to the description part of the SMI object macro.

SMI	NMF approach	proposed enhancements
ASN.1 type	ASN.1 type	ASN.1 type
object	attribute	attribute
access type	property list	property list
description	behaviour	behaviour
object status not mandatory	-	CHOICE { nonExistence NULL, value <type> }
object group	managed object class	package
SNMP MIB SNMPv2 module compliance	-	managed object class
SNMPv2 module identity	-	informal comment
table	name-binding	name-binding
table entry type	managed object class	managed object class
InstancePointer	InstancePointer	distinguished name
trap SNMPv2 notification	generic notification	notification or ignored (see runtime aspects)
SNMPv2 notification group	-	package

Figure 2: Basic mapping rules for the MIB translation

So far, the above mentioned algorithms can be adopted without changes from the NMF approach. In the following, additional translation rules and modifications of those given in the NMF approach are discussed.

To handle the issue of non mandatory SMI objects (where GDMO does not provide a similar construct), a mapping to ASN.1 Choice type is recommended. Furthermore, SMI prescribes the arrangement of objects in object groups. Each object group represents a characteristic part of the agent functionality. In GDMO, the main mechanisms for grouping attributes are managed object classes and packages. We propose to map each group of SMI objects into a GDMO package which contains one attribute for each SMI object. In contrast to the NMF approach, all mandatory and conditional packages of an MIB are therefore contained in one single GDMO object. A positive side effect of our proposal is the mapping of an SNMPv2 module compliance macro to a managed object class template, because in both cases the same contents are described.

In Version 2 of SNMP a macro for specifying informal text related to the MIB is defined. The mapping of this module identity macro to a GDMO template is impossible, because GDMO does not support this additional information element. Therefore we suggest, that they are translated into an informal comment.

The only mechanism in SMI for the definition of complex data types is the SNMP table, realised internally by an ASN.1 SEQUENCE OF type. Just keeping this ASN.1 type in GDMO might be one solution for translation. This, however, causes problems with the mapping of different SMI properties distributed over the objects in the table. As the naming tree in GDMO is a proper mechanism for structuring information hierarchically an object instance for each row of a table has to be created. This is why the translation of SMI objects, which are part of a table, is a special case. Each SMI table has to be mapped to an individual GDMO class containing attributes related to the SMI objects in the table entry. In SMI, an object value is addressed with an instance pointer. GDMO does not support a corresponding addressing between attributes. However, the instance pointer is only useful in SNMP in the context of

table entries. Therefore it is proposed to use the full distinguished name in GDMO which provides an equivalent functionality.
Both protocols support the ability to emit unsolicited messages. The message emitted by an SNMP agent is called a trap (or notification in version 2). In CMIP the GDMO object can emit a notification. We see problems in mapping all SNMP traps to CMIP event reports. Due to the different semantics of an SNMP trap and a GDMO notification problems concerning the consistency of the respective databases arise. As a CMIP-based management system is designed on the assumption of the existence of an underlying flow control, the database is considered to be up-to-date unless a notification reports a significant event. So, a single bit of information emitted by the SNMP agent, could possibly be lost in the network and will lead to inconsistencies between manager and agent. This important fact has to be taken into account for the translation of information models. For this reason it is proposed to avoid relying on traps in SNMP as far as possible. An alternative approach based on CMIP notifications solicited by a local polling process will be developed in section 4 dealing with runtime aspects.
Additionally in the purely syntactic translation SNMP specific behaviour needs to be defined in the GDMO template. In contrast to the NMF approach, where all object classes are inherited from the managed object class TOP, we propose to define a common generic parent TRANSLATEDSNMPMIB. This ancestor should model the common properties of an SNMP MIB. All translated SNMP MIBs are inherited from this class. The TRANSLATEDSNMPMIB class contains SNMP specific attributes, such as the time of the last successful SNMP operation (a rejected operation is successful in this context, because it was successfully transmitted). It also contains communications parameters, such as the IP-address of the agent, a timer threshold for an outstanding operation response or the maximum number of connections attempts.
In a similar way a generic class TRANSLATEDSNMPTABLE can described which defines an additional GDMO behaviour for an SMI table. This class is inherited directly from TOP. Due to a different functionality, most elements of the class TRANSLATEDSNMPMIB are not required in this ancestor.
Also both generic classes contain some generic notifications, like the attribute-value-change-notification. The detailed functionality of these notifications are described in detail in section 3.

2.2 Semantic translation

As opposed to the syntactic approach where a new GDMO information model is created, the focus of the semantic approach is to *recover already existing* GDMO definitions. Abeck [9] has proposed a mapping technique, which allows a new, CMIP-based model to be translated from an SNMP MIB and derived from a generic CMIP information model, covering the functionality of both models. In this approach it is very easy to ensure the adaptation to the TMN philosophy. Building a *new* GDMO-based information model, however, means that an *already existing* GDMO application has to be adapted to this model.
In order to overcome this problem, an alternative approach is proposed which is illustrated in Figure 3. This scenario of a semantic translation describes the definition of mapping rules between an MIB and an *existing functionally identical information model*. This scheme has to be applied, for example, when two information models describing the same functionality have to be integrated in a heterogeneous management environment. The general idea is to identify information elements which are common to both information models. Based on this common information, a

complete set of rules mapping the SMI version to the GDMO version has to be defined.

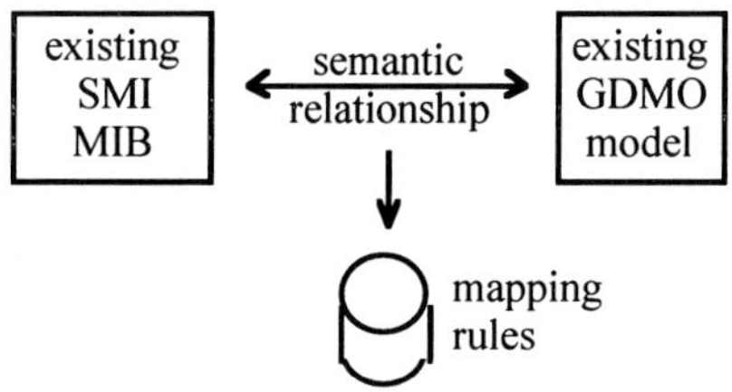

Figure 3: Basic principle of a semantic translation between two existing definitions

One very simple example which can already give an idea of the complexity of the problem is the observation of an error rate in a transmission line termination. The SMI management philosophy relies on polling rather than on evaluating notifications emitted by the agents. Due to this fact, SMI uses counter objects for the description of statistical events rather than statistical rates. The functionality of counting errors detected by the agent, for example, is described in an SNMP environment by an ERRORCOUNTER object. The value of this object has to be polled periodically by a manager application. The same management problem is solved within the CMIP approach in a totally different way. A TRANSMISSION object contains an attribute called ERROR-RATE. This attribute counts the errors per time period unit in the line termination. The value of an ERRORRATETHRESHOLD attribute, which is also included, can be adjusted by the manager application. As soon as the current error rate exceeds this threshold, a notification is emitted. The mapping rule for this small example is given by the mathematical derivation and interpolation of the error-rate from the number of counted events.

The semantic translation starts from the precondition that both the MIB and information model are not changed by the mapping. The task is then to find out a set of rules for the mapping that complies with this condition. These rules are determined in an iterative process where the result of the mapping and the inverse mapping is compared with the target MIBs, respectively. The main difficulty in this approach is to check the set of rules with respect to integrity and completeness. This problem can be reduced significantly by using the Managed Objects Conformance Statements (MOCS) as a means of structuring the mapping rules. To do so, the MOCS table is extended by one column containing the rules for the mapping of SNMP objects to the target CMIP MIB for each entry of the table. The integrity of the translation can be checked by comparing the SNMP object model with the inverse set of rules.

3 Runtime aspects of protocol mapping

The real-time access of an SNMP-based agent by a CMIP-based management system is realised by an entity which is called proxy in an SNMP environment and Q-adaptor within the TMN architecture.

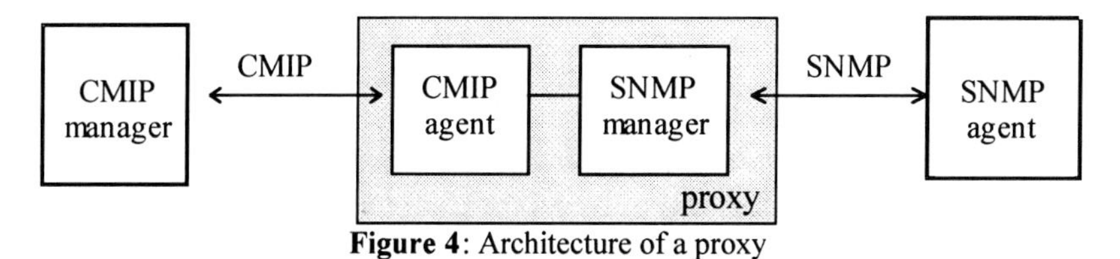

Figure 4: Architecture of a proxy

Figure 4 illustrates the basic principle of a proxy. A CMIP-manager seems to communicate with a CMIP agent whereas the SNMP agent seems to be managed by an SNMP manager. The adaptation between SNMP and CMIP is done internally in the proxy by applying the MIB translation rules. Corresponding to the two approaches for MIB translation, it is reasonable to distinguish between a syntactic and a semantic proxy. The properties of these two types of proxy will be discussed in the next two subsections, respectively.

3.1 Syntactic proxy

The syntactic proxy is based on a syntactic MIB translation. Figure 5 illustrates its architecture. A syntactical mapping of the management information between the CMIP agent and the SNMP manager part of the proxy is required. Receiving a CMIP operation it is translated into an appropriate SNMP function. Afterwards the information of a received SNMP response has to be mapped again to a CMIP which in turn has to be sent back to the CMIP manager. For this mapping the proxy needs information from the syntactic MIB translation. Therefore this mapping information is stored in a local database. This basic principle has already been explained in the NMF Document [11]. However, adaptations taking the modified mapping rules are required.

For a simple get or set request the contained management information has to be translated into the corresponding SNMP format. The proxy delivers this translated operation to the corresponding SNMP agents and waits a specific time for the response. If no answer is received, the proxy has to repeat the transmission several times. The exact parameters for the polling mechanism, such as the polling time interval, are stored locally and can be managed via the attributes inherited from the TRANSLATEDSNMPMIB class. Afterwards, the result of the operation has to be sent back to the CMIP manager. Therefore, the received SMI information has to be translated, in turn, into the GDMO format.

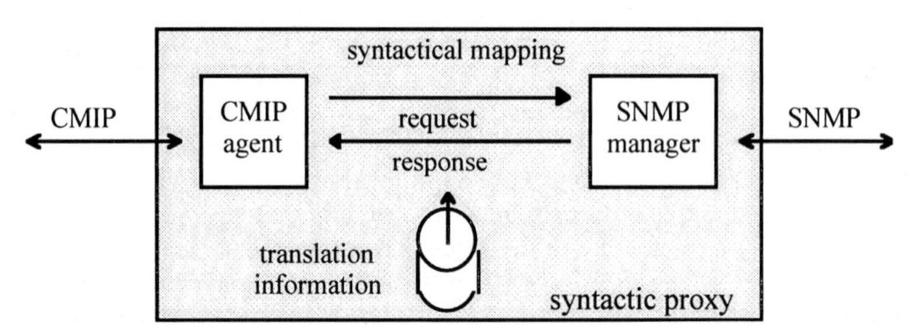

Figure 5: Architecture of a syntactic proxy

As an SMI table is mapped to a GDMO name binding (cf. Fig. 2), a CMIP create or delete request can only be performed when a corresponding table exists. In this case, the CMIP request will be translated into an SNMP set operation, which executes the corresponding table function.

In order to create state or attribute changed notifications, a value observation in the proxy is required. In the NMF approach, the proxy does not realise any polling mechanism. Therefore the CMIP manager has to poll for important values via the proxy. This functionality is very unusual for the CMIP and increases the transport overhead with get responses containing no information. A very simple but effective solution is to define an individual polling time for each SNMP object. The proxy then

automatically observes the value of each SNMP object and automatically submits an ATTRIBUTE-VALUE-CHANGE notification to the manager if necessary.
The implementation of the proxy could be specific or generic. In the specific approach, the management information is hardcoded with the proxy implementation. Since the set of rules for this kind of MIB translation is very rigid and most of the specific information is contained in the translation information, we propose to specify a generic implementation of the proxy. This means that one universal CMIP/SNMP proxy is used and the information of the specific information model is downloaded at runtime as "translation information".

3.2 Semantic proxy

The basis for a semantic proxy is the set of rules for semantic MIB translation. In Figure 6 the proxy is described using SNMP and CMIP terminology. In terms of an CMIP environment the combination of a semantic proxy with an SNMP agent is very similar to a managed open system. Furthermore, the proxy represents the function of the local system environment. Contained in the proxy is the CMIP agent which performs operations on the managed objects. From the viewpoint of the SNMP agent, however, these managed objects behave as independent SNMP managers. In CMIP terms this relationship between managed objects and managed resource is described by the managed object boundary. This principle of a semantic proxy is realised as follows.

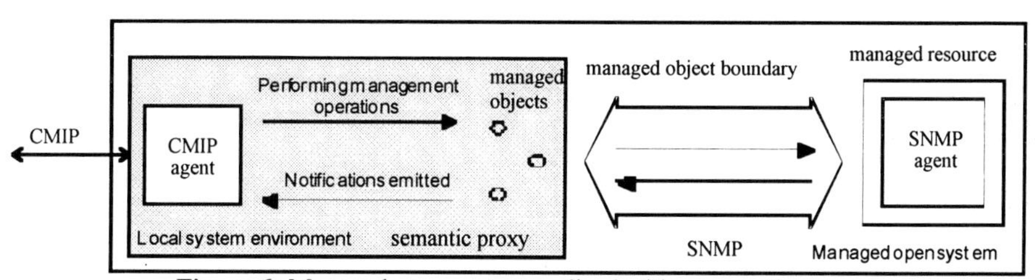

Figure 6: Managed open system realised with a semantic proxy

The proxy analyses a received CMIP request and performs the corresponding operation on the managed object. The result of this operation will be sent as a response backwards to the CMIP application in the manager role. The only difference between a CMIP agent and a semantic proxy is related to the distributed implementation of the managed object boundary. In case of a semantic proxy the mapping across the managed object boundary is realised by SNMP whereas for a CMIP agent, this mapping would be realised internally in the network element. In the case of a CMIP agent the managed object boundary could be realised for example by function calls, shared memory, or some proprietary interfaces optimised for this purpose.
Additional measures have to be taken in order to overcome functional deficiencies of SNMP. The main problem is the synchronisation between several SNMP operations. Due to this fact, a transaction oriented mechanism is required. This mechanism can be realised without enhancement of SNMP. For this purpose, all transmitted SNMP operations for a transaction are simultaneously stored in a local data base. In case an operation fails, related operations which have already been performed can be revoked therefore step by step.
One important aspect which must be recognised before implementing a semantic proxy is the complexity of this solution. In contrast to the case of the syntactic proxy,

it is impossible to design a completely generic solution for the semantic proxy. As in case of the semantic MIB translation, a great extent of functionality has to be implemented individually for each attribute. A single leak in the translation may render the complete proxy useless. Therefore it is important to analyse first whether a semantic proxy is really required and whether the management interface of the SNMP agent is really functionally identical to the target GDMO information model.

4 Migration of SNMP into a layered architecture

If a local area network grows very large, an SNMP approach can reach its limit in performance due to an increase in polling. One solution for this problem is to use CMIP instead. In the case of SNMP, the bandwidth required for management increases approximately linearly with the number of network elements. Due to the decentralised processing of management operations in a CMIP environment, this increase is sublinear when the number of network elements is large. Hence it may be advisable to migrate certain network elements to use CMIP. Therefore a solution based on an syntactic proxy is suggested.

In the architecture for this approach, a generic proxy is spread over the complete network. Thus it is possible to manage network elements with an SMNP interface in a centralised way with a CMIP based application. This application communicates with syntactic proxies which are located near by the SNMP agent. The advantage of this solution is that SNMP is only required in a small local area network between the proxy and the agent. Using the generic approach for proxy implementation (see section 3) the cost of implementation may be reduced significantly for this solution.

So far, one could be tempted to assume that the syntactic proxy is an universal solution in a large network. An MIB, however, which is derived by syntactical translation is not identical to an MIB which was designed in an object oriented manner. That is why a syntactic proxy is only a functional compromise between the use of cheap SNMP managed network elements and an efficient management solution.

5 Integration of SNMP and CMIP

At the level of data transmission, SNMP and CMIP managed networks have grown together in recent years and today interconnection represents no serious problem. In the context of a service oriented network the demand for a global management solution for these networks is strongly increasing at present. Therefore, the proxy techniques described above can be applied in different ways to the integration of SNMP into a TMN. These ways differ in the localisation of the mediation functions.

One way is based on a semantic proxy which is located near to the agent. In this approach the management efficiency underlies no restrictions, because a CMIP based application in the manager role can not identify any difference between an SNMP or a CMIP resource. The realisation of a semantic proxy, however, is more complicated than the re-engineering of a resource for the integration of a CMIP interface. Therefore, the semantic proxy is only useful, if a large number of legacy SNMP agents has to be integrated into a TMN.

The most common way for the integration of SNMP agents is to enhance the TMN management application by an SNMP interface. In this case, the management functions for a specific SNMP MIB of the network element are integrated into the TMN application. SNMP is still used for communication with the agent. Therefore this solution inherits all disadvantages of SNMP discussed above and can not be recommend for large networks.

An enhancement of the previous approach is to combine the specific merits of syntactic and semantic proxies. In this enhancement the syntactic proxy is located near to the SNMP agent in order to avoid the necessity of polling. Furthermore, the syntactic proxy performs scoping and filtering functions. The semantic proxy is located in a centralised way and performs a mediation function between the translated MIB in the syntactic proxy and the information model used in the management application.

6 Conclusions

In this paper we have considered several approaches for the integration of SNMP into an TMN using proxies. The technical requirements as well as the advantages and disadvantages of different approaches have been discussed in some detail.
In order to take the current Draft SNMPv2 Standard into account it is proposed to enhance the syntactic MIB translation described by NMF. In addition to that, semantic MIB translation is investigated thoroughly in this contribution. A new approach is described which allows a complete semantic translation of MIBs into GDMO information models without the need of changing either definitions. All translation intelligence is concentrated in the proxy in this approach. This allows the integration of SNMP agent into a TMN environment without changing existing management applications.

References

[1] "Simple Network Management Protocol", Case, Fedor, Schoffstall, & Davin, IAB Standard Protocol, RFC1157, May 1990.
[2] "Common management information protocol specification for CCITT applications", ITU-T, Recommendation X.711, Geneva 1991.
[3] "Hierarchical network management: a concept and its prototype in SNMPv2", M. R. Siegel and G. Trausmuth, Computer Networks and ISDN Systems 28 (1996).
[4] "CORBA-Based Telecommunication network management system", Object Management Group, Framingham Corporate Center, Framingham, Draft 2, January 1996
[5] "Structure of Management Information", Rose & McCloghrie, IAB Standard Protocol, RFC1155, May 1990.
[6] "SMI for SNMPv2", SNMPv2 Working Group ,IAB Draft Standard Protocol, RFC1902, January 1996.
[7] "Structure of management information: Guidelines for the definition of management objects", ITU-T, Recommendation X.722, Geneva 1992.
[8] "Translation of Internet MIBs to ISO/CCITT CMIP MIBs", Network Management Forum, Forum 026, Issue 1.0, October 1993.
[9] "Simply Open Network Management", an approach for the integration of the SNMP into OSI management concepts, Integrated Network Management, III submitted to the IFIP TC6/WG6.6 3rd International Symposium on Integrated Network Management, San Francisco, California, USA, 18-23 April, 1993.
[10] "Integrated Network and System Management", H.-G. Hegering and S. Abeck, Addison-Wesley 1994.
[11] "ISO/CCITT to internet management proxy", Network Management Forum, Forum 028, Issue 1.0, October 1993.

Managed Objects as Active Objects: A Multithreaded Approach

Rivalino Matias Júnior (rivalino@inf.ufsc.br)
Elizabeth S. Specialski (beth@inf.ufsc.br)
Curso de Pós-Graduação em Ciência da Computação
Universidade Federal de Santa Catarina
P.O.Box: 476 - Campus Universitário - Zip Code: 88040-900
Florianópolis-SC/Brazil
Phone: +55 (48) 231-9738 Fax: +55 (48) 231-9770

The computer network management requirements have been demanding that management platforms must be built according to the present technological advances, and also according to the large capability and variety of the available elements in these environments. In this paper, a multithreaded infrastructure for the managed objects implementation as active objects is proposed. The environment of this work is a network management platform project, which has been implemented following the OSI/ISO management model [1].
Keywords: Network Management, High Performance Agents Implementation, Active Objects

1 Introduction

The objectives of network management are immutable, however the process by which these objectives are reached has been changing in such a way that the current technological advances have been utilized. Since new technologies and services have been introduced, network management systems must be ready to attend such innovations, executing monitoring in real time and providing support to execute automatic corrective actions based on events and in current network conditions [2].
Automatic control and decision making, architectures for high performance, graphic resources utilization for the organization and visualization of many network elements, are some examples of new technologies which has been applied to the network management systems in order to make them efficient and capable of dealing with current management requirements.
Nowadays, due to the improvement in operating systems, programming languages and support environments to the applications development, multithreaded application implementation is becoming usual in many areas such as distributed systems, environments for parallel programming, operating systems, WWW browsers etc.
This paper presents a multithreaded infrastructure for managed objects implementation in agents of a network management platform, which is based on the OSI/ISO management model. The proposed infrastructure will allow managed objects to be implemented as active objects, providing the agent processes of the platform (under development) characteristics which allow higher performance through the execution of its managed objects and intra-agent activities.

2 OSI Network Management Agents

The standardization documents in the network management area, do not define the internal organization of management processes. It is a local matter of each implementation. Conceptually, management application processes are the applications which use the services provided by the system management application element (SMAE). This concept, defined in [3], can be illustrated in Figure 1.
Agent processes are part of the systems management which store and manipulate management information about a certain managed system. Hence, it can be asserted that agents are *management information servers*.

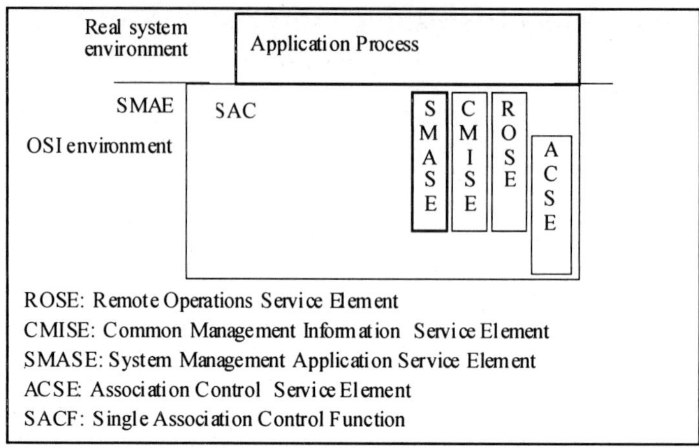

Figure 1: Management and application layer [3]

Thus, agents must monitor and control the system resources. Monitoring means the dynamic collection of information related to the objects (software or hardware) which are under observation [4]. Controlling can be taken as the execution of the operations issued by the manager over the managed objects.

In order to satisfy these requirements, agent processes must be structured in a way to minimize the answer time to the manager and always to keep its management information base updated and consistent. The internal organization of the agent processes and its implementation features are particular to each platform, since these questions are not defined in the standardization documents.

In this work, agent applications are defined and implemented as *dæmons*[5] processes of the UNIX operating system. As the MIB objects, the ACSE[6], ROSE[6], CMISE[7] service elements implementation and the LPP [6] protocol implementation are part of the agent code. The Figure 2 shows the structure of agent process in the platform.

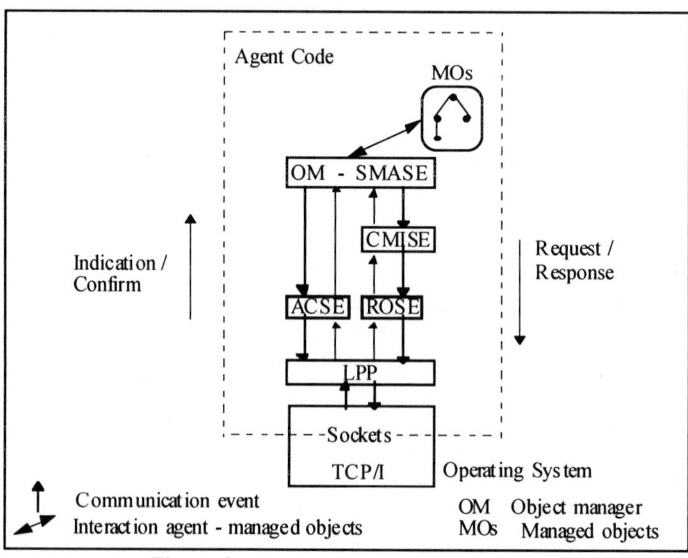

Figure 2: Agent structure in the platform

3 Agents and Managed Objects

When the agent application is loaded in to memory, it has in its code all the functions needed for its execution. Each agent process has definitions of the managed object classes, which will allow their instantiation at runtime. Managed objects are basically represented as instances of classes in C++[1] language.
Each agent managing its part of MIB, defines an internal structure for the MIT[8] representation. This structure will allow the objects representations to follow the *containment* hierarchy concept. All managed object classes are organized according to the inheritance hierarchy. Each managed object class has a method called "behaviour()" which implements the object conduct. This method will be more detailed in the section 4.3.
The MIT implementation is not part of the managed object class implementation. Its implementation is independent of object class which gives greater flexibility for its maintenance (creation, consult, deletion, etc.). A support object called OM (Object Manager) was defined to allow MIT management and maintenance of instance sets of managed objects supported by agent. A more detailed description of OM's behaviour will be presented in section 4.3.

4 The Multithreaded Infrastructure

Managed objects are a part of the agent processes code. Therefore, agent processes must support the basic requirements for the execution and maintenance of such objects. The multithreaded infrastructure which will support these objects, considers two fundamental aspects:

- Aspects of Management operation treatment and notification emission;
- Managed objects dynamic (behaviour).

Related to the management operation treatment and notification dissemination a policy was defined for the serialization of such operations. The operations treatment requested by the manager is done in a serialized way, according to the arrival order of these operations.
Since the multi-threaded infrastructure is inserted in the structure of the Figure 2, the management operation serialization occurs when PDUs come from the ACSE/CMISE. At this point, the OM input structure organization is in charge of grouping the operation according to a FIFO policy. The OM support object provides maintenance in MIT, distributes operations (which come from the manager) to its respective managed objects, makes the support operations to the multi-threading, and implements the access control to the MIB objects.
Concerning the treatment of external invocations of operations, some forms of organization of agent (server) processes were studied. One of the most utilized ones, which increases the performance of these processes, is similar to the RPC multi-threading server [9] implementation. In this approach, for each incoming operation a thread [10] is created to execute this operation. Such organization was *not* adopted, because the operations serialization is not guaranteed, and for those cases in which operations are made on the same object, the CMIP protocol requires the serialization of these operations. In case of non-synchronized operations on different objects, these operations can be executed without any serialization requirement.

[1] The language chosen for the platform implementation was C++, because of its support to object-oriented paradigm and its high portability in UNIX environments.

An example that nullifies the utilization of the adopted focus for the multi-threading RPC server, could be the execution of non-serialized operations on the same managed object. Suppose a manager requests a SET operation (*Replace-value*) on the attributes of several managed objects using scoping and filtering, and then issues a GET operation to recover the value of a modified attribute in one of the objects affected in the SET operation. The thread created for the SET operation could lose the processor before being finished. If the GET thread takes the processor it could recover the attribute value which has not been changed yet by the SET operation, creating an error situation.

According to the requirements of CMIP protocol serializations, it is considered that the management operations would be taken sequentially by the OM and by the managed objects. Such kinds of objects have operation queues following a FIFO policy. The Figure 3 shows the proposed infrastructure.

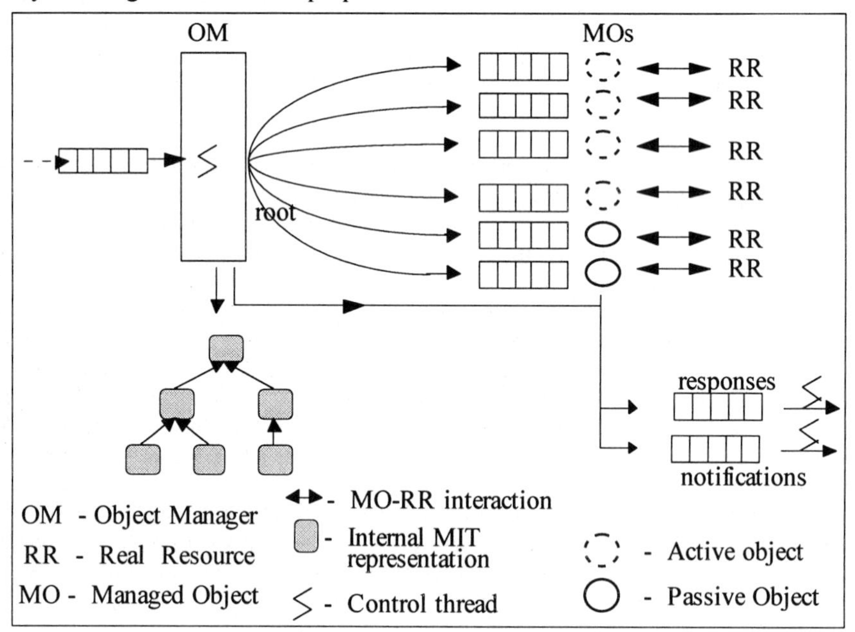

Figure 3: The Multithreaded infrastructure

After receiving messages in OM input structure, it is not required to invoke it, because it has its own control thread and it will be constantly consuming messages from its queue. When there are no messages to be consumed, the OM continues executing parts of its behaviour (priority changing between the threads, operations synchronization, etc.) related to the management of several managed objects.

By each object having its own control thread and operations queue, the execution is made into an independent and concurrent[2] form with other agent activities, increasing the agent performance.

Concerning the independence and concurrency between the managed objects, the infrastructure adopts the concept of active objects in the managed objects

[2] In multi-processed machines with shared memory, it can be seen as parallelism in the execution of several managed objects, by having each thread mapped to individual processors.

implementation. The Figure 4 shows the multi-threaded infrastructure localization in agent process code, according to the structure presented in Figure 2.

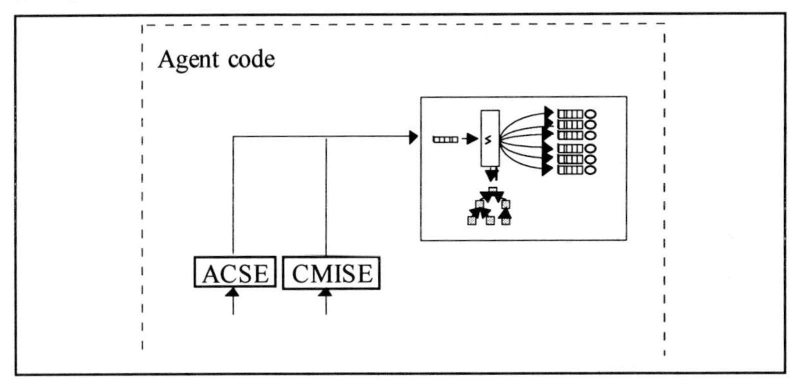

Figure 4: Multithreaded infrastructure localization in the agent process

4.1 Active Objects

In the object-oriented paradigm, the objects execution state can be basically classified as: Active or Passive [11]. According to Booch [12], concurrency in OOP is the quality distinguishing an active object from a non-active object (passive). When there is concurrency it is proper to visualize each object as an autonomous active entity which execute its behaviour independent of external invocations [13].

By using these concepts, it can be affirmed that active objects encapsulate their own control thread in order to have the autonomy to change their internal state without external invocations. Passive objects do not encapsulate their own control thread and just change their state by external invocation, which can be made by another active object or by the process control thread in which the passive object is located. After being invoked a passive object becomes active, makes some computation (its behaviour) and returns to the passive state.

Some object-oriented languages [14] provide support to active object implementation. However, the C++ language which has been used in the platform implementation, just implements the concept of passive objects. Due to this limitation, threads were used to allow the implementation of the active object concept using the C++ language.

4.2 Managed Objects as Active Objects

In order to address management requirements, the active objects where utilized to implement managed objects since the representation of real resources by active objects is more appropriate to the system resources abstraction (layer entity, transport connection, hubs, etc.).

The traditional focus (passive objects) limits the implementation of such abstractions, due to the lack of independent dynamics between the objects. Some matters concerning an object (e.g. failure in an interaction with an real resource) compromise other objects' execution, because the process control thread is dedicated to the object execution which failed causing the blockage of the entire agent process. To solve this problem, the platforms mainly restrain the objects′ implementation, because it is not allowed to block external events or even to execute a more complex computation. An example of such platforms is the OSIMIS [15] management platform, which defines that managed object implementations cannot block themselves in external communication points, causing the blocking of all agent processes. Restrictions like this limit the managed objects implementation related to interaction with system resources, especially loosely coupled real resources which demand a larger complexity in communication between the real resource and the managed object. If active objects were used in the previous example, the object that blocks itself would not affect other objects - not even the process, since its individual threads would be free to execute its computations (behaviour).

> *"Active objects reflect with greater fidelity the main function of managed objects; the representation of real resources."*

The proposed infrastructure (Figure 3) does not obstruct the utilization of passive objects, because in some cases, managed objects have a static behaviour and a passive object implementation is preferable. An example of managed objects implemented as non-active entities would be the log class, which just achieves its functions by external invocations (reading, writing, etc.). In these cases it is desirable that the object is implemented as passive, because its constant activity would just consume processing resources (CPU).

As pointed out previously, passive objects can block themselves and endanger the operation of other passive objects or the control thread that invoked it. A passive object implementation was proposed which allows that an eventual blockage given by its behaviour does not compromise the execution of other activities in the process. This implementation was possible due to the proposed infrastructure multi-threaded features and it is approached in the section 4.5.

4.3 Objects Dynamic: Behaviour

The managed objects concurrent execution provides to the manager system lower response time related to the requested operations and a more accurate view of MIB, referred to the managed objects state and the resource state that it represents.

Basically, managed objects can achieve information related to its resources accessing it in three basic forms:

- Upon manager requesting;
- Through periodic polling;
- Receiving clever resources alarms (with management function embedded).

1. In the first case, traditional organizations (single threaded) would cause the temporary blockage of all agent application, because at that moment the object is interacting with a resource, if this resource is loosely coupled to the system many later requisitions coming from the manager could be pending for a long period waiting for the communication to end between the object and the real resource. Using the proposed infrastructure all agent activities will be independent, providing more justice in its executions and not allowing that active computations in a given moment monopolize the processor utilization.
2. In the second case of interaction (periodic polling), traditional organizations and the proposed organization by the infrastructure multi-threaded provide a desirable support for their implementation. However, in the multi-threaded approach it is possible to provide to certain managed objects more priority related to their executions. This causes a larger frequency in its polling operations. This is desirable, because in some resources, more precise updating is needed in order to support real time features. In agent applications for fault management, this facility is desirable in the sense of providing more accuracy in problem detection which can occur with certain resources that are crucial to the perfect system functioning.

3. In the third interaction (traps) - similar to the previous case, some resources demand special attention (e.g., modem, routers) and can be privileged having their notifications satisfied with more priority than other managed objects that demand less attention.

The function of changing priority between managed objects is given to OM, which interferes directly in the set of threads priorities utilized by managed objects. The agent application designer is in charge of defining which priorities will be given to each object.

In order to further increase the performance of the agent process, objects can yield their execution turn to other system active entities (threads), verifying that their behaviour at a certain moment will not make computations. Some examples of conditions for which objects yield the processor to other activities in order to not consume all its execution time with no computation are empty queues, disabled operational state attributes, and real resources momentarily unavailable.

In the model presented in the Figure 3, there is a thread dedicated to the OM and two others to each queue of responses and notifications, which are dedicated dissemination of these messages. It is the function of OM to perform the maintenance (create, delete, suspend, reactivate, change priorities, etc.) of these threads and also of the threads of each managed object.

In the object creation, the OM associates a control thread to the created object, updates the MIT, initializes the attribute values of this object with any reference object (if specified in the name binding [16]), creates an operation queue for the object, and issues an event report notifying the creation and updates its internal state in order to support the new created object. The associated thread to the new object will execute the behaviour method of this object, where a part of this behaviour is defined by the platform and the rest will be the one defined by the template constructor (BEHAVIOUR) of the managed object class. The part of the object behaviour provided by the platform is related to the operation queue treatment of the managed object and also related to the interaction with synchronization objects (ψ) (section 4.4). The specification of how each operation (M-GET, M-CANCEL-GET, M-SET, M-ACTION, M-CREATE, M-DELETE, M-EVENT-REPORT) is implemented by OM is not in the scope of this paper, which is restricted to the presentation of the proposed multi-threaded infrastructure.

4.4 Operations Requiring Synchronization

The CMIP protocol allows two ways of synchronization:

- Atomic;
- Best effort.

In the best effort synchronization, the failure in the operation execution by some object of the group does not interfere in the execution of other objects. The atomic synchronization demands that every object of the group has success in the operation execution; in the case where some object fails, all the other objects involved abort the operation even if they are capable of executing it.

Each object having its own operations queue and executing independently and concurrently with other system objects, creates a problem in situations where an operation must be executed on multiple objects (scope) and it requires atomic synchronization. The problem with this kind of operation can be illustrated in the following example.

Suppose a management operation SET (S) is requested with scope and filter, and three objects are selected for this operation with atomic synchronization. The Figure 5 shows the scenario of the problem.
Since the object (1) has no operation messages in its queue, the SET operation will be the only one in the operation queue of this object. In the case of object (2) there are pending operations and the SET will be the fifth queue operation. The third object has just two operations in its queue and the SET operation will be the third to be taken.

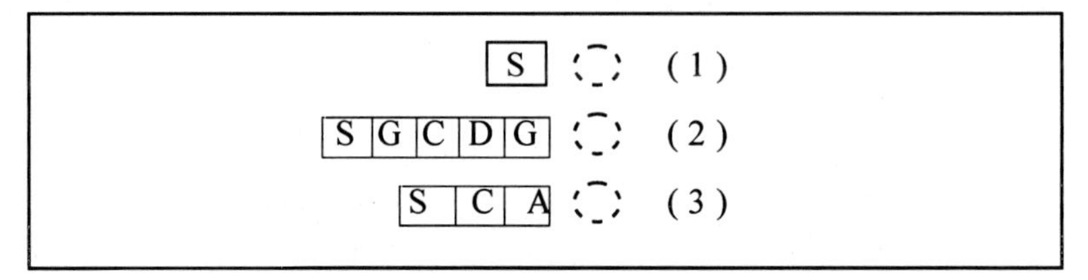

Figure 5: Scenario for the problem of atomic synchronization

In this scenario it is verified that the operation (S) could be executed by objects at different times. To solve this problem, an application support object called synchronizer object (ψ) will coordinate operations requiring atomic synchronization. The Figure 6 shows the solution.

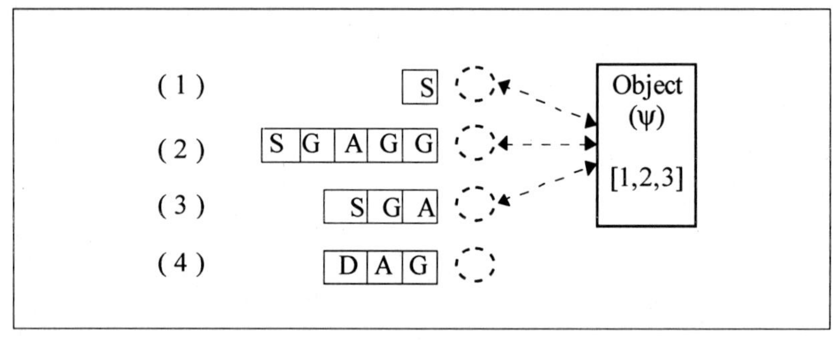

Figure 6: Synchronization with a centralized coordinator

When operations requiring atomic synchronization are submitted to managed objects, the OM will create an object (ψ) which will contain the object identifiers (OID) that will be in the operation group. In this example, just the objects 1, 2, and 3 are submitted to the atomic synchronization and its identifiers are contained in the object (ψ).
When a managed object takes an operation out of its queue which requires atomic synchronization with other objects, this managed object will notify its object (ψ) that it is ready to execute the operation. From this moment the managed object waits for a message from the object (ψ) which will allow its operation to be executed. This message will just be issued by the object (ψ) when all the objects, belonging to the group, are ready to execute the operation. Thus, new operations coming to the managed object queue are not be treated until it is liberated by the object (ψ). Other activities can be executed by the managed object in order to keep its real resource mirror always updated.
After the object (ψ) has received from all objects of the group notifications that all are ready, it issues a message for the objects to execute the operation. Once the operation is executed, the managed objects must issue to the object (ψ) a response specifying if

the operation was succesful or not, because in the operation on the real resources problems can occur in the interaction (managed object <-> real resource). If any failure response is notified the transaction is cancelled for all objects of the group, that after returned to the previous state are liberated to keeping serving its operations queues.

4.5 Non Blocking Passive Objects

In order to allow some managed object implementations (log, log record, etc.) to be modelled as passive objects without compromising other system activities (e.g, in case of these objects blocking themselves), it was defined that each passive object will have its own control thread. These objects will just be active in the presence of external invocations, preserving the passive objects' semantics. In this implementation, the OM does not access directly the passive object method, because if the method blocks itself for any reason the OM thread will be blocked and it will not attend to the new PDUs coming from the CMISE/ACSE. Therefore, the OM requests the passive object thread to invoke its methods to attend to its queue operations, not affecting the execution of other agent activities if the passive objects blocks itself. The Figure 7 shows conceptually the semantics of the active and passive objects' behaviour both with its own control threads.

As it can be observed, passive objects execute their behaviour by invocation from OM, which is represented by increment (sema_post()) of a semaphore variable (OM_Signal) utilized by the passive managed object. Active objects are constantly executing their behaviour.

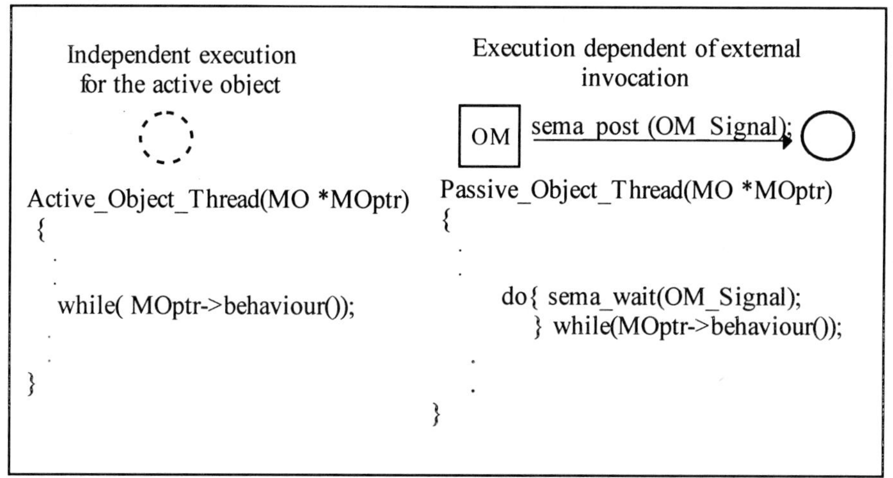

Figure 7: Threads behaviour in active and passive objects

5 Conclusions

The multi-threaded infrastructure presented in this paper, creates a basic core for the implementation of the OSI management agents, allowing the introduction of active objects concept in the implementation of managed objects. One advantage of this infrastructure over core implementations that use the single-thread approach, is the large independence between executions of managed objects' behaviour. This allows a high degree of fault tolerance for agent processes as a whole, providing protection from faults that may occur in the objects' scopes and affect other agent activities. The

greater performance in the execution of activities inside of agents is also another important factor present in the multi-threaded proposal.
This core has actually been implemented as part of an *Agent Toolkit*, whose objective is to automate the developing process of the agent code. The user has only to implement the managed objects' behaviour. This *Toolkit* belongs to a major project, which aims at the implementation of an OSI network management platform. The developing environment of the platform is based on Solaris 2 (UNIX SunOS 5.5), and many resources (Solaris Threads, LWP, mutex, semaphore,etc) have been used in this implementation. The thread manipulation is based on *pthreads library* [17] to make it conformant with the POSIX P1003.4a standard.
Future work will have as its objective the integration of this core in agents that use XOM/XMP interfaces, due to the large utilization of this APIs in the commercial management platforms.

References

[1] ISO/IEC 10040, Information Technology - Open Systems Interconnection - Systems management overview, 1992.
[2] Bean, A.; Wood, D.; Fairclough, W. "Specifying Goal-Oriented Network Management Systems", IEEE Communications Magazine, 1993.
[3] ISO/IEC 7498-4, Information Technology - Open Systems Interconnection - Management framework, 1992.
[4] Mansouri-Samani, M.; Sloman, M. "Monitoring Distributed Systems (A Survey)", Imperial College Research Report N°. DOC92/23, 1992.
[5] Bach, M. J. "The Design of the UNIX Operating System", Prentice Hall,1990, ISBN 0-13-201799-7.
[6] Rose, M. T. "The Open Book: a practical perspective on OSI", Prentice-Hall, 1990, ISBN 0-13-643016-3.
[7] ISO/IEC 9595, Information Technology - Open Systems Interconnection - Common management information service definition, 1991.
[8] ISO/IEC 10165-1, Information Technology - Open Systems Interconnection - Structure of management information: Management information model, 1992.
[9] Tanenbaum, A. S. "Distributed Operating Systems", Prentice-Hall, 1995, ISBN 0-13-219908-4.
[10] ISO/IEC 9596, Information Technology - Open Systems Interconnection - Common management information protocol - part 1: Specification, 1991.
[11] Tsichritzis, D.; Nierstrasz, O.; Gibbs, S. "Beyond Objects: *Objects*", IJICIS, vol. 1 no. 1, pp. 43-60, 1992.
[12] Booch, G. "Object-Oriented Analysis and Design with Applications - Second Edition", The Benjamin/Cummings Publishing Company, 1994, ISBN 0-8053-5340-2.
[13] Löhr, K. "Concurrency Annotations", OOPSLA'92, pp 327-340, 1992.
[14] Takashio, K.; Tokoro, M. "DROL: An Object-Oriented Programming Language for Distributed Real-Time Systems", OOPSLA'92, pp. 276-294, 1992.
[15] Pavlou, G.; McCarthy, K.; Bhatti, S.; Knight. G. "The OSIMIS Platform: Making OSI Management Simple",1994.
[16] ISO/IEC 10165-4, Information Technology - Open Systems Interconnection - Structure of management information: Guidelines for the definition of managed objects, 1992.
[17] Lewis B., Berg D. J., "Threads Primer: A guide to Multithreaded Programming", SunSoft Press,1996, ISBN 0-13-443698-9.

Internet - New Inspiration for Telecommunications Management Network

G. Bogler
Siemens AG, Germany

1 Introduction

As the complexity of telecommunication networks, and in particular carrier networks, has grown in the last two decades network management has changed for network operators from a speciality to a matter of being competitive and even of survival.
In the 1980s, standards organizations created the concept of a Telecommunications Management Network (TMN). TMN was seen in those days to be the solution to all interoperability issues. The vision of TMN - understood as a conceptual framework for network management - is still widely supported. Why do we think that TMN is in need of some inspiration? The answer may be that the situation of carriers and carrier networks has changed a lot since work on TMN started.
De-regulation and increasing competition between incumbent network operators and emerging service providers pose new challenges on carrier network management. Traditional TMN solutions reflecting the centralized network structures of the past monopolistic era are no longer viable in modern competitive multi-player networks. Answers to some of the main network management issues may now come from a direction that did not seem very promising a couple of years ago: the Internet.

2 The Main Challenges for Network Management

In a competitive environment network management is primarily driven by economic factors. Network management (including service and element management) is responsible for the largest portion of overall costs for a network operator. Therefore, efficient network management must focus on reducing operational cost.
Closely related to the goal of minimizing operational costs is the goal of minimizing expenditure for the deployment and upgrade of management systems. Open interfaces can be seen as the prerequisite for creating a market for network management products. This gives network operators a real choice when buying the management products best suited to their needs, thus saving money in the end. Suppliers benefit, too, as they can address a broader market and reduce their own costs through economies of scale. Obviously, these goals have been reached to only a small degree.
Today, carrier network management can be characterized by

- proprietary applications, interfaces (with the exception of the communication protocols), and procedures requiring individual solutions, and
- non-interoperability: current multivendor capabilities are usually very restricted in functionality.

This has meant and still means excessive costs for operators and suppliers. Today's market for carrier network management products is limited. Hard-coded, specialized, and thus expensive management applications are still the rule. In contrast, the market for management products for enterprise networks has traditionally been broader and more competitive. Internet and World Wide Web technologies have already given new impetus to developments in enterprise network management, and are also affecting carrier networks to an ever greater extent.

3 The Internet: Inspiration for Carrier Network Management?

The Internet and its technology can be seen as one of the main sources of innovative ideas today. While this is already well recognized for the management of enterprise networks (see [7]: several articles, or [15]), the Internet and its technology have had limited influence on carrier network management up to now. We believe this will change soon and the Internet and its technology will drive the evolution of carrier network management, too.
What are the lessons that can be learnt from the Internet? To answer this question it is worthwhile to have a look at some aspects contributing to the stunning popularity of the Internet:

- The Internet has not one, but many independent owners (or no owner at all). Nevertheless some important decisions are made at a global scale, e.g. addressing. This shows that two principles which are conflicting at the first glance are important: *co-operation* between organizations and *distributing control.*
- The Internet provides *global connectivity and interoperability* despite the fact that it is a conglomerate of many networks which differ in topology, size and underlying technology. This proves that interoperability is also possible to some degree in a deregulated, multi-provider, multi-technology environment.
- The Internet is an open network, it is based on a rather small set of basic standards. This has finally lead to a flourishing market for Internet related products, such as web-browsers. The Internet shows the importance of *pragmatic standards* which are strictly oriented at the market needs.

Obviously, carrier network management has to meet similar challenges:
How to achieve interoperability of network management in a multi-provider, multi-technology environment, how to distribute responsibility for network management (centralization vs. delegation of control) and how to make this reality, i.e. which standards, tools and methods shall be used to implement it.
In a nutshell, we think that carrier network management can benefit in two ways from the Internet: first, adopting some principles of its philosophy and second, using its technology where this is appropriate.

4 What can be Learnt from Internet Philosophy?

In this section we will discuss some conclusions for carrier network management which are not directly related to the technology of the Internet.

4.1 Delegation of Control: Customer Network Management (CNM)

Traditionally carrier network management has been organized in hierarchical and centralized structures. In most cases one network operator has been responsible to manage all types of services, network resources and network equipment. Presently we observe that this situation is changing rapidly. Value-adding service providers are offering services e.g. for business users on top of the public network, several service providers are competing for customers in one geographical area; the highly competitive field of mobile services may serve as an example.
As a consequence, management of carrier networks will have to be decentralized to a high degree. Decentralization brings the need for cooperation: to offer end-to-end services to the customer usually will involve more than one organization.

An important concept addressing these needs is Customer Network Management (CNM). CNM delegates responsibility for selected management tasks from the carrier, e.g. an incumbent Telco, to the carrier's customer, e.g. a large network user. CNM is already becoming increasingly popular, in particular in the U.S., see e.g. AT&T's CNM offering ("CNMS") or Sprint's "Sprint in Touch".
A typical example for the use of CNM is a VPN (Virtual Private Network) service. VPN allows an organization, e.g. an enterprise, to build an enterprise network from resources subscribed from a public network provider. Corporate customers do not have to buy all the network equipment to build their own physical enterprise networks. The beauty of VPNs is that their owners do not need to care about hardware maintenance, ordering of equipment etc. On the other hand any network owner - including the owner of a VPN - needs control about his network resources.
CNM gives the VPN owner management access to resources in the network provider's domain, subject to a bilateral contract. This management access may include reading status and performance information, customized billing, flexible allocation of bandwidth (e.g. between different sites of an enterprise) and other features. An example of a typical VPN user may be an Internet Service Provider. He may create a VPN consisting of ATM connections in a public backbone network connecting his Points of Presence (POP).
VPN is also a good example to highlight another requirement for CNM: in order to be useful, CNM has to be able to present the CNM user with an integrated view of the network resources: examples include end-to-end PVCs in an ATM network, summary alarms condensing information collected from a variety of network elements etc. CNM will achieve this by coordinated cooperation with management applications which provide gateways to technology and vendor specific management systems already existing in the network.
CNM is an emerging topic in standardization. CNM interfaces have been defined up to now only for ATM networks by the ATM Forum [4] and for frame relay. In addition to these quite successful specifications, ITU-T X.160 series addressing CNM of X.25 networks should be mentioned as a relevant effort in standardization bodies.
As CNM interaction gives one organization management control of resources originally under another organization's responsibility, CNM is very sensitive in terms of *security*. Clear contractual agreements between the partners involved in a CNM relationship and the supporting security technology are a pre-requisite for CNM to deliver its benefits.
CNM like mechanisms are quite common in the Internet and in commercial on-line networks. Self-subscription to services, inquiries about offered services and billing information are state-of-the-art. Despite the obvious security weaknesses of today's Internet, comprehensive security mechanisms are expected to be widely available soon. Basing CNM on Internet technology, i.e. understanding it as a specialized Internet-type service, will allow it to benefit from these emerging security mechanisms.

4.2 The Need for Pragmatic Standards: a Short Look at the IETF

Why are the IETF's Internet standards so successful?
One reason may be the way Internet standards are created by the IETF (Internet Engineering Task Force). The IETF applies a process which is significantly different from the one at established organizations like ITU or ETSI. This process has lead to many successful standards probably because of the IETF's philosophy of "running

code," which means that a proposed specification is only approved as a standard if it works in practice and is interoperable. The IETF's lean, market-oriented process may not produce the "best" standard in any case but certainly it produces standards which are implementable and have a good chance to be accepted by the market.

What are the consequences for carrier network management?
In short, rely on pragmatic standards in the future, i.e. use Internet standards for carrier network management where appropriate. This avoids solutions which are not accepted by the market and therefore expensive for carriers and suppliers. This statement is by no means an assessment that Internet standards are technically "superior", it simply acknowledges the fact that they tend to be widely accepted and are usually supported by numerous products from a variety of suppliers.
We think TMN standardization could benefit a lot from standardization in the IETF, with respect to process efficiency and applicability of the resulting standards. This discussion, however, would go far beyond the scope of this submission.

5 What can be Gained from Internet Technology?

Introducing Internet technology into carrier network management may involve a variety of technical measures, such as communication protocols, management data schemes, APIs and system architectures. It is most certainly more than just using IP for transport and WWW-browsers for the user interface.
This section is intended to show some of the principal possibilities of applying Internet technology, in particular WWW-technology to the management of carrier networks. It does not propose any specific solution.

5.1 Management Interfaces, Part 1: Protocols

Conventional TMN strictly demands the use of OSI protocols for its management interfaces (for reference see [3]). This is in contrast to the large majority of practical solutions especially in the area of enterprise networks which use the IP based protocol stacks specified by the IETF. As the layer 7 management protocol conventional TMN demands CMIP ([2]). The IT-community, however, relies - besides some proprietary solutions - on the IETF's Simple Network Management Protocol (SNMP, see [5]).
We expect that protocols for carrier network management will eventually follow the main stream in the IT world, i.e.:
The *IP protocol suite* will be used. This acknowledges the fact that the adoption of the OSI system management as the basis for the TMN specifications has not managed to make a real break-through.
An *Internet/WWW management protocol* will be used. Several possibilities on how to realize this are presently being discussed. This submission shall identify some of the more prominent options; it does not suggest any of these as the preferred solution for future carrier network management:

Keep the IETF's Simple Networks Management Protocol and (SNMP) respectively its enhanced version (SNMPv2).

HTTP is used as the transport protocol (instead of UDP). Keeping SNMP as the management protocol avoids complex converting to and from SNMP MIB structures in the managed system because the managing system still talks SNMP. The embedding of the SNMP protocol messages into HTTP messages has not been standardized yet.
This method can be extended to network elements with CMIP based agents.
Potential draw backs:

SNMP does not allow direct use of state-of-the-art information presentation using an off-the-shelf WWW browser. A way out may be the use of SNMP browser applets (e.g. written in Java) which enhance the capabilities of the standard WWW browser. Example: see [14].

Use the Hypertext Transfer Protocol (HTTP, see [6]) as the management protocol.
In this case the managing system includes a WWW client (normally an off-the-shelf WWW browser), the managed system includes a WWW server. The HTTP operations have to be mapped onto SNMP/CMIP operations to access the individual MIB objects in the managed system. This will be done by a proxy application in the network element or an intermediate system.
The method is principally applicable for any type of MIB (SNMP, CMIP, proprietary) and for all types of management interfaces, e.g. CNM interfaces and interfaces to the network elements. This supports smooth evolution from today's management solution to solutions based on Internet technology.
This approach is explored by a number of organizations, examples include [8], [12] and [16]. It is also utilized in ACTS project MISA [9].
Potential drawbacks:
HTTP does not directly support basic management operations, in particular not unsolicited messages (which are needed e.g. for reporting network alarms). In addition, HTTP implies use of a great number of individual TCP connections; this may result in considerable overhead and performance degradation.

Use an extension of the Hypertext Transfer Protocol (HTTP) as the management protocol between managing and managed system.
This means enhancing the capabilities of HTTP to bring it to a functional level comparable to SNMP, for example, by supporting event notifications (similar to SNMP TRAPs), a feature not provided by current HTTP.
A group of organizations, called Freerange, which includes Microsoft, Cisco and other important players in the Internet arena have started an initiative to standardize this kind of capability (see [11]). Two main components were proposed: the Hyper Media Management Protocol (HMMP) to the IETF and the Hyper Media Management Schema (HMMS) to the Desktop Management Task Force (DMTF).
Potential drawbacks:
As with any new standards proposal, it is quite difficult to forecast future support of HMMP (or any similar protocol that may emerge). For any carrier network management solution, interworking between this new protocol/data scheme and the legacy systems using SNMP, CMIP or a proprietary management protocol will be required. Network managers, especially in the traditional Telco environment, may hesitate before committing to a new approach without having explored the viability of smooth migration from their existing solutions.

5.2 Management Interfaces, Part 2: A Short Note About Information Models

While the choice of communication protocols is important for network management, the true complexity lies within the management information. This was mentioned already in some places when discussing management protocols. Also in an Internet technology environment standardized MIBs will continue to define a common level of management functionality. They constitute some kind of contract between the managing and the managed entities specifying which properties of which entities are subject to management.

From a purely technical point of view it would not matter whether these MIBs were accessed by SNMP, CMIP, HTTP, HMMP, or any other management protocol.
In the context of carrier network management however the installed base plays the most important role. Therefore, it is an absolute must to keep existing MIBs (in carrier management mostly proprietary MIBs, but also CMIP and SNMP based) and interwork with them.

5.3 Management Applications: Clients and Servers

Using Internet technology goes beyond using its protocols and standard MIBs. Internet technology also has a strong impact on the involved systems.
Internet technology means use of a *client-server model.* For carrier network management this means that the managing entity will be the client, the managed entity will take the server role. Figure 1 gives an overview:

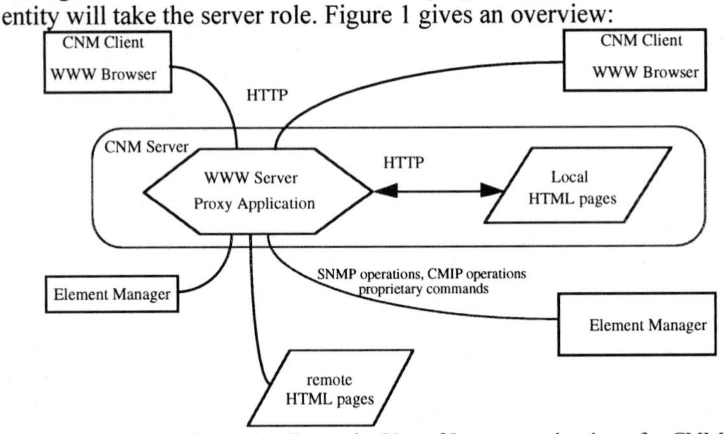

Figure 1: Example: Use of Internet technology for CNM

CNM Clients:

Core part of the client in carrier network management will be an industry-standard WWW browser. Using WWW-browsers has some distinct advantages:
WWW-like user interfaces can be seen as a de-facto industry standard. It can be expected that WWW browsers will replace most proprietary management user interfaces; additional features which should be available shortly in the WWW - like secure transaction support, authentication of users, etc. - will make them even more useful for network management.
Using WWW technology for CNM has extra benefits:
CNM is envisaged by the CNM users as their prime contact point to the network provider and their individual window into the CNM provider's network. Therefore CNM will be one of the major criteria for customer satisfaction. From the CNM users' point of view, CNM will look like a collection of WWW pages referring exclusively to management of their resources, having the same look and feel, and using the same software technology as other WWW services.

CNM Server:

Core part of the CNM server will be an industry-standard WWW server. Proxy applications will be responsible for interaction with other systems, e.g. legacy network elements and legacy element managers. Management operations will be supported by HTML pages which may be local (in the server library) or remote.

These HTML pages can be enhanced by executable application code (applets) loaded at run time from the server library or from external systems, e.g. the existing management systems or network elements. The rise of platform-independent languages like Java allows the creation of intelligent agents in the network elements which incorporate basic management capabilities of the respective piece of network equipment.
This may help to solve one of the most difficult problems in conventional TMN interface standardization in many practical cases: How to specify the dynamic behaviour of the managed objects in sufficient detail that manager and agent applications can interwork across the interface?
Experience has shown that efforts to standardize the dynamic behaviour of managed objects in complex systems were by far not as successful as originally expected. Mostly using plain English text behaviour descriptions tend to be too vague and leave too much room for interpretation. Downloadable applets import the exact knowledge of the behaviour of the managed objects to the managing system as this knowledge is incorporated in them. This eliminates the need to agree on any thinkable detail of a management interface in advance.
Two examples for the use of loadable applets:
SNMP MIB browser [14]: The SNMP MIB browser applet (including the protocol stack) is uploaded from a network element to the element management system. This reduces the software development effort for the management system and makes it more interesting for the suppliers to build management capabilities into their network equipment.
CNM: The CNM user (the CNM client) loads applets from the CNM server, e.g. for calculating statistics from raw performance and usage data collected in the network. This relieves the CNM user from the cumbersome work of writing his/her own applications dealing with system specific data formats. An example for WWW based CNM is the prototype described in [10].

5.4 Linking the Pieces: how to Promote Management Integration

Having discussed some of the benefits of WWW technology to achieve a widely accepted look-and- feel for network management we need still to explore the perhaps most exciting feature of the WWW - commonly known as "surfing".
What does it mean to "surf" in the context of carrier network management?
Surfing means following a chain of hyperlinks. Hyperlinks guide the user through the various operational procedures, e.g. "Create a new Subscription" or "Create a permanent ATM end-to-end connection". The (CNM) server hosts a library of interlinked pages which cover the most essential operational procedures. Hyperlinks may take these forms (see [12] for a similar approach):

- Links may lead to a selection menu page (e.g. select a management function)
- Links may lead to a parameter form page (e.g. fill-in the attributes of a new end-to-end ATM connection in a VPN)
- Links may lead to any kind of help information (e.g. help information explaining a management function, an operational procedure, giving details about a message from the network); note that this help information can be physically located anywhere, e.g. in an existing element management system, in a network provider's billing center or even at the support site of the respective equipment supplier.

- to metadata describing the syntax of management information (this kind of capability is also addressed in [8]).

Integrating help information and documentation into a WWW-style user interface, helps to provide management interfaces which offer a considerable degree of practical interoperability even for their proprietary parts. Clearly, the level of interoperability across these interfaces is different than that conventional TMN was aiming at: but without expensive upgrades to the network elements it eases the problem of cooperation with legacy systems in a surprisingly simple way because:

- Network operators and service providers have an interface to all types of systems in a heterogeneous network (both legacy and new systems) via WWW pages (i.e. via a widely known user interface of a common look and feel).
- Network operators and service providers can get context-sensitive help and system-specific documentation by just clicking at the appropriate hyper links in the respective WWW pages. Network operators and service providers are thus relieved from the tedious learning of cryptic command languages for various systems and from looking up information in voluminous operation manuals. Besides saving cost for staff training and day-to-day operation, this reduces the need to standardize all the numerous bits and pieces of a management interface.

6 Conclusion

This submission intended to give some evidence that the Internet and its technology are indeed a source, perhaps *the* source, of inspiration also for network management of carrier networks. This inspiration is not only a technical one: it has also to do with the way networks are engineered, operated and used.

We expect that recent developments for management of enterprise networks will also set the trends for the future management of public carrier networks. Distributing and delegating control by Customer Network Management (CNM) and applying Internet technology at all levels of network management can be expected to be on the agenda for future evolution of network management.

Many activities in particular for Internet-based element management are currently under way. Only a few could be mentioned in this submission. While they are important as such, they need to be complemented by Internet-based CNM to meet the management challenges of future networks. All involved parties will finally benefit:

- Carriers (network operators) can increase their revenue by cutting costs and using CNM as a service differentiator.
- Network and service users, e.g. value-adding service providers can be more competitive by tailoring services and management to their individual needs.

We do not expect that the Internet and its technology will incite a revolution in the TMN community. Although some of the fundamental assumptions of conventional TMN, in particular the goal of full semantic standardization of management interfaces, may need to be revisited, TMN will remain the framework for a systematic description of management functionality.

The Internet and its technology should be not be seen as a threat to the TMN vision, but understood as an unique opportunity to make TMN fit for the future. In the short term, Internet-based management will also not replace existing management solutions in the carrier environment. It will rather co-operate with legacy management systems and thus preserve carriers' investment in networks and their management.

References

[1] ITU-T: Recommendation M.3010 "Principles for a Telecommunications Management Network",

[2] ISO/IEC, ITU-T, Information Technology - OSI, Common Management Information Protocol (CMIP) - Part 1: Specification ISO/IEC 9596-1, ITU-T Recommendation X.711

[3] ITU-T, Lower Layer Protocol Profiles for the Q3 Interface ITU-T Recommendation Q.811 ITU-T, Higher Layer Protocol Profiles for the Q3 Interface ITU-T Recommendation Q.812

[4] ATM Forum, Customer Network Management Interface M3

[5] J. Case, M.Fedor, M. Schoffstall and C. Davin, "The Simple Network Management Protocol (SNMP)", RFC 1157, May 1990

[6] T. Berners-Lee, R. Fielding and H. Frystyck, "Hypertext Transfer Protocol HTTP/1.0", RFC 1945, May 1996,

[7] The Simple Times, Vol 4, No3, July 1996 : various articles about the use of WWW technology for network management

[8] L. Deri, Surfin' Network Resources across the Web, Proceedings of 2nd Intl IEEE Workshop on Systems and Network Management, Toronto, June 1996

[9] ACTS Project MISA, http://misa.zurich.ibm.com/World/

[10] S. Covaci, GMD Fokus, "Java and WWW-based CNM Interface for Global Broadband Connectivity Services", http://www.fokus.de/djjava/cnmi/

[11] Freerange, Enterprise Management Using Web-Based Technology, (Initial contributors: BMC Software Inc., Cisco systems Inc., Compaq Computer Corp., Intel Corp., Microsoft Corp.), http://wbem.freerange.com/wbem/

[12] Dossick/Kaiser, "WWW access to legacy client/server applications", 5th International WWW Conference, May 6-10, 1996, Paris, France

[13] Sun Microsystems, The Java Programming Language, Addison-Wesley, 1995, ISBN-0-201-63455-4

[14] Advent Network Management Inc., MIB Browser Applet Overview Version 1.0, http://adventnet.com/browser_overview.html

[15] Data Communication on the Web, M. Jander, "Welcome to the Revolution", Nov 21, 1996, http://www.data.com/

[16] Data Communication on the Web, A. Larsen, "Making the Web work for Management", Dec 1996, http://www.data.com/

Mobility

Roberta Gobbi
Italtel, I-20019 Settimo Milanese (Milano), Italy
Email: roberta.gobbi@italtel.it

Customer demands are closely related to the social environment in which the individual resides. One would expect a huge variety of trends according to the country lifestyle. Fortunately, some trends seems to be shared by different societies belonging to the group of industrialized countries, and may well serve as a basis for some predictions of customer demand in the telecommunications market. The main trend is individualization: people are becoming more and more aware of their individual liberty, rights, demands, desires and so on.

The most important consequence of the trend of "individualization" is the increasing demand for freedom of movement, that is *mobility*. This is probably the most typical way of expressing individuality: therefore user mobility and service portability are becoming the essential feature in the telecommunication world.

Nowadays, the principle means of mobile communication is still the portable telephone, but (as in fixed networks) computers are playing an ever increasing role. It is therefore merely a question of time before the networks and portable terminals (such as telephones, portable digital assistants, laptop and palmtop computers) will be expected to provide much more functionality. Up to now, the actual networks are limited to voice (and voice facilities) and very low bit rate data (e.g. maximum user data speed is 9.6 kbit/s in GSM networks). The increasing demand for multimedia applications is pushing operators to provide multimedia services anywhere and at anytime.

So what is needed to meet this demand? There can be no simple right or wrong answers, but a clear target vision is needed that will lead to a viable systems. A term coined by ETSI in PAC EG5 Report [1] for such systems is *Global Multimedia Mobility* (GMM). This phrase denotes the need for the networks to be Global and also accommodate full Multimedia applications and full Mobility. GMM communications will be supported not by one system but by a set of systems. Therefore the approach that should be followed is not to define a single unique architecture, but rather a framework within which different elements and sub-systems can work to provide GMM services.

In alignment with the GMM concept, the first paper of this section [*Towards Global Multimedia Mobility*] presents one view of the paths towards multi-mode services provision with the basic assumption that the future terminal equipment should be able to connect to several types of access networks. From a core network perspective IN-techniques provide the "glue" to support multiple access technologies and multi-mode terminals.

An example of how the IN platform can provide mobility functionality in GSM networks is presented in the paper [*GSM Evolution to an IN platform: offering GSM Mobility as an IN service*]. The approach proposed is a complete integration scenario of GSM and IN. The effect of moving from an architecture dedicated to mobility (GSM) to a general control architecture is to increase in the signalling traffic through the network. This is to be expected and is shown in the simulation results in the paper. Nevertheless, the initial simulation results indicate that the level of performance of GSM networks can be maintained in the integrated scenario.

The concept of "service interoperability" between fixed and mobile network is well addressed in the paper [*Integration of mobility functions into an Open Service*

Architecture: the DOLMEN approach] where a long term vision of how fixed and mobile technology can be integrated under a Common Service Architecture is proposed. Two approaches are here compared: the Service Engineering approach and the System Engineering one. They differ with respect to whether mobility is assumed to be provided by the Service Architecture itself or by a network infrastructure where, for example, Universal Mobile Telecommunication System (UMTS) functions are included.
The UMTS concept has been introduced by ETSI and it refers to a system that will be capable of offering mobile voice, messaging, data, image and video to all the users, using one easy-to-use terminal and will provide a seamless wireless access. It will use frequencies in the 2GHz range. The last paper of this section [*Consistency Issues in the UMTS Distributed Database*] focuses on the critical aspect of database systems in future mobile networks, under the assumption that IN is the vehicle to facilitate mobility and service interoperability. Here a Distributed Database (DDB) system is proposed as the most appropriate solution: a mechanism of enforcing mutual consistency between entries in the DDB is required and analysed.
To conclude it is becoming apparent that the business of providing mobility services and features is challenging not only the traditional mobile network operators, but also the fixed network operators. For these operators a new market is just within view. IN can serve both the fixed and the mobile networks to enable short-term seamless deployment of fixed and mobile services.

Reference

[1] Global Multimedia Mobility (GMM) A Standardisation Framework, ETSI PAC EG5 Report, October 1996.

Towards Global Multimedia Mobility

J. C. Francis (Swiss Telecom PTT)[1], B. Diem (Swiss Telecom PTT)[2]

In this paper we present a vision of the road towards Global Multimedia Mobility (GMM). Through evolutionary steps driven by market forces we expect future core networks such as the GSM-UMTS NSS, ISDN and B-ISDN to provide multimedia services to multi-mode terminals through multiple access networks. Starting from the DECT/GSM dual-mode service, we trace the path via CTM/GSM dual-mode towards multi-mode services. Our contention is that IN-techniques can provide the "glue" to unify different access technologies and provide support for services based on multi-mode terminals and multiple access technologies. Intelligent Networks, dual-mode and multi-mode terminals are seen as the enabling technologies that will make GMM a reality.

1 Introduction

Global Multimedia Mobility (GMM) is a term coined in the PAC EG5 Report [1] to denote the mobility aspects resulting from the convergence of telecommunications, information technology and entertainment services as envisaged by EII/GII [2]. A basic GMM assumption is that future terminal equipment should be able to connect to several types of access network. The choice of access will be made dynamically and will depend on a variety of factors such as the application service requested by the user, the service subscription, and the access networks available locally. A variety of access networks can be identified which include the UMTS RSS, the GSM BSS, DECT, satellite (S-PCN), LAN, ISDN and B-ISDN.

The GMM report indicates that the dynamic use of multiple access networks will enable high bit-rate services to be gradually introduced according to market demand. Dual-mode and multi-mode terminals are identified as key to the success of GMM, and it is predicted that in the near future there will be multi-mode terminals supporting GSM900, DCS1800, Satellite, DECT and PCS1900 radio interfaces. Compact multi-mode terminals are made possible as a result of technological advances in batteries, electronics, chip design, RF-techniques and satellites. Already, manufacturers are developing terminals which combine the capabilities of DECT and GSM.

2 DECT/GSM and CTM/GSM Dual-Mode

The standards for *Digital Enhanced Cordless Telecommunications (DECT)* were primarily elaborated for wireless telecommunications (including speech and data) in the private and business environment. More recently, the public environment including radio in the local loop and public mobile networks has gained increasing importance. The *Generic Access Profile* (GAP) ensures full interoperability between radio installations of different manufactures and the arrival of the *CTM Access Profile* (CAP) is imminent. The evolution towards wireless multimedia services with bit-rates of up to 552 kbit/s within pico-cells is foreseen.

The *Global System for Mobile communications (GSM)* denotes GSM900, DCS1800 and the North American derivative PCS1900. It was developed for mobile telecommunications (including speech, data and short messages) in cellular public networks. By 1996, GSM had reached a market acceptance of 12 million subscribers,

[1] Dr. J. C. Francis, Swiss Telecom PTT, Research and Development, Mobile Communications/FE423, CH-3000 Bern 29, Switzerland. Tel:+41 31 338 02 04, Fax:+ 41 31 338 51 74, E-Mail: francis@vptt.ch

[2] Dr. B. Diem, Swiss Telecom PTT, Research and Development, Mobile Communications / FE423, CH-3000 Bern 29, Switzerland. Tel: +41 31 338 4002, Fax: + 41 31 338 51 74, E-Mail: diem@vptt.ch

with some 100 networks in operation world-wide and subscriber growth exceeding 50% per year. Current developments focus on *Intelligent Network* (IN) support for GSM services (CAMEL); on a full-bit-rate voice codec pushing GSM speech quality towards that of fixed networks; on circuit switched bearers with bit-rates up to 38.4 kbit/s; and on packet switched data services with bit-rates reaching 100 kbit/s.
DECT is appropriate for mobile communications within indoor environments and for a limited range of outdoor applications. In outdoor environments multipath interference is a serious problem causing a rapid degradation of service quality with increasing cell radius. For digital cordless systems which are not equipped with equalisers, multipath interference cannot be compensated and as a consequence outdoor deployment is limited to Radio in the Local Loop and mobile service provision for pedestrians. DECT uses a dynamic radio channel selection scheme and supports traffic of up to 10000 E/km^2 for a cell radius of about 10 metres. The traffic capacity decreases with increasing cell spacing, and at a cell radius of about 100 metres the maximum traffic capacity of DECT is below that of GSM.
GSM900 and derivatives DCS1800 and PCS1900 are suitable for seamless wide area radio coverage, but they are not appropriate for indoor mobile service provision because of inter-cell interference and capacity limitations. With decreasing cell size, the capacity of GSM increases and reaches a maximum at a cell radius of about 100 metres. Below this radius, the repetition of frequencies is limited and traffic capacity can no longer be increased.

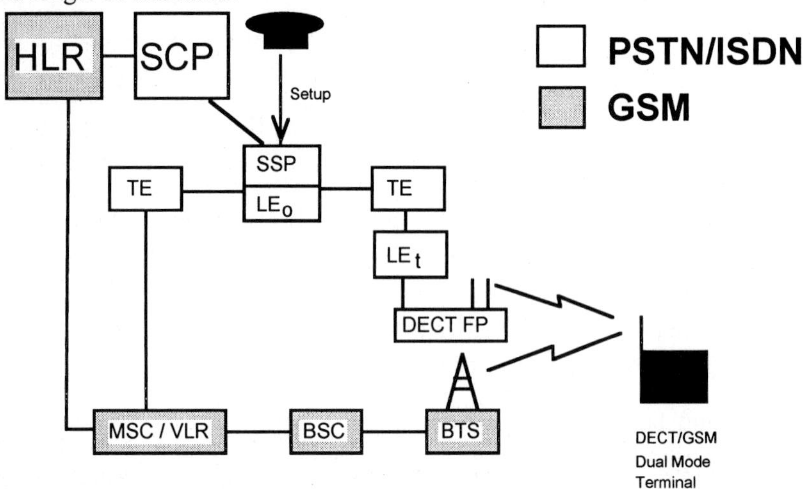

Figure 1: (Architecture of the DECT/GSM Dual-Mode Service)

Such properties of GSM and DECT reveal that both systems are complementary and that mobile communication services can be provided most economically by joint operation of DECT and GSM radio access systems using DECT/GSM dual-mode terminals. Combined operation of DECT and GSM can be based on GSM network functions only or can take advantage of Intelligent Network technology. For an operator such as Swiss Telecom PTT running both GSM and fixed networks, the latter option is beneficial because the local exchanges are equipped with IN service switching functionality but do not support functionality for GSM mobility.
Swiss Telecom PTT has scheduled a large user trial combining DECT and GSM by means of IN for Spring 1997. A DECT-GSM dual-mode handset will be used to

access the fixed network through a DECT base station when in the "home environment", and through the GSM network when out of range. The user is either located at home or in the GSM network, and as there is no mobility within the fixed network, the GSM mobility management procedures can be reused to locate the user. This is achieved by interworking the INAP and MAP protocols. For a user residing in the GSM service domain, the *Service Control Point* (SCP) retrieves a roaming number from the GSM *Home Location Register* (HLR) and routes the call directly to the dual-mode terminal. Since routing is determined on the basis of a roaming number, a common subscriber number can be used for the fixed network and GSM. This avoids undesirable service interaction in the CLIP service and the like. The architecture of the DECT/GSM Dual-Mode service is shown in Figure 1.

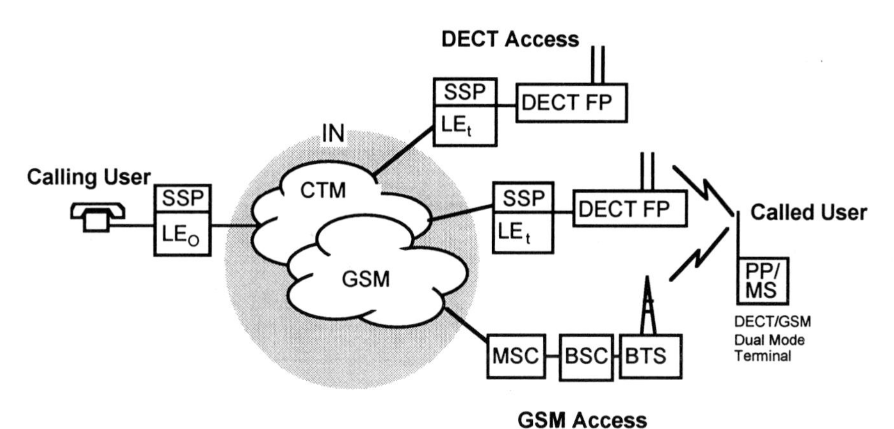

Figure 2:. (CTM/GSM Dual-Mode Service)

Swiss Telecom PTT envisages enhanced support of DECT/GSM dual-mode terminals so that the user can get access and be reached in one of several DECT islands or through the GSM network. This service is known as *CTM/GSM Dual-Mode*, and will use IN to combine *Cordless Terminal Mobility* (CTM) with GSM mobility management (Figure 2). It is more difficult to achieve than the DECT/GSM Dual-Mode service since two mobile networks (CTM and GSM) with different architectures must interact. A key to the solution is the maintenance of user location data in one single network point, which is represented conceptually in Figure 3 by the "SCP/HLR" box. The box may be composed of more than one physical node. It communicates with the fixed network using the INAP protocol and with the GSM network using the MAP or M-INAP (CAP) protocols.

The integration of DECT and GSM services will satisfy demand from the domestic and business market for support of low bit-rate mobile telecommunication services in large cellular networks in combination with medium bit-rate multimedia services in pico-cells. The feasibility of DECT-based mobile multimedia services within pico-cells of public networks will be demonstrated in the ACTS project EXODUS (Experiments on the Deployment of UMTS). The EXODUS experiments with DECT-based multimedia services are scheduled during 1998 in Switzerland and Italy. EXODUS uses an IN-based mobility management scheme that employs GSM concepts (e.g. the usage of a roaming number). Deviations from the GSM concept were adopted to minimise the signalling load between the visited and home networks.

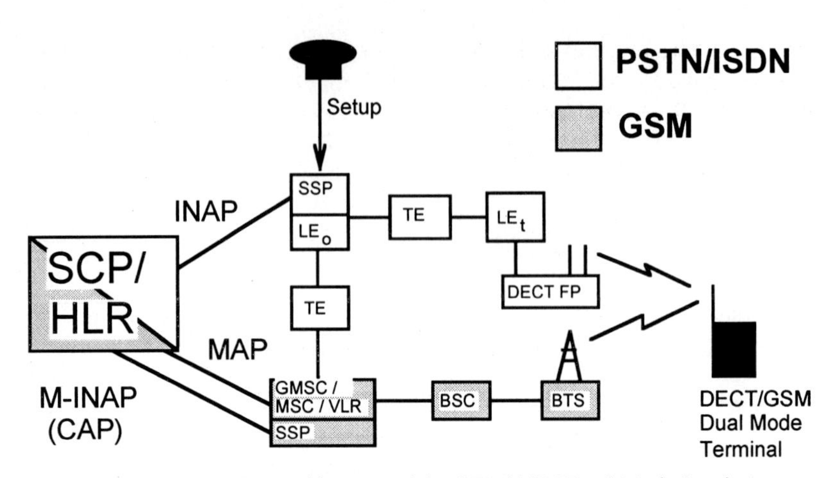

Figure 3: (Architecture of the CTM/GSM Dual-Mode Service)

3 From Dual-Mode to Multi-Mode Terminals

The demand for multi-mode terminals will be driven by the scarcity of spectrum, by new wireless multimedia services, by the need for the integration of wired and wireless access in one terminal, and by the desirability of global service provision.

No one radio interface is optimal, and each has its advantages. The appropriate wireless interface in each environment can optimise spectrum usage and minimise infrastructure costs. GSM900, for example, is a suitable basis for countrywide coverage with world-wide roaming, but it has limited spectrum allocation. DCS1800 is a way to increase capacity since greater spectrum is available with a higher capacity, but range is limited and cell size is smaller. As such it is an ideal solution for cities, but less economic than GSM900 for countrywide coverage. DECT offers very high capacity and easy implementation, but coverage is limited and cell sizes are very small. Wireless LAN is an appropriate solution for data communication in certain business environments.

The demand for mobile multimedia services with wide area coverage and spectral efficiency, will be met by a future radio interface optimised for multimedia services (UMTS). However, this multimedia wireless interface is unlikely to be optimised for speech and so a combination with native GSM and DECT will be advantageous.

The demand for the integration of wireless and wired access will be driven by high-bandwidth multimedia applications. With DECT, a maximum bit-rate of 552 kbit/s can be reached and with the future UMTS radio interface a maximum rate of 2 Mbit/s in pico-cells is envisaged. Higher bandwidth is achieved if the terminal can be connected to the access network with wires or fibres.

Only satellite systems offer truly global coverage, but with limited capacity. It is noted, however, that widespread GSM terrestrial radio coverage may reduce demand for satellite access, and it is therefore questionable whether satellite access will be an important driver for multi-mode terminals.

Based on market factors, multi-mode terminals will evolve to support services by use of the most appropriate wireless or wired interface. The evolution of such terminals is shown in Figure 4.

The same IN technology that is suitable for support of DECT/GSM dual-mode operation is also suitable for support of multiple access technologies with different wireless interfaces. The key is to maintain location data of a user in at least one single point within the network and to co-ordinate the service logics related to the different access technologies in an appropriate way. The feasibility of unifying the mobility management of wireline terminals and wireless terminals was studied in the ACTS project EXODUS and will be demonstrated in field trials.

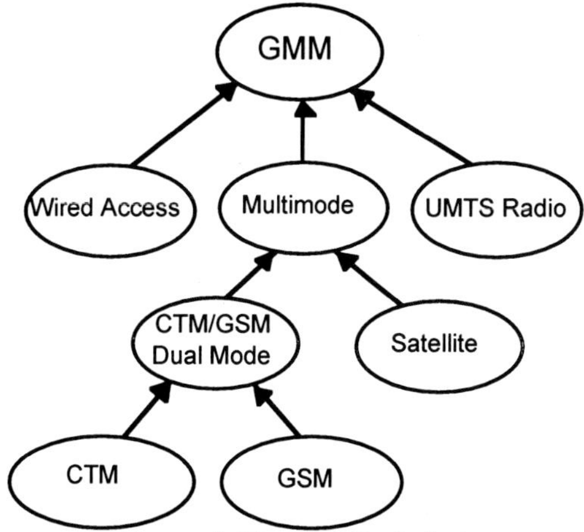

Figure 4: (Terminal Evolution)

Key issues to be addressed for multi-mode terminals include the seamless provision of services to the user. Services such as Calling Line Identification Presentation (CLIP) should behave in a consistent manner regardless of the access network. Likewise, call forwarding should function correctly and loops caused by the interaction of call-forwarding services in GSM and other mobile systems must be avoided. There is an important role to be played by standardisation bodies in ensuring the alignment and interworking of services.

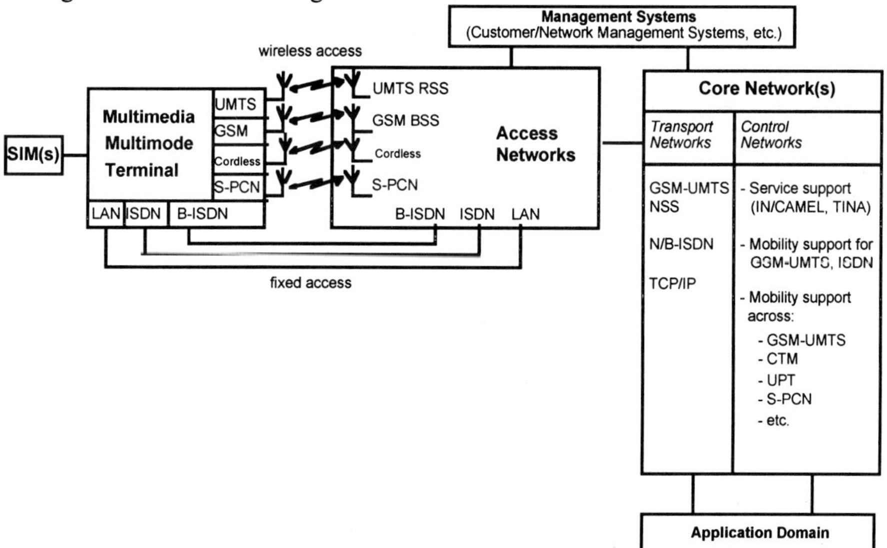

Figure 5: (Standardisation Framework for Global Multimedia Mobility)

4 IN & GSM Evolution Towards GMM

Within ETSI SMG2, a new base station subsystem using frequencies in the 2 GHz range is planned. GSM-core network enhancements by ETSI SMG3 for support of the new radio interface are also foreseen. These activities will lead to a GSM-variant of the Universal Mobile Telecommunications System (GSM-UMTS).

The UMTS Radio Sub-System (RSS) specified within SMG2 is expected to be designed in a manner which does not necessitate the use of the GSM-UMTS transport network. In line with the GMM report, this allows the emergence of UMTS-variants based on other transport networks such as the ISDN and B-ISDN switching networks. In the case where the ISDN or B-ISDN core networks support UMTS-access, IN-based techniques can play a role for mobility management. Where multiple access networks and multi-mode terminals are involved, IN will find an additional role providing support for the mobility management across various access networks. This will allow calls to be routed to the appropriate access network and provide handover between different access networks. The role of IN within the GMM Standardisation Framework is shown in Figure 5.

5 Conclusion

The GMM report states that evolutionary emergence from telecommunication networks and dual-mode GSM/UMTS radio capability will be instrumental for the successful introduction of UMTS. We consider the DECT/GSM dual-mode and CTM/GSM dual-mode services as necessary and pioneering steps, towards a multi-mode service supporting UMTS and other radio interfaces.

References

[1] Global Multimedia Mobility (GMM) A Standardisation Framework, PAC EG5 Report, Version 1.0.0, 27/03/96.

[2] Report of the Sixth Strategic Review Committee on European Information Infrastructure. Part B: Main Report and Annexes, June 1995.

Abbreviations

B-ISDN	Broadband ISDN
BSC	Base Station Controller
BTS	Base Transceiver Station
CAP	CTM Access Profile (DECT)
CAP	Camel Application Part (GSM)
CLIP	Calling Line Identify Presentation
CTM	Cordless Terminal Mobility
DCS1800	Digital Cellular System - 1800 MHz
DECT	Digital Enhanced Cordless Telephone
DECT FP	DECT Fixed Part
EII	European Information Infrastructure
E/km^2	Erlangs per square kilometre
ETSI	EuropeanTelecoms Standard Institute
EXODUS	(see below)
GAP	Generic Access Profile (DECT)
GII	Global Information Infrastructure
GMM	Global Multimedia Mobility
GMSC	Gateway MSC (GSM)
GSM	Global System for Mobile Comms
GSM900	GSM working at 900 MHz
GSM BSS	GSM Base Station Subsystem
UMTS NSS	UMTS Network Subsystem
HLR	Home Location Register (GSM)
IN	Intelligent Network
INAP	Intelligent Network Application Part
ISDN	Integrated Services Digital Network
LAN	Local Area Network
LE_o	Originating Local Exchange
LE_t	Terminating Local Exchange
MAP	Mobile Application Part (GSM)
MS	Mobile Station (GSM)
MSC	Mobile Station Controller (GSM)
PAC	Programme Advisory Committee
PCS1900	Personal Comms. Services 1900 MHz
PP	Portable Part (DECT)
PSTN	Public Switched Telephone Network
RF	Radio Frequency
SCP	Service Control Point (IN)
SIM	Subscriber Identity Module
SMG	ETSI Special Mobile Group
SSP	Service Switching Point (IN)
S-PCN	Satellite Personal Comms Network
TE	Transit Exchange
UPT	Universal Personal Telecoms
UMTS	Universal Mobile Telecoms System
VLR	Visited Location Register (GSM)

Acknowledgement: Part of the work cited has been carried out within the ACTS project EXODUS (EXperiments On the Deployment of UMTS) in which the authors are working. However, the views expressed are those of the authors and do not necessarily represent those of the project as a whole. EXODUS is partially funded by the EC and by the Swiss Government Office for Education and Research (BBW).

GSM Evolution to an IN Platform: Offering GSM Mobility as an IN Service.

Sutha Siva, Laurie Cuthbert,
Department of Electronic Engineering,
Queen Mary and Westfield College,
Mile End Rd, London E1 4NS, UK.
Tel: (44) 0171 415 3756
Fax: (44) 0181 981 0259
Email: {s.siva, l.g.cuthbert} @qmw.ac.uk

Malcolm Read,
Cellnet (UK),
260 Bath Road,
Slough SL1 4DX, UK.
Tel: (44) 01753 56 56 22
Fax: (44) 01753 56 50 21
Email: MREAD@cellnet.co.uk

1 Introduction

Work to date on the third generation *future, public land mobile* networks (known as FLMPTS or FPLMTS) and *Universal Mobile Telecommunication System* (UMTS) has assumed that mobility management within these networks will be delivered using *Intelligent Network* (IN) principles, that is the separation of the mobility *service* from the call control functions of the mobile switch. This separation is partially implemented within current *Global System for Mobile* communications (GSM) networks, but since these networks were defined prior to IN, the GSM mobility management functions (and similarly the GSM service components) are not defined as IN *Service Independent Building* blocks (SIBs). Also, because of the limitations of switch processing power at the time that GSM was specified, not all of the mobility functions were separated out from the GSM Mobile Switching Centre (MSC), to reduce the MSC signalling and processing load (within most GSM MSC implementations there are local mobility management functions - the GSM Visited Location Register).

GSM has been a major world-wide success, and the mobility functionality within GSM has been exhaustively tested, both within and between national GSM networks, and is a well proven standard. It is therefore highly likely that when the mobility functions for FLMPTS and UMTS are specified, they will be based on these well proven GSM principles, although the actual base protocol is likely to be derived from INAP. (It is currently proposed that mobility functionality is specified within Phase 2 of IN Capability Set 3).

At present the FLMPTS and UMTS architectures are only defined in functional terms, and the physical realisation of these networks, and network interfaces is yet to be defined. This paper presents an investigation of the performance issues involved in providing the GSM mobility management functions as IN SIBs, within a simulated, physical network, as a contribution towards the standardisation work for FLMPTS and UMTS.

The approach adopted within the study was firstly to model the signalling transactions within a GSM network, to validate the model against a known, operational implementation. Having validated the model, the GSM mobility functions within the model were then decomposed into mobility service components, modelled on IN SIBs, and the GSM mobility management functions realised within SCPs, separated from the Mobile Switching Centres. Two alternative IN scenarios were modelled, the first with a local temporary database (SDP_{temp}) associated with each MSC, and the second with centralised databases only.

This paper presents results from simulation studies carried out to compare the performance of the proposed architecture against GSM performance. Protocol based simulations models were developed on 'OPNET™' (a general purpose simulator), for

both the proposed architecture and the GSM architecture. From a signalling point of view, the location updating procedure is the most expensive and as such was the first investigated. Our results show the comparison between the two networks under different behaviour conditions and preliminary indications are that apart from the increase in signalling load on the core network, the IN approach does not significantly degrade the performance of GSM mobility procedures offered.
The opinions expressed here are those of the authors and do not necessarily reflect those of their employers.

2 GSM - IN : Integration Scenario

The GSM [8-11] network philosophy embodies elements of IN, such as the modularity of its architecture, distributed processing and the use of databases. These similarities combined with the use of ***common channel Signalling System No.7*** (SS7) [13-15] in both networks lead to two networks similar in their functionality. It is these similarities which make the proposals made here feasible.

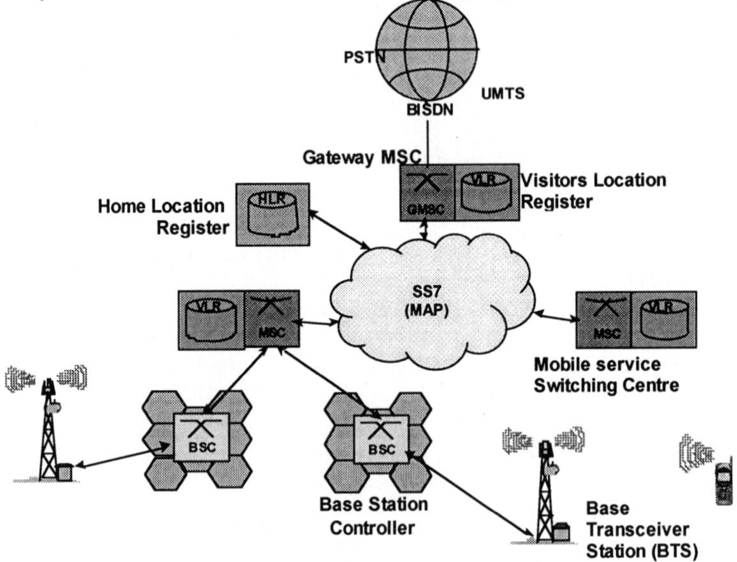

Figure 1: GSM architecture.

GSM mobility functionality is distributed between three GSM entities, the ***Mobile Switching Centre*** (MSC), ***Home Location Register*** (HLR) which is assumed to include the ***Authentication Centre*** (AuC) and the ***Visitors Location Register*** (VLR). Mobility functionality from these entities will need to be moved to an IN platform and offered as IN services. To be offered as IN services, they will need to be mounted on the ***Service Control Point*** (SCP). In IN, services are offered as a combination of ***Service Independent Building Blocks*** (SIBs).
From GSM mobility procedures a series of 'atomic elements' have been identified, which in effect are the SIBs necessary to offer GSM mobility procedures as an IN service. Using these GSM SIBs, call control, mobility management, service creation and service management in GSM will be offered from a single platform IN, eliminating the need for two control platforms. It would also use a single signalling protocol (INAP) for service control by absorbing MAP within INAP. Furthermore

this integration opens the door on a new set of services which are mobility based, such as Universal Personal Telecommunications (UPT).

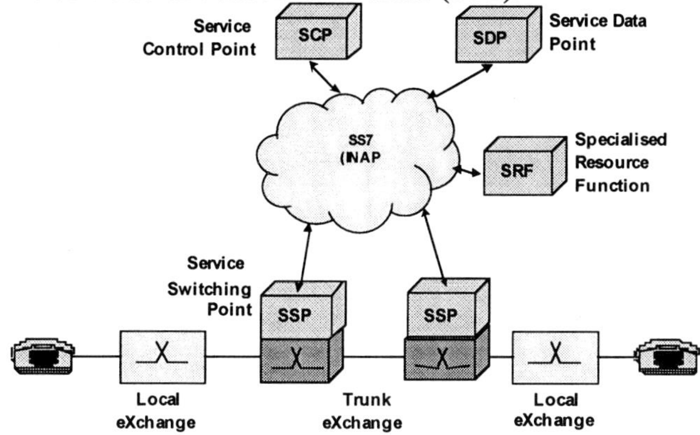

Figure 2: Intelligent Network Architecture.

2.1 GSM / IN Integrated Architecture

The resulting integrated architecture[12] with the following aims in mind is presented:

- Separate the radio access elements (BSS) and the mobility management elements (NSS) of the GSM network.
- Retain the radio access network with little or no change.
- Transform the mobility management and call handling network architecture to an IN architecture, as mobility will be offered from an IN platform.

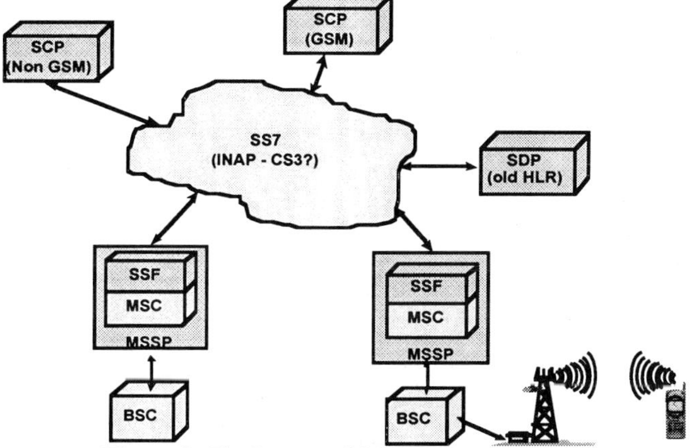

Figure 3: The Integrated IN / GSM architecture.

The result is the amalgamation of two networks, with GSM radio access network and IN control network. The point at which both networks meet, either the MSC or the ***Base Station Controller*** (BSC), will have to be fitted with functionality to recognise and manage IN service requests, i.e. with ***Service Switching Functionality*** (SSF). The MSC was chosen over the BSC for the following reasons:

- All mobility management messages and service messages from and to the mobile terminal are aimed at the MSC for forwarding or processing in the GSM network. Therefore it is in a central location to detect and process messages both from the mobile terminal and SCPs.

- The present GSM architecture will enable several BSCs to access a single SSF attached to a MSC.
- By not introducing IN elements to the radio access network, it will be possible in the future for the radio access network to evolve independently of intelligent networks and visa versa.

The node containing the MSC and SSF functionality is termed the ***Mobile Service Switching Point*** (MSSP). The switching point between the radio access network and the outside world will remain the MSC, as there is no reason to change this. The MSC will retain its radio channel and radio network control functionality. The functionality to handle inter BSC handovers will also be retained by the MSC. For inter BSC handovers, switching the call at any point other than the MSC serves to no advantage and the closest control point for inter BSC handovers is the MSC. The control of inter-MSC handovers will be by the SCP. The reason for the SCP handling inter MSC handovers is to eliminate the anchor MSC concept found in GSM networks and switch the call to the new MSC via the shortest path and not via the anchor MSC. Although this can easily be done for voice calls, it is not straight forward for data calls as the anchor MSC is the buffering point for the data traffic. The use of the shortest path for data calls is still under study.

The GSM user service profile (including roaming information) and the users IN service profile are combined to form a 'character set'. This character set will be maintained at the 'home' ***Service Data Point*** (SDP). Therefore any SCP with the appropriate service logic will be able to serve the user by downloading the necessary elements of the character set. This enables the local SCP to serve the user rather than the user being tied to a 'home SCP'. Having moved the mobility procedures from the HLR to the SCP and the database to the SDP, the need for a HLR no longer exists and we propose to remove it from the architecture.

The VLR in GSM serves two purposes. It provides mobility control functionality and serves as a temporary database for user information when the user is roaming the local MSC. This temporary storage reduces the signalling load on the core network by reducing the need to access the HLR. With the mobility control functions moved to the SCP, the VLR now only serves as a temporary database, hereafter referred to as SDPtemp.

In the integrated architecture, the SCP accesses the SDPtemp via the core network which will not necessarily reduce the signalling load on the core network. With the exception of the SCP and the MSC sharing a common physical platform. But what the SDPtemp does is, it reduces the load on the centralised databases, the SDPs. In a mobile environment this is critical as procedures such as periodic and intra MSC location updating will generate a considerable amount of traffic.

The effects of having or not having a temporary database (SDPtemp) attached to the MSC is investigated as part of the simulation studies. As a result two variations on the integrated architecture arise; with a SDPtemp attached to the MSSP and without a SDPtemp. The following sections look at the integrated architecture in detail.

2.2 Mobile Service Switching Point

The ***Mobile Service Switching Point*** (MSSP) will comprise of; MSC functionality for supporting the radio access network, ***MSC Call Control Functionality*** (MCCF) and ***Service Switching Functions*** (SSF). SSF will identify request for IN services, interact with other IN entities to invoke, execute and manage IN services and present the SCP with a view of MSC's switching functions controllable by the SCP. The existing MSC and SSF functionality will need the following modifications to form the MSSP.

- Traditionally IN services are requested during a call. To enable the processing of the service requested, the state and progress of the call is monitored by the IN call model. The IN call model is based on the ***Basic Call State Model*** (BSCM), which is a finite state model description of the ***Call Control Function's*** (CCF) call processing activities. The MSC in GSM has a call model which needs to be enhanced to cater for the IN environment as the MSC call model is currently oriented towards mobility[16]. The GSM call model will need to be enhanced such that it can interact with the SSF.
- The signalling between the MSSP and the SCP will be using INAP. Between the MSSP and the mobile terminal the ***Radio Interface Layer 3 Mobility Management*** (RIL3-MM) and ***RIL3-Call Control*** (RIL3-CC) signalling protocols are used. Therefore the MSSP will need to be installed with a translation function between these two signalling protocols.
- The MSC will now be in receipt of two types of signalling messages, messages related to the radio access network control and management, and messages related to IN services. These messages will obviously require different processing. Messages relating to IN services need to be passed on to INAP, while messages regarding the radio access network will need to be forwarded either to BSS Management Part (BSSMAP) or Direct Transfer Application Part (DTAP). The proposed modification to MSC's protocol stack is the addition of a further distribution layer above the existing distribution layer. The new distribution layer will separate between DTAP and INAP.
- In GSM request for services can be made in the presence or absence of a call. For example requests for the 'call waiting' service or handovers are made during a call. While location update requests or Short Messaging Service (SMS) requests may be made outside a call. The mechanism for the detection of service requests outside a call exist in GSM. This mechanism will then need to invoke the appropriate 'call model' for the service. The service detection mechanism is a CCF functionality and not a SSF functionality. Therefore in the IN / GSM integrated network, the GSM detection mechanism can continued to be used with enhancements to trigger the SCP.
- For the execution of certain services, the SCP needs to know the state of the switch and have control over it. Inter MSC handovers is such a service. At present the capabilities offered by CS-1 IN-Switching Manager (IN-SM) would seem adequate to meet these requirements.
- Services may be invoked in the absence of a call, during a call and in parallel with the execution of a current service. As such the question of compatibility and interaction between service becomes complex to solve for the ***Feature Interaction Manager*** (FIM). The CS-1 FIM approach of solving interaction and compatibility issues between services by maintaining a look up table of all possible interactions for all possible combination of services is not viable. As suggested in CS-1, a knowledge based approach is probably needed.

All GSM mobility control procedures have been transferred to a common IN platform (SCP) from the various GSM nodes (HLR, MSC, VLR), but none of the procedures have been modified. To the radio access network and the GSM user, this change is transparent. The mobile terminal does not see any changes in the network For all intends and purposes, the mobile terminal still believes its communicating with the MSC, as such is unaware of any IN components. As IN mobility control functionality imitates GSM functionality, there is no need to suspend call processing as in POTS to conduct IN processing. This is only true for GSM mobility procedures offered as IN services. When GSM mobility procedures offered as IN services are combined with other IN service, it may become necessary to suspend call processing and transfer control to the SCP. Figure 4 illustrates a possible combination of MSC and SSF functionality in the MSSP.

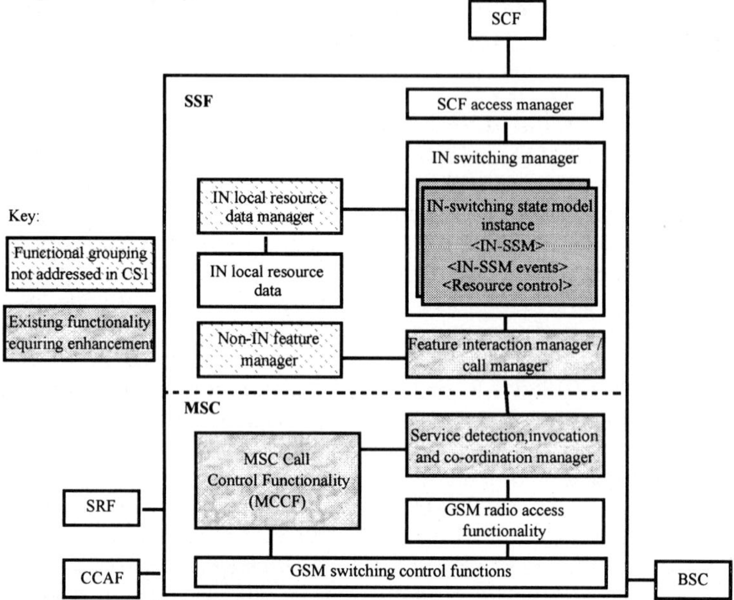

Figure 4: The MSSP functionality based on CS-1 SSF functionality.

2.3 Service Control Point

The integrated architecture does not require any modifications to the SCP. But in CS-1, SCPs cannot interact to share, negotiate or handover control between SCPs. Interworking between SCPs is being addressed in CS-3. This additional functionality will be of benefit for efficient provision of GSM services from an IN platform.

2.4 Mobile Service Independent Building Blocks

By modularising GSM mobility procedures (location updating, handovers, call set up, etc.), it is possible to identify commonality within the various procedures. The common sub procedures are authentication, issuing a new TMSI, paging, database enquiry and database updating. These sub procedures are self contained and identical immaterial of the GSM mobility procedure being used in. Each of these sub procedures will need to be converted to a SIB, i.e. ***Authentication*** SIB, ***Paging*** SIB and ***TMSI*** SIB (includes ciphering of channel). For database enquiry and database updating a derivative of the CS-1 ***Service Data Management*** (SDM) SIB can be used. The modified SDM SIB will be referred to as ***Mobile Service Data Management*** (MSDM) SIB. These SIBs will offer most of the functionality required for GSM mobility functions, but not all. Depending on the mobility procedure, further specialised SIBs will need to be introduced for inter MSC handovers, completion of incoming and outgoing GSM calls.

IN CS-3 [17] has called for the inclusion of several mobility based benchmark services into the IN network capabilities. These include services such as User Authentication (UAUT), Handovers, Terminal Location Registration, Terminal

Paging, Mobility Call Origination among others. These services can be composed using GSM SIB. GSM has a tried and tested mobility functionality which would save the wheel being reinvented. If GSM SIBs are inadequate for offering any of the CS-3 mobility services, then an enhanced set of mobility SIBs can be derived from the GSM SIBs.
The advantage is that a single set of standards will be used and it eliminates the need for multiple application protocols (MAP, INAP) as highlighted in [16,18-19], while maintaining backward compatibility.

3 Simulation Model and Results

This paper has presented a functional network architecture for offering GSM mobility services from an IN platform and has proposed modifications necessary to the GSM network to achieve it. In support of the proposal, the feasibility, performance and constrains of the IN/GSM integration scenario needs to be investigated. Furthermore it is prudent to know if the proposed network architecture will continue to maintain the GSM performance levels. Moreover the proposals made so far have been at a functional level, from which several physical implementations may be realised. As a result different physical implementation will lead to different levels of performance. In order to investigate these performance issues simulation models were built using 'OPNET™', a commercial general purpose telecommunication networks simulation tool.
Three simulation models were developed, first modelling the GSM network, the second modelling the IN/GSM integrated architecture with SDPtemp and finally the IN/GSM integrated architecture without the SDPtemp. These simulation models are based on signalling protocols and will focus on mobility procedures such as location registering/updating, mobile originating/terminating call set-up, handovers and gateway MSC functionality. Figure 5 and 6 (when a SDPtemp is present) and Figure 7 (in the absence of a SDPtemp) show how GSM location updating procedures for both inter and intra MSC cases were modelled in the simulation. When SDPtemp is absent, there is no difference in the procedures for intra and inter MSC location updates. The message sizes used are based on the equivalent GSM message size[20-21].
The aim of this simulation study is to present a relative analysis between the existing GSM network and the IN/GSM scenarios from a signalling point of view. The model uses simple queueing models for signalling links and access to node processing. Mobility is modelled using the fluid flow model [22-23].

3.1 Signalling Network

The IN/GSM integrated architecture makes use of three types of transmission mechanisms for signalling, just as the GSM network. The radio interface is used between the mobile terminal and cell site (BTS), ***Link Access Protocol for the ISDN 'D'*** channel (LAPD) a derivative of NISDN signalling for the Abis interface (between the BTS and the BSC) and SS7 between all other network nodes.

3.2 GSM Radio Access Network

Both the GSM and the IN/GSM integrated architecture use the same radio access network, i.e. from the MSC to the mobile terminal. Similarly both the GSM and IN/GSM simulation models use the same radio access network model. The radio access network is modelled as a black box which contains the mobile terminals, the cell sites (BTS) and the Base Station Controller (BSC).

For signalling over the radio interface, the Stand-alone Dedicated Control CHannel (SDCCH) is used for signalling outside a call and the Fast Associated Control CHannel (FACCH) during the call [9-10]. Over the Abis interface 64kbps links are used. The radio access network 'black box' is connected to the MSC via 64 kbps links. When compared to the effective rates of signalling links over the radio interface (about 1 kbps for SDCCH and 9 kbps for FACCH), the propagation delay over the radio interface is negligible. Between the BSC and the BTSs, a mean separation of 25Km is assumed.

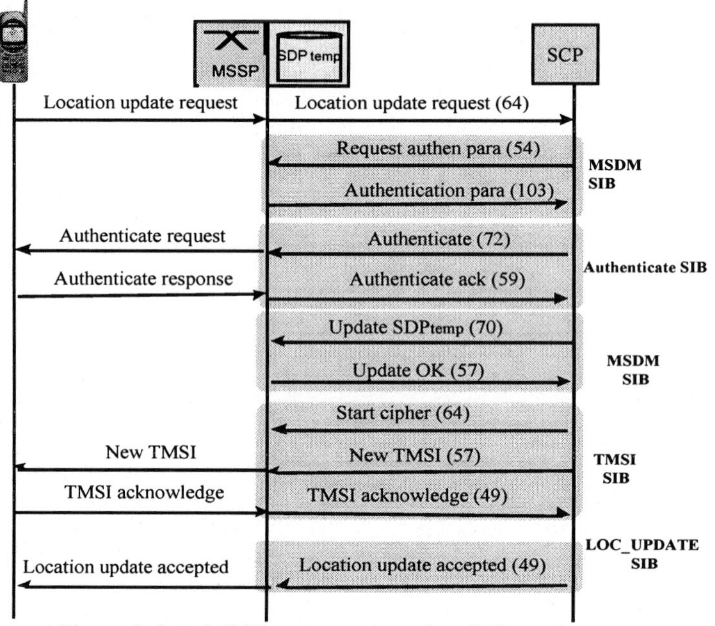

Figure 5: Intra MSC location update when SDPtemp is present.

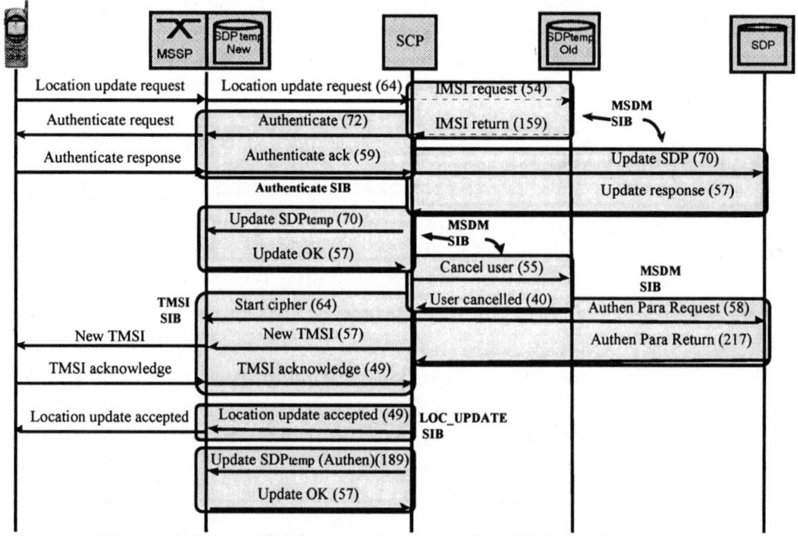

Figure 6: Inter MSC location Update when SDPtemp is present.

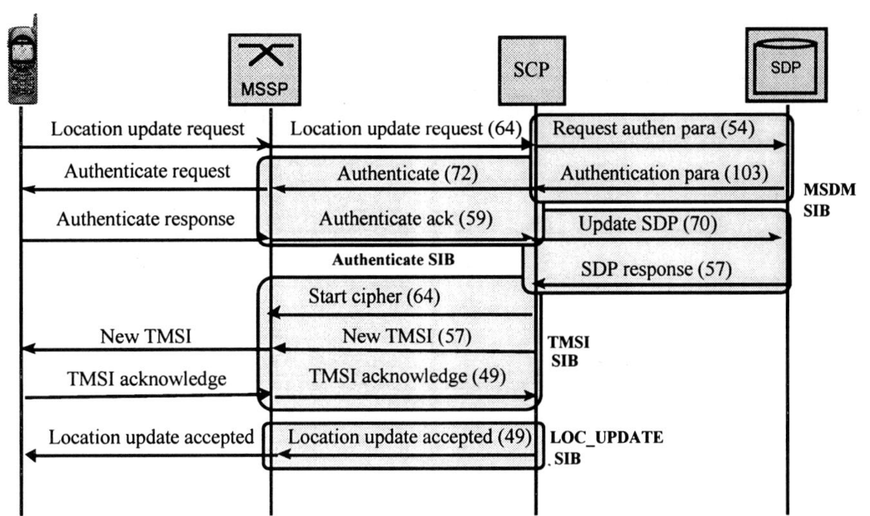

Figure 7 : Location update when SDPtemp is absent.

Node processing times within the radio access network are assumed to be 400μs for forwarding of messages by the BSC and BTS. A 800μs delay for composing of messages or for composing replies is used.
The request for mobility services are generated by the mobile terminal module. The rate at which services are requested (such as location updates), the holding time for services (such as call duration) and type of service (mobile to mobile call or mobile to fixed call) is determined using a combination of data collected from Cellnet (UK)'s networks and using the fluid flow model. This simple fluid flow model gives the number of boundary crossings out of an area with a perimeter of length L, where the user density (ρ) is uniform throughout the area. Users within the area are equally likely to move in any direction and have an average speed of V. The rate of boundary crossings is given by

$$M = \frac{\rho \, VL}{\pi} \text{ crossings per second.}$$

This equation is used to determine the rate of handovers and location updates.
Finally, each user on the network is not modelled individually, but each request for a service is given a unique ID. This ID is used by all subsequent signalling messages. Therefore it is possible to relate two service requests, such as a handover request to a particular call request.

3.3 Network Subsystems

MSC/VLR In The GSM Model

In the GSM simulation model, the MSC and the VLR were modelled as one node. Messages from the signalling network have access to the MSC processing via a single queue. A delay of 400μs is used for forwarding of messages to the VLR by the MSC. At the VLR 800 μs is used for composing a new message or replies. Furthermore for VLR (and the HLR) database access delays of 3ms, 6ms and 10ms is used for reading, updating and deleting of records.

The Mobile Service Switching Point

For the MSSP two variations were developed. The first with access to a temporary service data point (SDPtemp) and secondly without a SDPtemp. As in the GSM MSC model a single queue is used to access the MSC processing here. For messages from the radio access network, the MSC forwards the appropriate data to the service switching functionality (SSF) to compose the equivalent IN message to be sent to the SCP. A 400μs delay is used for forwarding the data and a 800μs delay for the SSF to compose the appropriate IN message. For messages from the SCP, the reverse is true. When a SDPtemp is present, access is via the MSC. SDPtemp access times are as those for a VLR.

The Service Control Point

When a service request is made to the SCP (parent process), an instance of the service is created (child process) form the library of services. Every request for a service will create a new service instance. The service instance is actually the service logic processing program (SLP) which determines the logical execution of service independent building blocks (SIBs) that make up the service and the flow of information between them. The SLP will create instances of SIBs (grandchild process) as required and it is the SIBs which generate the signalling messages. Each instance of a service execution or SIB has a unique ID and all signalling messages contain this ID. All replies to messages from the SCP will include the service ID and hence the SCP (parent process) can pass the information on to the appropriate service or SIB instance.

From a development point of view, having a library of services and SIBs allows for new services to be simulated easily. Having each request for service associated with an unique instance allows for a detailed monitoring of service progression. It also helps capture a better picture of the processing power that would be required of a SCP.

Signalling messages access the SCP node processing via a single queue. Figure 8 illustrates the SCP node model.

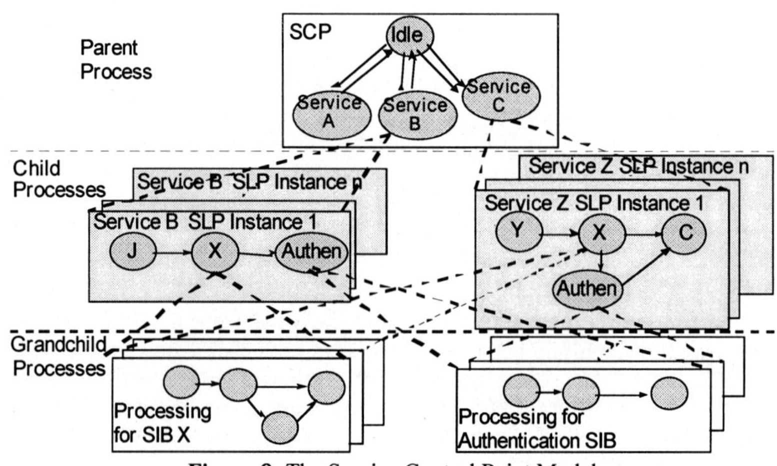

Figure 8: The Service Control Point Model

The SS7 Signalling Link

A simplified SS7 signalling link model is used, where signalling packets are received by layer 3 with all the MTP headers included. A delay 1ms is introduced for level 3 and level 2 processing. The access to the physical link is at 64kbps and the access delay is based on the MSU size. The propagation delay is calculated depending on the distance between nodes at half the speed of light. An error free transmission environment is assumed.

Routing of messages are carried out by MTP level 3 and a routing table is established at the beginning of the simulation.

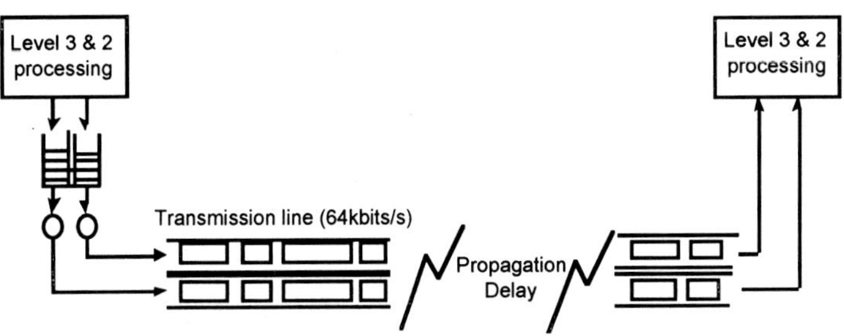

Figure 9: SS7 Signalling Network Model

3.4 Simulation Results

The initial set of simulations carried out were to compare the performance of the proposed architectures with the GSM network for conducting location updating. Location updating is the most expensive of all mobility procedures in terms of signalling. With no changes being made to the radio access network and the control elements being moved around in the core network, it is the volume of signalling that would have the biggest influence on performance.

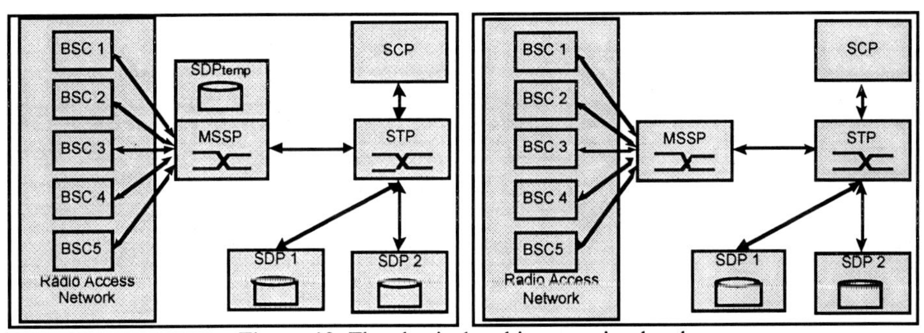

Figure 10: The physical architecture simulated.

Figure 10 shows the physical architecture used in the simulations. From a GSM mobility management point of view, it is the worst case scenario as the SCP is physically separated from the MSC (MSSP). Hence every single communication with the mobile terminal by the SCP has to be routed across the core network and so has the request for information from the SDPtemp.

The scenario simulated is for a MSC covering a circular area of 25 miles radius with 500,000 users. The average speed of the users is assumed to be 5 miles/hour. Only location updates as a result of user mobility were simulated. The results do not include periodic location updates.

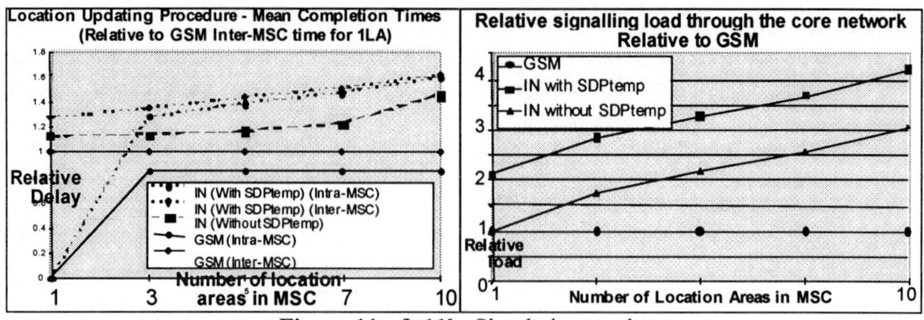

Figure 11a & 11b: Simulation results.

The mean time for completion of location update procedures are shown in Figure 11a. The times are relative to inter MSC location update in the GSM network (1.25s). The above results were obtained by using the same number of SS7 links between the various entities on all 3 models. For the GSM network the volume of traffic generated in the core network does not change as the number of location areas in the MSC varies. Figure 11b illustrates how the signalling load for the two IN / GSM architectures increase with increasing number of location areas in a MSC. As the number of links is fixed, increase in the signalling load will push the link utilisation levels up from 30% for 1 location area to 80% for 10 location areas for the IN / GSM architectures. Hence the delays associated with the links will increase and this is reflected in the time taken to complete location update procedures in Figure 11a.
When the SDPtemp is absent in the IN / GSM architecture, the signalling load is evenly distributed between the SCP-SDP and SCP-MSSP links. With the SDPtemp the load is more concentrated on the SCP-MSSP link. Hence the location update procedures take longer with SDPtemp. As stated this is the worst case scenario as the point of control (SCP) is separated from the MSSP (MSC). But even for this worst case scenario the performance of the integrated architecture with 30% loading on the links is acceptable. When a SDPtemp is present the average increase in completion time is 40% and 20% in the absence of SDPtemp. These results would suggest that performance of the integrated architecture is better when the SDPtemp is absent for this particular physical implementation. The obvious disadvantage of not having a SDPtemp is the load on the SDP as it needs to be interrogated every single time. For 3 location areas the load on SDP is doubled when the SDPtemp is absent and 3 times greater for 10 locations areas. Only after simulating different physical implementations and other mobility services can the case for having or not having a SDPtemp be fully made.

4 Conclusion

An architecture based on the integration of IN and GSM was presented as a possible path for the evolution of GSM to UMTS. From the standards point of view what is need is the absorption of MAP into INAP, resulting in a unified protocol. With a unified protocol the architecture presented here becomes viable. The decomposition of GSM services into SIBs would also enable greater flexibility in the creation of new services within GSM and a single IN service environment would enable potential conflicts between new services to be resolved easily. This integrated scenario also opens the door on new mobility based services such as UPT using existing GSM services.

A further advantage of evolving the mobility signalling procedures for UMTS from GSM, is that the UMTS core network is intended to support a number of radio access networks, including GSM. The proposals in this paper would enable this objective to be achieved.
The effect of moving from an architecture dedicated to mobility (GSM) to a general control architecture is the increase in the signalling volume through the network. This is to be expected and hence shown in the simulation results. But the initial simulation results would indicate that the level of performance of GSM networks can be maintained in the integrated scenario and the support of GSM mobility signalling procedures within an IN network is certainly feasible.
The authors will like to thank Cellnet (UK) for the funding of this study.

References

[1] Katoen J-P, Saidi A, Baccaro I, A UMTS network architecture, RACE Mobile Telecommunications Workshop Amsterdam, May 1994.
[2] Van den Broek W, Georgokitsos K, Impact of UMTS developments on IN standardisation, RACE Mobile Telecommunications Workshop Amsterdam, May 1994.
[3] Lobley N C, Intelligent mobile networks, BT Technology Journal, Vol. 13 No 2, April 1995.
[4] Th^rner J, Intelligent Networks, Artech House, 1994.
[5] Duran, J M, Visser J, International Standards for Intelligent Networks, IEEE Communications Magazine, Feb 1992.
[6] Abernethy T,W, Munday A,C, Intelligent networks, standards and services, BT Technology Journal, BT Labs, Vol. 13 No.2 April 1995.
[7] ITU recommendations Q1201 - Principles of Intelligent Network architecture.
[8] Laitinen M, Rantala J, Integration of Intelligent Network Services into Future GSM Networks, IEEE Communications Magazine, June 1995.
[9] Redl S M, Weber M K, Oliphant M W, An introduction to GSM, Artech House, London, 1995.
[10] Mouly M, Pautet M B, The GSM system for mobile communications, M.Mouly et M.B. Pautet, Palaiseau, France 1992.
[11] ETSI GSM recommendations 03.02 - Network architecture.
[12] Siva S, Read M, Cuthbert L, GSM/IN integration - offering GSM mobility procedures from an IN platform, 13th IEE UK Teletraffic Symposium, Glasgow, March 1996.
[13] Manterfield R J, Common Channel Signalling, Peter Peregrinus Ltd, 1991.
[14] Modarressi A R, Skoog R A, Signalling System No.7 : A Tutorial, IEEE Communications Magazine, pg.19-35, July 1990.
[15] Roehr W C, Inside SS No.7 : A Detailed look at ISDN's Signalling System Plan, Data Communications, pg. 120-128, Oct. 1985.
[16] Lilly N, Integrating GSM and IN architectures and evolving to UMTS, GSM MoU / ETSI Workshop on The Evolution of GSM towards IN, Brussels, 1-2 February 1995.
[17] ITU - WP 4/11, IN CS-3 Benchmark Services, Nov 1996.
[18] Lilly N, Integrating and evolving the INAP and MAP protocols, GSM MoU / ETSI Workshop on The Evolution of GSM towards IN, Brussels, 1-2 February 1995.
[19] Lilly N, The ETSI Framework for Evolving GSM and IN towards UMTS, GSM MoU / ETSI Workshop on The Evolution of GSM towards IN, Brussels, 1-2 February 1995.
[20] Meier-Hellstern K, Alonso E, Signalling System No.7 Messaging in GSM, Technical Report WINLAB-TR-25, Rutgers University, December 1991.
[21] ETSI GSM recommendations 09.02 - Mobile Application Part Specifications.
[22] Pollini G P, et al, Signalling Traffic Volume Generated by Mobile and Personal Communications, IEEE Communications Magazine, June 1995.
[23] Pollini G P, Goodman D J, Signalling System Performance Evaluation for Personal Communications, IEEE Transactions on Vehicular Technology, Vol. 45, No 1, February 1996.

Integration of Mobility Functions into an Open Service Architecture: The DOLMEN Approach

S. Palazzo[1] * , M. Anagnostou ** , D. Prevedourou *** , M. Samarotto * , P. Reynolds ****
* University of Catania, Italy ** National Technical University of Athens, Greece
*** Intracom S.A., Greece **** University of Plymouth, United Kingdom

This paper analyses the impact of mobility functions to an open Service Architecture for a mixed fixed and mobile environment like the one the ACTS DOLMEN project is currently developing, demonstrating, and promoting. In particular, the paper identifies the architectural means needed for the support of terminal and personal mobility functions in a telecommunication environment encompassing mobile networks, and outlines some possible approaches to the placement of these functions into the Reference Service Architecture adopted in DOLMEN.

1 Introduction

In the last years different Service Architectures have already been devised in order to facilitate the creation, deployment and maintenance of services over a variety of fixed networks. They address the needs of traditional telecommunication services, future interactive multi-media services, information services, operation and management services, etc., and provide the flexibility to operate these services over a wide variety of technologies. The main effort in this field was made and is still in progress within the TINA (Telecommunications Information Networking Infrastructure) Consortium [1], which is leading to an emerging *de facto* standard as a service architectural framework. Other examples of service architectural frameworks were developed in the past within the RACE Programme, for instance the so-called OSA (Open Service Architecture) defined by the CASSIOPEIA Project [2].
As interest for PCS (Personal Communications Services) is rapidly growing and new generation mobile systems like UMTS (Universal Mobile Telecommunication Systems) are anticipated as forthcoming [3], current Service Architectures must be upgraded to cover support of mobility. In fact, fulfilment of the PCS target, that is, the ubiquitous coverage enabling anyone to communicate instantly with anyone else anywhere in the world, implies that all the functions needed to manage mobility should be hidden from the users accessing the PCS services. This can be effectively achieved through the adoption and the use of a reference Service Architecture encompassing support of mobility.
Within the ACTS Programme, the DOLMEN Project has the main objective to develop, demonstrate, assess and promote an Open Service Architecture for a fixed and Mobile environment (OSAM) in which PCS services are deployed. The Service Architecture developed in DOLMEN is intended to feature conceptual, computational and engineering support to mobility, basing upon TINA architectural framework and upon results from the previous RACE projects on service engineering and mobility.
In this paper we present the methodology adopted in DOLMEN to identify which parts of a Service Architecture should be affected by mobility functions. The methodology used follows two different approaches, which reflect two complementary views of the provision of mobility services. The two approaches, named *Service Engineering* and *System Engineering* respectively, differ with respect to whether mobility is assumed to be provided by the Service Architecture itself or by

[1] Corresponding author: Prof. Sergio Palazzo, Istituto di Informatica e Telecomunicazioni, Università di Catania, V.le A Doria, 6 - 95125 Catania, Italy.
Tel:+39 95 339449 Fax:+39 95 338280 email: palazzo@iit.unict.it

a network infrastructure where, for example, UMTS functions are included. These also constitute two different ways (*recipes*) to implement a system compliant to the Service Architecture. The pros and cons of the two approaches are analysed. In order to illustrate how the methodology can be applied, a case study regarding a significant personal mobility function is finally presented.

2 The Reference Service Architecture

A *Service Architecture* dictates concepts, rules, guidelines and prescriptive models for the design of advanced telecommunications services and of the systems supporting their provision. The notion of Service Architecture is an advance on the Intelligent Network (IN) principle of separating physical-connection-oriented call processing from service-oriented call processing [4]. The notion of call is replaced by the notion of a service instance since in the long-term services call orientation is not necessarily among the key criteria. As compared with the IN approach, the vision adopted in the TINA-based approaches, like the one taken by DOLMEN, represents a new foundation in Service Engineering. In fact, intelligence for control and management of services and resources is distributed among network nodes and user/terminal nodes. In contrast, the IN approach of open service provisioning is based upon creation of newer and newer capability sets as needed by new classes of services, while keeping an already standardised INCM (Intelligent Network Conceptual Model) where all the intelligence is in the network elements.
The key design principles of the *Reference Service Architecture* adopted in DOLMEN are :

- system-independent modelling of services ;
- nested services for defining and providing new services ;
- coherent design of management and control ;
- customisation and personalisation of services ;
- separation of media services from their control and management ;
- separation of service access from service core ;
- separation of application and session as well as resource and communication-oriented problems.

Conformance to these principles results in good and flexible design of open-ended systems offering a multiplicity of (customised) quality services. Such systems also exhibit service flexibility allowing public and proprietary services to be integrated and combined in a modular way.
As illustrated in Figure 1, the main elements of the DOLMEN *Reference Service Architecture* are :

- An *Application stratum*, which includes all elements specific to applications fully in the realm of end-users (clients and servers) and totally transparent to Telecommunication Actors [A];
- A *Service Machine*, which is an abstraction of the whole set of software entities involved in service creation, provision and management ; it includes:
    - communicating *control and management service components* that can be invoked to support various applications [C],
    - a service execution environment providing an *open distributed processing platform* for interaction and exchange between entities possibly residing in different nodes, both fixed and mobile [P], and

- *network resource adapters*, aiming to provide standard ways to access, control and manage the various elements of the network resource infrastructure [N].
- A *Network Resource Infrastructure*, which comprises a variety of network technologies (narrowband and broadband, fixed and mobile, access and core domain) and elements to support them (protocols, interfaces, dedicated computing devices, databases, etc.) [I].

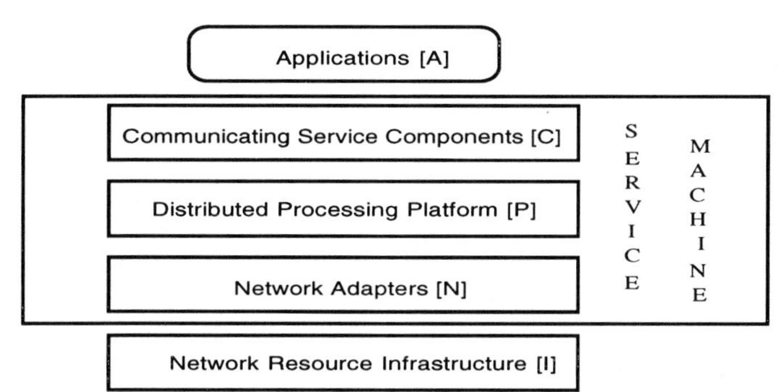

Figure 1: The DOLMEN Reference Service Architecture

The Service Machine, which is the core of the Reference Service Architecture in DOLMEN, can be regarded as a useful alias for the whole set of TINA compliant entities belonging to its overall framework. In particular, according to the TINA model, the above defined elements, [C], [P] and [N], are equivalent to the concepts of TINA Service Architecture, DPE (Distributed Processing Environment) and KTN (Kernel Transport Network), respectively. The categorisation of the Service Machine entities into several strata, which was introduced in DOLMEN, adds a new useful dimension to the TINA framework, in that it helps to separate concerns for service modelling and service reuse.

The Service Machine is concerned with the provision of computational and engineering support for service definition and deployment.

The computational model of the Service Machine addresses:

- Management aspects, which refer to software modules implementing Fault, Configuration, Accounting, Performance and Security management of the services and the resources needed;
- Session Control aspects, which provide software modules and their relationships or associations, whose goal is to control service provisioning to the end-user (establishment, usage and release of sessions are supported and three types of sessions are defined, i.e., access, service and communication sessions.)
- Connection Control aspects, which model transport network connection establishment (comprising software modules for negotiating, setting up, maintaining and releasing telecommunications connections on a connection-oriented basis.)
- Provision Support aspects, which mainly refer to a set of software modules implementing services that are commonly needed in the support of service provision and can be (re-)used for the definition and provision of other services

(these include billing accounting, call logging, event notification and handling, service directory, etc.)

- Resource aspects, which mainly refer to software modules modelling services of the resource infrastructure together with their service metrics (services of the resource infrastructure can refer to transport network elements or legacy databases or even already existing software)
- The engineering model of the Service Machine, which is often referred to as the *Service Network*, addresses:
- *Distribution Support Services*, which comprise a software platform providing distribution transparencies (distribution transparency is defined as the property of hiding from a particular user the potential behaviour of some parts of a distributed system : distribution transparency may involve access, failure, location, migration, persistence, relocation, replication and transaction transparency, as well as engineering objects implementing trading, binding, etc).
- *Service Nodes*, which are the actual hosts of the Service Machine software modules and of the Distribution Support Services (the structure of a Service Node follows the structure proposed in the Engineering Viewpoint of the ODP-RM [5]).
- *Kernel Transport Network*, which enables the interconnection of and

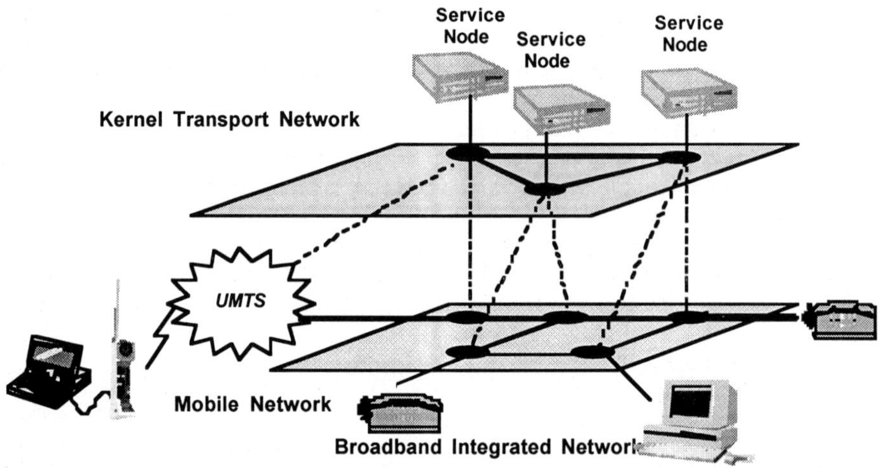

Figure 2: Engineering View of the Service Machine

communication between Service Nodes.

- *Resource Adaptation*, which enables access to and usage of the services of the Resource Infrastructure (the concept of Resource Infrastructure refers to whatever is available from a system which is not part of the Service Network but can be used, controlled and managed by it ; the Resource Infrastructure is seen as a set of interconnected Resource Nodes : a Service Node can be related to any number of Resource Nodes and vice versa.).
- An overall view of a Service Network is shown in Figure 2.

3 Mobility background

In order to evaluate the main impact of mobility on the Service Architecture the following steps have to be followed:

- to specify and classify the functions and requirements a system providing mobility must offer and satisfy,
- to identify which of these functions have impact on the Service Architecture,
- to decompose mobility functions into elementary components and, for each of them, to determine which parts of the Service Architecture are significantly affected.

With respect to the first issue, an analysis of the mobility functions to be considered has to be carry out at quite an abstract level, irrespective of any possible system architecture or network environment configuration. The classification of mobility functions has to be made in order to separate concerns, thus facilitating identification of architectural requirements.

The mobility functions which should be considered can be straightforwardly derived from the context of the current technical and standardisation activities, such as UPT [6, 7] and UMTS [8]. In principle, the functions under study can be distinguished in two main groups: the terminal and the personal mobility functions.

Functions belonging to the first group provide basic feature for a mobile network and have already been implemented in many analog and digital cellular systems of the past and current generations such as NMT, TACS, AMPS and GSM. *Terminal mobility* refers to the "ability of a terminal to access telecommunication services from different locations and while in motion, and the capability of the network to identify and locate that terminal" [8]. Examples of such functions are handover, location and domain updating, attach, detach, etc.

The second group of functions supports the user with new capability in telecommunication networks. *Personal mobility* refers to the "ability of a user to access telecommunication services at any terminal on the basis of a personal telecommunication identifier, and the capability of the network to provide those services according to the service profile of the user. Personal mobility involves the network capability to locate the terminal associated with the user for the purpose of addressing, routing and charging of the user's calls" [8]. Among the functions belonging to this group we can mention for example user registration and de-registration, user access and authentication, etc.

Without going through in detail into the description of all these functions and into their decomposition into elementary functional entities, we will include in the following a description of the two approaches, named *Service Engineering* and *System Engineering* respectively, which have been established in DOLMEN to evaluate the impact of the mobility functions on the Service Architecture. These approaches also constitute two different *recipes* to implement a system compliant to the Reference Service Architecture. In the Service Engineering recipe the assumption is that mobility functions are implemented as part of the Service Machine, whereas in the System Engineering recipe the assumption is that mobility is mainly taken care of by the Network Resource Infrastructure. A detailed example of the analysis carried out for identifying the impact of mobility functions according to the two different approaches will be shown in the case study presented at the end of this paper.

4 The Service Engineering Approach

According to the Service Engineering approach the complete Service Architecture will be exercised, applied and evaluated in the creation and provision of services (components) which support mobility. The assumed Network Resource Infrastructure involves both mobile and fixed transport networks *deprived* of intelligence to control and manage resources : such intelligence is considered to be part of the Service Machine. In this approach the evaluation of the impact of mobility on the Reference Service Architecture can be carried out by considering mobile communication services as any other telecommunication service: most of the functions related to mobility should be explicitly provided by and included in the Service Machine.
The mobility support constituent of the Service Machine must be taken into account as a means for providing mobile users with network (and service) access, in a transparent way. In other words, the service user should not be bothered with mobility aspects and should not be able to notice mobility in general. The only observable impact of mobility on the user is a certain limitation on the quality of service imposed by the limited radio interface bandwidth and by the specific type of terminal, which is designed for portability.
According to the Service Engineering approach, for example all the Personal Mobility procedures should be included in the Service Machine and designed according to Service Architecture principles. In this approach, because "Personal number", "User Profile" and "Terminal Profile", are key concepts strongly related to the Personal Mobility provision, their definition and management should be supported by the Reference Service Architecture. As a consequence, the main impacts of Personal Mobility on the Service Architecture concern the capability to define abstract data structures capturing the semantics of such concepts, and to create components to manage such data structures irrespective of their real implementation and location.
The result of this approach is summarised in Figure 3, where all the mobility functions and the resource adapters to the mobile network are identified as generic Mobility Components.
The Service Engineering approach guarantees requirements of generality, reusability and flexibility which can be useful for future development of new added value services. It also makes possible a direct control over the infrastructure in order to provide the required Quality of Service (QoS).

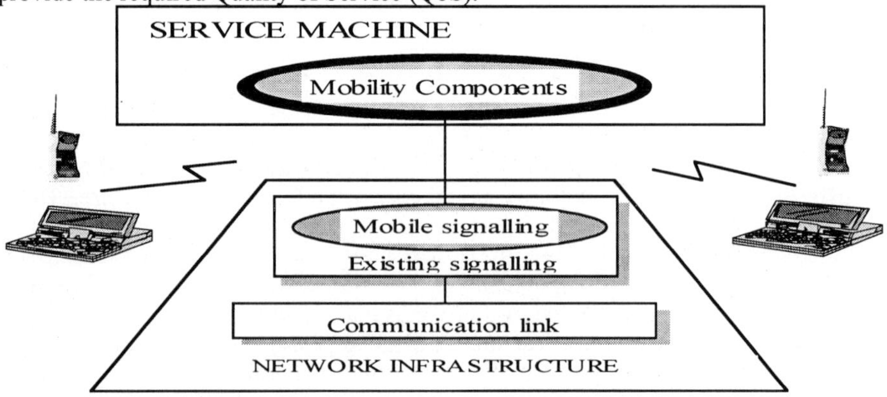

Figure 3: Mobility in the Service Engineering approach

5 The System Engineering Approach

The System Engineering approach starts from the assumption that mobility functions are provided by existing mobile systems, which is realistic in the case of UMTS. For this reason they can be considered as part of the Network Resource Infrastructure, like other basic telecommunication services, and hidden behind suitable adapters in the user/network interface. In this case the Service Architecture should be minimally affected by mobility functions and the Service Machine will include an interface to these systems as it does to other basic telecommunication services.

In this sense the main difference between the two approaches is that the System Engineering approach simplifies the Service Machine by considering mobility functions inside the Network Resource Infrastructure, as illustrated in Figure 4. The existing mobile system is assumed to provide terminal mobility and to support personal mobility in its most general standardised concepts.

In accordance with the above assumptions, and considering that mobility functions are already provided in a Network Resource Infrastructure that includes new generation mobile systems, the main impact on the Reference Service Architecture is that the Service Machine has to deal with the interface provided by the underlying mobile system and has to enhance its quality. In other words, the Service Machine only has indirect control over mobile system functions, in contrast with the solutions presented in the Service Engineering approach. In particular, the Service Machine only has control over the quality parameters of the telecommunication service provided at the interface between the Service Machine and the mobile system. The main task of the Service Machine is, therefore, to protect the quality of the services it supports by properly handling the quality and performance parameters associated with the telecommunication services provided by the underlying mobile system, according to what has been stated in this field by the current standards for the provision of advanced telecommunication services set by the B-ISDN community. Unfortunately, the user network interface has not been defined in sufficient detail so far, neither in B-ISDN nor in UMTS terms. Therefore, a precise description of the transformation of telecommunication quality into the final quality of service (performed by the Service Machine) is not possible. However, this problem is also met in the relationship between the Service Machine and fixed networks and is expected to have minimal implications on the Service Architecture definition.

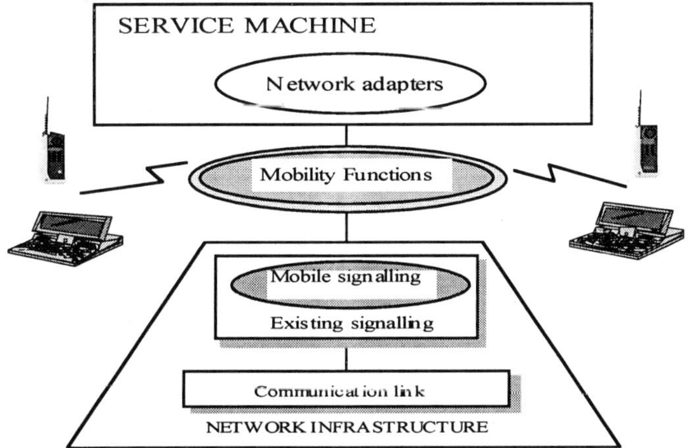

Figure 4: Mobility in the System Engineering approach

Concluding, in the System Engineering approach the factors to be taken into account in evaluating the impact of mobility on the Service Architecture are:

- The user control over functions and features supporting terminal mobility is filtered and therefore limited by the user-network interface, which will hide most of the mobility-supporting internal functions of a mobile network.
- The only possibility to interact with the mobile system is provided by the protocols supported by the user-network interface.

6 Comparison between the two approaches

In the previous section, two approaches have been discussed to evaluate the impact of mobility on a Service Architecture.

The Service Engineering approach assumes that all Personal Mobility and most of the Terminal Mobility functions should be included in the Service Machine. The consequence of this is easy and flexible reuse of the functions in the definition of new services. The System Engineering approach shows the mobility environment as an already available set of services provided by an advanced system supporting mobility, for example UMTS.

The implications of each of the two approaches on the Service Architecture can be summarised as in Figure 5: while the control of the mobility functions resides in the Service Machine in both cases, their support is up to the Service Machine in the Service Engineering approach alone. In other words, the two approaches considered can be regarded as two opposite solutions in which mobility functions are supported by either the Service Machine or the Network Resource Infrastructure.

Both approaches have drawbacks. Choice of the Service Engineering approach could require a full redesign of already defined mobility services. Choice of the System Engineering approach has the disadvantage of being strongly dependent on the progress of the developing third-generation mobile systems. One might conclude that the long-term preferred and natural solution would be the Service Engineering approach, provided that maintaining legacy of current mobile system implementations or specifications does not constitute an issue.

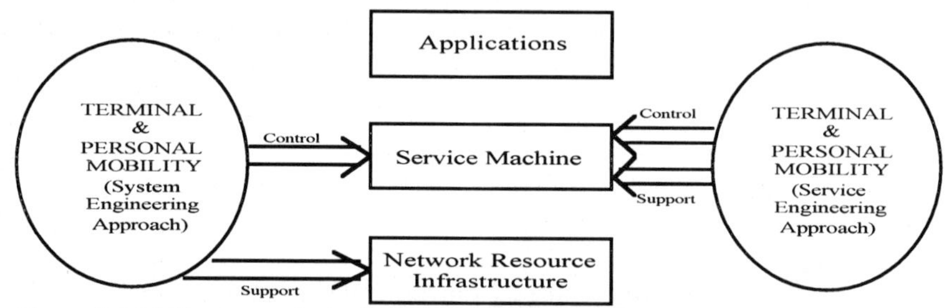

Figure 5: Mobility impact in the Service Engineering and System Engineering approaches

There could be a benefit in looking at "trade -off" solutions which reside in between the two poles. In these solutions the Service Machine is defined in a form that lends itself to implementation with currently available technology and constitutes a significant input for designers of trials and demonstrators of the Service Architecture.

One "trade-off", illustrated in Figure 6, is the following. All the Personal Mobility functions are placed in the Service Machine. Conversely, most of the Terminal Mobility functions are allocated in the Network Resource Infrastructure, because it is part of the existing environment, while only appropriate Network Adapters have to be designed inside the Service Machine to interface them.

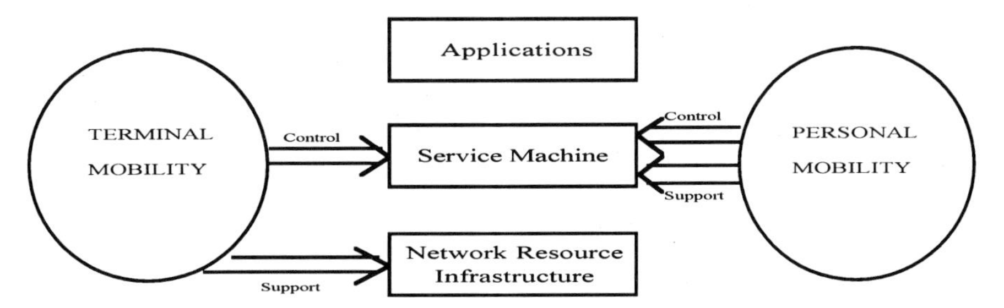

Figure 6: Architectural impact in the "trade-off"

This solution guarantees that Terminal Mobility functions already available in existing systems can be reused, whereas Personal Mobility ones can be designed according to the Reference Service Architecture, thus guaranteeing principles of generality and reusability and allowing easy development of support for new applications. In addition this trade-off does not prevent components providing Personal Mobility from being reused to (re)design Terminal Mobility functions. For example, components providing user-location transparencies can easily be reused for (re)designing terminal location transparencies.

7 A case study: the User Registration procedure

In this section we present a case study which concern a specific personal mobility procedure: the user registration. It has been chosen because it represents an essential aspect of personal communications since it provides the user with the capability to dynamically associate himself with different terminals for different services. The analysis carried out in the case study leads to the mapping of the elementary components of the function onto the various strata of the Reference Service Architecture. The impact of the this procedure on the Reference Service Architecture is evaluated according to both approaches, in order to compare the two alternatives.

In a Personal Communication Environment, in order to receive or activate services, the user must associate himself to one or more terminals. The registration procedure informs the service provider about the terminal from which a user wants to receive or activate services. The procedure can be decomposed into three different phases : a Request phase, when the user forwards the registration request to the service provider ; a Decision phase, during which the user requests are checked against the terminal capabilities ; and an Execution phase, which provides the profile updating. The main elementary functions involved in this procedure are described in the following.

Request Phase

UIH : User Interface Handling. This function has in charge handling of the terminal-user interface by supporting the user with friendly access to the provider environment.

TPL : Terminal Profile Location. This function has in charge to find where the Terminal profile is located by using the identity of the terminal the user has selected for registration.

Decision Phase

UPA : User Profile Access. In order to check if the user is allowed to register for the selected service, the User Profile has to be accessed. As the User Profile is a database under the service provider control, this specific function has to be performed to guarantee a protected access.

UPC : User Profile Checking. Before the registration can be carried out, a check must be performed against the user subscription data maintained in the User Profile, in order to verify whether the user is allowed to register for the selected service.
TPA : Terminal Profile Access. In order to check if the user is allowed to register on the selected terminal, the Terminal Profile has to be accessed. As the Terminal Profile is a database under the service provider control, this specific function has to be performed to guarantee a protected access.
SCC : Service Capability Checking. The terminal profile has to be checked to control if the user is allowed to register on it and if the terminal capabilities are compliant with those given to the user at subscription time for the service he wants to be registered for.
Negotiation. If a difference between the terminal capabilities and the capabilities given to the user at subscription time is detected, a negotiation with the user has to be carried out to define the characteristics of the service for the specific registration.

Execution Phase
TPU-UPU : Terminal and User Profile Updating. After the characteristics of the service have been agreed, both the Terminal and the User profiles have to be updated in order to maintain the service registration.
Table 1 shows the mapping of the above described functions on the different strata of the Reference Service Architecture, according to both the approaches.
It can be observed that, in the Service Engineering approach, the Communicating Service Components of the Service Machine related to Session Control aspects have to provide the association between user and terminal and have to support the user with the capability of a negotiation that could ensure service characteristics compliant with the ones given to the user at subscription time. This requires the development of specific new components to support the registration procedure, which can be considered as a fundamental facility for PCS. The problem of locating the Terminal Profile in the network, in order to provide its checking and updating, has been supposed to be solved by the Distributed Processing Platform which will support the applications and the other components of the Service Machine with distribution transparencies, i.e. with the capability of accessing distributed data in the system while abstracting from their location. Proper Adapters will provide a single interface with storage devices and will have to take into account the modelling of user and terminal profiles data to allow access and updating with the registration information.

Service Engineering Approach			
C	N	P	I
• User interface handling • User profile checking • Service capability checking • Negotiation	• User profile access • Terminal profile access • Terminal and User profile updating	• Terminal profile location	

System Engineering Approach			
C	N	P	I
• Service capability checking • Negotiation	• Mobile system interface		• User interface handling • User profile access • User profile checking • Terminal profile location • Terminal profile access • Terminal and User profile updating

Table 1: User registration procedure in the Service and System Engineering approaches

In the System Engineering approach mobility is assumed to be supported by the Network Resource Infrastructure. The main role of the Service Machine is therefore to provide proper Network Adapters that enable control of the necessary functions in the underlying infrastructure. In the registration procedure, aspects of negotiation are very important to give the user the option to accept the service or not, depending on the service conditions offered on that terminal in that place. To this purpose the Service Machine has to support the interface which gives the user the capability to control the characteristics of the requested service by negotiation. These aspects will have to be considered in the Provision Support Services of the Communicating Service Components of the Service Machine. As far as location, access and updating of the data stored in the user and terminal profiles are concerned, the mobile system already handles this data and only proper Network Adapters need to be provided to interface the mobile system elements in the Network Resource Infrastructure. If the underlying mobile system does not satisfy the requirement of data location transparency, the Distributed Processing Platform must provide this for User and Terminal profiles.

8 Conclusions

In this paper we have presented the work done in the ACTS DOLMEN project about the evaluation of the impact of mobility to a Service Architecture that is devised to encompass deployment of Personal Communication Services in a mixed fixed and mobile network environment. First we have briefly outlined the foundations of the Reference Service Architecture adopted in DOLMEN, which is based on the long-term vision also pursued, in the global standardisation arena, by the TINA-C consortium. Then we have described the methodology used to evaluate the impact of both terminal and personal mobility functions in terms of their allocation into the main constituents of the Reference Service Architecture. The methodology can be applied according to two distinct approaches, named Service Engineering and System Engineering, which also constitute two different ways (or recipes) to implement systems compliant to the Reference Service Architecture. Pros and cons of the two approaches have been discussed and an insight to possible trade-off solutions has been given. Finally, in order to assess the applicability of the methodology, we have introduced and analysed a case study, which refers to a function that is classically used to demonstrate the general principles for personal mobility, that is, the user registration procedure.

Acknowledgements: The authors want to thank the whole team of the DOLMEN Project and especially all the people in the WorkPackage on "Mobility and Personal Communications Aspects" who contributed to the deliverable from which this paper has been derived.

References

[1] TINA-C: "Overall Concepts and Principles of TINA", February 1995.

[2] CASSIOPEIA: "Open Services Architectural Framework for Integrated Service Engineering, March 1995.

[3] R. Pandya, "Emerging Mobile and Personal Communication Systems", *IEEE Communications Magazine*, June 1995.

[4] S. Trigila, A. Mullery, M. Campolargo, J. Hunt, "Service Architectures and Service Creation for Integrated Broadband Communications", *Computer Communications*, vol. 18, no. 11, Nov. 1995.

[5] ITU-T Recommendation X.902: "Open Distributed Processing - Reference Model - Part 2: Foundations and Part 3: Architecture", 1995.

[6] ITU-T SG1 Rec. F.851 (Feb. 95) - Universal Personal Telecommunication (UPT) - Service Description (Service Set 1).
[7] ETSI DTR/NA-072206 "Universal Personal Telecommunication (UPT) - Phase 2 - User Procedures and User States " - Final Draft June 1995.
[8] MONET-R2066, Deliverable R2066/BT/PM2/DS/P/113/a4, UMTS System Structure Document (Revised), 1995.

Consistency Issues in the UMTS Distributed Database

Eleftherios P. Adamidis
NTUA, Greece
eadami@cc.ece.ntua.gr

Gary Fleming
TELTEC, Ireland
gary.fleming@ul.ie

Efstathios D. Sykas
NTUA, Greece
sykas@central.ntua.gr

In Universal Mobile Telecommunication System (UMTS) a Distributed DataBase (DDB) is responsible for storing and manipulating user and terminal related data. Many of the simple operations and system procedures in the DDB act on distributed and interdependent entries. In this paper, in order to preserve consistency in the DDB after the execution of an operation or procedure, information flows based on the Two Phase Commit Protocol (2 PC) principles are proposed for the DDB communication protocols.

1 Introduction

The infrastructure of Mobile Telecommunication Systems includes a database system which is responsible for storing and manipulating user and terminal related data. In third generation mobile telecommunication systems like the Universal Mobile Telecommunication System (UMTS), a Distributed DataBase (DDB) system is proposed as the appropriate solution.

A DDB to support mobile users has some novel characteristics. The entries of the UMTS DDB are not always independent, but in many cases interdependent entries occur. Reference pointers can be used in remote nodes to allow entries to be easily and efficiently located. In addition, due to the transient nature of the data entries for mobile users (i.e., temporary Terminal Data, temporary Registration Data etc.) a much higher frequency of creates and deletes is perceived than in existing DDBs (e.g., X.500). Due to the aforementioned characteristics there is a high probability of *consistency* breach in the DDB. Thus a mechanism of enforcing consistency in the DDB is required.

The mechanism used to enforce consistency must ensure that two or more values must be in agreement with each other in some way (mutually consistent): for example, two copies of the same data must have the same value or a reference must exist only if the referenced data exists (and vice versa). This is reflected in the fact that the data contained in the database must be correct, i.e., there must be a coherent vision of the world they represent.

In this paper, the UMTS DDB architecture, data organisation and locating strategies are presented in section 2. In section 3 the UMTS DDB consistency requirements are analysed. The properties and the variations of the Two Phase Commit (2 PC) protocol and their applicability in the UMTS DDB are discussed in sections 4 and 5. The impact of the proposed mechanisms on the DDB communication protocols is discussed in section 6, whereas in section 7 some conclusions are drawn.

2 UMTS Distributed Database Architecture

The Intelligent Network (IN) technology is considered as the vehicle that will offer mobility and service interoperability in UMTS. The implementation of the UMTS DDB will be based on the IN functionality of the Service Control Function (SCF) and Service Data Function (SDF).

The proposed functional model (FM) for the UMTS DDB, shown in figure 1a along with a possible mapping into IN FM, consists of three levels: a *processing* level, a *directory service* level and a *data storage* level [4]. The respective functional entities on these levels are: Query Handler FE (QH), Directory Service FE (DS) and Data storage FE (DA).

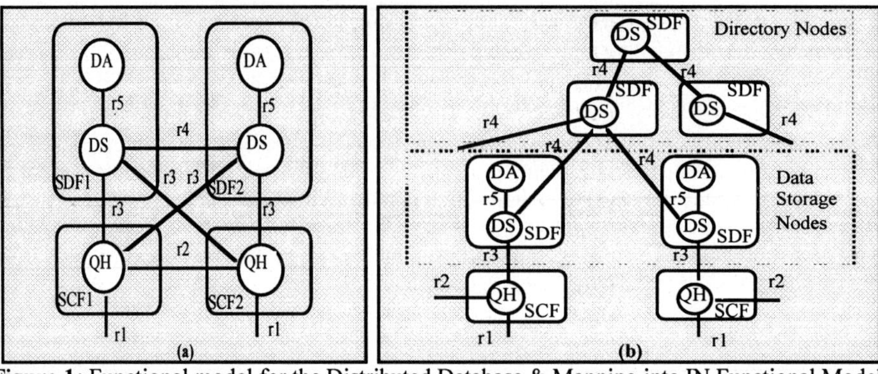

Figure 1: Functional model for the Distributed Database & Mapping into IN Functional Model.

Service specific queries, received via relation r1 from other entities in the system, are analysed at the processing level and decomposed into simpler generic directory requests. Such directory requests are transferred via relation r3 to the directory service level. If the QH cannot resolve the query locally it can forward the query (or a part of it) to another Query Handler via relation r2. The Directory Service level provides access to the requested data. If data is not available locally the Directory Service FE may communicate, via relation r4, with other Directory Service FEs, possibly belonging to different DDB networks. Finally, the Storage level simply stores the data. At this level data operations, requested by the Directory Service FE, are performed locally.

In [1], a logically hierarchical tree structure is proposed for the DDB architecture, as shown in figure 1b. The lower nodes of the tree, referred to as *Data Storage* nodes, store user and terminal related data whereas the upper nodes of the tree constitute the *Directory* part of the database, containing pointers (static or dynamic) to the lower nodes of the tree used to efficiently locate a requested data item. Among several possible options, the SDF-SDF relationship is used for handling data distribution inside the domain of a single network operator whereas the SCF-SCF relationship is used for inter-network communication [1].

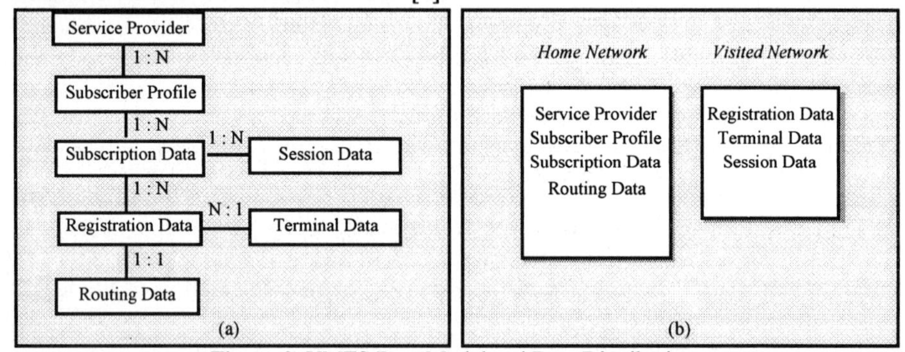

Figure 2: UMTS Data Model and Data Distribution

2.1 Data in the UMTS DDB

The database in UMTS contains data required to support the UMTS services and mobility procedures. The UMTS data is organised as entities in the data model shown in figure 2a [1]. The entities are linked by several relations to fully structure the data held in the database. In figure 2b a basic data distribution scenario considering a user roaming outside his/her home network is shown.

The *Service Provider Data* entity contains all the information concerning the service provider. This information is common to all the subscribers of a service provider. *Subscriber Profile Data* contains all the information agreed between the subscriber and the service provider at subscription time.
Subscription Data contains all the information concerning a user in a subscription including the User Profile and data required for user authentication. This type of information applies to all the registrations made by the user. *Session Data* is a replicated subset of the Subscription Data containing data needed to establish and maintain a user session.
Registration Data describes the attributes of a visiting user at a terminal and is service specific. A registration is required to associate a user with a terminal. In this entry the necessary part of the User Profile, required for service delivery, is replicated. *Terminal Data* keeps the details of terminal location. This entry is required as more than one user might be registered on different terminals for different services. The user location is found indirectly through Registration Data, which stores the identity of the corresponding terminal for each registered service. *Routing Data* are created in the home network of the user whenever performs a user registration. Routing data maintains the necessary information to locate a user registered for a service. This information may be kept either as an address of the visited node or the visited network where the user is registered.

2.2 Data Locating Strategies in the UMTS Database

The manner in which the DDB is searched for a particular data item depends on the different data locating strategies applied (e.g., *R*, *V* and *RV*) [1]. Locating of a data record is performed either through the use of the directory pointers, which could be static or dynamic, or/and through the use of references stored in the data storage nodes pointing the location of the data. For example, in the data locating strategy *R* the directory structure should first guide the request to the Resident node (R-node), of the requested data entry. At this node there is always either the requested data entry or a reference to the Visited node (V-node) where the user is currently roaming, which finally contains the requested data. In the data locating strategy *V* the directory structure should dynamically guide the request directly to the V-node without involving the R-node. Finally, in the *RV* locating strategy a mixture of the *R* and *V* locating strategies is applied.

3 Consistency Requirements in the UMTS DDB

In the UMTS DDB two types of consistency requirements can be discriminated:

- consistency requirements when several actions are performed as a result of a single database operation (e.g., update of a data entry, deletion of a data entry);
- consistency requirements when several database operations are performed as a result of a UMTS procedure execution.

DDB operations always act on a single data entry and obviously they are restricted in the same network. The execution of a DDB modification operation (Create, Delete, Update) might require several actions to be performed in the data storage nodes and probably in the directory nodes depending on the locating strategy adopted. For example, a delete operation might require deletion of an entry, update of reference and update of directory.

On the other hand UMTS procedures do not always act on just a single DDB entry. Instead, in many cases, a procedure results in more than one DDB entry being acted upon. These aspects are further complicated by the fact that the effects of mobility procedures are not always limited to a single network, since data entries may be distributed in different networks. Table 1, presents the procedures that require consistency in the UMTS DDB.

Procedure	Operation	Acts on (data)
User Registration	Create	Registration
	Delete	(old) Registration
	Update/Create	Routing
User De-Registration	Delete	Routing, Registration
Location Registration	Create	Registration
	Update	Routing
	Delete	(old) Registration
User Profile Modification	Update	Subscription, Registration
Attach/Detach	Update	Routing, Registration

Table 1: Consistency Requirements in the UMTS Procedures

To maintain consistency in the DDB after the execution of an operation or procedure some mechanism is required to guarantee that the involved actions are either all successfully completed or else non of the actions is performed. That is, the involved actions should form an *atomic action* [2].

Considering UMTS procedures, the most important consistency requirements arise from the procedures that are strongly related to the quality of service offered to the user. These are the User Registration, Location Registration, User Profile Modification and Attach/Detach.

In the sequel we will discuss the implications in case not all operations succeed during the execution of a system procedure while at the same time we will try to identify cases where inconsistency could be tolerated.

When a user performs user registration or location registration, it should be guaranteed that the user is reachable for the selected service. The following generic approach is used for user locating. Firstly, the registration or routing data are searched for locally where the query originates and if no information is found, the routing data in the home network is queried. This way, it is clear that the crucial requirement is to keep the registration data stored in the currently visited network and the routing data stored in the home network aligned, otherwise locating of the user will not be possible.

Let us now consider that a user already registered for a service, performs a user registration in a *new* visited network. In this case old registration data will exist in the old visited network. If the old registration data is not deleted, we have inconsistent information in the old visited network. That is, for calls originating in the *old* visited network the user will be paged unsuccessfully and considered as unreachable. The above discussion shows clearly that these three operations should form an atomic action.

Location registration is performed when a user registered in a mobile terminal for a number of services enters a new visited network. Location registration can be regarded as consisting of the following two steps: retrieval of the registration data

from the old visited network and user registration in the new visited network. Thus, the remarks made for the registration apply also for this case.
When a terminal attach/detach is performed, it is important that the routing data are consistent with the registration data in terms of the terminal status flag. Two cases of inconsistency can occur:

- Home network: *attached*, Visited network: *detached.* This situation can be accepted from the user point of view however, network resources are wasted since interrogation to home network continues with interrogation to the visited network where the terminal is detached.
- Home network: *detached*, Visited network: *attached.* This situation is not acceptable from the user point of view since the user is reachable only from the currently visited network. Thus, the terminal status should be consistently updated in the home and visited network.

Considering user profile modification, it is essential that the modifications in the replicated user profile made locally should result in an update of the corresponding parameters of the user profile stored in the home network and vice versa, otherwise registered services cannot be handled properly.
Finally let us consider user de-registration. When a user roams outside the home network, registration and routing data exist in different networks. If deletion of routing data succeeds and deletion of registration data fails or vice versa, then the visited or the home network of the user, respectively, will point to a non-existent registration. In this case, the user might be considered registered for a while, and this might waste network resources however from the user point of view this does not cause a serious problem. Thus, it could be argued that for user de-registration there is no need to define these two operations as an atomic action. However, for completeness in the following sections a consistent version of this procedure will be demonstrated.

4 The 2 Phase Commit Protocol

To ensure that all actions involved during the execution of a DDB modification operation or a UMTS procedure are treated as an atomic action, *transaction* based protocols should be introduced in the UMTS DDB. A transaction is a logically unique unit of work consisting of a sequence of operations [2]. If the transaction can complete its task successfully, we say that the transaction *commits*. If, a transaction cannot complete its task, we say that it *aborts*. When a transaction is aborted, its execution is stopped and all of its already executed actions are undone and the database is returned to the state before their execution. This is also known as rollback. To support these properties in distributed transactions the widely known and used Two Phase Commit (2 PC) protocols were developed [2,3].

4.1 Variants of the 2 Phase Commit Protocol

There are two interesting variations of 2 PC which are of relevance to the UMTS DDB. The first is the centralised approach shown in figure 3a. In this figure, site 1 is the co-ordinator of the action, while sites 2-5 are simple participants which perform operations under the instructions of site 1. Upon receiving a *prepare* message the participants reply whether they are able or not to perform the requested operation. In case all participants vote *ready* the operation is committed and an acknowledgement is sent to the co-ordinator. If any of the participants replies *refuse,* indicating that it is not able to perform the operation, the transaction is aborted. It can be seen that a large quantity of signalling is needed between the co-ordinator and its participants. In fact four messages are required for each participant.

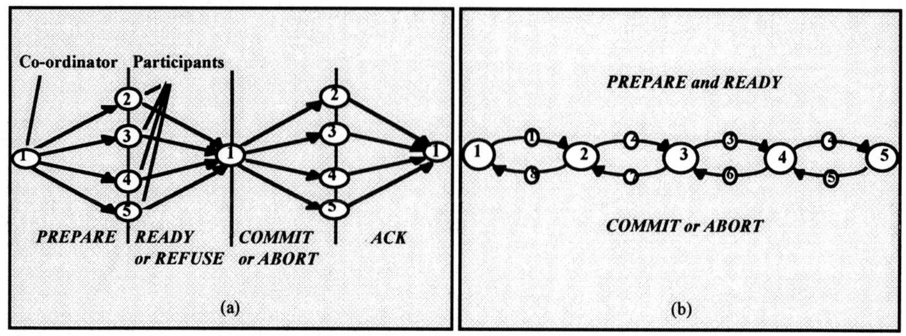

Figure 3: Centralised and Linear Atomic Action.

A simplification of the centralised approach can be achieved by imposing an ordering of sites. In this mechanism each site, except the first and the last one, has a predecessor and a successor. Instead of sending the *prepare/commit* messages from the co-ordinator to all the other participants, the message is passed from each participant to its successor. In this mechanism there is no need for separate *prepare* and *ready* messages. A single message indicates that the site is ready and also informs the successor to prepare. In this way the combined *prepare/ready* message propagates along a chain of sites as shown in figure 3b.

A further extension on the centralised approach is to allow some participants to also be co-ordinators. This leads to a hierarchical structure for the atomic action, where some operations also lead to sub-transactions. This approach will be discussed in the following sections.

4.2 Assumptions

One of the first optimisations of the 2 PC protocol for UMTS is the removal of the Commit Acknowledge message. Since a reliable communication network is assumed for the DDB, it is not considered necessary. The main purpose of the usage of transactions in UMTS is to ensure consistency in the DDB in the case of node errors or failures, but not link failures. Considering that only node failures can occur, the following assumptions are made:

- *Reliable transfer of messages.* It is assumed that the underlying transport mechanism is capable of transferring messages without loss.
- *No usage of the Ack message in the flows for consistency.* In the UMTS DDB we will assume that once a node receives a commit message it is able to commit the changes.
- *Reliable backups exist of both the SDF and the SCF nodes.* Thus the state of any node is assumed to be recoverable, irrespective of node failures.
- *Time-outs are used to encompass node failures.* If a participant node fails during an operation the co-ordinator will re-send the operation request following a time-out.
- *The response of a node to a transaction time-out is a local rollback.*

- *Duplication of Operations is not considered an error situation.* When an operation is performed twice (e.g., trying to create a data entry which already exists) it is not considered an error situation.

5 Applying the 2PC Protocol in the UMTS DDB

5.1 The Linear Approach

The linear approach is very suitable, due to the reduced messaging it incurs, but it does not support operations being performed in parallel. The linear approach can be applied, in the UMTS DDB, to transactions or sub-transactions which exhibit an explicit ordering of participant sites. The hierarchical organisation of the DDB nodes belonging in the same network provides a logical ordering of the DDB nodes.

A general scheme of the message flows between the DDB nodes in the linear approach of the 2 PC protocol is shown in figure 4, assuming that the *R* locating strategy is applied. These flows can be applied in all the modification operations (Create, Update, Delete). A similar scheme can be applied also in case the *V* locating strategy is used.

In figure 4, the name of the query originating in the O-node is simply indicated with the abbreviation "Op" indicating any of the modification operations.

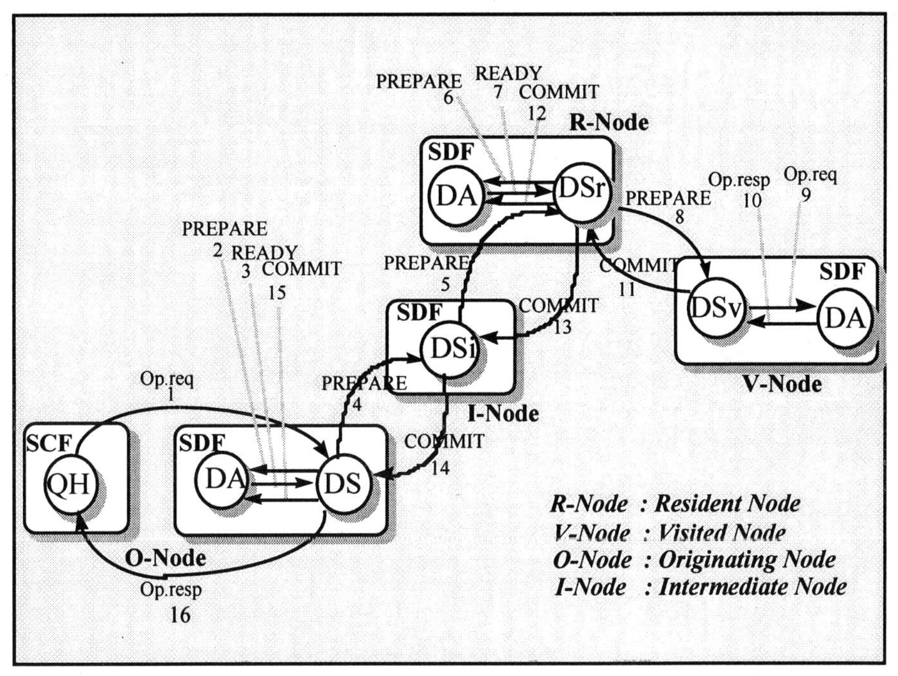

Figure 4: Modification Operations using the Linear Approach of the 2 Phase Commit Protocol

The query is originated in the O-node and propagates up and down in the directory, until the R-node is found (there can be several interactions, indicated in the figure by means of the I-node). Then, depending on the operation a further V-node can be involved. For example, if we consider a create operation, data will be created locally (O-node) and the reference to this entry will be created in the R-node. In comparison with the normal flows only two additional message (namely, the Commit 12, 15 messages) are required and since they are internal in the involved nodes no additional signalling load is required.

One strong advantage of the linear approach is that it necessitates minimal changes to the existing flows for the DDB. While *prepare/ready* and *commit* messages were explicitly shown in the messages in figure 4, they are not really necessary. The *prepare/ready* can be implicit in a normal operation request, while the *commit/abort* can be determined from the status flag of an operation response. Thus normal request/response type messages can be used to support linear 2 PC.

Considering the implications of node failures on the technique shown in figure 4, inconsistency can occur if any of the involved nodes fail after the operation is performed in the R or V-node. In such case, the commit message will not propagate back to the O-node and the transaction will time-out. If we assume that rollback is always the response to a time-out then there will be inconsistent data in the DDB. However, the drawbacks of this inconsistency are not very critical. The inconsistency will be removed by the QH invoking the operation again. In this case the R and/or the V-node will try to modify the data a second time. The DS service logic should be able to detect these situations and consider it normal (i.e., not an error situation).

5.2 The Centralised Approach Applied in UMTS

When a transaction involves two operations, and no ordering of participant sites exists, then the linear approach is not applicable. For situations like this the centralised approach, shown in figure 5a, can be applied. It provides a high degree of protection against node failures but incurs additional overhead in terms of signalling. However, when only two operations are involved, it is not necessary to use the full centralised approach. Instead the messages of the full 2 PC protocol, with separate *prepare, ready* and *commit* messages, are only required for one of the operations. The other operation can be performed using request/response messages, in which the *prepare, ready* and *commit* are implied. This is shown in figure 5b, where the "*semi-centralised approach*" is compared with the centralised approach.

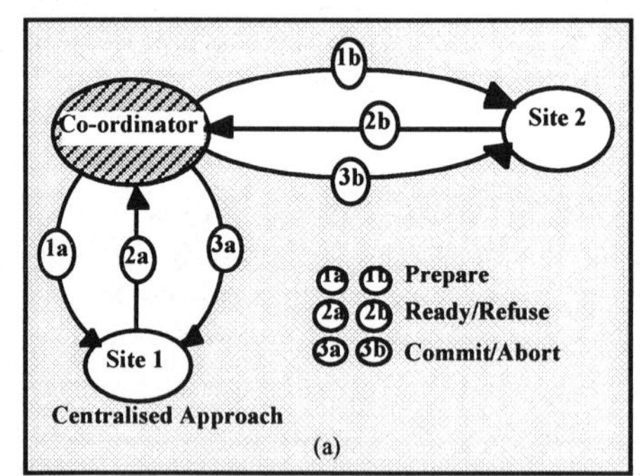

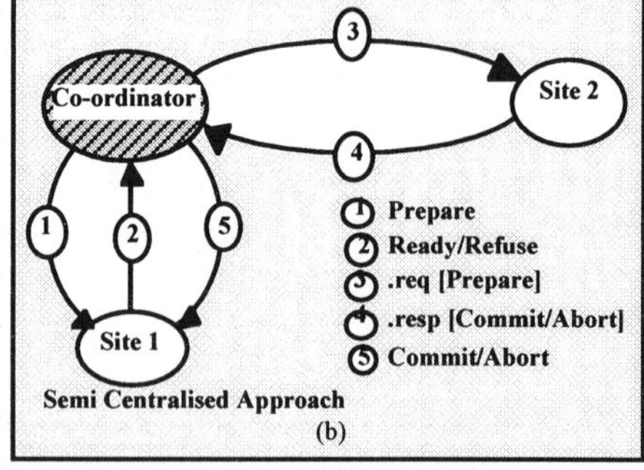

Figure 5: Transaction containing two operations - Centralised Approach.

In figure 5b, the second operation is only performed after a *ready* has been received from the first operation. Then simple request/response messages are used to perform the second operation. If the response message indicates success the co-ordinator commits the first operation, otherwise if the response indicates failure the first operation is rolled back.

Example: User De-Registration, User Profile Modification, Attach/Detach.
In this example of a User De-Registration, the most general case is considered where the user roams outside the home network.

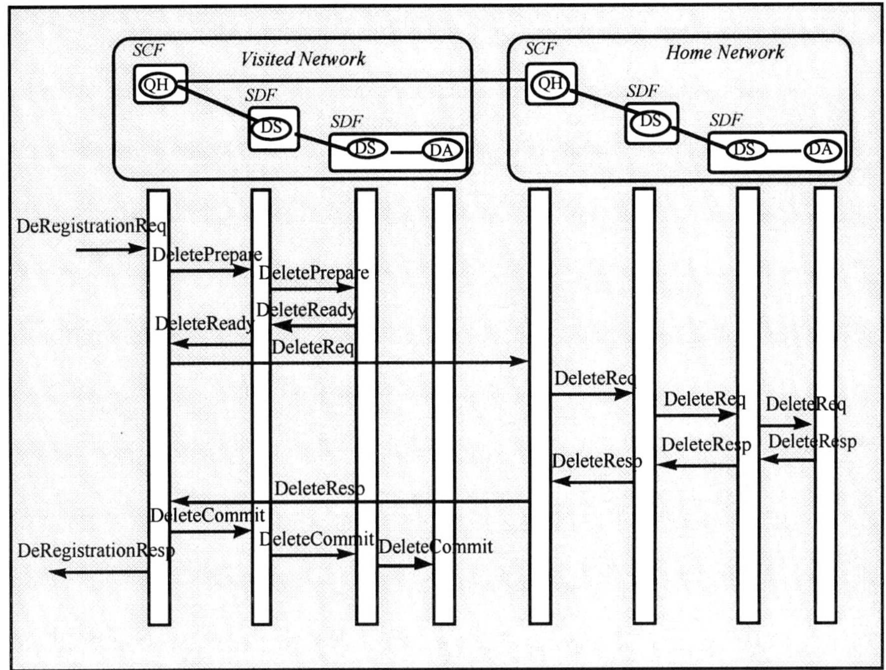

Figure 6: User De-Registration using the semi-Centralised Approach of the 2 PC Protocol

In figure 6, on receipt of a DeRegistrationReq, the SCF of the visited network issues a DeletePrepare to the local SDF. The message propagates through the directory to the SDF which contains the registration entry. Only after the receipt of a DeleteReady from this SDF, which indicates ability to perform the Delete operation, is a DeleteReady forwarded to the SCF. The SCF now begins the second operation to delete the routing entry. If the outcome of this second operation is successful, the SCF commits the first operation by sending a DeleteCommit to its local SDF. The local SDF forwards the DeleteCommit to the SDF which stores the actual registration entry and the delete is made permanent.
Similar flows can be applied for the Attach/Detach and User Profile Modification procedures by replacing the Delete messages with the relevant Update operations.

5.3 The Hierarchical Approach Applied in UMTS

When more than two operations are involved the centralised approach is still applicable. However, since the operations often occur sequentially rather than simultaneously, the hierarchical approach, illustrated in Figure 7a, could be preferable.
With this approach, one of the participants in the transaction has now become a co-ordinator for a sub-transaction consisting of two separate operations.
The hierarchical approach can be optimised in the same way as the semi-centralised approach was optimised previously. The full 2 PC messages are not necessary for the sub-transaction, instead normal request/response messages can be used. Similarly one of the operations in the sub-transaction does not need the full 2 PC messages. The

"*semi- hierarchical*" approach is shown in figure 7b. Note that the hierarchical approach can be mixed and combined with the linear approach to meet the specific needs of UMTS operations and procedures. For example, in figure 7b, the operation occurring at site 3 is a sub-transaction which exhibits an ordering of participant sites. Thus the linear approach is used between site 3 and its participants, as shown by the dotted lines.

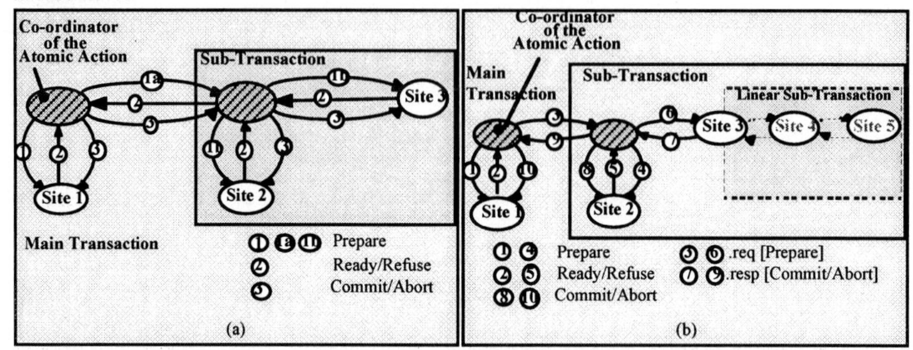

Figure 7: Composition of a transaction with three operations - Hierarchical Approach

Example: User Registration, Location Registration.

In this example the most complicated case where the user roams outside the home network is considered. Figure 8 shows the information flows for User Registration, using the semi-hierarchical approach of the 2PC protocol. Similar flows can be applied for the Location Registration procedure as already explained in section 3.

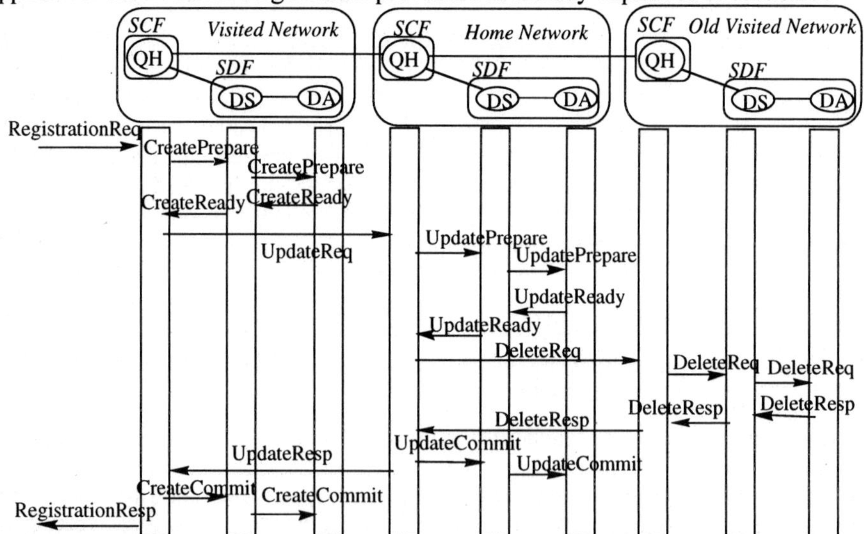

Figure 8: User Registration using the semi-Hierarchical Approach of the 2 PC protocol

According to this approach the originating SCF becomes the co-ordinator of two operations, namely the create and update operations. The second operation, the update of the routing data, has become a sub-transaction consisting of two separate operations. This sub-transaction has as co-ordinator the SCF of the home network. This way, the use of the full 2 PC protocol is avoided in the SCF-SCF interface.

Firstly, on receipt of the RegistrationReq the SCF issues a CreatePrepare message to its local SDF. Only after the receipt of a CreateReady, which indicates that the SDF is able to perform the operation, the SCF begins the second operation to update the routing entry. The SCF in the home network after receiving the UpdateReq issues an UpdatePrepare message to its local SDF. The response UpdateReady of the home SDF, indicating ability to perform the operation, carries the address of the old registration information. Using this information, the SCF of the home network initiates the delete operation in the old visited network. After the successful deletion of the data in the old visited network, the SCF of the home network makes permanent the update of routing data by issuing the UpdateCommit message to its local DS and returns an UpdateResp in the originating SCF which makes permanent the creation of the registration data.
As an alternative, the fully centralised implementation of the 2 PC protocol could be applied. According to this option the originating SCF has the overall control of the transaction. That is, the originating SCF co-ordinates the three operations create, update and delete. Note that this requires knowledge of the location of the old registration data and thus the originating SCF should wait for the outcome of the update operation containing the address of the old visited network before issuing the delete operation. Moreover, this way the full 2 PC flows are required now on the SCF-SCF interface.

Effects of Node Failures in semi-Centralised and semi-Hierarchical Approaches
We will consider the effect of node failures using the example of figure 8, since the semi-hierarchical approach is the most general case. The discussion applies also for the centralised approach.
Two cases of inconsistency can occur: (i) Any of the participant nodes fails after the delete operation has been performed. Then a transaction time-out will result in the originating SCF performing the transaction again. (ii) The originating SCF fails after the delete or update operation has been performed. In this case after recovery the originating SCF will invoke the transaction again. In both cases, as already discussed, the duplication of the operations should not be treated as an error situation.
Note that failure of a node being at a *Ready* stage does not cause problems, since after recovery the node will query its co-ordinator node whether the result of the transaction is aborted, committed or unknown yet and will resume accordingly.

6 Impact of consistency on the protocols

The use of a transaction based mechanism for maintaining consistency in the UMTS DDB generates additional requirements on the protocols used for handling data. The mechanisms described in this paper necessitate support of transactions in both the database access protocol (i.e. the SCF-SDF protocol) and the intra-network database protocol (i.e. SDF-SDF protocol).
In current IN standards (IN CS-1, and CS-2), a reduced version of the X.500 Directory Access Protocol (DAP) is proposed for the SCF-SDF protocol [5, 6]. The DAP protocol however, is not a transaction based protocol that is, only single request/response operations are supported whereas the 2 PC flows require a transaction mechanism. If full consistency is required in the DDB, additional functionality should be added in the SCF-SDF protocol in order to be able to support 2PC protocol transactions.
The SDF-SDF protocol responsible for handling data distribution inside a UMTS DDB network has not been considered so far in IN CS-1 and CS-2 however, the

development of this protocol is foreseen for the IN CS-3 phase. One of the candidates for the implementation of the SDF-SDF protocol interface is the X.500 Directory Service Protocol (DSP) [1]. The X.500 DSP does not explicitly support consistency enforcement mechanisms and thus the X.500 DISP and DOP protocols are also being considered (in combination with DSP) for support of SDF-SDF interface. These protocols could be used to support data consistency through the use of a "shadowing" agreement between a data copy and the original in case replication of data is applied.
However, as shown earlier, an equally critical aspect for the UMTS DDB is procedural consistency, which would guarantee that the UMTS DDB maintains a consistent representation of the mobile network after the execution of a system procedure. None of the X.500 protocols supports this type of consistency. The X.500 DAP and DSP protocols require additional messages to include the 2 PC like flows required for this.
As far as the SCF-SCF protocol is concerned, in IN CS-2 a number of messages, which are not yet stable, have been proposed for this protocol. Support of consistency does not pose any significant additional requirements in this interface. That is, with the application of the appropriate variants of the 2 PC protocol normal request/response messages can still be used on the SCF-SCF interface.

7 Conclusions

Due to the large number of interdependent operations which occur in the UMTS DDB some mechanism of enforcing mutual consistency between entries in the DDB is required. To support this, transaction based protocols are necessary. Thus, variants of the 2 PC protocol were evaluated for their applicability to UMTS namely the linear, the centralised and the hierarchical approach. Examples of information flows for consistent versions of the DDB modification operations (i.e. Create, Delete, Update) as well as for UMTS procedures were given and the effects of node failure on these flows were considered. The use of the 2 PC variants impacts the DDB communication protocols. The introduction of transaction mechanisms was found to require additional functionality to be added in the SCF-SDF and SDF-SDF protocols. It is anticipated that transaction based DDB protocols would be extremely powerful in the IN Database. They would facilitate the efficient integration into IN of currently emerging and evolving telecommunications systems. Many of the emerging mobile systems and services will require interdependent data entries and operations as well as data distribution.
It is worth noting that this paper concentrated more on highlighting the consistency issues in the UMTS DDB and exploring alternatives to tackle these issues than making hard decisions. This is due to the fact that there are still many unknown aspects of both the UMTS DDB and system to allow informed choice to be made yet.

Acknowledgements: The authors are working in the ACTS-013 Project EXODUS. Part of the work is based on the work done in RACE 2066 MONET in which the authors participated. The paper does not present the views of the project as a whole, but these of the authors.

References

[1] RACE R2066 MONET, R2066/BT/PM2/DS/P/113/b1: UMTS System Structure Document (Revised), 1995.
[2] R. Sharp, *Principles of Protocol Design*, PrenticeHall Int. Series in Computer Science 1994
[3] S. Ceri, G. Pelagatti, Distributed Databases Principles and Systems, McGraw-Hill 1984
[4] ITU-Baseline document - text on DDB, WD KEN-5 - Geneva, 5-23 September 1994.
[5]: ITU-T Rec. Q.1218: Interface Recommendation for Intelligent Network CS-1, 1995.
[6] ITU-T Rec. X.500: The Directory - Overview of Concepts Models and Services, 1993.

Security

Jaime Delgado
Logic Control, Barcelona, Spain.

This section has the general subject of "Security". Currently, this is a very broad term that may refer to different concepts depending on the context in which it is used. Since this Conference is on Intelligence in Services and Networks, the scope is smaller. However, it is still too wide.

Three papers dealing with Security issues have been selected for this section of the book. However, there are other papers that, although not dealing specifically with security, consider security aspects in the contexts of other topics. The three papers approach the security aspects from very different points of view, *Security architecture*, *Integrity* and *User authentication*, so they represent different examples on security issues on intelligent services and networks.

- The *Security architecture* is defined for inter-TMN management. Security services are integrated in existing systems by defining sub-architectures for the different security services, that are common in distributed systems, such as authentication, access control, data-integrity, confidentiality and non-repudiation.
- Concerning *Integrity*, this is considered from the general point of view of general telecommunication services and networks. A formal methodology is proposed to guarantee the integrity of the system in the face of unexpected changes. This problem of integrity is not related to the common security services in distributed systems, but applies to the general security of a complex system.
- *User authentication* is a well known security issue. Here it is considered in a particular scope: using voice biometrics to enhance the security of a smartcard in a GSM system.

In the rest of this introduction, a few details are given on how the three papers approach the mentioned topics.

The first paper, *A Security Architecture for TMN Inter-Domain Management*, presents a security architecture intended to fulfil the needs of the security policies for inter-TMN management initially defined by the ACTS TRUMPET project. Three different policies have been formulated: minimal, basic and advanced. The security architecture specifies the identified security services in terms of security components, their placement and their relationships.

Some decisions have been taken, such as to use public key cryptography, to use TTPs in the role of Certification Authorities (CA), and to use symmetric encryption for confidentiality and data integrity protection.

The goal of the security services is to protect inter-domain management between management systems based on X interfaces as defined by the TMN model, taking into consideration the concerns of each actor with respect to security and privacy.

Sub-architectures are defined for authentication, access control, data-integrity, confidentiality and non-repudiation.

Integration of security services with the TMN communication capabilities strongly depends on the openness of the management platform: if open, security can be added at for example the application, presentation or transport layers; if not, all security must be added on top of the service interface provided by the stack. The second case is the one with most commercial platforms.

Security services and mechanisms are then placed in relation to the TMN architecture and to management applications. Security will for the most part be transparent to the applications, but some interfacing is necessary.

The second paper, *Maintaining Integrity in the Context of Intelligent Networks and Services*, discusses the need for a formal and systematic framework to assess the degree of integrity of telecommunication networks and services, based on their vulnerability to the introduction of changes or unexpected perturbations.
The paper outlines a methodology that can be developed to determine the cause, extent and outcome of integrity violations, thus assisting the design of robust systems and the management of "live" networks services.
Techniques for modelling and simulation are proposed, also using SDL (Specification and Description Language) and MSC (Message Sequence Charts), both from ITU-T.
Future work will include more complex scenarios, i.e. multi-service and multi-network, thus allowing the examination of issues such as services interactions and interconnections.

The last paper, *User Authentication in Mobile Telecommunication Environments using Voice Biometrics and Smartcards*, presents the work of the ACTS ASPeCT project on the implementation of security services in general for UMTS (Universal Mobile Telecommunication System) and authentication of users to the network in particular. This includes the authentication protocols of the network and also the authentication of a user to the network. These features will be implemented by the project in a real trial.
The paper considers how the authentication between the user and a smartcard, acting as a security module, can be enhanced by means of voice biometrics. It explains how such authentication is carried out in the GSM system by means of a PIN. The key features desired of a new authentication mechanism and how well biometrics approaches meet these requirements, and the approach being further researched in the ASPeCT project, are also explained.
Employing vocal biometrics techniques may allow the enhancement of the user authentication mechanism in the mobile telecommunications environment. This environment is atypical and places special constraints on the implementation of the authentication mechanism because the service provider is not able to trust all the equipment being used.

A Security Architecture for TMN Inter-Domain Management

Francois Gagnon, ASCOM Monetel (email: gagnon@ascom.eurecom.fr)
Dominique Maillot, Telis (email: Maillot@sqy.tel.telis.fr)
Jon Ølnes, Norwegian Computing Centre (NR) (email: Jon.Olnes@nr.no)
Lars Hofseth, TELSCOM (email: hofseth@vptt.ch)
Lionel Sacks, UCL (email: l.sacks@eleceng.ucl.ac.uk)

This article presents a security architecture suitable for implementation of the security services and mechanisms selected by the security policies defined by the ACTS TRUMPET project for an appropriate protection of the external interfaces of TMNs. These security policies define the protection measures to use for communication in the inter-domain management case. These measures need to be implemented within each individual domain. Each security service is discussed separately with respect to the architecture specification.

1 Introduction

This article presents a security architecture appropriate to fulfil the needs of the security policies for inter-TMN management as they have been defined by the initial work of ACTS TRUMPET project [AC112] in its second deliverable [D2B]. Three different policies have been formulated: minimal, basic and advanced, in order to meet security requirements of different severity. The security architecture specifies the identified security services in terms of security components, their placement and their relationships. It should be able to support a larger variety of policies than those specified by TRUMPET. However some policy decisions inherently form the architecture, like the decision to use public key cryptography. TRUMPET's security policies prescribe the use of public key mechanisms for authentication, non-repudiation, and key management (of secret session keys). Use of TTPs in the role of Certification Authorities (CA) is anticipated. Symmetric encryption is used for confidentiality and data integrity protection. Management of security is internal to the individual management domains, with the exception of some part of the management of some public cryptographic keys. Key management ensures that the keys generated have the necessary properties, are distributed to the parties that will use them and are protected against disclosure and substitution.

The goal of the security services is to protect inter-domain management between management systems based on X interfaces as defined by the TMN model [M.3010], taking into consideration the concerns of each actor with respect to security and privacy. The inter-domain security depends to a certain degree on the internal security of each TMN domain. This intra-domain security is out of the scope of this article.

In this article, each security service is treated separately, and sub-architectures are defined for *authentication*, *access control*, *data-integrity*, *confidentiality* and *non-repudiation*. The sub-architectures are described in sections 3.2-3.5 of this article.

Integration of security services with the TMN communication capabilities strongly depends on the openness of the management platform. If the internals of the protocol stack can be accessed, security can be added at for example the application, presentation or transport layers. If not, as is the case with most commercial platforms, all security must be added on top of the service interface provided by the stack, and is limited by the functionality provided by this interface.

With respect to the inter-domain management applications, security for most parts is viewed as parts of the platform. Interfacing between the applications and the security services needs to be defined, particularly for authentication and access control, but the internals of the security architecture should be transparent to these applications.

2 Global Architecture

The security architecture implemented in TRUMPET is designed to secure the management interactions between two management entities (manager-agent) belonging to different domains and communicating across a single X interface. More precisely, the communication to be secured is the communication between a Management Application Entity (MAE) belonging to an OS from one domain to another MAE belonging to an OS in another domain. There may be several MAEs belonging to the same OS. Each MAE may use different security profiles and the choice of the security profile is made during the initialisation of the security context. The choice of the security profile may be constrained by the mechanisms supported by the architecture, by internal policies or by target TMN policies.
The security architecture presented in this article is designed to be scaleable, so as to support the various security profiles (minimal, basic, advanced) proposed by TRUMPET. The security architecture will be validated in trials devoted to the management of connections across multiple domains.

2.1 Integration of Security Components to the OSI Model

The security architecture is defined for management applications based on OSI principles and the corresponding OSI architecture. In this context, the management applications can be defined above CMISE, ACSE and possibly other Application Service Elements (ASE). An important issue for TMN security is the accessibility to the OSI protocol stack. Typically, security mechanisms need to be integrated on several layers. In this article, the term *research platform* is used when the source code of layers inside and below the application layer can be modified wherever necessary. The term *commercial platform* is used when only a well-defined interface above CMISE is available, and none of the layers below that can be modified.
The security architecture presented in this article deals with the accessibility issue by proposing security architectures for both research and commercial platforms[1]. However it is not possible to fully implement all security services on a commercial management platform.

Research Platform Security Architecture

Research management platforms, such as OSIMIS, and their OSI stack allow addition of protocols and ASEs. In such environments, a complete security architecture can be implemented. To provide full support for integrity, confidentiality and security negotiations, some security functionality must be introduced below or in the bottom part of the application layer of the OSI stack.
Figure 1 shows the architecture for a research management platform and open OSI stack. The shadowed boxes refer to existing components expected to be available in TMN management platforms. The white boxes refer to new or modified components.

---

[1] A third, short-term, solution is to place parts of a security architecture in front-end separate software / computers.
The most common front-end approach is the use of "trusted routers" that operate at the network layer. Confidentiality and data integrity, as well as peer-entity authentication with respect to other routers, are provided by use of the Network Layer Security Protocol (NLSP) [X.273].

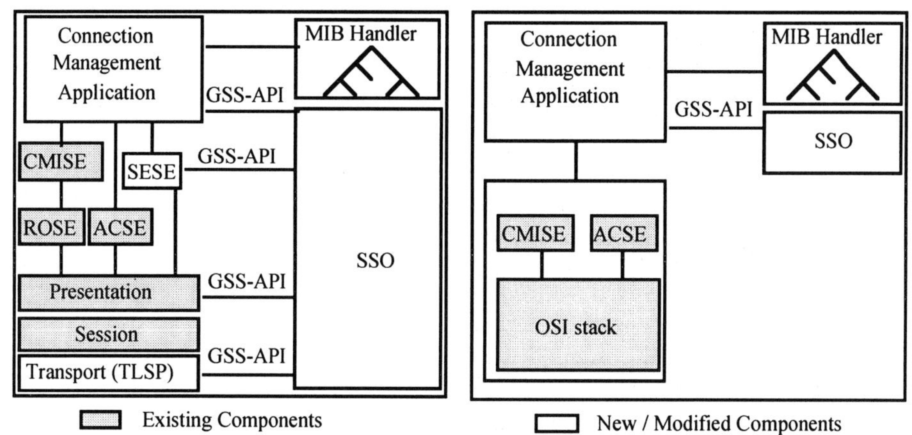

Figure 1: Research Platform Architecture **Figure 2**: Commercial Platform Architecture

The Generic Upper Layer Security (GULS) [X.830-3] specifications can only be applied to a research management platform because of the need to interface directly to the presentation layer, which in turn must be enhanced with additional functionality to support security transformations. The Security Exchange Service Element (SESE), part of GULS (X.831-X.832), supports arbitrary security negotiations and is also used for key establishment. Connection integrity and confidentiality security components are introduced in the transport layer of the OSI stack according to the Transport Layer Security Protocol (TLSP) [X.274]. Selected field integrity / confidentiality and non-repudiation are added at the presentation layer.

The security architecture is intended to be as independent as possible from particular security mechanisms. To support this goal, the security architecture makes use of the GSS-API [RFC 1508][2] which specifies a generic security programming interface. The GSS-API interfaces to a Security Support Object (SSO). The SSO is used to establish a security context between the communication parties and to perform security transformations on the application data.

Commercial Platform Security Architecture

Figure 2 shows the architecture for a commercial OSI management platform. The security components are integrated strictly to the application, above CMISE and ACSE, or the interface to those ASEs, like XMP.

Although this architecture is based on a commercial management platform, there is some minimal support required for the transfer of security data between communi cating parties. The specific requirements that must be satisfied are:

- The authentication field of ACSE must be supported to establish the security context, perform authentication and, possibly, establish session keys.
- The access control field of CMIS management operations must be supported to transfer security related information.

The security architecture also requires that agents have control over accesses to the MIB. This is necessary to enforce access control to MOs. The implementation of agents may be restricted by tools provided by the platform provider. For example, the code generated by a GDMO compiler may not be compatible with the introduction of access control mechanisms.

---

[2] The discussions in this article are based on the first version of [RFC 1508].

The architecture shown in Figure 2 is able to support a subset of the security services required by the TRUMPET policies. The security services that cannot be fully supported by integrating security in the top part of the application layer are *data integrity, data confidentiality, non-repudiation and security negotiations*.
Integrity of application data can be provided by application security. The data must be transferred to an application independent encoding, like ASN.1 BER, before the Integrity Check Value (ICV) is calculated, to ensure that the receiver will validate the ICV using exactly the same representation. This implies an extra encoding / decoding step. Integrity of data introduced by the protocol stack cannot be provided. ICVs may be transferred via the CMIS access control parameter, or through separate fields of the CMIS operations requested.
Confidentiality can only be partially supported for selected application data. The encrypted data must be inserted into one of the fields of the particular CMIS operation being requested. However, most of the fields have specific pre-defined types that cannot accommodate an encrypted data type. Fortunately, the fields used to carry the attribute values to and from the target MIB can accommodate encrypted data, which means that encryption of attribute values is possible by first encoding them according to BER and then encrypting them.
Non-repudiation cannot be fully supported by application layer security for the same reasons as for integrity. That is, non-repudiation can be supported for selected application data, but not for data introduced below the application.
Security negotiations involving a maximum of two exchanges can be performed during association establishment using ACSE fields such as the authentication or the user information field. However, in some cases, it is desirable to perform more than two exchanges (e.g. 3-way authentication) or to support security exchanges (e.g. re-negotiation of a session key) after the association has been established.

3 Architecture Components

3.1 Security Context Establishment

In order to successfully establish a security context with a target peer, it is necessary to identify appropriate mechanisms which support both initiator and target. The definition of a mechanism embodies not only the use of a particular cryptographic technology, but also definition of the syntax and semantics of data element exchanges which that mechanism will employ in order to support security services.
Unless SESE is available for context negotiations, the security context must be negotiated at association establishment time, by transferring GSS-API tokens in the authentication field of ACSE. This enables only a two-way negotiation procedure, which anyway is assumed to be sufficient in most cases.
Appendix B of [RFC 1508] specifies a mechanism-independent, encapsulated representation for the *initial* token of a GSS-API context establishment sequence, incorporating an identifier of the mechanism type to be used on that context. The security tokens are created and validated by the SSOs, which also hold the information that constitutes the security context.

3.2 Peer-entity Authentication

The peer-entity authentication mechanisms that are required by the security policies formulated by TRUMPET are weak authentication, strong-unilateral authentication and strong-mutual authentication. Weak authentication is based on passwords while both types of strong authentication services are based on public key technology. The

security architecture is based on [X.509], and is common for research and commercial platforms. Weak authentication is intended as an interim solution until strong mechanisms can be assumed to be generally available.
Peer-entity authentication will take place during the security context establishment. The initiating MAE requests from the Security Support Object (SSO) an authentication token as a part of the initial GSS-API token, and transmits this via the ACSE authentication field. For mutual authentication, the receiving MAE must return an authentication token as a part of the GSS-API token in the ACSE response. A security context will only be established if authentication has succeeded.
The weak authentication procedure requires a single authentication token sent from the initiator of the association to the target entity, i.e. one-way authentication. The schemes defined by [X.509] for protection levels 1 and 2 are used. Hashing, random numbers and time stamps are used to preserve the confidentiality of the password and the integrity of the information. The hash functions are selected by the initiating MAE, presumably following an inter-domain policy agreement. Identification of the hash algorithms is specified using a GSS-API-like mechanism type. If the algorithm cannot be processed by the receiver, the authentication will fail. Negotiation of the mechanism type cannot be performed using two-way context negotiation.
The strong authentication procedures are based on public key technology. A signed authentication token is sent from the initiator to the target, which must return a similar token if mutual authentication is requested. [X.509] does not mandate a particular cryptosystem. As for hash functions in the previous section, identification of the particular algorithm is specified using the GSS-API mechanism type.

3.3 Access Control

The access control architecture is based on [X.812] (access control framework) and [X.741] which specifies a model for controlling access to management information and operations. The access control profiles AOM24322 (Access Control Lists with Item Rules) and AOM24326 (Capability List) [ISP] are also used.
The actual access control is carried out within each individual domain. However, the domains need to agree on the nature of the Access Control Information (ACI) and how this is exchanged over the X interface.

Global Access Control Architecture

Figure 3 shows the global access control architecture. Initiator domain: access control is applied to outgoing management association requests, outgoing management operation requests and outgoing management notifications (event reports). Target domain: access control is applied to incoming management association requests, incoming management operation requests and incoming management notifications. There is no difference between research and commercial platforms.
For incoming access control in the target domain, the access control services may require the support of TTPs. Use of Privilege Attribute Certificates (PAC), where a privilege attribute server in the initiator domain (or perhaps even a TTP) signs a certificate for the access rights granted to the requesting entity, is for further study.

Access Control to Management Associations

Access control to the establishment of management associations requires an Access Control Enforcement Function (AEF), an Access Control Decision Function (ADF) and a security MIB (S_MIB) in both domains. The AEF enforces access control and is always part of the access path between the initiator and the target as defined in

[X.812]. In practice, the AEFs will in most cases be integrated with the Message Communication Functions (MCF) on both sides.
Initiator domain: we assume that the initiating person/entity/process (illustrated by the Human Machine Adaptation (HMA)) has a group or role identity assigned by the initiating system. The MAE requests association establishment through a MCF, specifying the identity of the target MAE, its own identity, and possibly the identity of an entity it is acting on behalf of (for example a human operator). The AEF invokes access control in the initiator domain, to decide whether the association establishment request shall be allowed to leave the domain. If access is granted, the MCF carries out the association request to the target domain under the identity of the initiator domain (or for the advanced profile a group or role). After processing of the association request in the target domain, the initiator domain receives a reply to the management request. The response may be a failure report if access control in the target domain was not successful. The activity concerning access control will be forwarded by the Event Forwarding Discriminator (EFD) and written in a log.

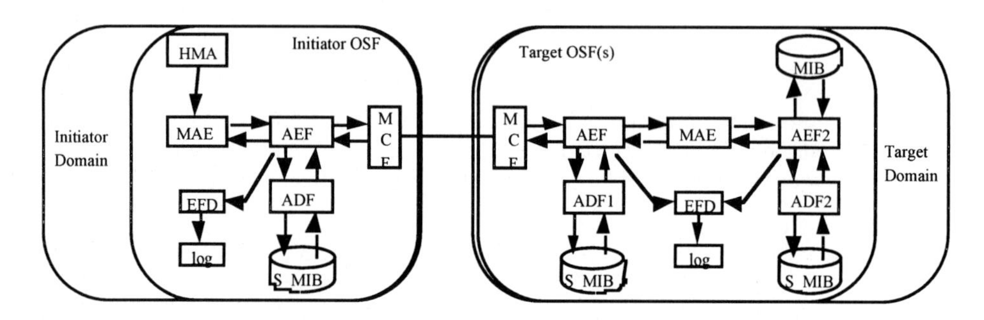

Figure 3: Global Architecture for Access Control

Target domain: upon reception of the management request, the AEF will initiate access control using the initiator ACI (initiator domain or role identity) contained in the request. The establishment of the management association is granted or denied by ADF1. The activity is taken note of and forwarded by the EFD to the log.

Access Control to Management Operations

Access control to management operations is performed in both the initiator and the target domains. As specified in [D2B], the outgoing access control to management operations uses the access control profile AOM24326 (capability based) while the incoming access control is based on an identity scheme (identity of initiator domain or a role, as discussed above) using Access Control Lists (ACL).
Initiator domain: outgoing access control to management operations is granted or denied based on identity of the initiating HMA (the assigned group or role is correlated to a capability based scheme) and whether the requested operations and targets are within the constraints of the capability.
Target domain: the MAE receives the management operation request from the initiator domain. AEF2 passes the decision request to ADF2 containing the following information: initiator ACI, requested management operation, ACI related to the data, identification of the target object class and instance, action identifier, attribute identifier. ADF2 bases its decision on the above information together with the retained ACI, the target ACI, the access control rules and contextual information.

ADF2 uses an ACL to determine whether the initiator may access the particular target in the inter-domain MIB ([D2B], section 11.2.3.5). To do this, the <Initiator, Target> pair is mapped to the access rights (of the Initiator) listed in the ACL. If access to the target is granted, the management operation is applied to the target.
The target could be the MIB itself, a MO, a group of attributes or even an attribute of a particular MO. TRUMPET has decided to support controlled access down to the attributes and attribute values on the target side. Choosing this fine-grained access control on the target side can cause performance problems. In the near future, TRUMPET assumes that the initiator TMN domain is expected to have a limited number of managers accessing a target TMN domain. However, this situation might change when customers are given access to TMN.

Access Control to Notifications

Notifications are forwarded to managers that have previously subscribed to receive these notifications. This is enforced in the agent by EFDs. Because subscribing to receive notifications from an agent from a remote domain is a management operation that is submitted to inter-domain security measures, only legitimate managers will have registered to receive notifications and only those managers will receive notifications. Therefore, there is no need for extra outgoing access control mechanisms in the agent domain.
The above guarantees that notifications will not be emitted to illegitimate managers, it does not stop misbehaving agents from emitting notifications to managers. In the manager domain, it is possible that unwanted notifications are received from properly authenticated (notifications are received only after a secured management association has been established), but somehow misbehaving agents. For example, an agent may not have properly secured its access to EFDs.
Figure 4 shows the access control architecture for notifications. A notification is issued by a MO and passed to the EFD. If the notification passes the discriminator filter, an event report is generated and sent to the target manager. Manager side; the AEF and ADF verifies whether the incoming notification should be processed by the manager.

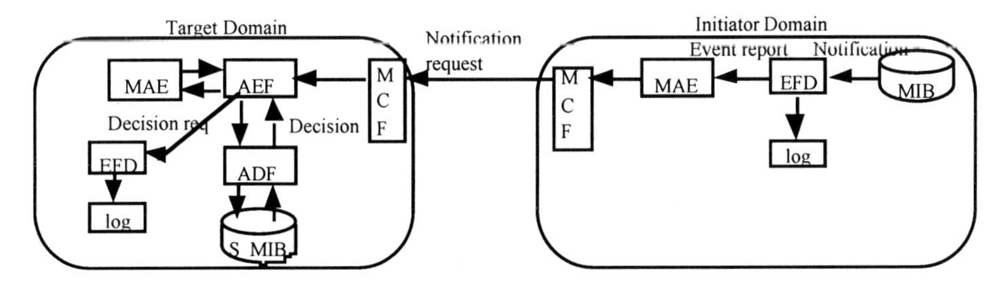

Figure 4: Access control to Management Notifications

3.4 Integrity and Confidentiality

Research TMN Platform

Connection Integrity / Confidentiality
Connection integrity and confidentiality are provided at the transport layer using TLSP [X.274]. This is the preferred solution, but it can be implemented only for a research TMN platform.
As illustrated in Figure 1, TLSP can access the SSO which performs the security transformations on the data, using session keys that have previously been established during the security context negotiation.
The security transformations take place at the bottom of the transport layer. That is, the TPDU to be protected contains the usual transport layer header. If only integrity protection is requested, the Integrity Check Value (ICV) is sent in clear text, and the algorithm for the ICV must be a cryptobased Message Authentication Code (MAC). If connection confidentiality is selected as well, a non-cryptobased Manipulation Detection Code (MDC) can be used. The ICV will subsequently be encrypted along with the data. The result of the security transformations is a Security Encapsulation TPDU (SE TPDU) as specified by [X.274].
The decision to use connection integrity / confidentiality is taken at the time of establishment of the management association. The transport layer entity may be instructed to use TLSP through QoS parameters, or through management operations. All security transformations are done by calls over the GSS-API from TLSP, which will normally be a separate entity placed between the transport and network layers.

Selected Field Integrity / Confidentiality
Selected field integrity and confidentiality are provided at the presentation layer by use of the Protecting Transfer Syntax (PTS) of ASN.1 [X.833]. Note that this syntax is only defined for the 1994 version of ASN.1.
All transformations are done by calls from the presentation layer over the GSS-API interface. If non-repudiation is to be used, use of a public key algorithm is necessary. This will also provide integrity protection, and additional use of a symmetric algorithm for integrity protection is then redundant.
The application is responsible for selection of the fields which need protection. This selection must be signalled over the CMIS interface by some means, and the selection must be communicated to the presentation layer from CMISE. The CMIS interface needs enhancements to be able to handle parameters for this purpose.

Commercial TMN Platform
For commercial TMN platforms where the internals of the OSI stack are not accessible, integrity and confidentiality services will be implemented above CMISE. We are aware that this approach has several disadvantages:

- Connection confidentiality cannot be provided, and only some fields (e.g. attribute values) may be selected for confidentiality protection (see section 2.1).
- The security services become much more application dependent.
- For both services, extra encoding / decoding of application data is required. It is assumed that the overhead imposed by these operations is significant.
- For both services, the security transformations may have to be applied to several fields for a single CMIS service request. The optimisation of the security transformations for several fields increases the complexity of the algorithms.

- For confidentiality, a modification of application data types is required.

Figure 5 shows the integrity and confidentiality procedures for sending data. Security transformations are performed through calls over the GSS-API from the application.

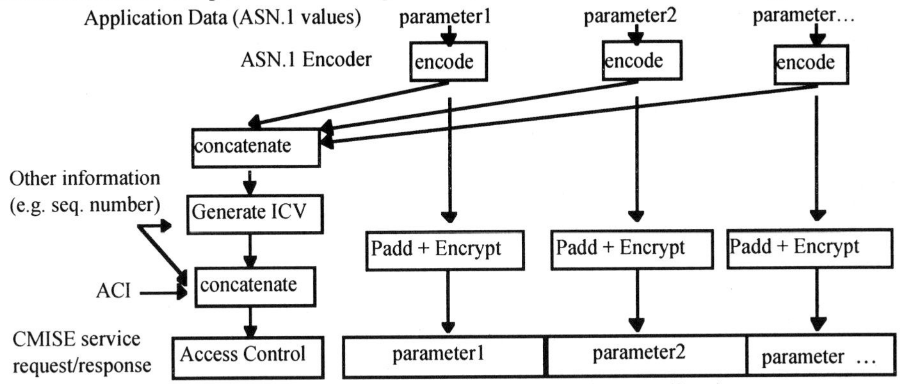

Figure 5: Integrity and Confidentiality Architecture (Send)

The application data corresponds to a set of parameters for a CMISE service request or response. The parameter values are encoded using an ASN.1 encoder. For integrity, the encoded parameters are concatenated and the ICV is generated using the concatenated data and other information. Because integrity is performed separately from confidentiality, the ICV algorithm is a cryptobased MAC. The ICV is concatenated with ACI (if applicable) and other information. The result is transferred to the peer entity using the access control field of CMIS.

For confidentiality, each encoded parameter is padded (if necessary) and encrypted separately (possibly sharing encryption algorithm information) and then transferred using the appropriate CMIS field. The parameters are not concatenated before encryption because we do not consider it a reasonable alternative to put the result of this encryption into a single parameter and put dummy values in other parameters.

Encryption is illustrated using dotted lines because it can only be performed on CMIS parameters which are able to accept a data type suitable for an encrypted value. For example, the Invoke Identifier parameter cannot accommodate an encrypted value.

3.5 Non-Repudiation

Non-repudiation consists of the following facilities: generation of evidence, recording of evidence, verification of generated evidence, retrieval of the evidence, re-verification of the retrieved evidence. The last two are typically the tasks of a dedicated application, and may be performed an "arbitrary" time after the events that caused the evidence to be stored. Non-repudiation may be applicable to both the initiator and receiver of a management request, demanding non-repudiation of delivery and non-repudiation of origin respectively.

Weak non-repudiation, i.e. generation of evidence provided by authentication, integrity and logging of the operations performed, will be sufficient in all but a few cases. The logging should be done by both TMN systems, providing both (weak) evidence of origin and delivery. For weak non-repudiation of delivery, logging of acknowledgements should be used in addition to logging of outgoing management requests. According to standard security terminology, this can strictly speaking not be called a real non-repudiation service.

Strong non-repudiation may only be attained by asymmetric cryptographic mechanisms or use of a TTP. [D2B] prescribes use of signatures produced by a private key

paired with an authentication certificate issued by an off-line CA. This may be the same as the key pair/certificate used for peer-entity authentication, or a separate pair if for example different algorithms are used for authentication and for signature.
At present, an architecture for strong non-repudiation is left for further study by TRUMPET. On a commercial management platform, signing of management operations suffer from the same problems as cryptographic integrity protection, as discussed earlier. On a research management platform, non-repudiation may be implemented at the application or presentation layer, by signatures on the PDUs transmitted. This level of granularity is clearly not very convenient with respect to later retrieval and re-verification. Nevertheless, requirements for strong non-repudiation (at least of origin) have been identified, and the service should not be discarded completely.

4 Conclusion

This article presented an architecture for implementation of security policies for inter-domain management as formulated by TRUMPET. The architecture consists of sub-architectures for each security service. Security services and mechanisms are placed in relation to the TMN architecture and to management applications. Security will for most parts be transparent to the applications, but some interfacing is necessary. Partly defined protocol specifications show how to communicate security information and establish a security context. Use of TTPs is anticipated at least for authentication (certification of public keys, and distribution of public key certificates), which means that TTPs are necessary components of the infrastructure.
The placement in relation to the TMN architecture and management applications strongly depends on the platform on which the security will be implemented. Most commercial platforms do not grant any access to the communication protocol stack, which means that all security must be implemented above the stack. Use of a commercial platform necessitates deviations from the security policies formulated by TRUMPET. If the internals of the communication stack are accessible, security may be integrated at the application, presentation and transport layers.

References

[AC112] ACTS TRUMPET Project AC112, “Inter-Domain Management with Integrity”.

[D2B] TRUMPET Deliverable 2B, “Inter-TMN Security Policies”, Jul. 96. http://ascom.eurecom.fr/Trumpet/

[ISP] ISO/IEC DISP 12060-9, "International Standardised Profiles AOM243 - OSI Management - Mngt Functions- Part 9: AOM2432n - Access Control", Dec. 96.

[M.3010] ITU-T M.3010, “Principles of a Telecommunications Management Network”.

[X.273] ITU-T Rec. X.273 | ISO/IEC 11577, “OSI - Network Layer Security Protocol”, 94.

[X.274] ITU-T X.274 | ISO/IEC 10736, “OSI - Transport Layer Security Protocol”, 94.

[X.509] ITU-T X.509 (1995) | ISO/IEC 9594-8: 95, “OSI - The Directory - Part 8: Authentication Framework”, Revision 1, 95.

[X.741] ITU-T X.741 | ISO/IEC 10164-9, "OSI - Systems Management - Part 9: Objects and Attributes for Access Control", 95.

[X.812] ITU-T X.812 | ISO/IEC 10181-3, “Security Frameworks for Open Systems - Part 3: Access Control”, 94.

[X.830-3] ITU-T X.830-833 | ISO/IEC 11586, "OSI - Generic Upper Layers Security”, 95.

[RFC1508] RFC 1508, J. Linn, “Generic Security Service Application Program Interface”, 93.

Maintaining Integrity in the Context of Intelligent Networks and Services

Victoria Montón, Keith Ward
Department of Electronic and Electrical Engineering, University College London, Torrington Place, London WC1E 7JE, U.K.
email:vmonton@eleceng.ucl.ac.uk

Mark Wilby
The Network Design House, 31 South Parade, Ossett, West Yorkshire WF5 OEF, U.K.
email: mark@NDH.co.uk

Modern, high functionality networks support an ever increasing range of sophisticated services. But this strength is also a weakness due to the fragility of the complex distributed processing capability. Unexpected perturbations can cause widespread network outages i.e. 'brownout'. The risks to network integrity are exacerbated by regulatory policies that demand open network interconnection between competing network operators and service providers; and the increasing complexity where single experts cannot comprehend the whole problem space. This paper proposes a design methodology for a formal and systematic framework to assess the risk to network integrity and hence minimise network and service failure probability.

1 Introduction

Modern telecommunication networks are experiencing a dramatic evolution and increase in complexity. The introduction of processor controlled digital systems with message addressed common channel signalling permit the introduction of the Intelligent Network (IN) [1, 2]. An increasing degree of intelligence is being introduced in telecommunication networks, causing a dramatic increase in control complexity and fragility. Services have to co-exist and co-operate across platforms, giving rise to undesirable feature interactions [3, 4, 5, 6]. Unexpected perturbations may also ripple through the network to give widespread outages i.e. the, so called, 'brownout'.

These problems are exacerbated by regulatory policies that demand the open interconnection of networks of competing operators and service providers together with internetworking of services [7]. But the process from conception to launch of a new service embraces many specialised activities from the specification of customer requirements to pre-service testing. Such an expertise base is well beyond the comprehension of one individual and should be incorporated into a meaningful and manageable structure, easily accessible by a variety of specialists.

This paper discusses the need for a formal and systematic framework to assess the degree of integrity of telecommunication networks and services, based on their vulnerability to the introduction of changes or unexpected perturbations. This framework can be used to assess the risk of failure, to avoid and prevent failure situations and to assist rapid recovery in the case of failure. Actions to be taken in order to construct this framework, include:

- assessment of the end-to-end process to identify areas for improvement,
- adoption of new design methodologies,
- use of risk analysis methods to assess the risk introduced by limitations in the design, implementation and testing of systems,
- use of real time monitoring to quickly detect integrity violation,
- construction of knowledge based systems to capture information regarding previous experiences in a way that can be accessed by a variety of people.

Modelling and simulation techniques are proposed to gain a better understanding of the intelligent services and networks, together with the use of formal methods such as

SDL (the ITU-T Specification and Description Language [8]), in conjunction with Message Sequence Charts MSC [9]. A methodology is presented to assess the end to end process and also, using SDL and MSC, to identify and categorise possible sources of errors and their effects in services.

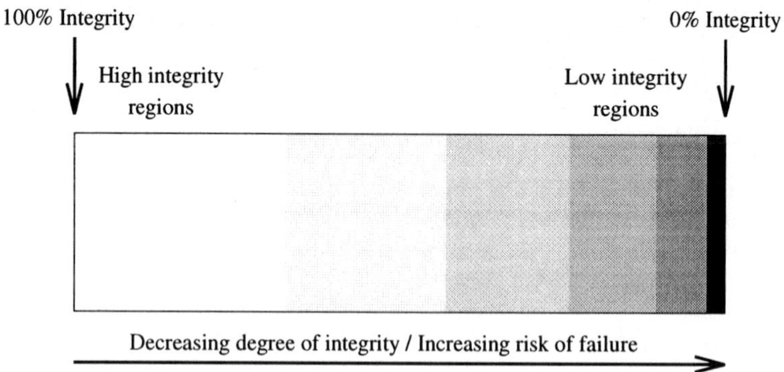

Figure 1: Integrity definition

2 Network Integrity Definition

A definition of network integrity is proposed as "the ability of a network to retain its specified attributes in terms of performance and functionality". This definition considers integrity in terms of robustness, invulnerability and incorruptibility. The degree of integrity of a system is inversely related to the risk of failure, i.e. the higher the degree of integrity, the easier it is for the system to maintain its attributes, and hence the less likely it is that unexpected perturbations can result in failure. Figure 1 shows that the degree of integrity can be divided in a collection of states of integrity [10]. At one end there is 100% integrity, which means that the system is absolutely robust and will not be affected by unexpected problems. At the opposite end, there is 0% integrity, meaning that any perturbation will cause failure. The size and shape of the different integrity bands will depend on the particular system and the criteria adopted by the operator to define regions of risk.

Loss of integrity means moving from one band to another band of lower integrity. This will result in failure only in one of the following cases:

- if the system was initially in a band of very low integrity, because any minor perturbation would put it in a situation of failure; hence, it is important to operate in a high integrity band;
- if the loss of integrity is very high, causing the system to move from a band of acceptable integrity to the region near 0%; it is desirable to minimise such instances by identifying possible sources of failure and designing them out, or monitoring the relevant parameters to take avoiding action.

3 Methodology Framework

The methodology framework has been divided into two categories, namely static and dynamic actions.

Static Actions – previous to the launching of the services, i.e. actions taken during the design and implementation stage.

- Improvement of design methodologies, taking account of factors effecting integrity, to build systems that are robust to loss of integrity.

- Risk analysis, using the information provided by the integrity definition framework, to assess the consequences of possible design limitations, testing limitations and errors.

Dynamic Actions – taken in real time, i.e. during the operation of the services.

- Monitoring and control of the integrity parameters to identify the band of integrity in which the network is operating a any time.
- Restoration mechanisms in case of integrity violation detection.
- Risk analysis to asses the severity of detected integrity violations and to decide the appropriate restoration policy to be taken.

4 New Services Process

The process, from inception to launch of a new service (Figure 2) starts with the feasibility study on which the requirements specification of the overall system are based. The architectural design sets the parameters within which the detailed design is carried out and the different elements of the architecture and their inter-operability are considered. Then the most detailed level of unit design and coding. Each of these processes requires a different level of testing. At the most detailed level, testing must ensure that the individual coding works. Interaction testing is necessary to ensure that when two or more components are put together they perform as expected. When the overall system design is tested, the acceptance testing is carried out to ensure that not only does the system work but it also conforms to the requirements. The launch phase requires not only testing in an operational environment, but also trailing of the in-service support processes.

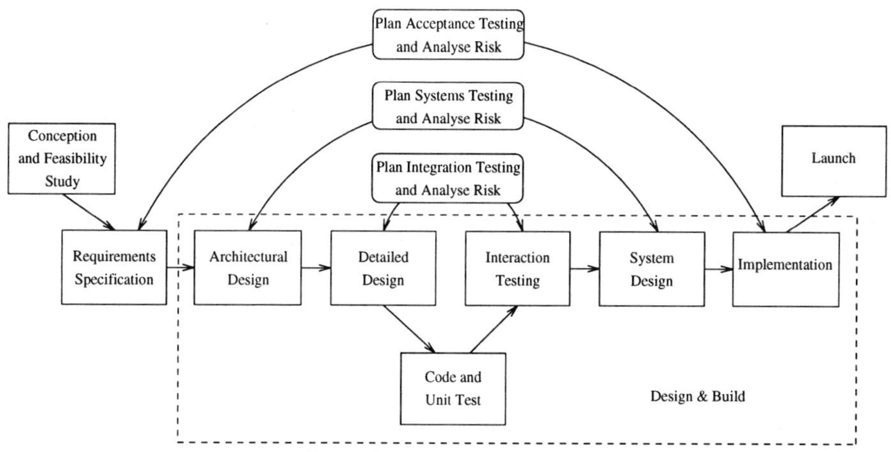

Figure 2: New services process

At the customer requirements definition stage, consideration should be given to how the service will be used by customers, and the potential for undesirable feature interactions with other services.

Early in the design stage, modelling should be carried out to assess the risk of feature interaction and integrity violation. The objective being to remove problems at the early stages of the design cycle rather than the time consuming and costly method of identifying problems when the development is completed.

Testing should be carried out at appropriate stages in the development as shown in Figure 2. Acceptance testing should embrace not only testing of the completed design on laboratory models but also intra-network field testing of the installed equipment and/or software. Where the service is to be internetworked via the networks of other operators, comprehensive non-intrusive testing should be carried out between captive models followed by intrusive testing of the internetworked services, in the operational environment, before they are used by customers.
Due to the high complexity of the services, it is not possible to test everything. Testing must focus in those areas where problems are more likely to arise. The level of testing carried out must also be subjected to risk assessment, in order to understand the consequences of the limited amount of testing. For this reason, a risk assessment methodology that helps to assess the probability, impact and consequences of actions must be developed and integrated in the development of systems. As shown in Figure 2, it should be applied at the different levels of systems development, in order to determine the amount of testing that needs to be performed for different levels of risk. Increasing the amount of testing decreases the risks, but increases the costs and delays so a balanced approach must be taken.
Part of this risk assessment activity would consist of keeping records of any constraints decisions taken during the development of the systems, and an assessment of the effects that these limitations would have in adverse conditions. Such an activity would:

- provide a better understanding of the systems operations;
- make the information accessible to different people, instead of being only in the heads of those involved in the design process;
- help to identify weak points in the systems and where problems are likely to arise;
- help in the diagnosis of failure and to a more rapid identification of the actions required to fix the problems;
- help to evaluate consequences of failure, for example, in terms of lost calls, damage to equipment, cost, etc.

Other activities and factors that should be considered in order to tackle the network integrity problem are:

- *Population and user models* – the behaviour and operation of a service is effected by the way the service is used by the customers. These factors can be embraced in so called *customer models.* Customer models can be decomposed in two main areas, called *user models* and *population models.* User models represent how an individual user would interact with the service, e.g. what actions they are likely to take, possible mistakes they would make and how they would react in different situations. Population models represent the characteristics of all users of the service e.g. number of customers, geographical distribution, degree of mobility, and usage patterns – call holding times, average and peak calling rates etc.
- *QoS* – quality of service represents how the service is perceived by the customers. In a competitive environment, where customer satisfaction is a primary objective, special care must be taken to understand those parameters that measure customer perception and to maintain the appropriate values for them. A major problem is how to map the users QoS requirements into network

parameters, and a great deal of work is being done in this area, e.g. the RACE project TOMQAT [11], and the ACTS project MISA [12].

- *Performance* – the behaviour of a service can be heavily influenced by network performance characteristics. Therefore, the study of possible erroneous behaviour must include the relationship to performance. The performance objectives of the design should also map on to the QoS requirements.
- *Signalling* – this factor refers to the general information flows within the system and the associated processing, as opposed to specific signalling protocols or architecture. Signalling is the area of most growth in complexity, and hence highest risk to integrity.
- *Monitoring* – comprises the checking of progress throughout the development cycle and in-service where it is important to detect potential integrity violations before they become customer effecting.
- *Automated records* – as systems evolve and become more complex, the amount of information which is required to be understood increases. Hence it becomes necessary to store this information in an structured and automated way that makes it accessible, meaningful and useful to different people.
- *Risk analysis* – part of the problem of maintaining network integrity is being able to identify areas of risk, and predict the consequences of actions such as limitations and constraints of the developed systems. Hence, risk assessment methodologies should play a main role in the integrity problem.
- *Modelling* – modelling and simulation techniques are needed to improve the understanding of the new systems before they are developed. This is particularly important in the telecommunications environment, characterised by fast changes and increasing complexity.

	Population Models	User Models	QoS	Perform -ance	Signal -ing	Monitor -ing	Automated Records	Risk Analysis	Model -ing
Req.	!	√	?	?	–	√	?	!	–
Spec.	!	√	?	?	–	–	?	!	√
Design	!	√	?	?	√	–	?	!	√
Implem.	!	√	?	?	√	–	?	!	!
Testing	?	√	?	√	√	√	!	√	!

Table 1: New services process

√ Activity carried out ? Lack of information
! Activity not carried out – Not applicable

The cross relationships between the above factors and the main processes in development processes can be presented in the form of tables, representing the relationship between the different work areas and the factors relevant to the integrity problem, as shown in Table 1. The content of each square shows whether the integrity factors on the top have been applied to the process areas on the left. To facilitate the investigation, the development of a new service is considered in terms of five sub-processes, namely: requirements capture, service specification, design, implementation and testing.

Typical results of such an analysis are illustrated in Table 1. Application of the analysis methodology to a the design processes of a “Telco” has revealed a number of areas requiring action. In particular, the lack of predictive modelling to identify areas of potential risk to integrity.

5 Modelling Methodology

In the complex intelligent network scenario, modelling becomes an activity of paramount importance. An adequate modelling exercise helps to detect and resolve potential problems prior to the detailed design phase, thus preventing expensive retrospective design modifications. Furthermore, telecommunications markets are moving towards new services where there is no expertise, and appropriate modelling can provide a better understanding of how these new services operate services.

As quoted in [13], in order to build a good model, the issues of abstraction, completeness and simplicity of the models need to be carefully considered. The models must capture all the aspects of interest of the problem they are trying to solve, but at the same time maintain a level of simplicity that allows people to understand the models and interpret the results. Due to the large and increasing complexity of services, it is necessary to apply some decomposition paradigm that allows to break down the problem into smaller areas. A top down approach [10] is used, that borrows heavily from the IN conceptual models, consisting of four main layers corresponding to different levels of detail, namely service layer, functional layer, network layer, operational layer and customers layer. The important point of this decomposition is that different aspects of the models would access only those layers that are required, thus providing a framework that allows the treatment of different aspects of the problem with the level of detail needed.

A specification language is needed, and SDL (the ITU Specification and Description Language [8] is already in widespread use. The advantages of SDL are that it is an international standard, it is easy to use, it has proven to be the most used method in telecommunications applications (currently used by a large number of operators and manufactures), and there are advanced software tools available to build and test SDL systems. Moreover, there is a strong international research activity in SDL, and modifications are introduced to the language to respond to new demands. Hence SDL is a likely candidate to be used as the modelling language.

6 Proposed Approach

6.1 Overview

The initial approach focuses on 'signalling'. The approach is from a services perspective, where a service is viewed as a collection of distributed entities performing different tasks and interacting with each other to provide the required functionality. The aim is to identify the possible states of the system, the transitions between them and their probabilities; both for normal behaviour and possible error states.

The behaviour of a system can be represented as a extended finite state machine, as is done in SDL. A system run or trace is a specific path of behaviour, i.e. a set of states and the transitions between them. A path of correct behaviour consists of a sequence of acceptable states. Under adverse conditions, the system might leave this path and move to an erroneous state. Some errors will not produce a strong deviation from the path of correct behaviour and therefore can be considered minor errors, since the system remains within an area of acceptable behaviour. Conversely, major errors or combination of errors can cause 'control mutations', i.e. can make the network follow a path divergent from that of the correct behaviour. The framework must provide the probability of erroneous behaviour and should permit identify the risk of control mutations which can take the system out of its stable condition towards catastrophic breakdown. There is a need to detect deviations and their size, classifying regions of risk and categorise errors according to their consequences. Thus minor errors can be treated differently to those leading to a catastrophic situation.

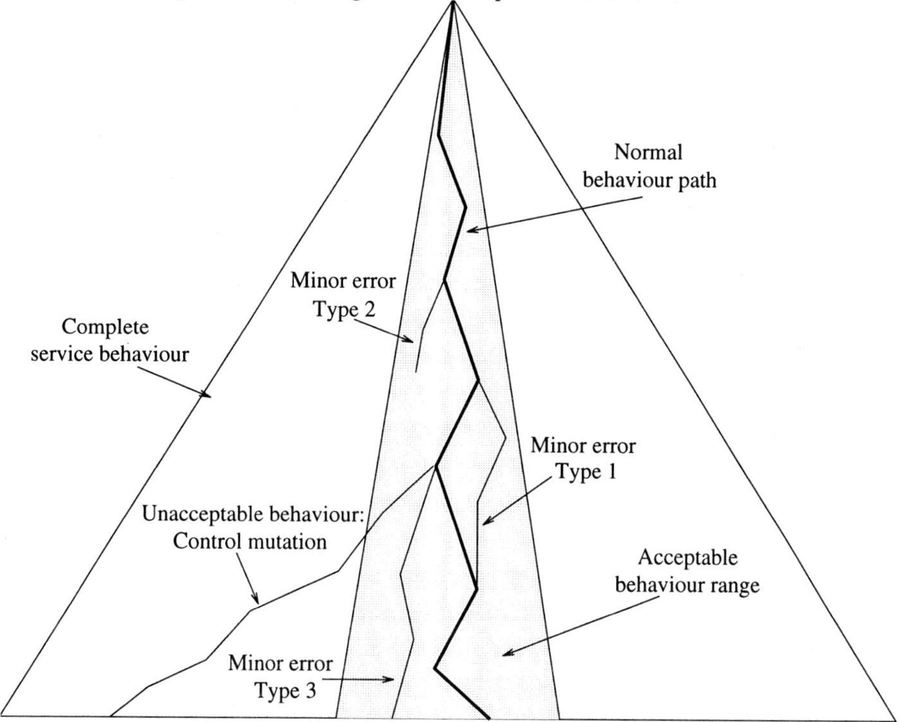

Figure 3: Path of correct behaviour and deviations from it

The concepts are illustrated in Figure 3, where the large triangle represents the whole domain, i.e. all the possible paths that the system can follow when the service is running, the shaded area indicating the region of acceptable behaviour. The path of correct behaviour, starting from the initial state (the top vertex), is shown, together with minor deviations that return to the correct path (Type 1), or are detected and cause the process to stop (Type 2), or terminate in a final state out of the normal path but still within the safe region (Type 3). A control mutation takes the behaviour path outside the safe region.

6.2 Stages

The stages to develop the framework, for the simplified case of one single service, are as follows:

Stage 1. Modelling and understanding of service behaviour – A formal service description that defines the behaviour and operation of the service is needed, e.g. SDL can be used to describe the overall behaviour, whereas MSC [9] can be used to visualise selected traces and identify message interchanging between communicating entities.

Stage 2. Find path of normal behaviour – focus on one path of desired behaviour and analyse the transactions involved, e.g. signalling, state transitions, etc. from the MSC description.

Stage 3. Identification of possible sources of errors – identify possible error events, e.g. perturbations such as invalid messages that may trigger a violation of integrity, and study how the introduction of these errors affects the behaviour of the service. Two inputs are needed in order to identify errors, namely

- *Performance models* – performance characteristics can have a great impact in the behaviour of services. Two types of errors can be differentiated. *Static errors* are due to bad definition of services, poor logic or incorrect programming, inherent to the services, which can be minimised with adequate specifications, precise definitions and carefully developed programming. *Dynamic errors* can occur due to performance limitations from delays, signalling overloads, etc. and they are more difficult to detect and correct because they are caused by numerous event combinations. In order to identify sources of dynamic errors performance models need to be built and coupled to the overall model. One of the major limitations of SDL in this context is that it does not include performance evaluation (some research is being done in that direction [14]). There is a need to study how to couple the performance models to the SDL description of the service behaviour.
- *Customer models* – user an population models (discussed in section 4) are needed as another input to the overall model.

Sensitivity analysis must be carried out on performance and population models to find the limits for which they are valid. After studying standard cases, effort must concentrate in stressful situations that might cause service malfunction.

Stage 4. Introduction of errors in the model and identification of error states – the MSCs that describe a normal path of behaviour need to be modified to introduce errors. Analysis of this information will permit the identification of error states that can be gradually incorporated to the model in order to build paths of erroneous behaviour.

Stage 5. Find probabilities for the different error states – error events must be matched to rates, i.e. the probability of errors occurring per time unit, taking into account performance characteristics and customer models. Initially, error rates must be assigned based on external information, e.g. previous experiences, and there will be an associated degree of uncertainty. This uncertainty will always exists, but it will be reduced as the methodology progresses and provides more information about errors and their probabilities.

Stage 6. Construct overall probability tree – identify possible sequences of error states that will progressively take the system away from the normal behaviour. The correlation between different error states, i.e. the probability that one error state leads to another, forms a large and complex decision tree. A simple example is demonstrated by the tree structure illustrated in Figure 4. Each node represents a state and the interconnecting branches represent transitions between states and their probability. This will show the overall probability of moving from the initial state (A) to each of the possible final states.

In complex systems, where the number of possible paths and final states is extremely large, it would be unfeasible to find the probabilities to reach each of the states. Therefore, the work must concentrate in two cases, namely

- the most likely states (such as state C1), which describe the normal operation of the system, and, predominantly,
- the most dangerous states, i.e. those that fall out of the safe region, even if their occurrence is very unlikely, because they need to be quickly identified in order to prevent catastrophic failures.

Stage 7. Organise information – extract useful information from the data obtained in the previous stages, so that in can be easily accessed and utilised as a pre-emptive framework. There is the need for adequate documentation of all the acquired information, possibly in the form of a new kind of expert system where it can be extracted in an intelligent way.

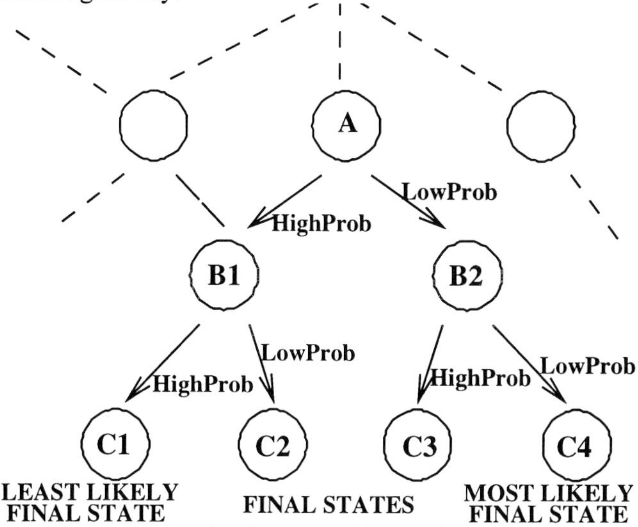

Figure 4: Example of state transitions and probabilities

7 Conclusions

The ever increasing complexity of telecommunications networks and services progressively increases the probability of service outages due to loss of integrity. The paper outlines a methodology that can be developed to determine the cause, extent and outcome of integrity violations. Thus assisting the design of robust systems and the management of "live" network services. The assessment of the development process of a telecommunications operator has been carried out and identified a number of areas for improvement. The modelling methodology is being tested for the simplified case of one service and one network. SDL models of the service are being developed. Future work will include more complex scenarios, i.e. multi-service, multi-network, thus allowing the examination of issues such as services interactions and interconnect.

References

[1] ITU-T, *Recommendation Q.1211:Introduction to intelligent network capability set 1*, May 1995.

[2] ITU-T, *Draft Recommendation Q.1221:Introduction to intelligent network capability set 2*, November 1995.

[3] E. J. Cameron, N. Griffeth, Y.-J. Lin, M. E. Nilson, W. K. Schnure and H. Velthuijsen, "A feature interaction benchmark for IN and beyond," *IEEE Communications Magazine*, vol. 31, no. 3, pp.64-69, 1993.

[4] L. R. Brothers, E. J. Cameron, Y.-J. Lin, M. E. Nilson and E. Silverstein, "Feature interaction detection," *IEEE International Conference in Communications*, vol. 3, pp. 1553-1557, 1993.

[5] E. J. Cameron and H. Velthuijsen, "Feature interactions in telecommunications systems," *IEEE Communications Magazine*, pp. 18-23, August 1993.

[6] B. Kelly, M. Crowther and J. King, "Feature interaction detection using SDL models," *IEEE Global Telecommunications Conference*, vol. 3, pp. 1857-1861, 1994.
[7] K. Ward, "Impact of network interconnection on network integrity," *British Telecommunications Engineering*, vol. 13, pp. 296-303, January 1995.
[8] ITU-T, *Z.100 (1993). CCITT Specification and Description Language SDL*, June 1994.
[9] ITU-T, *Z.120 (1993). Message Sequence Chart (MSC),* September 1994.
[10] V. Monton, "Risk assessment methodology for network integrity," *BT Technology Journal*, vol. 15, pp. 223-234, January 1997.
[11] RACE Project 2116, *TOMQAT: Total Management of Service Quality for Multimedia Applications on IBC Infrastructure.,* 1995.
[12] ACTS Project AC080, *MISA: Management of Integrated SDH and ATM Networks. Deliverable 3*, October 1996.
[13] K. C. Woollard, "What's IN a Model? Modelling IN services using formal methods," *BT Technology Journal*, vol. 13, pp. 43-50, April 1995.
[14] M. Diefenbruch, E. Heck, J. Hintelmann and B. Müller-Clostermann, "Performance evaluation of SDL systems adjunct by queuing models," *SDL'95 with MSC in CASE, Proceedings of the Seventh SDL FORUM*, pp. 231-242, Elsevier Science Publishers B.V. 1995.

User Authentication in Mobile Telecommunication Environments Using Voice Biometrics and Smartcards

Martine Lapère
Lernout & Hauspie Speech Products
Koneng Albert 1-Laan 64, 1780 Wemmel
Belgium

Eric Johnson
Giesecke & Devrient GmbH
Prinzregentenstraße 159, D-81607 München
Germany

Biometric authentication mechanisms offer both new and improved possibilities for the authentication of users of mobile telecommunication systems. However, the nature of such systems is different in character to those where biometric mechanisms have been traditionally employed. This is primarily because the service provider can only trust the smartcard rather than the mobile terminal equipment and this smartcard has limited capabilities. This paper examines these issues and discusses a potential solution being implemented in the ASPeCT project.

1 Introduction

The emergence of extensive, affordable, accessible and effective global telecommunications services is resulting in an explosive growth in the variety and volume of transactions conducted by electronic means. The safeguarding of the integrity and security of such transactions has a high priority and is being addressed by the developers of such services.

In particular, the secure access to telecommunications services is being regarded as a high priority by ETSI during the development of UMTS (Universal Mobile Telecommunication System).

Under the auspices of the European Community's ACTS programme the ASPeCT project is investigating the implementation of security services in general for UMTS and authentication of users to the network in particular. This includes the authentication protocols of the network and also the authentication of a user to the network. The project will first demonstrate these features and in the final part of the project implement then for a trial involving real users.

This paper considers how the authentication between the user and a smartcard, acting as a security module, can be enhanced by means of voice biometrics. It first explains how such authentication is carried out in the GSM system by means of a PIN, it then goes on to highlight the key features desired of a new authentication mechanism and how well biometric approaches meet these requirements. Finally, it explains the approach being further researched in the ASPeCT project.

An important theme of the paper is how it is possible to accomplish the user authentication despite the restrictions imposed by the mobile telecommunication environment. These restrictions – principally the non-local nature of the authentication and the fact that it must use non-trusted equipment, do not usually apply to biometric authentication systems.

2 User Authentication in GSM

In the GSM system every subscriber is supplied with a smartcard called a SIM (Subscriber Identification Module) by his service provider. The SIM contains secret information from the Service Provider and which the SIM uses in a unilateral authentication protocol to prove to the network that it is a valid SIM. The SIM will only prove its validity to the network if it, in turn, has already authenticated the user.

The user authenticates himself to the SIM by presenting a secret number of between 4 and 8 digits to the SIM. This number is called the PIN (Personal Identification

Number) and its value can be changed by the user at his discretion following a successful authentication with the current PIN. The PIN is protected by a counter which will block the card if the correct PIN is not entered in a limited sequence of consecutive trials. Typically, a SIM will be blocked if an incorrect PIN is entered on three consecutive attempts.

The SIM has a second secret number called the PUK (Personal Unblocking Key) which can be used to unblock a card and change the PIN. The PUK also has a false presentation counter but if the incorrect value is repeatedly entered for the PUK then the SIM will become permanently blocked. The PUK is sometimes kept in the possession of the Service Provider rather than being distributed to the subscriber.

This mechanism is both simple to implement and simple for the user to comprehend but it does have several shortcomings:

- users often find the system too inconvenient to use and disable the PIN checking entirely
- users change the PIN to easily guessed values, or values which are used on other systems. Breaking these other systems or correctly guessing the PIN allows access to the network
- the PIN verification is static – the same value is entered every time, thus making compromise of the PIN possible by observation as it is being entered or by a terminal recording the value for later replay.

One other feature is that PIN verification is a binary process – either the PIN has been entered correctly or it has not, there is no room for doubt and the checking mechanism is straightforward.

An important feature of the GSM system is that the subscriber is free to use any handset he chooses and so the only influence the Service Provider can have on the remote equipment is through the SIM. This also means that the Service Provider cannot trust any processing in the remote equipment except that carried out by his trusted component, the SIM.

3 Biometric Authentication Mechanisms

Biometrics is an exciting area of recent technology development that deals with user-friendly automated methods of verifying a person's identity from one or more behavioural or physiological characteristics. An extensive variety of biometric techniques are being actively developed and researched including fingerprints and palm-prints, hand geometry, retinal and iris scans, dynamic and static signature capture, keystroke dynamics, gesture analysis, odour, and vocal characteristics.

Biometric mechanisms are typically used as a means of access control because they do not rely on a user simply possessing a token but are based on properties of the individual himself. This means that the authentication verifies the actual user and not simply the presence of a token.

Biometric authentication mechanisms can provide high levels of security, however, these come at a price:

- they often require expensive sensors
- the input signals typically require a large amount of processing to perform the authentication
- the authentication criteria are not precise – there is always a trade-off between rejecting valid users and accepting invalid users.

In all cases of biometric authentication the actual authentication is based on a set of calculations which use the input biometric data and some previously processed data which has been stored as a user template. The actual authentication process varies between the different mechanisms – it may simply be checking that characteristics of the input data lie in a certain range or it might involve a neural network.

4 User Authentication in Mobile Telecommunication Environments

In considering how to improve the user authentication in a mobile telecommunications environment, specifically for UMTS, by employing biometric mechanisms it is important to consider the desired goals and relevant restrictions:

- The new mechanism should be more convenient to use than a PIN so that the user does not choose to disable the mechanism
- The new mechanism should work in tandem with the PIN mechanism. Depending on circumstances, it may not always be convenient to use the new mechanism. That is, the user should be able to choose whether to use the new mechanism or the PIN. For example, lending the SIM is feasible by establishing a temporary PIN but would not be practical for a biometric mechanism.
- In a roaming environment communication back to the Service Provider is expensive. Therefore it must be possible to perform the biometric authentication locally in the terminal.
- The mobile terminal and any remote network cannot be trusted to perform security related operations.
- A mobile terminal already contains both speech input sensors and a substantial digital signal processing capability for use in the biometric processing.
- UIMs will contain limited memory for template storage and possess limited processing power to perform the comparison
- The authentication mechanism must be portable between different mobile terminals by simply transferring the UIM from one terminal to another. This means that all user specific data must be stored on the UIM.
- Since a PIN approach is trivially vulnerable to a replay attack, the mechanism can improve the security level by making a replay attack more difficult to mount. Note also, that with a 4 digit PIN the security of the PIN mechanism is not greater than 1 in 10^4.
- A security mechanism with variable thresholds could play a useful rôle. For example, a low security level could be used to allow access to low cost services such as local telephone calls. International calls or more expensive services would require a higher security threshold.

Based on these observations it is apparent that vocal (or speaker) biometric authentication looks potentially the most attractive for the mobile communication services. It can make use of the existing speech sensors and signal processing power and it allows a user interface which naturally fits in with the way the system is used.

Note that the processing of the data derived from the sensors can be split between the terminal and the UIM. It is, however, important that the UIM does not disclose information which would allow a fraudulent terminal to achieve a successful authentication. This parallels the fact that a UIM does not output the PIN but can test an input value to see if it is the correct PIN.

5 Capabilities of the UIM

The GSM SIM is implemented using an integrated circuit card commonly referred to as a smart card. These are special security versions of micro-controllers and are produced by a range of semiconductor manufacturers. The physical and electrical characteristics of the SIM have been standardised along with the logical interface that they provide to the mobile terminal equipment. This standardisation has been very successful and any approved SIM should operate correctly in an approved terminal. Despite some comments from ETSI suggesting otherwise it seems very unlikely that UMTS will use a different format for the UIM. The increasing uptake of smart cards by the banks and the potential for multi-application cards makes the idea to use a different format even less likely.

As highlighted in the previous section the format of the UIM clearly imposes some restrictions on the exact form of biometric authentication that can be used in UMTS. Current smart cards are almost exclusively 8 bit processors with up to 512 bytes of RAM, 16 Kbytes of ROM and 8 Kbytes of EEPROM (non-volatile memory). These memory resources will certainly increase in the next few years but past trends do not suggest that this will be that rapid. Furthermore, the price of the components is largely controlled by the size of their EEPROM and this is in a non-linear fashion. Unless there is a real perceived need for large memory capacities customers are unlikely to be prepared to pay for this resource.

Newer smart cards support a co-processor but this is specially designed to process large integers for public-key cryptographic computations and is not likely to be of use in biometric type calculations which typically involve many floating point arithmetic operations. It seems most improbable that smart card manufacturer's would add a floating point co-processor but simpler special purpose hardware for biometric processing could be an option.

Floating point operations can be implemented using the 8 bit processor found in smart cards but they will necessarily be slow and could render the authentication impractical.

The most promising way forward at present is to partition the processing between the terminal and the UIM. This partition must be carefully arranged because the terminal cannot be entirely trusted – thus we must seek to minimise the possibility of the terminal to defraud the UIM. Since a UIM without its own biometric sensors must rely to some extent on the terminal, we cannot prevent all attacks but we should at least ensure it is not too easy to determine the secret information stored on the card.

6 Possible Approaches to Vocal User Authentication

There are two obvious ways to perform vocal authentication – the processing can be performed either locally or remotely.

As pointed out previously, the second of these is not appropriate for the mobile environment. However, it is worth noting that it may be appropriate if a user wished to perform high security transactions over a telecommunications link. For example, authentication could be carried out remotely by a user's bank before carrying out a fund transfer. In this circumstance, the bank is unlikely to trust the mobile telecommunications Service Provider to carry out the authentication and so remote authentication will be required.

In the remainder of this paper we shall assume that the authentication used to access mobile telecommunications services is performed locally. There are three different approaches as to how the authentication could be achieved:

- *Free Speech Input.* In this case the subscriber simply uses the mobile equipment and the authentication takes place in the background using his speech as the input. This is the most complicated form of authentication and requires large amounts of data storage and processing.
- *Prompted Text.* In this variant the terminal prompts the user with some text and the characteristics of his response is compared to the responses expected from the correct user.
- *Passphrase.* This mechanism relies on only one type of utterance being supplied by the user. It has the advantage that the knowledge of this utterance can be used as part of the authentication.

7 Vocal User Authentication in the ASPeCT Project

As part of the ASPeCT project we are investigating the possibility of using a Passphrase system with small storage requirements – the password-based voice prints must be stored locally on the UIM. More precisely we are aiming at a system requiring only a small enrolment session; consisting of a simple 3 times repetition of the desired password. This enables the user to change his password locally in an autonomous and flexible way in order to avoid possible misuse by a third party.

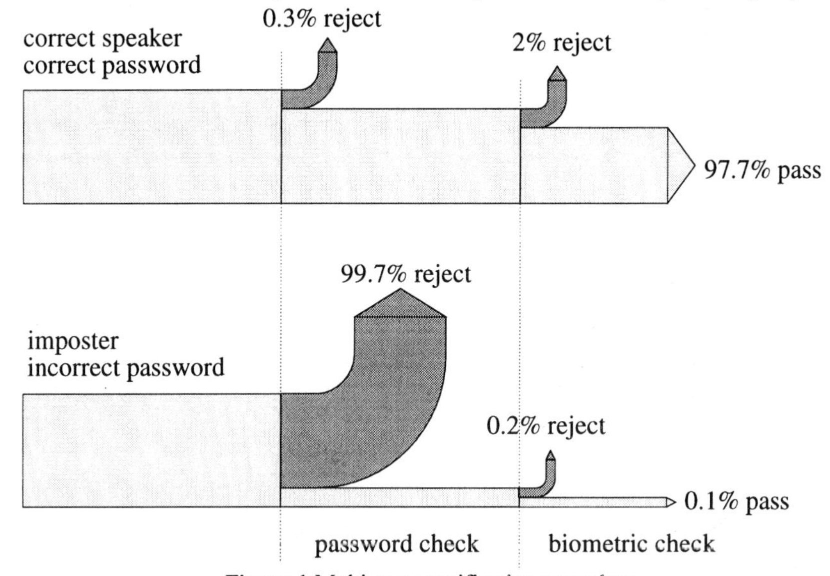

Figure 1:Multi-stage verification procedure

As illustrated in Figure 1, the verification process will be a two stage process. The gross protection of the system should come from the first stage: the password check. In a normal situation an impostor is not aware of the right password. The first step or *Password check* will validate the correctness of the spoken password. Figure 2 shows preliminary results on the systems capacity to reject false passwords. The obtained *equal error rate* (EER), or the point where the ratio of false rejection of bona fide speakers equals the ratio of false acceptance of impostors was measured to be 0.3%. Passwords of an average duration of 1 second were used in this test. Only the utterances which passed this first *Password Check* will be submitted to the second *Biometric Check*. This will be 99.7% of bona fide trials against only 0.3% of false trials. Preliminary results have revealed an EER of 10% for this second or biometric check. This is the discrimination of valid users and impostors uttering the same passwords. The acceptance/rejection curves for this biometric check have a similar behaviour to the curves for the first password check of Fig. 2. This means that if we put a threshold such as to reject not 10, but only a 2 of bona fide users, the false acceptance ratio will increase from 10 to 40%. With these preliminary results however, we will accept 97.7% of bona fide users (98% of 99.7%), while accepting only 0.1% of impostors (40% of 0.3%).

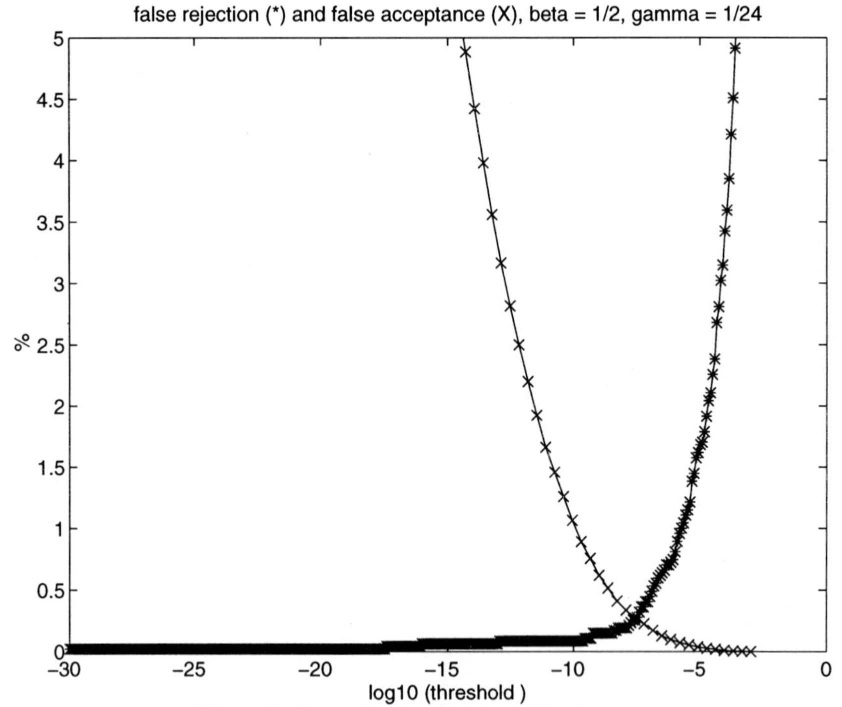

Figure 2: Separation on *Password Check*

During the training session, the required password models will be synthesised out of the simple say-in of the passwords. Since no graphemic input is required, this process can run in a complete hands-free voice controlled manner. This hands-free control makes the system less vulnerable to remote camera surveillance than is the input of a PIN on a keyboard device. Due to the need to enable on-line adaptation to new passwords, the password transcriptions should be stored on the UIM itself. The memory constraints of the UIM then place a severe restriction on the number of parameters that can be retained for the authentication.

The spoken utterances are mapped onto predefined and speaker independent acoustic codebooks which are stored in ROM memory. This mapping limits the required non-volatile storage for a single password transcription to a few tens of bytes. The coded password transcriptions are complemented by additional parameters in order to make the password discrimination feasible.

For the biometric check the distribution of the long time spectral envelope is used. Approximately 10 parameters are stored for one single speech segment. The

separation power of this biometric check is again dependent on the resolution of the stored template. The resolution will be defined as the best trade-off of needed and available storage capacity, and the required separation power.
Multiple passwords shall give increased protection, similar to the way the protection of a PIN code is based on the length of the digit code.
The main advantage of the underlying multi-stage approach of single or multiple password and biometric checks is that any service provider making use of the UIM card would be free to choose the complexity level in correspondence to the required security for their services. This will definitely enhance the ease of use and acceptance by the user.
An important part of the work will be the decision about how to partition the processing of the passwords between the terminal and the UIM. Preliminary work in this direction has already been carried out.

8 Conclusion

We have shown that there is an opportunity in the mobile telecommunications environment to enhance the user authentication mechanism by employing vocal biometric techniques. This environment is atypical and places special constraints on the implementation of the authentication mechanism because the service provider is not able to trust all the equipment being used.
The ASPeCT project is investigating these constraints and an appropriate one is being implemented using a prototype UIM.

Human Factors

Gérard Lacoste lacoste@vnet.ibm.com
Centre d'Etudes et Recherches - IBM France

Homo faber, the inherent characteristic of mankind for creating and building artificial objects, is still at work today. Man will continue to pursue this action, as ever more powerful tools are in great demand. In the distant past he cut and polished flints, but rapidly realized that he could support his struggle for life against hostile environments with other useful objects. He became extremely prolific in creating countless tools to produce these objects, including the tools themselves. The 18th. Century saw homo faber inventing machines that could substitute, in part, for human physical processes and generating ever more powerful means of production. With the information age, men are transgressing the physical limits of their artificially constructed environment. They have begun to dream of an abstract world, with virtually no physical barriers. A world that will support their relentless search for power. The world of thought is now receiving the careful attention of innovation and research.

During their development, tools and machines were assigned more and more power, but, at the same time, they lost much of their simplicity. Tools, machines and systems required more and more knowledge to create and to master them. They segmented the human world into narrower fields of specialization, in stark contrast to the natural simplicity and wonderful versatility of human hands and brains. In a way, machines tend to make men their slaves. Men have to employ increasingly more effort and knowledge to maintain control over their "creatures". This situation is currently intensified by the existence of information systems which can challenge the human brain. Mankind's freedom is seriously at stake, and one of the battles to preserve this freedom is raging at the interface between the automaton and its user. In other words, men will only stay in control if the ever more powerful and intelligent machines are equipped with man-machine interfaces which are well suited to human physical and intellectual abilities. Through their interfaces, machines must bring the complexity of the world, virtual or not, down to an easily understandable human level, without compromising their power and without limiting human expressiveness. They must not be allowed to seriously challenge existing human principles and the concepts which govern our advanced societies, such as human rights, equal opportunity, public health, etc. This is especially important, as fewer humans will be able to operate the machines of the information age if too much knowledge, skill, and training are necessary.

Together with power and simplicity, reliability is, undoubtedly, the third fundamental property required if the man-machine interface is to serve man's quest for freedom. Men must have sufficient confidence in machines before they will accept delegating their power to them. Confidence in the information processing machines operating in the virtual world is of paramount importance, as people will see, feel, and act on virtualities through man-machine interfaces alone. Such virtual situations require that counterparts to the physical world emerge at the man-machine interface, counterparts on which confidence can grow. The semantics of both commands and communications travelling via the interface must be carefully delineated, so that user requirements can be fully executed without delay and without errors. Likewise, machine communications must be faithfully transcribed in human terms and without distortion.

These are the conditions under which man-machine interfaces should operate. Needless to say, we are still far from achieving these ideal conditions. In the

information age, the man-machine interface will be the sole visible part of the machine. Therefore, the focus should now be directed towards that interface and every effort made to enhance the relationship between men and machines, for the benefit of the majority of people. Otherwise, co-operation between human beings and machines will not happen and men will be denied the true contribution which machines could make to their lives
The three papers which follow illustrate three significant research efforts to help people to benefit from the virtual world. The first paper investigates users acceptance of security features in the world of electronic commerce. Security techniques are inherently complex, and so is the networked world. Despite their complexity, they must appear to providers and consumers of electronic commerce as simple and as convenient as the security mechanisms of the real world. The second paper deals with information access by diverse classes of users. The goal is to ease user access by having the machine learn users' preferences, ability, and experience. The third paper demonstrates that application of existing concepts of the real world to the virtual world of network management can make the analysis of information as easy as inspecting the state of a road network. All three articles advocate the need for powerful and easy-to-use man-machine interfaces, which users can rely on in the upcoming information society.

End User Acceptance of Security Technology for Electronic Commerce

Dale Whinnett dalew@iig.uni-freiburg.de
Albert-Ludwigs-University Freiburg, Germany
Institute for Computer Science and Social Studies

This paper examines the current advantages and limitations of the developing Global Information Infrastructure (GII) for commerce from the point of view of today's players. It is based on the interim results of an Expert Survey being carried out as part of the ACTS project SEMPER (Secure Electronic Marketplace for Europe)[1]. The findings can be broadly categorised as 'network' and 'non-network' influences on participation in electronic commerce, i.e. those directly related to the use of computer technology for connecting to and navigating in open networks and the non-technical influences on the willingness or ability to conduct business in this environment. The analysis is primarily based on use of the Internet because it is currently the most widely used precursor of the GII.

1 Introduction

Secure communication over open networks cannot be achieved by technological innovation alone. Its success will also be determined by the ability of end-users (business, government, or the private consumer) to both appreciate the significance of security and make intelligent use of it. Further, the willingness of the end-user to conduct business in an advanced, networked environment will be determined not only by the performance of the enabling technology, but also the legal, financial and regulatory issues surrounding this activity.

Those who are professionally involved in the design and implementation of information technology are well aware of both the opportunities and risks of doing business in an advanced networked environment. The research on which this paper is based[2] indicates that the non-professional user is not fully aware of either the opportunities or risks related to network use. Restrictions on the infra-structure and continuing problems of inter-operability bar today's end-user from discovering many of the potential advantages of the "Information Highway". Misunderstanding or a lack of knowledge regarding the technologies involved, on the part of both users and regulators, prevents most users from conducting business securely.

A sketch of current Internet use and type of users is followed by a discussion of the research results, with particular emphasis on implications for the implementation or acceptance of security technology. Finally, suggestions are made regarding measures to support the development of secure commerce.

2 Internet Use and Users

Truly global use of the Internet is still very much a vision of the future. The USA is leading the way with roughly 2/3 of all host computers in the world. Taken collectively, the countries of the European Union follow with 20%[3]. Estimates of

---

[1] The results of the Expert Survey form part of the document D05 First Year Surveys *Requirements and Trials* available at <www.semper.org>

[2] Research included a review of existing network user surveys and a series of extended interviews with experts in the field of electronic commerce.

[3] Data taken from Network Wizards Internet Domain Survey, January 1996. Method pinging 1% of all known hosts. Data available at http://www.nw.com.

Internet usage vary from 26 million[4] to as little as 6 million[5], however, due to the unreliability of all currently available user surveys, host count figures provide a conservative basis for estimating usage.

Recent studies [CNN95, HKN96] indicate that despite the growing numbers of total users, the *active* base of regular Internet users is, in fact, relatively small. In addition, it is estimated that 15% of all Internet users comprise 50% of all usage [CNN95]. The fact that this small group of relatively sophisticated computer users is still experiencing some very basic problems indicates areas which must be improved if electronic commerce is to become more widespread.

Today's active network users are young, well educated and well equipped. On average, they spend at least an hour a day on-line. Roughly half access the Internet from home, the other half from work, or an educational establishment. In spite of the fact that 28.8 Kbps modems are rapidly becoming the norm and most respondents have the hardware to support graphic-intensive content [GVU96], speed of connection to the Internet remains the greatest limiting factor for today's users (most common problem for 80.9% of the GVU survey).

Although e-mail is the number one Internet application, used by 3/4 of all PC users with Internet access[6], browsing the WWW is clearly the fastest growing application. The number of PC users accessing Web services has shown a 250% increase compared to 1994.[7] Most users are aware of potential security risks when browsing the Net, but a surprising number are unaware that their e-mail travels via this medium and are, subsequently, also unaware of the potential threat to the privacy of the communications which they send in this way. Business customers, on the other hand, frequently avoid sending email via the Internet by dialling into their office networks in order to protect valuable company information.

The most popular computing platform for today's users is Windows. This broad base of non-technically oriented PC users places very high value on ease-of-use and they want security mechanisms to mirror the "drag and drop", "cut and paste", "click and go" environment of Windows. They view the management of multiple passwords and phrases, currently required for the use of subscription services, as inconvenient, and frequently take unnecessary security risks in care and use. They lack confidence in their own ability to install and configure security interfaces.

In spite of positive signs of growth in electronic commerce, roughly half of Web users [GVU96] have never purchased anything on-line and few purchase regularly. Security fears appear to be the reason why. "As people learn more about the Web they become less likely to trust it with their financial transactions. People are less likely now than they were just six months ago to post credit-card information on-line."[8] User reluctance is not restricted exclusively to the potential loss of financial information. It is expected that data privacy generally will become increasingly important as the Internet becomes a part of many people's daily life. "While the majority of users understand the basic information that can be recorded per transaction, many do not know some of the advanced features like

---

[4] International Data Corporation, World Wide Web Surfers, May 1996 in CyberAtlas published by I Pro last updated 18 July 1996.

[5] O'Reilly & Associates, US users with direct access, July 1995 in CyberAtlas published by I Pro last updated 18 July 1996.

[6] CI (Computer Intelligence) Consumer Technology Index, July 1996. <http://www.ci.zd.com/news/ctinet2.html>

[7] Ibid.

[8] *BYTE*, March 1996.

cookies. Additionally, the current HTTP specifications do not enable the user's email address to be logged, thus indicating that 45.2% of the users hold a false belief about what is loggable. Yet, given the recent implementation bugs (enabled the user's email address to be sent to whomever) of certain browsers that implement scriptable languages like Javascript, this result may be a bit ambiguous." [GVU96]

The ability to *control* the use of demographic information collected and to visit Web sites anonymously are considered by users as essential aspects of their right to data privacy. Over a quarter of the GVU respondents reported having provided false demographic information when registering with Web sites and there was also strong support for the statement "ought to be able to take on different aliases/roles at different times on the Internet." [GVU96]

3 Network Influences

"Customer feedback, has been that they find the security in general an annoying feature."[9]

3.1 Transmission Speeds

As mentioned above, connection speed is currently considered to be the greatest limiting factor for Internet users. One might argue that this is an infrastructure problem, but the ways in which users compensate for this restriction can have important implications for the success of security mechanisms. According to the FIND/SVP survey[FD/SVP95] more than a third of Web users (34%) turn off graphics in order to speed up browsing and most refuse to download any files over 25k to 35k. As a result, important transaction or security related information cannot depend on graphics being viewed online and the size of files a user is expected to download in order to install a security interface is likely to influence acceptance. Insufficient speed also discourages on-line shopping and creates the danger that users will interrupt transactions if they are not implemented quickly enough. There must always be a way to discover the status of a partially completed communication.

Transmission speeds also result in the at-work volume of Internet use being considerably higher than home use, indicating that security mechanisms must be designed for use in both the business and private environment, preferably for use in both environments by the same user. At work use is primarily within a company network which requires taking into account multiple users and possible firewall interference, but even private access is frequently shared. Recent surveys of German users [FOC96, IST96] indicated that 40 to 60% of users share their Internet access with others, with an average of 3.7 persons using an Internet connection.

3.2 Ease of Use

"What do you do to your customer base and about a retail marketplace developing on the Net if you say that it's a normal thing to expect of a customer, that he is capable of going to a different site and ftping Netscape in zipped or unzipped format and then putting those executable files on his hard disc and then putting them in a directory and then unzipping them and loading them and running them. What have you said about the type of person you expect to be able to access your Web site?"[10]

A simulation study on the use of digital signatures carried out by the GMD [ROS94] also identified ease of use as a key factor for user acceptance. GMD test participants avoided using digital signatures as much as possible because they found them much more inconvenient than using their handwritten signature. Even when it was made

---

[9] Financial Services Provider.

[10] Business Development Director, Financial Services.

possible for them to use the PIN for a group of documents or a work session this wasn't sufficient because the signing function wasn't integrated in the word processing programme. GMD concluded that the signature function has to be integrated and implemented by the click of a mouse.
Even mechanisms which are less sophisticated than the digital signatures tested by the GMD meet with user resistance. If they are to be accepted security mechanisms must be as simple as possible, virtually automatic. However, some products which contain features to assist users in coping with multiple passwords, such as the password file included in Windows 95, actually create security weaknesses, e.g. anyone gaining access to the computer also gains access to this file. In respect of Win 95 it was also felt that many users will ignore the product advice to disable shared files and printers before accessing the Internet. Most recommendations for increasing security by correct use of passwords or phrases are ignored by users. Even well-informed experts admitted to laxness in security principles, e.g. using the same PIN or pass phrase for all electronic accounts, simply for convenience. "Human short term memory is limited to a certain amount of information (seven plus or minus two chunks)..." [LUSI96]
There is some evidence that consumers will only take proper care in creating and protecting security information if forced to do so. However, this approach conflicts with the ease of use requirement discussed above and supports those who argue that, in order to be successful, security options should be independent of the computing hardware which would facilitate both ease of use and user control.
It was suggested that consumer education in respect of security mechanisms could be carried out in stages. "A service dialogue that is structured to support users new to the service might become frustrating over time, and users will want to speed up the interaction and reduce the number of procedures required to complete their tasks." [LUSI96]
Although it may be possible to simplify certain aspects regarding security procedures once the user has become familiar with them, it should never be possible to disable a warning at the point where the user takes an action which cannot be revoked. In addition, mistakes should be easy to correct. "One of the best ways of achieving this is to allow the user to undo the previous action or to step back. If there is no chance of easy recovery then it is good practice to make the user confirm the action." [LUSI96]

3.3 Incompatibility

> "...the superhighway should provide a "seamless" web of features and services to users, with thousands of systems and components interacting, or interoperating, in a way that is transparent to users. Achieving interoperability will require manufacturers to co-operate with standards-setting bodies to establish common interfaces and protocols."[11]

The second greatest technical problem (apart from speed of connection) currently facing service providers and end users alike, is browser incompatibility. Browser incompatibility is not only an impediment to the development of electronic commerce in general, it can also have serious implications for the success of security mechanisms. According to respondents one of the principal features of the Internet, its universal accessibility, is being threatened by the introduction of incompatible extension sets.
The fact that Netscape and Microsoft have extended the HTML command set in ways that only their own browsers can read creates formidable problems in the virtual

[11] United States General Accounting Office Report to the Congress, *Information Superhighway: An Overview of Technology Challenges,* GAO January 1995.

world.[12] There are also some who don't think the competitive battle will be restricted to browser software. "It's not just technology. I'm sure they (Microsoft) will start playing games with integration with their TCP/IP stack. I'm sure they'll start bundling the browser with the operating system."[13]

For the Internet user this incompatibility renders Websites that have been designed for a browser other than the one they have installed, from simply difficult to use, to totally inaccessible. It also presents the possibility that they will either have incomplete access to valuable consumer information (product descriptions, conditions of sale, etc.) which an electronic enterprise has included in its Website and that important transaction related information could be overlooked due to poor visual presentation (warnings may be illegible, instructions misunderstood). In the worst case the user is barred from using the site at all.

Companies currently operating in the Internet are forced to deal with the problem by 1. restricting Website design to the lowest common denominator, 2. supplying their customer base with the appropriate browser for their site or 3. modifying their website to offer different levels of service depending on the features of the customer's browser.[14] "The Web should be open to all access providers, interoperability is the key."[15] The same principle must be applied to the design of security services and emphasises the necessity for non-proprietary solutions.

4 Non-Network influences

> "Security problems are more problems of awareness than a technical issue. People don't take the care that they need to take, either because they don't know, or because it's too difficult."[16]

In general, the experts consulted felt that consumer, or end-user perceptions of security threats currently result more from media coverage of the issue than personal knowledge, or experience. As a result, awareness of a threat tends to result in *"non-use"* rather than *"care in use"*.[17]

Experts felt that very few end-users understand Internet technology well enough to appreciate where security threats occur, or to prevent their abuse by personal intervention. Misunderstanding of Internet technology is seen to lead to resistance on the part of both endusers (business or private users) and the official bodies or agencies which must co-operate if electronic commerce is to be successfully implemented. This applies particularly to legal issues, but also has implications for the way services can be made available. Respondents referred to a frequent misconception that the Internet is a broadcast medium, similar to radio or television. They felt the fact that that information in the Internet is essentially *made available* for retrieval is widely misunderstood, i.e. that it is essentially a *pull* rather than a *push* medium.

4.1 Existing Laws are Inadequate for Regulating Electronic Commerce

> "The jurisdiction applying to contracts will depend on the facts of the situation, the choice of law clauses in the various contracts, the location of the parties and the location of the equipment attached to the network." [MC96]

---

[12] *Convergence*, Mar. '96

[13] Founder of a company which makes Web authoring software.

[14] online aktuell, April 1996.

[15] Hakon Lie, INRIA programmer in *Convergence*, Mar. 1996

[16] Consultant to the Printing Industry.

[17] According to [GVU96] users are now less likely to enter credit card information via the Internet than they were a year ago.

Legal experts identified a number of key issues regarding the legality of business conducted electronically. These included the inadequacies of existing legislation for application to the new communication medium, the question of establishing legal jurisdictions, the legal acceptance of digital signatures and the burden of evidence for electronic contracts. At the same time they were careful to point out that until new legislation is passed, existing laws will be applied to electronic transactions and it is the responsibility of legal advisors to create as much certainty as possible.
Parties to traditional contracts are generally free to specify the jurisdiction which will regulate the contract. It can be assumed that contracts negotiated electronically will also include a clause of this nature, but the *global* nature of electronic commerce will make disputes more difficult than ever to resolve. In addtion to the changes in technology and business practice which are currently taking place, an equally important factor is the change of participants in the global market. Whereas previously, with high level international transactions, you had businesses taking advantage of expert (and expensive) legal advice, the fact that, via the Internet, for the first time small businesses and ordinary consumers are becoming involved in international small scale transactions raises an entirely new set of concerns.
If the consumers are members of a defined group, e.g. because they sign a subscription agreement, (the closed user group model which is being used for much of the electronic commerce which is currently taking place), then that agreement should specify the jurisdiction in which future transactions will take place. In the absence of such a provision, such as in the commercial situation addressed by SEMPER, where no pre-established relationship between the buyer and seller is assumed, the consumer should be warned, at the point of sale, that as a condition of sale a certain jurisdiction will be used.

4.2 Acceptance of Digital Signatures

"They shouldn't call it a signature, they should call it something else." 18

Digital signatures are currently viewed as the most viable option for concluding legally binding contracts in a global network environment, but their use is also subject to criticism and their legal status remains uncertain. One of the greatest legal problems accompanying the concept of a digital signature appears to be the very specific legal interpretation of a *signature.* A fresh approach is required to resolve the legal issues related to electronic commerce. Taking too narrow a view, or attempting to mirror traditional contracts, is viewed as an impediment to finding the solutions required.

Possible Solutions

A solution which is currently being used for electronic commerce is to use a traditional paper contract to establish the conditions and terms for business, which will subsequently be carried out electronically. This has the disadvantage, of course, that the consumer has to establish a relationship to the vendor and mitigates against spontaneous transactions. The legal experts within the SEMPER consortium aim to develop an initial legal framework for electronic contracts which can be used in the spontaneous business context.[19]

---

[18] Financial Services Provider.

[19] The initial proposals for this can be found in Chapter 4 - Framework for Electronic Commerce in D05 *First Year Surveys Requirements and Trials,* available at <www.semper.org>.

In order to achieve some level of clarity many businesses now construct their web sites in such a way that the consumer is forced to view and acknowledge their terms and conditions of sale before proceeding with a transaction, e.g. before entering order details, confirming, or sending an order. This method is frequently employed by software manufacturers which allow their products to be downloaded via the Internet. The consumer has to acknowledge (by clicking) acceptance of the licensing agreement, before the download can proceed.

> "A clearly visible hyperlink should guide the consumer from the product offering to the terms of condition and sale and the ordering procedure should be structured in such a way that the consumer must view these terms and conditions before it is possible to place an order." [Zor96][20]

It appears that many Internet entrepreneurs have enough confidence in the bindingness of electronic negotiations to conduct business make the *electronic acknowledge* of their licensing agreements, or terms and conditions of sale, a prerequisite for obtaining their product. It may well be, however, that this results more from the lack of an alternative (e.g. easy to implement digital signatures) and it is usually restricted to products where the financial gains of an Internet presence outweigh the financial losses resulting from occasional abuse.

In traditional commerce and electronic commerce alike, not all contracts require a written signature. Virtually any commercial transaction constitutes a contract being formed between the two parties and what is lacking is clarification of the legal status of these contracts when they are concluded electronically. A distinction was made here between the need for legal recognition of digital signatures and the need for electronic contract law.

Companies are already doing business over open networks, in spite of the current restrictions and anything which has the potential for increasing the certainty of their situation is welcome.There is widespread willingness to trial security solutions.

> "They're holding up the electronic thing to much higher standards than the manual system. People are saying the whole world has to agree on this before we can move forward. I guarantee it won't happen if that's the approach we take. What we need is...this little geography has certificates and they work and they spread their experience to other people. If we let governments and politicians and standards people dictate for the Internet, it ain't gonna happen."[21]

Electronic Notary Services

> "It (the notary's stamp) is a formality which delays things, doesn't particularly prove anything and mostly is used to satisfy governments which like to see a lot of stamps, red tape and sealing wax. If there's an electronic way to speed up that process, where somebody in one country can push a button and have it verified that, yes, they're a real company, that would be extremely useful."[22]

Respondents felt that two services currently provided by notaries could not only be achieved well in an electronic environment, but would, in fact, offer a considerable improvement over the current situation. These were: 1. the neutral authentication of identity and/or other information regarding a person or business and 2. acting as a trusted third party by holding funds, deeds, etc. The notary's role is also frequently viewed as one of selling trust which extends beyond the individual or company.

---

20 Advice of a German lawyer to companies wishing to do business on the Internet.
21 Director of Advanced Technology, software company.
22 Legal advisor to the software industry

> "In Germany, the notary has a lot to do with personal trust in a person. A digital signature is nothing more than a substitute for a normal signature. The notary is at a much higher level and, in addition to his signature, he also adds a seal."[23]

The comments above seem to indicate that in certain business situations there is a perceived requirement for an *electronic seal*, which, if necessary, could be added to the digital signature. In view of current discussion regarding certificates, this might be viewed as support for different *levels* or *degrees* of certification to cater to the requirements of various business situations.

Legislation Governing Electronic Business/Contracts

> "I think the biggest problem at the moment is that judges don't understand electronic anything."[24]

Legal experts stressed the fact that technology is much more advanced than the laws which must be applied to settle disputes arising from its use. There is a need for clear rules for the exchange of electronic contracts. Some form of proof of contract will be required, as well as the legislation which says this is a valid way of concluding a contract.

According to one expert, although there are number of factors which are seen to be legitimately holding back the development of legislation for electronic commerce, among these, the need for more detailed research to discover the extent to which current definitions (of *document, writing, signature, record* and *instrument*) produce legislative barriers. In his opinion, however, there are also some areas where legislation could provide an immediate improvement of the legal situation:

1. "There is no justification for imposing different requirements for the admissibility of computer records as evidence from those imposed for other types of documentary evidence.
2. It could remove some of the uncertainty about how computer records are to be authenticated in civil proceedings. This could be achieved by setting up a certification scheme, under which particular technical methods of authentication would be certified as providing sufficient proof of the accuracy of computer records. Legislation would then provide that such records were prima facie presumed to be accurate, on proof that the certified authentication method had been used, whilst of course leaving open to the parties the possibility of adducing evidence to the contrary. The certification (and decertification) of authentication methods would allow the law to respond rapidly to technical change." [Ree 94][25]

A number of respondents stressed the point that much traditional business is conducted on the basis of mutual trust and unwritten business relationships which are built up over time. As one respondent put it, great deal of business has been conducted on a handshake. In the situation where the negotiations between these two established business partners take place electronically what may be more important than the question of a legally binding signature, is simply that the *electronic handshake* can be authenticated as actually coming from a known business partner. This indicates further support for a requirement for *levels* of security, depending on the parties involved in the particular business application, or the value of the transaction. At the same time, however, there is also evidence that much of the business being currently being conducted electronically (on an insecure basis) results from the lack of an alternative.

---

[23] Manager Research & Development, German Federal Printing Office.

[24] Legal advisor to the software industry.

[25] The author of these recommendations felt that the most efficient approach would be for certification to be carrried out by independent, non-govermental bodies.

"It is tempting when faced with these problems to throw up one's hands in horror and say, 'We must wait for Parliament to reform the law'. This is not a realistic option for a commercial lawyer. If a transaction can be carried out using computer technology, it is certain that at some time or other a client will decide to do so." [Ree 94]

Government Restrictions on the Choice of Technology

"Critics of federal involvement argue that current federal initiatives represent a danger to civil liberties, and that individuals should be free to choose the technical means for achieving information security." [GAO95]

In the minds of most respondents the protection of intellectual property, a requirement for secure data transmission and questions regarding the use of encryption are all closely linked, so there will be a certain amount of overlap in the discussion of these issues.

"Giving away the fruits of intellectual labor without fair and equitable compensation is a policy not destined to survive the rigors of a marketplace economy." [Bra 95]

The ease with which "digital" products can be copied and distributed presents enormous challenges for electronic commerce. The information industry, which previously functioned on the basis of recognised arrangements between authors, publishers and libraries, is now facing a completely new economic environment. Most of the current solutions for secure delivery of software via the Internet are seen to be complicated by the conflict between facilitating legitimate enduser use of the software and protecting copyright, e.g. solutions which require the physical delivery of a dongle, delivery of code discs, placing up to 10 products on a single CD with separate encryption keys to open each one, delivering keys separately by email, supplying trimmed versions of products, etc.

Respondents from the printing industry were particularly concerned about the secure transmission of their products. In the pre-press industry, for example, the electronic transmission of data is extremely efficient, but the value of the products is also very high, which means that high priority is placed on the fact that the data must be transferred intact and cannot be intercepted, by competitors. Encryption is generally viewed as the most efficient form of protection.

"Headers or electronic envelopes, encryption and other tools will be essential to maintaining the integrity of works."[26]

There are any number of examples of other types of high value or sensitive data which requires the same protection, e.g. patents, passport or health information. For most business users secure email is a top priority. There was a virtually unanimous demand among the experts interviewed for this facility. Most were reluctant to send mail containing sensitive company information over the Internet and few had found a solution which was both practical and secure.

"For us secure email is almost more important than secure payment. We tried PGP and I think the technology is probably very good, but it's public domain software which consists of a lot of individual programmes and it's much too complicated. It should be integrated in a programme like Eudora or Netscape."[27]

"Failure to implement strong encryption on-line could even open on-line service companies to lawsuits from clients and customers who find their business secrets laid bare to competitors. People in the financial services sector owe certain duties of confidentiality to their clients."[28]

---

[26] Ibid.

[27] Internet Access Provider.

[28] Robert Carolina, Attorney from London law firm Clifford Chance, in *Convergence*, Mar 96.

According to respondents the problem is not merely a lack of suitable products, but also the effect of US export controls on strong encryption.

> "Business needs strong encryption just to function, not 'it would be nice if'. I know of at least one major company which has told its government that it will move certain aspects of business out of the country if strong encryption is prohibited."[29]

In view of the rapid advances in computing power it is also assumed that encryption key length will continuously have to be adjusted, i.e. what is referred to as a C.O.C.A. (cost of cracking adjustment) - a predetermined increase in the bit length of encryption keys every few years - will have to be included in new developments. It is important that any legislation regulating the use of encryption take this factor into account if the law is to keep pace with technological advances.

> "Because many kinds of information must be kept confidential for long periods of time, assessment cannot be limited to the protection required today. Equally important, cryptosystems - especially if they are standards - often remain in use for years or even decades. The life of a cryptosystem is likely to exceed the lifetime of any individual product embodying it." [BDR96]

5 Conclusion:

There is a pressing requirement for clear, non-technical information regarding network technologies, in particular, those which employ controversial, or restricted technologies, such as strong cryptography, or where security protection relies on unfamiliar and technically complex mechanisms, such as digital signatures.

This information should be generated for two specific audiences; for those persons who will influence and implement new legislation for the regulation of their use and for those who influence public opinion (journalists/media). The aim should be to promote public discussion which is based less on the sensationalism which is prevalent in today's media coverage, and more on an informed evaluation of the existing options. The challenge is to "translate" technological achievements into unintimidating layman's language, with the aim of promoting not only understanding, but co-operation.

The role of the end-user for the success of security technology cannot be emphasised enough. The greatest danger for the future of secure electronic commerce is the current lack of security solutions which are easy to understand and simple to use. There is a risk that the credibility of security technology will be undermined if users develop false confidence in security products and conduct business insecurely because the technology is insufficiently transparent.

The goal of the project, SEMPER (Secure Electronic Marketplace for Europe), is to develop the fundamentals for secure electronic commerce by providing the first open and comprehensive solutions for secure commerce over the Internet and other public information networks. SEMPER will integrate existing architectures, tools and services where appropriate, with the aim of ensuring compatibility and interoperability of different services and service implementations.

Electronic commerce is not a vision of the future. It is currently taking place, in spite of technological problems, insufficient security mechanisms and legal uncertainty regarding the transactions which are conducted. The question is no longer "Is there a business case for electronic commerce?", but rather, how can this business case be best supported?

---

[29] Legal Advisor to the software industry.

Acknowledgements: This work was supported by the ACTS Project AC026, *SEMPER*. However, it represents the view of the author. *SEMPER* is part of the Advanced Communication Technologies and Services (ACTS) research program established by the European Commission, Directorate General XIII.
This description is based on joint work of the *SEMPER* consortium. It is a pleasure to thank all of them for their co-operation and contributions. The SEMPER homepage is at <http://www.semper.org>.

References

[BDR96] Blaze, M., Diffie, W., Rivest, R.L., Schneier, B., Shimomura, T., Thompson, E., Wiener, M., *Minimal Key Lengths for Symmetric Ciphers to Provide Adequate Commercial Security*, <http://www.bsa.org/policy/encryption/Cryptographers.html>, January 1996.

[Bra95] Branscomb, Anne W., *Common Law for the Electronic Frontier*, Scientific American, (The Computer in the 21st. Century) Special issue, 1995.

[CNN95] CommerceNet/Nielsen Internet Demographic Survey, 1995, Executive Summary available at: <http://nielsenmedia.com/commercenet/exec_sum.html>

[D0596] D05 *First Year Surveys Requirements and Trials* available at <www.semper.org>

[FD/SVP95] FIND/SVP, HSF Consulting, C+C Data, *American Internet User Survey*, Sept. - Dec. 1995, <www.findsvp.com>

[FOC96] Focus Online Nutzerbefragung, Will & Partner Marktforschung, Augsburg Burda Medien Forschung/Studienleitung New media, March 1996. Available from: Focus, Arabellastrasse 23, 81925 München.

[FOR95] Forester Research, *Conducting Business on the Internet*, O'Reilly & Associates, <http://www.ora.com/survey/> 1995.

[GAO95] United States General Accounting Office Report to the Congress, *Information Superhighway : An Overview of Technology Challenges*, GAO January 1995.

[GVU96] GVU - Fifth WWW User Survey, April - May 1996, <http://www.cc.gatech.edu/gvu/user_surveys/survey-04-1996/>

[HKN96] Hoffman, Kalsbeek, Novak , *Internet Use in the United States: 1995 Baseline Estimates and Preliminary Market Segments*, <http://www2000.ogsm.vanderbilt.edu/baseline/1995.Internet.estimates.html> 12 April, 1996.

[IST96] IST On-line Umfrage, Nov. 1995- Jan. 1996, Frauenhofer Institut Systemtechnik und Innovationsforschung, Südwest Funk, Telecooperation Office Universität Karlsruhe. <http://www.teco.uni-kar> and <http://swf3.de>

[LUSI96] Clarke, Anne M.,Editor, *Human Factor Guidelines for Designers of Telecommunication Services for Non-Expert Users*, Volume 1, Published by HUSAT Research Institute (for LUSI Consortium), 1996.

[MC96] Millard, Christoper, Carolina, Robert, *Commercial Transactions and the Global Infrastructure: A European Perspective*, in The Marshall Journal of Computer & Information Law, Winter 1996.

[NOP95] NOP Internet User Profile Survey, Nov.-Dec. 1995, <http://www.nopres.co.uk/inet/proposal.html

[ROS95] Roßnagel, A , *Die Similationsstudie Rechtpflege: Eine neue Methode zur Technikgestaltung*, provet/GMD, Berlin: Ed. Sigma, 1994.

[SRI95] SRI International, *Exploring the Web Populations's Other Half*, 06/15/1995 <www.future.sri.com/vals/vals-survey.results.html>

[TM95] Times Mirror Americans On-line Survey, sample size: 4.005, June 1995, (US).

[Zor96] Zorn, Prof. Dr. Ing. Werner ,Universität Karlsruhe, *Visionen zur Zukunft des Internet und elektronischer Dienste*, CW Nr. 3/96.

Personalized Hypermedia Information Provision through Adaptive and Adaptable System Features: User Modelling, Privacy and Security Issues

Josef Fink, Alfred Kobsa, Jörg Schreck
GMD – German National Research Center for Information Technology
Institute for Applied Information Technology (FIT)
Human-Computer Interaction Research Division (HCI)
D-53754 Sankt Augustin
+49 2241 14 {2729, 2315, 2859}
{josef.fink, alfred.kobsa, joerg.schreck}@gmd.de

Users of publicly accessible information systems are generally heterogeneous and have different needs. The aim of the AVANTI project is to cater to these individual needs by adapting the user interface and the content and presentation of WWW pages to each individual user. The special needs of elderly and handicapped users are also partly considered. A model of the characteristics of user groups and individual users, a model of the usage characteristics of the system, and a domain model are exploited in the adaptation process. This paper describes the detected differing needs of AVANTI users, the kind of adaptations that are currently implemented to cater to these needs, and the system architecture that enables AVANTI to generate user-adapted web pages from distributed multimedia databases. Special attention is given to privacy and security issues which are crucial when personal information about users is at stake.

Keywords: Adaptive hypermedia, individualization, personalization, disabled and elderly users, adaptivity, adaptability, user modelling, user model server, privacy, security

1 Introduction

The aim of AVANTI [1], a collaborative R&D project partially funded by the European Commission within the ACTS programme, is to develop and evaluate a distributed information system which provides hypermedia information about a metropolitan area (e.g., about public services, transportation, buildings) for a variety of users with different needs (e.g., tourists, citizens, travel agency clerks, elderly people, blind persons, wheelchair-bound people, and users with (slight) forms of dystrophy).
In order to develop an information service which is able to take the aims, interests, experiences, and abilities of its different users into account, AVANTI will take advantage of:

- methods and tools developed in the context of adaptive and adaptable systems during the last few years,
- standardized software components in the area of the World-Wide Web (WWW), and
- the widespread availability of computers interconnected in metropolitan-area networks.

The AVANTI system can be accessed from offices, public information booths, people's homes, and appropriate mobile computing devices (e.g. message pads and palmtops) throughout the world. Internal models of both user groups and individual users will help adapt the content and presentation to each user's individual needs.

2 User Needs

Our investigations of the AVANTI user groups have shown that their needs are considerably heterogeneous. Moreover, individual differences in user needs have also been encountered. Some examples might illustrate this:

- For users who have never used the AVANTI system before, the topography of the hypermedia space should be kept simple (e.g., restricted to a sequence, grid, or tree [25]) in order to reduce the efforts necessary for building an appropriate mental model [7] [24]. Likewise, links to other hypermedia pages should be augmented by a label, or a short comment. Both adaptations can, however, be redundant (or even cumbersome) for citizens who use the information system of their home town frequently.
- For users interested in a specific subject, interesting details should be provided, e.g. an assessment of each painter in a web-based virtual museum. If the user lacks this specific interest, such detailed information should not be presented in order to reduce the efforts for building a mental model of the current hypermedia page [19, 24].
- For laypersons like tourists in a travel booking scenario, a technical term like 'check-out time' should be supplemented by an explanation. This is normally not necessary for domain experts like travel agents.
- For users with low-bandwidth network access (e.g., via a slow modem), information that requires high bandwidth (like videos and high-resolution pictures) should be replaced by less demanding but nevertheless appropriate equivalents.
- For blind users, the modality of the presented information must be changed in the case of tactile and/or audio output. Moreover, additional orientation and navigation aids (e.g., table of contents, indices) are helpful for this user group [14].
- For wheelchair-bound users, information concerning the accessibility of premises (e.g., the availability and the dimensions of ramps and elevators, the type and width of doors) is important and should therefore be provided.
- For users with (slight) forms of dystrophy, the man-machine interface (i.e., the interaction objects and associated manipulation techniques) should be adapted accordingly, i.e., should be made less sensitive to erratic hand movements.

When implementation issues are considered, it becomes obvious that all these needs can hardly be addressed within the scope of a single project. Consequently, we focused the further investigation on mainly mobility-related user requirements in the metropolitan areas of Sienna (Italy), Rome (Italy), and Kuusamo (Finland) and consolidated the findings.

3 Scope of Adaptivity and Adaptability

In order to cater to different user needs, information systems can be tailored manually by the user or automatically by the system. Systems that allow the user to change certain system parameters, and adapt their behaviour accordingly, are called adaptable [20]. Systems that adapt to users automatically based on their assumptions about them are called adaptive.

Both features, adaptivity and adaptability, will be provided by the AVANTI system:

- *Adaptivity and adaptability within the user interface*
  We integrate and implement (special) I/O devices (e.g., macro mouse, Braille display, speech synthesizer), visual and non-visual interface objects, and associated interaction techniques [23].
- *Adaptivity and adaptability within hypermedia pages*
  We implement the adaptation of the information content, information modality, information prominence, orientation and navigation aids, search facilities, and links to other hypermedia pages [4].

Whereas the first group of adaptations aims at enabling and improving the overall access to the information system, the second group of adaptations aims at individualizing one specific hypermedia system.

4 User and Usage Modelling

In order to provide user-oriented adaptivity, a so-called 'user model' has to be set up and maintained by the AVANTI system. A user model contains explicitly modelled assumptions which represent relevant characteristics of an individual user, like preferences and interests, domain knowledge, physical, sensorial, and cognitive abilities. Different methods for acquiring assumptions about the user have been discussed in the literature [6].
In AVANTI, assumptions will be acquired from the following sources of information:

- An initial interview provides the basis for primary assumptions about the user and is therefore a valuable source of information for initially assigning the user to certain user subgroups (see the 'stereotypes' below).
- Certain user actions can be exploited for the acquisition of primary assumptions. For instance, if the user requests an explanation for a technical term then he or she can be assumed not to be familiar with it [15].
- Based on primary assumptions about the user and additional information about the application domain, the system can draw inferences in order to acquire further assumptions about the user. For instance, if the user is interested in paintings, and being a tourist in Florence, has a special interest in the famous 'Galleria degli Uffizi' in Florence, we can predict the user's interest in Botticelli's 'The Birth of Venus'.
- So-called 'stereotypes' [22] contain assumptions about relevant characteristics of user subgroups (e.g., tourists, blind users). If certain preconditions are met, a stereotype can be activated for a specific user which means that the assumptions contained in the stereotype become assigned to the user.

In order to support technically motivated or usage-oriented adaptivity, a subcomponent of the user model, the so-called 'usage model', contains relevant characteristics of the environment (e.g., terminal location, user interface characteristics) and the user's interaction with the AVANTI system (e.g., history of visited pages, frequently requested pages, most likely future hypermedia page requests). Apart from information that is available *a priori*, such as about the environment of a specific terminal, most information in the usage model is elicited at run-time, either directly from hypermedia page requests via the HTTP [3] protocol or indirectly by employing statistical methods like regression analysis.

5 System Architecture

The following figure shows the architecture of the AVANTI system:

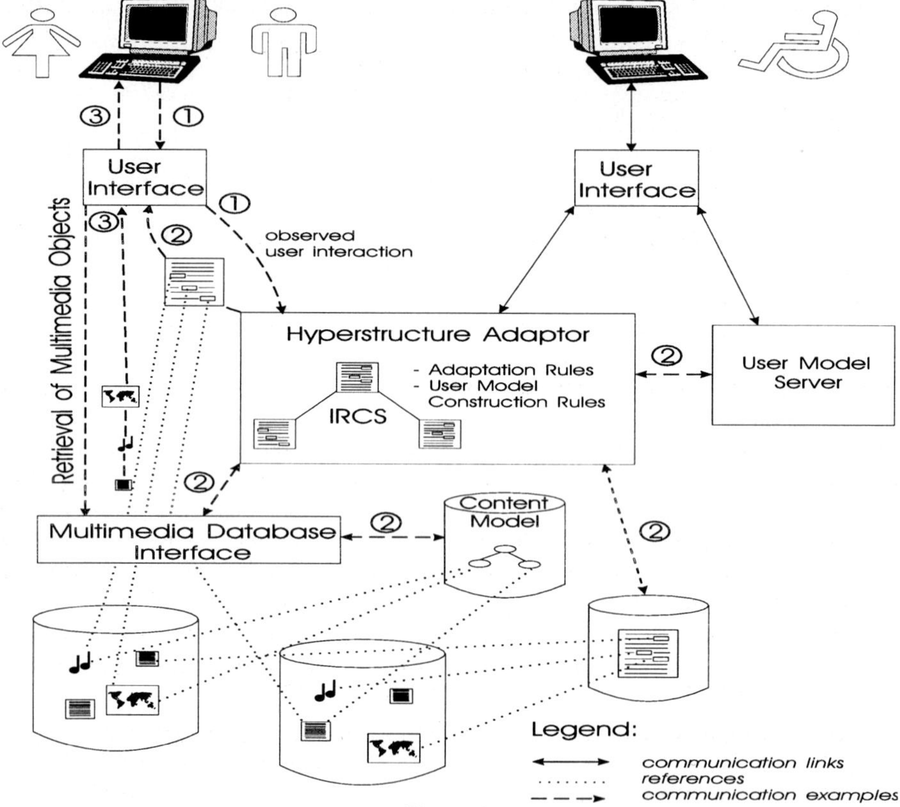

Figure 1.

In the following, we will focus on the functionality of, and the co-operation between, the main architectural components of the AVANTI system, namely the *User Interface* (UI), the *Hyperstructure Adaptor* (HSA), the *User Model Server* (UMS), and the *Multimedia Database Interface* (MDI) within the scenario of a request for a hypermedia page. The numbers refer to those in the figure:

① The user requests a hypermedia page. The UI forwards this request to the HSA.

② The HSA fetches the requested hypermedia page from secondary storage. The mark-up language used within this page is a subset of and an extension to HTML [28] named 'Information Resource Control Structure' (IRCS). Apart from static elements, an IRCS page may contain optional and alternative hypermedia objects, and also groups of hypermedia objects with an associated layout like a page header, toolbar, etc. An example for an optional element is supplementary information on wheelchair accessibility. Examples for alternative elements are technical vs. non-technical descriptions and an image of a painting vs. its textual description.

The processing of these optional and alternative elements is controlled by *Adaptation Rules*, which can take information from other system components into account, namely assumptions about user characteristics (e.g., knowledge, interests, preferences) from the UMS, and content-related information about multimedia objects

from the *Content Model* (CM) via the MDI. Information about the current user's session (e.g., previously requested IRCS pages, previously provided input) is available as well. A second group of rules that may be contained in this IRCS page are *User Model Construction Rules*. They control the formation of so-called primary assumptions about the user (i.e., assumptions which are directly derived from the user's interaction with the hypermedia page). Primary assumptions are directly reported from the HSA to the UMS.
The HSA interprets the requested hypermedia page and the Adaptation Rules, generates an adapted page (which is compliant to standard HTML) and hands it over to the UI for presentation.

③ The UI interprets the hypermedia page, retrieves multimedia objects from the AVANTI databases transparently via the MDI, and finally presents the requested hypermedia page to the user.[1]

The communication between all active components is carried out via the HTTP protocol. On top of it, a restricted and slightly enhanced version of KQML (Knowledge Query and Manipulation Language [8]) for user modelling purposes is used for communication with the UMS [17].
The main advantages of this architecture include the following:

- Already existing software in the area of the WWW (e.g., communication libraries, browsers, servers, proxies, web development environments, and database gateways) can partially or fully be used for the development of AVANTI components. This allows the developers to focus on adaptivity and adaptability, and on the evaluation of these concepts in several field tests.
- Most WWW browsers available today can access the AVANTI system and take advantage of the customization features that are based on user-oriented adaptivity and adaptability.
- All active components within AVANTI can be fully distributed according to organizational and technical requirements. This is achieved by employing an HTTP-based name service for resolving symbolic references at run-time [13].
- Certain content adaptations may be dynamically delegated from the HSA to the UI, if the necessary environment for the execution of (mobile) Java code [5] is present there. Delegated adaptations relieve the server-based HSA and allow for a more scalable architecture, avoiding the inherent limitations of a purely server-based approach.

The HSA and the UMS are central constituents of the AVANTI architecture. Their development does not have to be started from scratch since already available software can be employed as a basis, including 'WebObjects' [18] for the HSA and 'BGP-MS' (Belief, Goal and Plan Maintenance System [16]) for the UMS.

6 Security and Privacy Issues

The distributed architecture of AVANTI implies that its constituents communicate via network connections. The distribution and the fact that the system is shared between the user and the information provider poses challenges regarding the security and privacy of the users being modelled. In the following, we briefly discuss technical

---

[1] As pointed out before, the UI is able to perform additional adaptations (e.g., use alternative I/O devices, visual and non-visual interface objects and associated interaction techniques) which are not further discussed here.

means for ensuring secure and private communication between the constituents of the AVANTI architecture. Moreover, we will outline various user modelling policies.
Encryption techniques provide the basis for secure information flow. *Link encryption* based on the hardware of network connections (i.e., on layer 1 or 2 of the ISO/OSI reference model [12]) provides a simple and transparent means for keeping transport data private (see e.g. the proposed Internet Protocol standard 'IPv6' [11]). In order to take advantage of link encryption as an end-to-end service, all physical nodes within a communication channel have to support this kind of service. At the moment, this requirement is not generally met.
Software solutions offer various opportunities to shield data transparently on the transport layer (i.e., on layer 3 and 4 of the ISO/OSI reference model). For example, particular implementations of TCP/IP establish protected communication channels. In order to take advantage of these encryption services on the transport level, all (potentially heterogeneous) operating systems that are hosting AVANTI components must employ compatible implementations of secure TCP/IP. Up to now, this pre-condition is normally not met.
Recent developments like 'Secure Socket Layer' (SSL, [10]) and 'Personal Communication Technologies' (PCT, [2]) reside above the transport layer and allow for safe communication between applications. The only requirement that has to be met is that the communicating parties actually use SSL (or PCT). This precondition is normally met since these implementations are available on many platforms and are interoperable. On the presentation layer (i.e., layer 6 of the ISO/OSI reference model), high-level protocols can be employed in order to tailor the security mechanisms to the respective needs and to realize *end-to-end encryption.* Secure HTTP (S-HTTP, [21]) allows for different modes of protection depending on the kind of transported data. The 'Protocol Extension Protocol' (PEP, [27]) and the 'Security Extension Architecture' (SEA, [26]) of the World-Wide Web Consortium offer mechanisms to communicate that transportation security and authentication is required. A certification authority has to be added to the overall infrastructure in order to authenticate the communicating parties.
For the AVANTI system, a dual approach is appropriate. S-HTTP is recommended for the safe exchange of hypermedia pages between the HSA and the UI. This would enable end-to-end encryption and authentication between these components using already established WWW standards. The KQML-based communication between the UMS, the HSA, and the UI should also meet these security requirements. An extension to KQML like the one proposed in [9] would allow the security aspects (i.e., encryption techniques) to be negotiated within a communication that can take the sensitivity of the transported data into account.
Confidentiality must not only be guaranteed for data exchange but also for data storage since personal information about users resides in the UMS. These data include usage records with time stamps, data that the user supplied, and assumptions that were inferred from the user's data and usage behaviour. Privacy issues arise if a user accesses the system by revealing his or her identity rather than remaining anonymous. If the user provides information on disabilities and interests, this data is not only person-related but possibly even sensitive. Several options should be offered by the system in order to accommodate user's privacy expectations:

- If possible, users should be given the option of accessing the system anonymously (e.g., with a pseudonym) if they do not want to reveal their identities (such an option is even an obligatory provision in the current draft of the forthcoming German multimedia law.).

- In an (optional) initial dialogue the user should be able to choose between
  - no user modelling,
  - short-term user modelling, e.g. for the current session only,
  - long-term modelling using persistent user models that are augmented with information from the current session.
- At the end of each session the user should be asked if his or her model is to be deleted or stored for subsequent sessions.

These measures are taken to meet legal regulations regarding systems that process personal information and to increase user acceptance by making the system transparent. The fact that data about the user are gathered and processed should be pointed out to the user at the beginning of each session.

7 Related Projects

The main motivation for developing adaptive hypertext and hypermedia systems is the overwhelming growth of many hypermedia spaces (e.g., the WWW) in terms of size, complexity, and heterogeneity. Likewise, the user population which is confronted with these hypermedia spaces is growing, also both in size and heterogeneity. In order to keep pace with these trends, at least twenty adaptive hypertext and hypermedia systems have been developed in the last few years in order to provide more sophisticated tools for orientation, navigation, and search (for an overview and a brief description of most of these systems we refer to Brusilovsky [4]). While AVANTI shares characteristics with some of them, there are also several important distinguishing features including the following:

- In AVANTI, user-oriented *adaptations* take place *within hypermedia pages* as well as *within the user interface*. This seems to be especially beneficial for disabled users (e.g., blind users), since they can often take advantage of adaptations on both levels. The extension of hypermedia adaptation techniques for catering also to perceptual and motor disabilities is a unique characteristic of AVANTI.
- AVANTI puts an *extendible set of adaptation techniques*, namely adaptive presentation of multimedia elements, direct guidance, adaptive sorting, hiding, and annotation of links at the disposal of the hypermedia author.
- The *adaptations* offered in AVANTI *address heterogeneous user needs*, *usage patterns*, and *environmental conditions* (e.g., technical capabilities of the user interface, available network bandwidth), and provide therefore a more holistic approach to the challenges mentioned at the beginning of this chapter.
- In AVANTI, user models are entirely hosted by a *central user model server* which offers the user and other AVANTI components location-independent access to the most recent user-related information. Moreover, synergetic effects with respect to user-related information can be expected (e.g., the HSA takes advantage of assumptions acquired by the UI and vice versa). Another advantage of this centralized user modelling approach is that the other components of the AVANTI system become totally relieved of user modelling tasks and can take advantage of sophisticated run-time services of the UMS.

References

[1] AVANTI Home Page. Available at http://www.gmd.de/fit/projects/avanti.html
[2] Benaloh, J.; Lampson, B.; Simon, D.; Spies, T.; Yee, B.: Personal Communication Technologies, Microsoft Corporation, 1995. Available at http://www.lne.com/ericm/pct.html
[3] Berners-Lee, T.; Fielding, R.; Frystyk, H.: Request for Comments 1945, Hypertext Transfer Protocol - HTTP/1.0, Category: Informational, 1996. Available at http://www.csl.sony.co.jp/cgi-bin/hyperrfc?1945
[4] Brusilovsky, P.: Methods and Techniques of Adaptive Hypermedia, User Modeling and User-Adapted Interaction 6(2-3), 1996, pp. 87-129.
[5] Campione, M.; Walrath, K.: The Java Tutorial, 1996. Available at http://www.javasoft.com/tutorial/index.html
[6] Chin, D. N.: Acquiring User Models, Artificial Intelligence Review 7, pp. 185-197, 1993.
[7] Conklin, J.: Hypertext: An Introduction and Survey, IEEE Computer, September 1987, pp. 17-41.
[8] Finin, T. W.; Weber, J.; Widerhold, G.; Genesereth, M.; Fritzson, R.; McKay, D.; McGuire, J.; Pelavin, R.; Shapiro, S.; Beck, C.: Specification of the KQML Agent-Communication Language, 1993. Available at http://www.cs.umbc.edu/kqml/papers/kqmlspec.ps
[9] Finin, T. W.; Mayfield, J.; Thirunavukkarasu, C.: Secret Agents - A Security Architecture for the KQML Agent Communication Language, Intelligent Information Agents Workshop (CIKM'95), Baltimore, 1995.
[10] Freier, A. O.; Karlton, P.; Kocher, P. C.: The SSL Protocol, Version 3.0, 1996. Available at ftp://ietf.org/internet-drafts/draft-ietf-tls-ssl-version3-00.txt
[11] Huitema, C.: IPv6 - The New Internet Protocol, Prentice-Hall, Englewood Cliffs, New Jersey, 1996.
[12] Hunt, C.: TCP/IP Network Administration, O'Reilly & Associates, Sebastopol, California, 1992.
[13] KAPI, 1996. Available at http://hitchhiker.space.lockheed.com/pub/aic/shade/software/KAPI/
[14] Kennel, A.; Perrochon, L.; Darvishi, A.: WAB: World-Wide Web Access for Blind and Visually Impaired Computer Users. New Technologies in the Education of the Visually Handicapped, Paris, June 1996 and ACM SIGCAPH Bulletin, June 1996. Available at http://www.inf.ethz.ch/department/IS/ea/blinds/
[15] Kobsa, A.; Müller, D.; Nill, A.: KN-AHS: An Adaptive Hypertext Client of the User Modeling System BGP-MS, Proceedings of the Fourth International Conference on User Modeling, Hyannis, MA, pp. 99-105, 1994.
[16] Kobsa, A.; Pohl, W.: The User Modeling Shell System BGP-MS, User Modeling and User-Adapted Interaction 4(2), pp. 59-106, 1995.
[17] Kobsa, A.; Fink, J.; Pohl, W.: A Standard for the Performatives in the Communication between Applications and User Modeling Systems (draft), 1996. Available at ftp://ftp.informatik.uni-essen.de/pub/UMUAI/others/rfc.ps
[18] NeXT Corporation: WebObjects, 1996. Available at http://www.next.com/WebObjects/Products.html
[19] Nielsen, J.: The Art of Navigating through Hypertext, Communications of the ACM, vol. 33, no. 3 (March 1990), pp. 296-310.
[20] Oppermann, R. (Ed.): Adaptive User Support - Ergonomic Design of Manually and Automatically Adaptable Software, Lawrence Erlbaum Associates, Hillsdale, New Jersey, 1994.
[21] Rescorla, E.; Schiffman, A.: The Secure HyperText Transfer Protocol, Enterprise Integration Technologies, 1995. Available at http://www.eit.com/creations/s-http/draft-ietf-wts-shttp-00.txt
[22] Rich, E.: User Modeling via Stereotypes, Cognitive Science, 3, pp. 329-354, 1979.

[23] Savidis, A.; Stephanidis, C.: Developing Dual User Interfaces for Integrating Blind and Sighted Users: the HOMER UIMS, Proceedings of the CHI'95 Conference on Human Factors in Computing Systems, Denver, Colorado, May 7-11, 1995.
[24] Schaumburg, H.; Issing, L. J.: Lernen mit Hypermedia: Verloren im Hyperraum?, HMD - Theorie und Praxis der Wirtschaftsinformatik, No. 190, pp. 108-121, 1996.
[25] White, B.: Web Document Engineering, Tutorial Notes, Fifth International World-Wide Web Conference, Paris, May 1996, O'Reilly & Associates, Sebastopol, California, 1996.
[26] World Wide Web Consortium: SEA: A Security Extension Architecture for HTTP/1.x, W3C Working Draft, 1996. Available at http://www.w3.org/pub/WWW/TR/WD-http-sea.html
[27] World Wide Web Consortium: HTTP/1.2 Extension Protocol (PEP), W3C Working Draft, 1996. Available at http://www.w3.org/pub/WWW/TR/WD-http-pep.html
[28] World Wide Web Consortium: HyperText Markup Language (HTML), 1996. Available at http://www.w3.org/pub/WWW/MarkUp/

VRML: Adding 3D to Network Management

Luca Deri
IBM Zurich Research Laboratory[1]
and University of Berne[2]

Dimitrios Manikis
IBM Zurich Research Laboratory[3]

The increasing number, complexity, and heterogeneity of network management resources have pushed industry and research to find new ways to visualise information that go beyond classic 2D solutions. Network topologies and layered information models have exposed the limitations of these solutions, demanding more powerful 3D visualisation techniques.

This paper describes a novel approach to the visualisation of network management resources using VRML, a 3D modeling language. VRML has been successfully applied to selected network management problems, showing that network management information can often be presented profitably in 3D format. The ability of modern Web browsers to handle both HTML and VRML allows a simple yet powerful and flexible system for two and three-dimensional network management visualisation to be created.

Keywords: Network Management, 3D Visualisation, VRML, World Wide Web.

1 Introduction

Over the past few years, graphical computer capabilities have been improved significantly. Today, computers come with a high-resolution video board, and graphical user interfaces are the standard way to interact with software applications. Pushed by the game industry and by vertical markets such as computer graphics and medical visualisation, computer manufacturers are producing faster and faster chips and video systems able to draw and animate realistic images and mathematical models. This quick evolution in computer hardware enabled the move from 2D to more realistic 3D representations. After some attempts to create APIs and modeling languages for 3D, Silicon Graphics, a pioneer of computer graphics, released a graphics library called OpenInventor [9], available on many platforms. Its wide acceptance in the industry contributed to the unification of the various existing APIs for 3D visualisation, apart from being used to originally implement VRML, acronym for Virtual Reality Modeling Language [10]. VRML is a modeling language, hence it describes a set of 3D elements usually called a *3D virtual world*. With VRML people can build virtual rooms, towns and landscapes specifying how such virtual worlds look. A software application called *VRML viewer* interprets VRML to render the world, allowing people to explore and navigate it. One of the most interesting features of VRML is its ability to link virtual worlds with the Web associating an HTML anchor with each VRML element. This allows users to jump to other VRML worlds and HTML documents and vice-versa, exploring the Web as if wandering through a vast universe.
Besides some rare exceptions, network management visualisation is still limited to 2D. In many cases, classic 2D representations are too limited and do not allow complex information to be represented easily. Recently, the great diffusion of the Web

---

[1] IBM Research Division, Zurich Research Laboratory, Säumerstrasse 4, 8803 Rüschlikon, Switzerland. Email: lde@zurich.ibm.com, WWW: http://www.zurich.ibm.com/~lde/.
[2] Universität Bern, Institut für Informatik und angewandte Mathematik, Software Composition Group, Neubrückstrasse 10, CH-3012 Bern, Switzerland. Email: deri@iam.unibe.ch, WWW: http://iamwww.unibe.ch/~deri/.
[3] Current address: National Technical University of Athens, Telecommunications Laboratory, 15773 Zographou, Athens, Greece. Email: jman@telecom.ntua.gr, WWW: http://www.telecom.ntua.gr/~jman/.

promoted the development of Web-based network management systems, one of which developed by one of the authors, shows how greatly management systems can benefit from their integration into the Web.
The aim of this paper is to demonstrate how 3D visualisation based on VRML can be effectively applied to network management and combined with HTML to build a simple yet powerful and modern management system. This paper covers the design and the implementation of a VRML-based network management visualisation system including its advantages and disadvantages. In addition it demonstrates how selected network management problems can be solved by combining technology developed by the authors with the power of VRML and the Web, whilst avoiding both the need to purchase expensive/proprietary toolkits that run only on specific platforms, and to use specialised tools for each task.

2 The Virtual Reality Modeling Language

A VRML file is a textual description of a VRML world. It describes how to draw shapes, where to place, and how to display them. Each VRML file is composed of: a) VRML header, b) a set of nodes, and c) fields and comments (optional). The VRML header specifies the VRML version. The nodes describe the shapes and their properties in the world being defined. Each node contains the type of the node and a set of optional node attribute fields. Some node types are Cube, Cylinder, and Sphere, whereas attributes are radius, width, and height. VRML files can also contain a camera object which defines the position and the characteristics of the default view of the world. Additionally the LOD (Level of Detail) VRML tag allows VRML viewers to handle big worlds efficiently. LOD specifies which elements have to be displayed and at what level of detail according to the distance of the camera into the current VRML world.

3 VRML and Network Management

This section describes how network management can benefit from 3D visualisation, why VRML has been selected for this purpose and how it has been applied to selected network management problems.

3.1 Why VRML?

The idea to apply 3D visualisation techniques to selected network management problems is derived from the need to represent management information in a way that is as close as possible to reality. Conventional 2D visualisation systems have many limitations, some of them being (c.f. section 3.2):

- a lack of realistic representation of the information whenever such information is implicitly in 3D format since it has to be flattened in order to be represented in 2D;
- a lack of expressiveness whenever a large quantity of sparse information has to be combined in order to build a compound view of it;
- an inability to display topological information as it is in reality.

Besides all this, 3D visualisation offers several interesting advantages. It allows people to represent the information in a way very similar to reality: to change perspective, to move the viewpoint, and to add or eliminate details by getting closer to the information. Beyond these benefits, it is not straightforward to identify how to apply 3D to network management and where to prefer it to 2D. 3D is significantly more computation-costly than 2D and it is usually not platform-independent, in the sense that applications written using standard 3D graphic libraries cannot run unmodified on different platforms. Additionally, the cost to write an application for 3D visualisation is high in terms of development time and expertise. The solution to all these problems has been the adoption of a modeling language because it allows different worlds to be represented easily, leaving to the language viewer the task to visualise the information. VRML has been selected because it is at the moment the standard modeling language which is also well integrated with the Web since a) Web browsers usually come with a VRML viewer, b) VRML files are retrieved using HTTP, the same protocol used by the Web to retrieve documents, and c) it is possible to jump transparently from HTML files to VRML and vice-versa.

The integration of the Web with network management is becoming more and more important since it allows network resources to be managed in a simple, cheap, platform-independent way directly from within a standard Web browser. This seamless integration is also important in the context of a research project called *Webbin' CMIP*. In this project one of the authors has developed a software application named *Liaison* [5], which allows CMIP/SNMP resources to be managed through the Web. In this view VRML has been preferred over more powerful modeling languages like 3DMF [1] because:

- HTML is a simple, platform-independent and elegant way to display information that can be represented in 2D. The ability to jump from HTML to VRML and vice-versa enables developers to use the best visualisation format for each situation;
- VRML can be visualised efficiently on standard PCs without the need to purchase additional custom hardware;
- VRML is a very simple yet powerful language that can be learnt quite easily, hence it enables developers to create new VRML worlds or enhance existing ones without technical knowledge of 3D visualisation because the VRML viewer is responsible for this.

3.2 Applying VRML to Network Management

The use of VRML in the context of network management is derived from the need to represent management information in an effective way, employing either 3D or 2D depending on the situation. The ability of Liaison to handle different file formats makes it easy to use HTML to display 2D information, whereas VRML is used exclusively for 3D visualisation. The following sections cover the some situations where VRML has been applied in order to overcome the limitations of 2D.

Network Topology

Network topology deals with visualisation of network elements. Topology can be either logical or physical. Logical topology is used to visualise how elements are interconnected and what the connections are, without any constraint about element physical location or topological distance. Physical topology instead requires that elements are placed where they are really located and that distance constraints are satisfied.

In the case of logical topology, the goal is to display the information in the most readable way. This is in order to show human operators the current network status without adding additional information like element size or distance, which are not meaningful in this context and may confuse operators.

ATM (Asynchronous Transfer Mode) topology is an example where logical topology is used. In ATM the PNNI (Private Network-Network Interface) [8] database can be queried for retrieving the status of connections and additional information such as ATM addresses. This information is useful not only to retrieve the actual network topology but also to determine the load of the active connections, and hence to optimise the global performance by rerouting the overall network traffic.

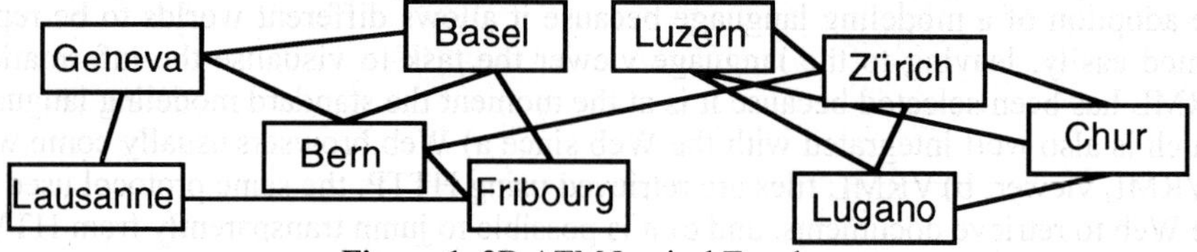

Figure 1: 2D ATM Logical Topology

The picture above displays a typical 2D ATM topology. While it shows what the operator is supposed to see, it does not highlight important information like the fact that Zurich and Bern are the two main ATM centres in Switzerland. Additionally, in order to display such topology inside a HTML browser, there are two nontrivial solutions: an HTML imagemap or a Java [7] applet. VRML instead allows HTML anchors to be associated with each basic element, hence providing a high degree of control with very low granularity. The following picture shows how the previous ATM logical topology has been implemented using VRML.

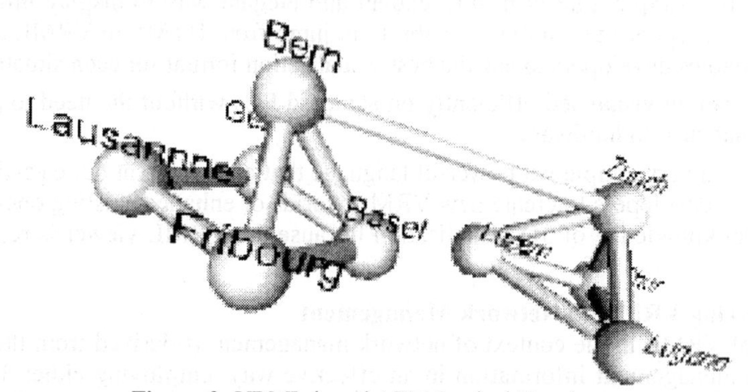

Figure 2: VRML-based ATM Logical Topology

The Z axis allowed Zurich and Bern to be placed higher than the other ones, hence to highlight the fact that such centres are the main ones in the country contributing to the elimination of line crossings. Each node has a corresponding 3D label and an HTML anchor which enables users to jump directly to an HTML page containing detailed information about it. Connections are depicted using a different colour according to the current link load and it has a corresponding HTML anchor just like the nodes. Users can freely rotate and walk through the world by exploiting the facilities offered by the VRML viewer, hence going beyond vertical/horizontal scrolling available for 2D representations. These advantages become much more evident when a large topology having multiple connections among nodes is to be displayed. LOD allows big worlds to be manipulated efficiently since the VRML viewer does not have to render elements that are too distant from the current camera or represent objects at a high level of detail which is useful when the camera is close to an object. Additionally, LOD allows subnetworks to be represented as one node when the camera is distant, and the architecture of the subnetwork to be visible when the camera is close enough to it.

Hierarchical Information

Quite often in network management, information is aggregated in hierarchical structures. In the OSI world, object instances are stored in a so-called containment tree. This means that an object instance can contain one or more subinstances in a tree structure. In SNMP the same structure can be obtained splitting the information according to the MIB groups, which are identified using nested object identifiers. The TCP/IP protocol identifies hosts assigning addresses of the type X.Y.W.Z, where X, Y, W and Z are integers; host addresses are then grouped in subnets according to a subnet mask. These examples show some simple cases of hierarchical information just to demonstrate that this kind of information structure can be encountered quite often in network management.

While 2D could be used to represent hierarchical data, it has the disadvantage that a) it does not respect the native structure of the information whenever there are connections between the different information levels, making the picture difficult to read, and b) it creates trees that may become very wide or tall if the tree contains many nodes. VRML, instead, can represent the information in its native hierarchical format. This is very convenient whenever the quantity of information to be represented is high, making it possible to create real 3D topologies instead of flat ones. Also, by using the LOD facility, the information about the subinstances in the containment tree can be hidden, shown only when the camera is close to the object.

Another situation to which VRML can be applied is to represent subnetwork connection-termination points. As described in [6], important concepts needed to describe networking are layering and partitioning. They allow a network to be partitioned into a hierarchy of subnetworks with successive levels of abstractions. The following figure shows the partitioning of a network connection into link connections and subnetwork connections (SNC) as for modeling according to the [5] way.

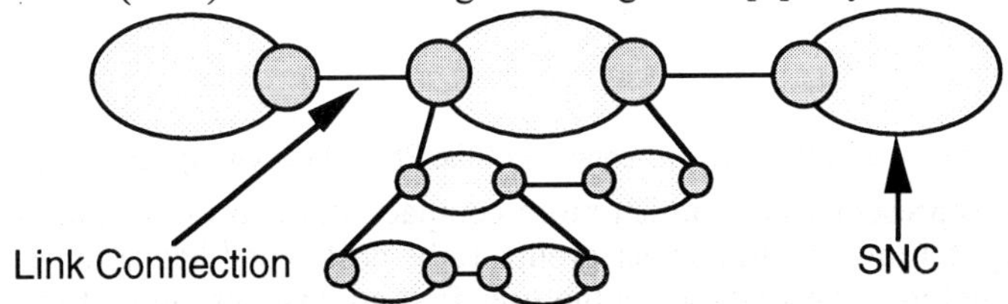

Figure 3: Partitioning of a Nw Connection into SubNw. Connection

When the number of nodes and SNCs is large, this technique produces complex 2D representations of multiple layers, which are difficult to read and to navigate. In this context, VRML allows layer navigation to be performed in a natural way.

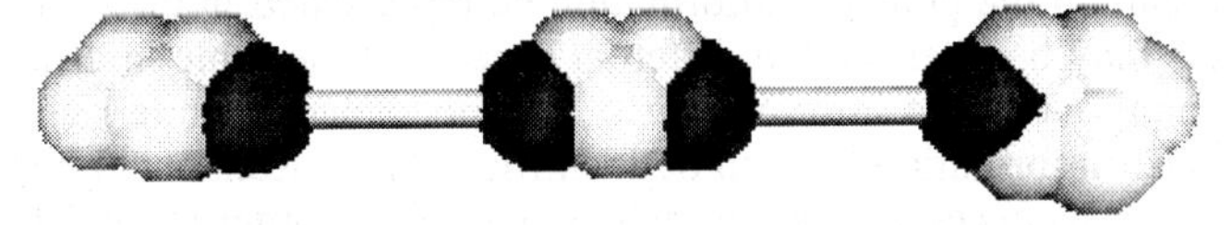

Figure 4: VRML Representation of the Previous Figure

Figure 4 above shows a VRML representation of SNCs. Subnetworks are represented as clouds whereas link connections are represented as pipes. When the camera is distant from the subnetworks, only the top level is displayed although the subnetworks contain further elements. Thanks to the VRML LOD tag, users can see the contained subnetworks and link connections by moving the camera towards a subnetwork. This operation is recursive hence contained subnetworks can be exploded by moving the

camera closer towards them. Each element contains a hotlink, which allows users to jump to other pages containing detailed information about the element.
VRML can also be applied to represent hierarchical views of ATM networks. The PNNI routing protocol views the world as a collection of peer groups. At the bottom, a peer group is formed by an administrator logically drawing a circle around a collection of switches. These switches must be a set of devices which are connected like a graph. Given several peer groups connected to each other by border switches, one can build a hierarchical structure by drawing a circle around this collection of peer groups, and declare that to be a peer group, the parent of each of the constituent peer groups. As shown before, it is possible to query the PNNI interface in order to build a hierarchical view of an ATM network.

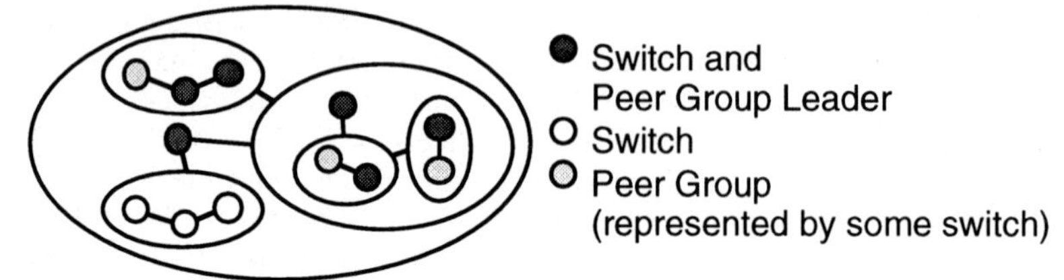

Figure 5: PNNI Peer Group Hierarchy

The figure above represents how a collection of switches might be aggregated into a hierarchical topology, which can be represented in VRML as a sphere which represents a peer group containing the other peer groups, switches, or peer group leaders.

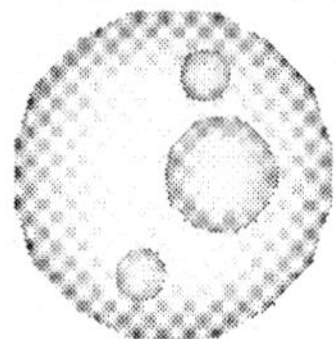

Figure 6: VRML Representation of the PNNI Peer Group Hierarchy

The VRML representation is much more compact and expressive than the 2D one. Additionally it allows multiple peer groups to be nested and hence to be represented in a single picture. The LOD gives the user first an abstract view of the whole network and then allows a more detailed view of the peer group of interest by moving the camera close to it.

Compound Information View

Very seldom can all the potential information be represented in the same view since different users may be interested in different aspects of the same information. Additionally, in many cases the information to be represented is quite rich, so ways to "compress" such information have to be identified. This means that a representation has to be as clear as necessary and as rich as possible in order to depict most of the information in a single picture. Compound information means that several aspects of the same information have been combined to produce a simple representation that removes or hides any information that is irrelevant for a given representation. VRML can help in this respect because it allows a great amount of information to included in a single virtual world.
The following example shows how this concept has been applied to a real situation. The European ACTS project MISA (Management of Integrated SDH and ATM Networks), deals with the management of integrated ATM and SDH networks. Members

of the MISA consortium, who are located in different European countries, provide access to resources relevant for the project. X.700 agents keep track of these resources: they contain object instances that represent locations, computers, and services provided. When this information is represented, it is important to show as much information as possible on a single screen to avoid having to walk through too many different screens. VRML facilitates this because it allows logic containment (country, location, computer, services) to be represented using real 3D containment and also because it allows views to be manipulated easily by moving the whole world and walking through it. The way to represent information, retrieved dynamically from the X.700 agents, is the following: a map of Europe has been wrapped over a 3D flat surface, partner locations are identified by building (boxes) located where the partner really is and labelled with the flag of the country. Services are depicted in a colour representing their state. Each element has an HTML anchor that links it to the corresponding resource details or that allow to jump to other VRML/HTML pages.
Agents contain all the information needed to represent the view stored inside object instances that contain topology information. When the VRML world is built, the topology instances are located and agents are no longer accessed dworld uring the manipulation. The object containment tree is used to produce the 3D containment whereas information about the precise element location, maps and anchors to detailed information, as well as links to VRML element views are contained in X.700 object attributes as part of the object instances. The following picture shows a simple VRML world representing MISA partners and the services they provide.

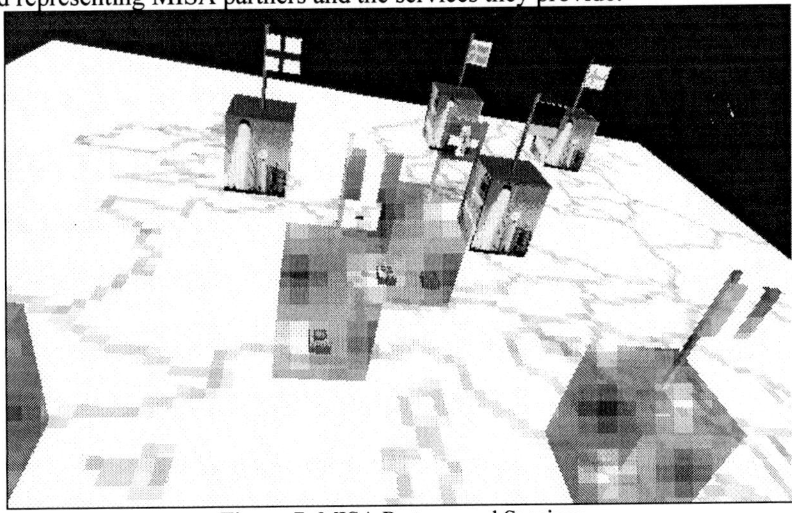

Figure 7: MISA Partners and Services

The use of the LOD contributes to simplify the exploration of the world since the VRML viewer removes/adds objects to the view according to the current camera position. When the user is far away from a location where a MISA partner is located, a box with a flag on top of it is displayed. The closer the user approaches a box, the more details are shown. It is like navigating the containment tree with a camera: users do not have to enter boxes to see what is inside and also because it is possible to represent much information in one view without demanding users to click several times in order to arrive at the object of their interest, although this possibility is provided as well.

4 Liaison: VRML Extensions

The examples shown so far have been implemented as VRML extensions to Liaison. Liaison is based on software components called *droplets* [3] that have the ability to be replaced and added at runtime allowing the behaviour of the application that contains them to be modified and extended at runtime. Each of these VRML extensions have been implemented using droplets.

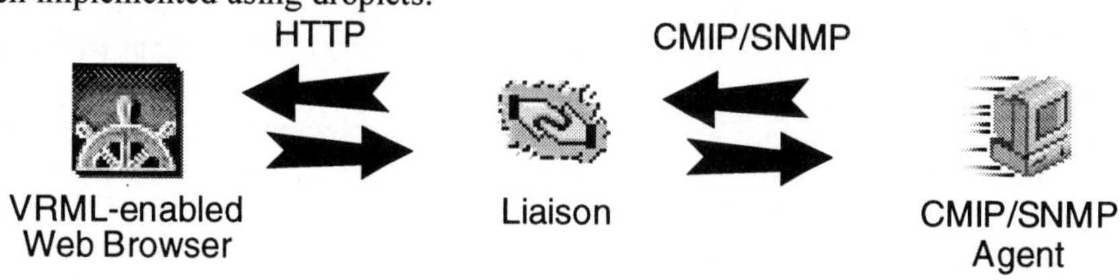

Figure 8: Liaison's VRML Extensions

When an HTTP request is received, Liaison routes it to the droplet responsible for handling it. The later issues one or more CMIP/SNMP requests and builds an internal representation of the virtual world. Since VRML specifies the location of each element, the droplet has to layout each element by assigning a 3D location to it. At this point the droplet returns an HTTP response and a file containing the VRML world whose type is `x-world/x-vrml`. The VRML-enabled Web browser (or the VRML viewer) receives the file and shows it properly according to the file type, which is VRML in this case. Once the user clicks on one of the HTML anchors contained in the VRML world, a request is issued and a new file is returned. It is possible to jump from HTML to VRML and vice-versa by using the file type, that in the case of HTML, is `text/html`.

For each object, the location and the name of the VRML file used to render the object are retrieved from the OSI agent. This file is basically a VRML file with certain parameters, replaced at runtime with their actual value, which represent translation coordinates, scale, rotation, anchors, etc. These parameters, which are not part of the VRML language, are defined as special strings contained in the VRML code. The following is a VRML code fragment which shows how these parameters are used:

```
WWWAnchor {
 name  "$anchor"                     # Object Name (URL)
 description "$title"                # Information about the object
 Translation { translation $X }      # Object position
 Texture2 { filename "$fn" }         # Texture to be mapped on the obj.
 Cube { }                            # Cube (default size)
} # WWWAnchor
```

The droplet responsible for VRML replaces the parameters at runtime with the actual values retrieved by the X.700 agents. This technique has additional advantages:

- if not all the objects have to be shown, the VRML location attribute can be set to null in order not to generate VRML code for those objects;
- it is possible to design an object using a sophisticated VRML editor and then to embed the parameters into the VRML code;
- the VRML file can be changed dynamically while the Liaison is running;
- the object instance information in the X.700 agent can be manipulated separately from the graphical VRML aspects;

- droplets greatly reduced the development time and code size since most of the services were already available and have been exploited thanks to the facilities provided by Liaison.

The only situation in which VRML is not suitable is whenever the management information changes frequently. VRML files, like HTML, are static, so they cannot be used in very dynamic situations because every time the file changes then the VRML viewer has to parse the file.

5 Conclusion

This paper showed how 3D visualisation can be used effectively for network management. It covered a novel technique developed to display management information using VRML.

The main characteristics of this technique are:

- Internet-ready, simple, and platform-independent based on VRML, which is a portable, compact, and widely established modeling language;
- ability to display 3D data using their native format instead of flattening them to 2D;
- ability to mix 2D and 3D data.

VRML has to be considered a novel and promising solution for Web-based 3D network management visualisation due to its wide range of appliance and its native compatibility with HTML. Its platform-independence, high flexibility and wide acceptance make VRML probably the most reasonable solution for 3D visualisation.

Acknowledgements: The VRML-based visualisation system described in this paper has been designed and implemented at the IBM Zurich Research Laboratory. The authors gratefully acknowledge partial funding by the European Commission, European ACTS project AC080 MISA, and by the Swiss Federal Office for Education and Science. Moreover, the authors would like to thank R. Akolk and D. Gantenbein for their suggestions and valuable discussions, in addition to users of *Webbin' CMIP*[4], who have greatly stimulated our work with all their comments and suggestions.

References

[1] Apple Computer Inc., QuickDraw 3D Metafile Format (3DMF), October 1995.
[2] J. Case, M. Fedor, M. Schoffstall and J. Davin, A Simple Management Protocol (SNMP), Network Working Group, RFC 1157, May 1990.
[3] L. Deri, Droplets: Breaking Monolithic Applications Apart, IBM Research Report RZ 2799, September 1995.
[4] L. Deri, Surfin' Network Management Resources Across the Web, Proc. of 2nd IEEE Workshop on Systems and Network Management, Toronto, June 1996.
[5] European Telecommunications Standard Institute, Telecommunications Management Network (TMN), Generic Managed Object Class Library for the Network Level View, I-ETS 300.653, DI/NA-043316, May 1996.
[5] ITU-T Recommendation G.805, Generic functional Architecture of Transport Networks, November 1995.
[7] Sun Microsystems, The Java Programming Language, Addison-Wesley, 1996.
[8] The ATM Forum, Private Network-Network Interface Version 1.0 (PNNI 1.0), AF-PNNI-0055.000, March 1996.
[9] Silicon Graphics, OpenInventor C++ Reference Manual, Addison-Wesley, 1994.
[10] G. Bell, A. Parisi and M. Pesce, The Virtual Reality Modeling Language (VRML), Version 1.0, May 1995.

[4] An on-line demo and a version available for public download (supported platforms: AIX, OS/2, MacOS, Linux, Win95/NT) can be found at http://misa.zurich.ibm.com/Webbin/.

List of Authors

Springer and the environment

At Springer we firmly believe that an international science publisher has a special obligation to the environment, and our corporate policies consistently reflect this conviction.

We also expect our business partners – paper mills, printers, packaging manufacturers, etc. – to commit themselves to using materials and production processes that do not harm the environment. The paper in this book is made from low- or no-chlorine pulp and is acid free, in conformance with international standards for paper permanency.

Lecture Notes in Computer Science

For information about Vols. 1–1169

please contact your bookseller or Springer-Verlag

Vol. 1170: M. Nagl (Ed.), Building Tightly Integrated Software Development Environments: The IPSEN Approach. IX, 709 pages. 1996.

Vol. 1171: A. Franz, Automatic Ambiguity Resolution in Natural Language Processing. XIX, 155 pages. 1996. (Subseries LNAI).

Vol. 1172: J. Pieprzyk, J. Seberry (Eds.), Information Security and Privacy. Proceedings, 1996. IX, 333 pages. 1996.

Vol. 1173: W. Rucklidge, Efficient Visual Recognition Using the Hausdorff Distance. XIII, 178 pages. 1996.

Vol. 1174: R. Anderson (Ed.), Information Hiding. Proceedings, 1996. VIII, 351 pages. 1996.

Vol. 1175: K.G. Jeffery, J. Král, M. Bartošek (Eds.), SOFSEM'96: Theory and Practice of Informatics. Proceedings, 1996. XII, 491 pages. 1996.

Vol. 1176: S. Miguet, A. Montanvert, S. Ubéda (Eds.), Discrete Geometry for Computer Imagery. Proceedings, 1996. XI, 349 pages. 1996.

Vol. 1177: J.P. Müller, The Design of Intelligent Agents. XV, 227 pages. 1996. (Subseries LNAI).

Vol. 1178: T. Asano, Y. Igarashi, H. Nagamochi, S. Miyano, S. Suri (Eds.), Algorithms and Computation. Proceedings, 1996. X, 448 pages. 1996.

Vol. 1179: J. Jaffar, R.H.C. Yap (Eds.), Concurrency and Parallelism, Programming, Networking, and Security. Proceedings, 1996. XIII, 394 pages. 1996.

Vol. 1180: V. Chandru, V. Vinay (Eds.), Foundations of Software Technology and Theoretical Computer Science. Proceedings, 1996. XI, 387 pages. 1996.

Vol. 1181: D. Bjørner, M. Broy, I.V. Pottosin (Eds.), Perspectives of System Informatics. Proceedings, 1996. XVII, 447 pages. 1996.

Vol. 1182: W. Hasan, Optimization of SQL Queries for Parallel Machines. XVIII, 133 pages. 1996.

Vol. 1183: A. Wierse, G.G. Grinstein, U. Lang (Eds.), Database Issues for Data Visualization. Proceedings, 1995. XIV, 219 pages. 1996.

Vol. 1184: J. Waśniewski, J. Dongarra, K. Madsen, D. Olesen (Eds.), Applied Parallel Computing. Proceedings, 1996. XIII, 722 pages. 1996.

Vol. 1185: G. Ventre, J. Domingo-Pascual, A. Danthine (Eds.), Multimedia Telecommunications and Applications. Proceedings, 1996. XII, 267 pages. 1996.

Vol. 1186: F. Afrati, P. Kolaitis (Eds.), Database Theory - ICDT'97. Proceedings, 1997. XIII, 477 pages. 1997.

Vol. 1187: K. Schlechta, Nonmonotonic Logics. IX, 243 pages. 1997. (Subseries LNAI).

Vol. 1188: T. Martin, A.L. Ralescu (Eds.), Fuzzy Logic in Artificial Intelligence. Proceedings, 1995. VIII, 272 pages. 1997. (Subseries LNAI).

Vol. 1189: M. Lomas (Ed.), Security Protocols. Proceedings, 1996. VIII, 203 pages. 1997.

Vol. 1190: S. North (Ed.), Graph Drawing. Proceedings, 1996. XI, 409 pages. 1997.

Vol. 1191: V. Gaede, A. Brodsky, O. Günther, D. Srivastava, V. Vianu, M. Wallace (Eds.), Constraint Databases and Applications. Proceedings, 1996. X, 345 pages. 1996.

Vol. 1192: M. Dam (Ed.), Analysis and Verification of Multiple-Agent Languages. Proceedings, 1996. VIII, 435 pages. 1997.

Vol. 1193: J.P. Müller, M.J. Wooldridge, N.R. Jennings (Eds.), Intelligent Agents III. XV, 401 pages. 1997. (Subseries LNAI).

Vol. 1194: M. Sipper, Evolution of Parallel Cellular Machines. XIII, 199 pages. 1997.

Vol. 1195: R. Trappl, P. Petta (Eds.), Creating Personalities for Synthetic Actors. VII, 251 pages. 1997. (Subseries LNAI).

Vol. 1196: L. Vulkov, J. Waśniewski, P. Yalamov (Eds.), Numerical Analysis and Its Applications. Proceedings, 1996. XIII, 608 pages. 1997.

Vol. 1197: F. d'Amore, P.G. Franciosa, A. Marchetti-Spaccamela (Eds.), Graph-Theoretic Concepts in Computer Science. Proceedings, 1996. XI, 410 pages. 1997.

Vol. 1198: H.S. Nwana, N. Azarmi (Eds.), Software Agents and Soft Computing: Towards Enhancing Machine Intelligence. XIV, 298 pages. 1997. (Subseries LNAI).

Vol. 1199: D.K. Panda, C.B. Stunkel (Eds.), Communication and Architectural Support for Network-Based Parallel Computing. Proceedings, 1997. X, 269 pages. 1997.

Vol. 1200: R. Reischuk, M. Morvan (Eds.), STACS 97. Proceedings, 1997. XIII, 614 pages. 1997.

Vol. 1201: O. Maler (Ed.), Hybrid and Real-Time Systems. Proceedings, 1997. IX, 417 pages. 1997.

Vol. 1203: G. Bongiovanni, D.P. Bovet, G. Di Battista (Eds.), Algorithms and Complexity. Proceedings, 1997. VIII, 311 pages. 1997.

Vol. 1204: H. Mössenböck (Ed.), Modular Programming Languages. Proceedings, 1997. X, 379 pages. 1997.

Vol. 1205: J. Troccaz, E. Grimson, R. Mösges (Eds.), CVRMed-MRCAS'97. Proceedings, 1997. XIX, 834 pages. 1997.

Vol. 1206: J. Bigün, G. Chollet, G. Borgefors (Eds.), Audio- and Video-based Biometric Person Authentication. Proceedings, 1997. XII, 450 pages. 1997.

Vol. 1207: J. Gallagher (Ed.), Logic Program Synthesis and Transformation. Proceedings, 1996. VII, 325 pages. 1997.

Vol. 1208: S. Ben-David (Ed.), Computational Learning Theory. Proceedings, 1997. VIII, 331 pages. 1997. (Subseries LNAI).

Vol. 1209: L. Cavedon, A. Rao, W. Wobcke (Eds.), Intelligent Agent Systems. Proceedings, 1996. IX, 188 pages. 1997. (Subseries LNAI).

Vol. 1210: P. de Groote, J.R. Hindley (Eds.), Typed Lambda Calculi and Applications. Proceedings, 1997. VIII, 405 pages. 1997.

Vol. 1211: E. Keravnou, C. Garbay, R. Baud, J. Wyatt (Eds.), Artificial Intelligence in Medicine. Proceedings, 1997. XIII, 526 pages. 1997. (Subseries LNAI).

Vol. 1212: J. P. Bowen, M.G. Hinchey, D. Till (Eds.), ZUM '97: The Z Formal Specification Notation. Proceedings, 1997. X, 435 pages. 1997.

Vol. 1213: P. J. Angeline, R. G. Reynolds, J. R. McDonnell, R. Eberhart (Eds.), Evolutionary Programming VI. Proceedings, 1997. X, 457 pages. 1997.

Vol. 1214: M. Bidoit, M. Dauchet (Eds.), TAPSOFT '97: Theory and Practice of Software Development. Proceedings, 1997. XV, 884 pages. 1997.

Vol. 1215: J. M. L. M. Palma, J. Dongarra (Eds.), Vector and Parallel Processing – VECPAR'96. Proceedings, 1996. XI, 471 pages. 1997.

Vol. 1216: J. Dix, L. Moniz Pereira, T.C. Przymusinski (Eds.), Non-Monotonic Extensions of Logic Programming. Proceedings, 1996. XI, 224 pages. 1997. (Subseries LNAI).

Vol. 1217: E. Brinksma (Ed.), Tools and Algorithms for the Construction and Analysis of Systems. Proceedings, 1997. X, 433 pages. 1997.

Vol. 1218: G. Păun, A. Salomaa (Eds.), New Trends in Formal Languages. IX, 465 pages. 1997.

Vol. 1219: K. Rothermel, R. Popescu-Zeletin (Eds.), Mobile Agents. Proceedings, 1997. VIII, 223 pages. 1997.

Vol. 1220: P. Brezany, Input/Output Intensive Massively Parallel Computing. XIV, 288 pages. 1997.

Vol. 1221: G. Weiß (Ed.), Distributed Artificial Intelligence Meets Machine Learning. Proceedings, 1996. X, 294 pages. 1997. (Subseries LNAI).

Vol. 1222: J. Vitek, C. Tschudin (Eds.), Mobile Object Systems. Proceedings, 1996. X, 319 pages. 1997.

Vol. 1223: M. Pelillo, E.R. Hancock (Eds.), Energy Minimization Methods in Computer Vision and Pattern Recognition. Proceedings, 1997. XII, 549 pages. 1997.

Vol. 1224: M. van Someren, G. Widmer (Eds.), Machine Learning: ECML-97. Proceedings, 1997. XI, 361 pages. 1997. (Subseries LNAI).

Vol. 1225: B. Hertzberger, P. Sloot (Eds.), High-Performance Computing and Networking. Proceedings, 1997. XXI, 1066 pages. 1997.

Vol. 1226: B. Reusch (Ed.), Computational Intelligence. Proceedings, 1997. XIII, 609 pages. 1997.

Vol. 1227: D. Galmiche (Ed.), Automated Reasoning with Analytic Tableaux and Related Methods. Proceedings, 1997. XI, 373 pages. 1997. (Subseries LNAI).

Vol. 1228: S.-H. Nienhuys-Cheng, R. de Wolf, Foundations of Inductive Logic Programming. XVII, 404 pages. 1997. (Subseries LNAI).

Vol. 1230: J. Duncan, G. Gindi (Eds.), Information Processing in Medical Imaging. Proceedings, 1997. XVI, 557 pages. 1997.

Vol. 1231: M. Bertran, T. Rus (Eds.), Transformation-Based Reactive Systems Development. Proceedings, 1997. XI, 431 pages. 1997.

Vol. 1232: H. Comon (Ed.), Rewriting Techniques and Applications. Proceedings, 1997. XI, 339 pages. 1997.

Vol. 1233: W. Fumy (Ed.), Advances in Cryptology — EUROCRYPT '97. Proceedings, 1997. XI, 509 pages. 1997.

Vol 1234: S. Adian, A. Nerode (Eds.), Logical Foundations of Computer Science. Proceedings, 1997. IX, 431 pages. 1997.

Vol. 1235: R. Conradi (Ed.), Software Configuration Management. Proceedings, 1997. VIII, 234 pages. 1997.

Vol. 1238: A. Mullery, M. Besson, M. Campolargo, R. Gobbi, R. Reed (Eds.), Intelligence in Services and Networks: Technology for Cooperative Competition. Proceedings, 1997. XII, 480 pages. 1997.

Vol. 1240: J. Mira, R. Moreno-Díaz, J. Cabestany (Eds.), Biological and Artificial Computation: From Neuroscience to Technology. Proceedings, 1997. XXI, 1401 pages. 1997.

Vol. 1241: M. Akşit, S. Matsuoka (Eds.), ECOOP'97 – Object-Oriented Programming. Proceedings, 1997. XI, 531 pages. 1997.

Vol. 1242: S. Fdida, M. Morganti (Eds.), Multimedia Applications, Services and Techniques – ECMAST '97. Proceedings, 1997. XIV, 772 pages. 1997.

Vol. 1243: A. Mazurkiewicz, J. Winkowski (Eds.), CONCUR'97: Concurrency Theory. Proceedings, 1997. VIII, 421 pages. 1997.

Vol. 1244: D. M. Gabbay, R. Kruse, A. Nonnengart, H.J. Ohlbach (Eds.), Qualitative and Quantitative Practical Reasoning. Proceedings, 1997. X, 621 pages. 1997. (Subseries LNAI).

Vol. 1245: M. Calzarossa, R. Marie, B. Plateau, G. Rubino (Eds.), Computer Performance Evaluation. Proceedings, 1997. VIII, 231 pages. 1997.

Vol. 1246: S. Tucker Taft, R. A. Duff (Eds.), Ada 95 Reference Manual. XXII, 526 pages. 1997.

Vol. 1247: J. Barnes (Ed.), Ada 95 Rationale. XVI, 458 pages. 1997.

Vol. 1248: P. Azéma, G. Balbo (Eds.), Application and Theory of Petri Nets 1997. Proceedings, 1997. VIII, 467 pages. 1997.

Vol. 1249: W. McCune (Ed.), Automated Deduction – Cade-14. Proceedings, 1997. XIV, 462 pages. 1997.

Vol. 1250: A. Olivé, J.A. Pastor (Eds.), Advanced Information Systems Engineering. Proceedings, 1997. XI, 451 pages. 1997.

Vol. 1251: K. Hardy, J. Briggs (Eds.), Reliable Software Technologies – Ada-Europe '97. Proceedings, 1997. VIII, 293 pages. 1997.

Vol. 1253: G. Bilardi, A. Ferreira, R. Lüling, J. Rolim (Eds.), Solving Irregularly Structured Problems in Parallel. Proceedings, 1997. X, 287 pages. 1997.